东亚中高纬度环流与季风及其区域气候变异论文选

廉　毅　沈柏竹　刘　刚
李尚锋　杨　旭　王　宁
编著

气象出版社
China Meteorological Press

内容简介

本书收集了廉毅教授及其团队近几十年来发表的25篇学术论文，研究成果揭示了北半球中高纬度大气环流系统（包括遥相关型、阻塞高压、东北冷涡等）与海洋外强迫的不同配置及海-气相互作用、东亚-西北太平洋夏季副热带季风在东北区域的建立、季内活动规律等对东北区域气候异常的影响机制，以及为应对气候变化和生态省建设开展的东北生态环境与区域气候变化的监测和反演，对粮食生产安全影响等。本书可供大气科学、环境科学等领域的相关专业研究人员和业务工作者阅读和参考。

图书在版编目（CIP）数据

东亚中高纬度环流与季风及其区域气候变异论文选 / 廉毅等编著. -- 北京 : 气象出版社, 2024. 8. -- ISBN 978-7-5029-8273-7

Ⅰ. P423.3-53

中国国家版本馆CIP数据核字第2024FV5614号

东亚中高纬度环流与季风及其区域气候变异论文选

Dongya Zhong Gao Weidu Huanliu yu Jifeng jiqi Quyu Qihou Bianyi Lunwen Xuan

出版发行：气象出版社

地　　址：北京市海淀区中关村南大街46号　　**邮政编码**：100081

电　　话：010-68407112（总编室）　010-68408042（发行部）

网　　址：http://www.qxcbs.com　　**E-mail**：qxcbs@cma.gov.cn

责任编辑：胡育峰　高菁蕾　　**终　　审**：张　斌

责任校对：张硕杰　　**责任技编**：赵相宁

封面设计：艺点设计

印　　刷：北京建宏印刷有限公司

开　　本：889 mm×1194 mm　1/16　　**印　　张**：20

字　　数：620千字

版　　次：2024年8月第1版　　**印　　次**：2024年8月第1次印刷

定　　价：200.00元

廉毅先生简介

廉毅 男，1945 年 11 月生，天津市人，1965 年毕业于北京气象专科学校（另于北京大学进修一年），中共党员，二级教授。曾任吉林省气象科学研究所所长、吉林省省委省政府咨询委员，先后获得吉林省劳动模范、全国“五一”劳动奖章等荣誉，并享受国务院政府特殊津贴。在中高纬度环流系统、东亚季风和生态环境等方面研究有一定建树，与多所院校协同培养博士、硕士研究生 15 名。获批 13 项国家自然科学基金项目，发表论文 100 余篇，获省部级科学技术进步奖一、二、三等奖 14 项。

序 一

坚持奋斗,团队优秀,为我国省级气象科学研究所发展作出卓越贡献

廉毅教授是北京气象专科学校1965年大专班毕业生,1977—1978年又在北京大学地球物理系暴雨进修班深造,学习了新的大气科学基础理论、知识和方法。通过这次学习,廉毅教授深深地体会到:北京大学培养出来的学生,大都具有明确的政治抱负、坚实的数学物理知识、较强的外语能力和良好的科学研究训练。一年的进修,是他人生的重要加油站。

我与廉毅是在北京大学暴雨进修班认识的,他是班长。他曾在我指导下撰写了“东北冷涡暴雨中分析”的毕业论文,从此开始加速了他的气象科学研究事业,在单位逐渐形成了自己的研究团队。他在担任吉林省气象科学研究所所长期间,比较敏锐地从吉林省所处的地理位置出发,不断地研究北半球中高纬度环流系统,积极开展东北冷涡、东北亚和亚洲乃至北半球阻塞高压活动气候变异及其对我国北方天气气候影响的研究。不仅如此,他与团队还注意研究东亚—北太平洋地区夏季副热带季风进退的特征,提出除南北向的低频振荡变化外,还存在西、中和东方向三种不同路径的变化。针对影响吉林省农业产量的东北地区夏季低温气候变异的研究,提出了相应的预测方法,据此准确地预测出2009年的夏季低温,得到吉林省主要领导的批示与采纳。还开展了生态环境和区域气候联系的研究,为吉林省在国务院获得生态试点省建设作出重要贡献。他荣获吉林省劳动模范和全国“五一”劳动奖章称号,并享受国务院政府特殊津贴。1997年,廉毅敏锐地看到,省级研究所如果不积极争取国家级的科研项目,很难获得生存与发展机遇。在他的带领下,1998—2020年,共获批13项国家自然科学基金项目,其中1项是同中国科学院大气物理研究所布和朝鲁教授合作,为国家自然科学基金重点项目。这些项目取得多方面的创新成果,获省部级科技进步奖14项,其中一等奖1项、二等奖5项。

廉毅在退休后,继续从事科研工作,于2006年2月20日成立了中高纬度环流系统与东亚季风研究开放实验室,担任主任;我和王会军、陈隆勋、何金海教授等为其学术委员会的主任和副主任。廉毅还聘请了国内著名学者20余位,请他们提供指导和帮助。在吉林省气象科学研究所同事们的积极努力下,他们争取到国家自然科学基金项目11项,撰写论文百余篇,为中国气象局和省级研究所科学研究事业作出重要贡献。在中国气象局的三次省级研究所评估中,均获全国第一,其中两次都是廉毅主持研究所工作时所取得的。

我同廉毅为师生关系,大他7岁,但也是挚友,感情深厚,多年交往,彼此关心。廉毅教授年近八十,仍

致力于省级研究所走“努力与国内外先进的大气科学理论、技术和方法接轨，向国内专家请教，以小求精，不断创新促发展之路”。正如他自己总结的所走过的道路：决策正确，奋斗不止，业绩催人，团队优秀。

在《东亚中高纬度环流与季风及其区域气候变异论文选》出版与中高纬度环流系统与东亚季风研究开放实验室成立18周年之际，我表示热烈祝贺！祝廉毅团队蒸蒸日上，长江后浪推前浪，永不停歇，取得更好的业绩！

丁一汇

（丁一汇：中国工程院院士）

2024年2月20日

序 二

我和吉林省气象科学研究所/中高纬度环流系统与东亚季风研究开放实验室主任廉毅教授是在中国科学院大气物理研究所1999年“973”项目年会上认识的，廉毅关于利用遥感陆地卫星解译东北西部荒漠化的发展研究受到广泛关注。他同时建议选择吉林省通榆县作为国际地-气交换的农牧交错带代表通量站，被采纳。从他的身上我深深感到，虽为省级研究所所长，但志存高远，坚持应用基础研究实践，眼光敏锐，能及时抓住时机，积极参与大院大所的国家项目研究，从中培养了人才，也锻炼了队伍。

2006年2月，在他的努力下成立了中高纬度环流系统与东亚季风研究开放实验室，我应邀担任其学术委员会副主任，同丁一汇院士和其他20余位专家受邀组成学术委员会。廉毅与同事们努力奋斗，获准11项国家自然科学基金项目，并取得了一批创新研究成果。他们提出极涡、NPO(北太平洋涛动)和阿留申低压，与北太平洋西风漂流区和赤道中东太平洋海温，组成北太平洋重要的海-气耦合系统；通过数值试验发现，衰减的ENSO位相可对中高纬阻塞和冷涡异常有强迫作用；也发现中高纬度环流系统对ENSO春季位相发展的重要反馈；他们在极端天气与气候事件研究方面也颇有建树。廉毅多次参加中国科学院大气物理研究所博士答辩会，还积极参与中国科学院大气物理研究所主持的项目研究，从而锻炼了他们的青年骨干，为日后的发展积蓄了能量。

吉林省气象科学研究所虽然不是“国家队”，但是他们坚持应用基础研究实践，以创新求发展，走出了省级气象研究所的发展新路，其开放实验室也有了一定的知名度，并为当地国民经济发展和生态环境建设作出重要贡献。

我和廉毅交往20余年，彼此关心，友情深厚。廉毅虽年近八十，仍具有勃勃生机，其团队蒸蒸日上。在《东亚中高纬度环流与季风及其区域气候变异论文选》即将出版，以及中高纬度环流系统与东亚季风研究开放实验室成立18周年之际，祝愿他们取得更优异的业绩，也祝愿廉毅活力永驻。

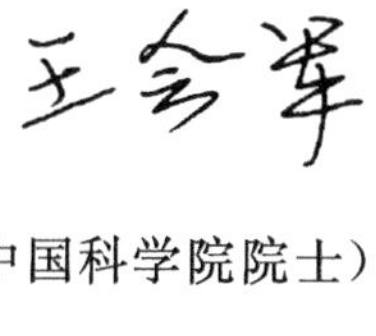

(王会军：中国科学院院士)

2023年4月

贺　词

甲辰新春，迎来《东亚中高纬度环流与季风及其区域气候变异论文选》即将出版与中高纬度环流系统与东亚季风研究开放实验室成立18周年，恰逢廉毅教授八十大寿，表示热烈的祝贺！我有他为知己，深感荣幸和由衷的高兴。

我和廉毅教授是于1997年2月24日在印尼巴厘岛参加第一届WMO(世界气象组织)国际季风会议相识的，彼此交谈，均有相见恨晚之情。自那以后，他也多次邀请我到吉林作学术报告，几十次相逢在国内的学术讲坛上。廉毅曾任吉林省气象科学研究所所长，总是自称为“游击队”，但常常展现出攻坚克难的“突击队”形象，在大气科学基础应用研究方面取得了有创新性的成果。特别在退休后，他仍然踔厉奋发，闯出了人生学术的又一高峰，堪为我等的榜样。该论文选中相当部分论文发表在国内外知名的学术期刊上，涉及北半球中高纬度极涡、NPO、阿留申低压、阻塞高压和高空切断冷涡等，其研究成果均有新的认识和创新，在极端天气与气候事件及其机理研究方面也颇有建树。在东亚—北太平洋夏季副热带季风研究方面有独到之处，廉毅应邀参加了WMO第一届(印尼)、第三届(杭州)和第四届(北京)国际季风会议。关于东北西部生态环境与区域气候的联系研究，采用多学科交叉也取得明显进展，为吉林省委、省政府领导的决策咨询发挥了重要作用。

2004年11月2日，WMO在杭州召开的第三届国际季风会议上，我同陈隆勋教授向参会的廉毅建议，在长春市成立一个北方的季风研究开放实验室，廉毅欣然接受。2006年2月，廉毅在北京召开了第一届中高纬度环流系统与东亚季风研究开放实验室学术委员会会议，我应邀担任开放实验室的副主任，从此我们的联系更加密切。

我们已经走过了相识27年的光景，彼此相互关心、相互支持，心灵相通，相互理解；人生几何？能与廉毅成为知交，实为人生幸事，其乐无穷！我衷心祝愿中高纬度环流系统与东亚季风研究开放实验室、廉毅及其团队再创辉煌！

（何金海：南京信息工程大学教授）

2024年2月20日

前 言

1999 年，因参加中国科学院大气物理研究所主持的“973”项目，我有幸结识了吉林省气象科学研究所廉毅所长。初见面时，他“行头”清廉，目光坚毅，透着一股坚韧不拔的精神劲头。一位省研究所所长，如此执着与国内大院大所加强联系和交流，其科学研究精神令人感动。从此我们结下了不解之缘，20 余载建立了父子般深厚的情感，他对我和课题组的学生们的扶持与关心倾注了大量的心血。

自担任中高纬度环流系统与东亚季风研究开放实验室的学术委员会委员以来，我们充分利用开放实验室这个平台，加强和国内外专家的交流与合作，获益良多。如今，在中高纬度环流系统与东亚季风研究开放实验室成立 18 周年之际，恰逢廉毅教授八十大寿，可喜可贺！廉毅带领“省级游击队”始终坚持“不断与国内外大气科学的理论、方法、技术接轨”的科研路线，取得了诸多傲人的科研成果，全国皆知。将这些成果结集成册，不只是为了系统整理关于东亚中高纬度环流与季风及其区域气候变异的相关理论成果，更好地指导实际业务，也是为了弘扬他们奋斗不止的精神，为我国省级科研团队树立标杆，用他们蒸蒸日上催人奋进的业绩，激励和鞭策青年一代。

《东亚中高纬度环流与季风及其区域气候变异论文选》收集了廉毅及其团队成员，包括沈柏竹、刘刚、李尚锋三位研究员和杨旭高级工程师等发表的 25 篇学术论文，集中在中高纬度环流系统与 ENSO、东亚—西北太平洋夏季副热带季风(包括东亚冬季风)和东北生态环境与区域气候变化三部分。

第一部分主要揭示了北太平洋极涡、阿留申低压和 NPO 等环流系统与西风漂流区和赤道中东太平洋 SST 之间的密切联系，以及上述北太平洋海-气耦合系统的形成机理，特别指出 NPO 异常位相不仅是夏季气候变化的重要前期信号，还是大气中除了天气尺度混沌分量外可提取的一种行星尺度稳定分量；系统地分析了北半球 5—6 月阻塞高压(简称阻高)活动特征，发现东北亚阻高活动天数具有明显的独立性，且 6 月跃升第一位，唯有 5 月北太平洋阻高是其 6 月的变化的稳定影响源，揭示了大气中冰晶相变过程引起非绝热加热的局地变化和垂直输送等对阻高强度及维持时间的影响机制，以及 1977 年前后乌拉尔山—鄂霍次克海双阻活动发生显著年代际突变的成因，还论证了初夏在贝加尔湖以南、青藏高原以北阻塞高压的形成过程；识别并研究了夏季东北冷涡的时空分布特征及其与北半球大气环流系统间的联系，指出前冬 NPO 正(负)位相有利于夏季东北冷涡活动偏弱(强)和中国东北降水偏少(多)。分析了初夏东北冷涡活动异常与北半球环流低频变化特征，系统总结了国内外近 80 年有关东北冷涡的形成和活动异常机理研究进展，近两年被国际学术期刊引用十几次；科学定义了冬季东北极端低温事件并将其划分为东北类和全国东部类极端低温事件，发现东北在这两类极端低温事件中都处于 300 hPa 波动能量辐散中心，并且全国类极端低温事件发生时罗斯贝波波动能量传播弱于东北类极端低温事件。在研究东北冷涡与东北夏季低温事件的机理的基础上，准确地预测出持续 15 a 高温的背景下 2009 年夏季低温的发生，其决策咨询报告被吉林省主要领导批示与采纳，还准确预测出 2010 年开始出现的冬季暂时变冷的趋势；揭示了春季 Niño 4 区 El Niño(La Niña)位相和西风漂流区冷(暖)位相配置模态强迫的非对称性，有利于 NPO 正(负)位相和初夏东北冷涡活动弱(强)；春季赤道中太平洋 El Niño 衰减位相与西风漂流区海温强迫产生后退的非绝热波并激发出热带大气流场椭圆形低频变化，增强了热带—副热带—中高纬度地区间流场呈 C-A-C-A 低频变化波列，是影响北太平洋副热带和中高纬度初夏环流异常的重要物理机制；春季赤道中太平洋 La Niña 衰减位相可以造成初夏鄂霍次克海至日本海上空产生异常高压遥响应，还间接触发了对流层低层东亚 EAP 正位相低频波列。在非线性分析方法上，采用了 WKBG 和信息流因果论，为研究中高纬度环流系统与外强迫之间的联系打开了“两扇门”，揭示了东北亚区域春夏转换时空特征与影响机制。

第二部分主要介绍有关东亚—西北太平洋夏季副热带季风的研究成果。尽管东北地区地处我国纬度较高的东北部，廉毅团队基于 20 世纪 60—70 年代夏季低温发生频率较高与东亚夏季风年代际偏弱的联系，给出了影响东北地区夏季风建立的标准、日期和主要特征，确定东亚—西北太平洋地区夏季副热带季风(EASSM)和活动范围候平均等时线的客观计算方法：季风总体以 110°—115°E 为轴心，向北凸起的折线表明同纬度中国大陆先建立，中国近海和西北太平洋地区后建立，向北凸起等时线的舌状如位于 110°E 以西为偏西路径，110°—115°E 为中间路径，115°E 以东为偏东路径，2022 年就属典型的偏西路径，造成长江中下游地区夏季干旱与持续高温；揭示了 EASSM 北进沿纬圈的东西向年际振荡特征，提出小于 35% 概率等值线的地区为其影响的边缘地带，北边界始于黑龙江省北部 48°N 附近，向东南经由黑龙江省南部和俄罗斯滨海区，穿过日本海止于日本北海道南部，其西边界由东北向西南沿中蒙边境止于青藏高原；提出了东北地区夏季受冷涡和副热带雨季影响的划分标准，以及各自的主要影响系统；研究了影响东北地区春、夏季降水的水汽输送特征，探讨了夏季风热带季风指数的定义等科学问题。

第三部分涉及东北生态环境与区域气候变化科学问题。参加中国科学院大气物理研究所主持的国家“973”项目“我国生存环境和北方干旱化趋势预测研究”，为廉毅团队进行气候与陆地卫星和下垫面变化多学科交叉研究提供了机遇。在吉林省省长几百万元基金项目资助下，采用了当时国际上先进的“3S”技术系统，解译了 20 世纪 80 年代与 90 年代后期的 Landsat TM 遥感信息，并与 20 世纪 30 年代吉林省土地利用调查数据对比，发现中国东北西部的区域增暖表现为现代土地覆盖类型上的荒漠化纬向扩展和区域气候变化暖干倾向，从三个方面逼近中国的商品粮基地——松辽平原，对其构成严重威胁。荒漠化加剧还表现为盐渍化和沙化土地覆盖面积增加，这不仅与该区气候暖干倾向联系显著，还与人类活动密切相关，具体表现为吉林省西部耕地在 1996 年比 1930 年增加了 3 倍，而林草地减少了 30%，原哲里木盟的新开河与乌尔吉木伦河间的撂荒地与沙化土地面积相当。相关研究成果呈送给吉林省委、省政府高层次决策部门，并成为省委、省政府被国务院办公厅批准为生态试点省建设的主要依据之一；1999 年 11 月，根据中国气象局、原国家林业局与荷兰 EARS 公司签署的“建立中国的荒漠化及粮食保障的中国能量与水平衡系统”合作项目协议，在吉林省松原市乾安县建立了大孔径闪烁仪(LAS)数据采集系统。LAS 是一种监测大气显热通量的国际成熟仪器，通过测量光束在传播过程中由空气引起的偏移脉冲，得出折射指数的结构参数 C_n^2，然后计算 H 值，再根据能量平衡方程，结合实测的土壤感热通量算出蒸散量值，提供了吉林省西部有代表的盐碱地表和大气间的感热通量参数，为荒漠化的成因以及陆地和气象卫星测量荒漠化提供了校正数据，也为当地盐渍化土地的治理提供了科学依据。此外，廉毅团队研究了吉林省粮食产量与气候变化的联系，并代表吉林省气象科学研究所与荷兰瓦格宁根大学签署了合作协议。

本文选于 2023 年 1 月开始选编，由廉毅统稿，王宁搜集、编排和校对文章，在电子化早期论文等方面做了大量工作。气象出版社胡育峰主任审阅全书，为本文选提出很多宝贵意见和建议。

特别感谢孙力、薛东友、裘善文、刘宗秀、杨崧、布和朝鲁、陆日宇、钟中、姜大膀、林中达、高枞亭、任红玲、张文哲、支克广、王杰、安刚、唐晓玲、赵滨、杨青、涂钢、刘实、姚耀显、王雅君、马梁臣、苏丽欣、曹玲、邵琪多、徐士琦、石晨、王冬妮等为文选所做出的贡献。我和团队中的龚志强、王启光、侯威、支蓉、张世轩、乔少博、赵俊虎、王晓娟、颜鹏程等也参与了部分研究工作。在此，对文选作者一并表示衷心的感谢！

文选的编写凝聚了廉毅及其团队多年的研究成果，记载着廉毅以气象工作者之身兢兢业业数十年的绵绵奔赴。观天测云，栉风沐雨，毫不倦怠。廉毅为气象事业奉献、孜孜不倦地探索之精神，以传承气象科研之火之匠心，是我等晚辈学习的楷模。借廉毅八十大寿之际，愿廉教授鹤寿松青，携气象精神与秋月明！衷心祝愿中高纬度环流系统与东亚季风研究开放实验室、廉毅及其团队创新进取、蒸蒸日上，以不懈的努力，取得更加丰盛的成果。

（封国林：国家气候中心研究员、扬州大学教授）

2024 年 2 月 20 日于北京

目 录

第一部分 中高纬度环流系统与ENSO

第二部分　东亚—西北太平洋夏季副热带季风

第三部分　东北生态环境与区域气候变化

第一部分

中高纬度环流系统与ENSO

Impacts of Polar Vortex, NPO, and SST Configurations on Unusually Cool Summers in Northeast China. Part I: Analysis and Diagnosis*

LIAN Yi [1,2]　SHEN Baizhu[1,2]　LI Shangfeng[1,2]　ZHAO Bin[3]　GAO Zongting[1,2]
LIU Gang[1,4]　LIU Ping[1,4,5]　CAO Ling[5]

(1 Laboratory of Research for Middle-High Latitude Circulation Systems and East Asian Monsoon, Changchun 130062; 2 Institute of Meteorological Sciences of Jilin Province, Changchun 130062; 3 National Meteorological Center, Beijing 100081; 4 Jilin Meteorological Science and Technology Service Center, Changchun 130062; 5 Center of Weather Film in Jilin Province, Changchun 130062)

Abstract: This study unveils the evolution of two major early signals in the North Pacific atmosphere ocean system that heralded abnormal high pressure blockings and cold-vortex activities across Northeast China, based on an analysis of the configurations of major modes including the polar vortex, the North Pacific Oscillation (NPO), and sea-surface temperature (SST) in the preceding winter and spring and atmospheric low-frequency disturbances in Northeast China. We analyzed these aspects to understand the atmosphere-ocean physical coupling processes characterized by the two early signals, and here we explain the possible mechanisms through which dipole circulation anomalies affect the summer low temperature processes in Northeast China. We further analyzed the interdecadal variation background and associated physical processes of the two early signals.

Keywords: polar vortex; NPO; North Pacific SST; atmosphere-ocean coupling; summer low temperatures in Northeast China; dipole pattern

1 Introduction

The interaction between the ocean and the atmosphere is an important subject of global climate dynamics. The North Pacific Ocean and the Eurasian continent have the most extensive ocean-land connections in the world. In this context, the atmosphere-ocean coupling processes and their associated teleconnections, as well as their impacts on climate variations, have attracted much attention. Bjerknes (1966, 1969) first proposed the concept of El Niño-Southern Oscillation (ENSO) that shows a major link with global climate anomalies in the form of atmosphere-ocean teleconnections. Rogers (1981) pinpointed a North Pacific Oscillation (NPO), similar to the North Atlantic Oscillation (NAO). Wallace et al. (1981)

* 本文原载于 *Advances in Atmospheric Sciences*, 2013, 30(1): 193-209

summarized the Western Pacific (WP) teleconnection pattern, an upper level teleconnection version of the NPO. Some studies have revealed a teleconnection wave train that travels in a large circular path between the North Pacific and the North American continent (PNA) (Namias, 1951; Klein, 1952; Wallace et al., 1981). Knowledge of atmosphere-ocean teleconnections in the North Pacific region has improved, especially since Madden et al. (1971, 1972) discovered the 40-day low frequency oscillation (LFO). Some regional circulation anomalies might be the result of a string of dynamic processes, including the forcing both in and outside of the region and the nonlinear interactions between waves and currents. The presence of the Pacific Japan Oscillation (PJO) is closely associated with convective activities over the western equatorial Pacific or ENSO events. Huang et al. (1988) held that the Pacific Japan (PJ) wave train that collided with North America via East Asia is directly associated with the convection heating in the Philippines area. They also proposed an East Asia-Pacific (EAP) teleconnection pattern, in an attempt to demonstrate the intervariation within the western Pacific subtropical high, the Meiyu trough, and the Northeast Asia high (Huang et al., 1987, 1992; Yang et al., 2005a, 2005b; Liang et al., 2011).

Attentions have been given to the possible linkage between the North Pacific atmosphere-ocean system and the unusually cool summers in Northeast China since the 1980s. Li (1989) reported a negative anomaly center above Northeast China in an El Niño summer, which contributes to the lower-than-normal temperature in the region. Zhang (1990) also indicated that El Niño results in the lower-than-normal temperature in the Northeast region. Other studies, however, have shown that in the 1950s—1970s, Changchun has an unusually cool summer if El Niño warming commences from the first half of the year (Lian et al., 1998). However, in the period 1980—1995, Changchun reported a higher or slightly higher than normal temperature pattern in summer, which had to be explained by the differed interseasonal oscillations staged by El Niño and outgoing long-wave radiation (OLR) (Lian et al., 1998). Zheng et al. (1999) pointed out that the unusually cool summer in Northeast China had a decadal connection with ENSO events. Sun et al. (2006) also confirmed this finding. In addition, scientists have begun to investigate the possible impacts of a polar vortex and NPO on climate in China, assembling a range of facts to make the teleconnections evident between the polar vortex and unusually cool summers in the Northeast, and between the NPO phases in the preceding winter and the summer atmospheric circulations, including Hadley circulation and precipitation distribution in China (Zhang et al., 1985; Zhao et al., 1992; Peng et al., 1998; Liu et al., 2002, 2003; Lian, 2007; Zhou et al., 2008).

In 2009, both Jilin and Heilongjiang, two provinces in Northeast China, reported an unusually cool summer (cooler than any in the prior 15 years), with most of the two provinces reporting an averaged temperature anomaly exceeding −0.5 ℃ (National Climate Center, 2010). The cool temperatures jeopardized local food production, especially the rice yield, which drew additional public attention again (Shen et al., 2011).

This study investigated the configuration of the polar vortex and NPO with North Pacific westerly drifts and ENSO phase, from the perspective of the interactions between the spring polar vortex and high-latitude atmosphere-ocean system, based on the two typical Northeast China summer low temperature incidences that occurred in 1964 and 1993. In this study, diagnosis analysis was applied to analyze the fact that the opposite early signals given by the two examples in the spring surprisingly yielded a remarkably similar sophisticated story, where high blocking and diversified abnormal cold vortex activities prompted the occurrence of an unusually cool summers in the Northeast China. Herein, the so-called "spring forecast threshold" is explained, and the physical mechanisms of high-low incidences that allow the mid-and high-latitude atmosphere-ocean coupling processes to stage a sustained cool summer decadal variation period in the Northeast China are revealed. The numerical modeling results will be presented in Part 2 of this study.

2 Data and calculation method

In this study, the 1948—2009 global reanalysis data was acquired from the National Centers for Environment Prediction/National Center for Atmospheric Research (NCEP/NCAR), the 1958—2002 reanalysis was provided by the European Center for Medium-Range Weather Forecast (ECMWF), and the 1881—2005 SST data from the Meteorology Office of the Hadley Center for Climate Change. According to the time scale defined by Shen et al. (2011) for classifying the extreme summer low-temperature events in Northeast China, a grade A cool summer event (i. e., serious) was defined as $\overline{T}_{6-8} \leqslant -1.0\sigma$ ($\overline{T}_{6-8}$ stands for the summer mean temperature of Changchun observatory station, σ represents standard deviation) observed at a weather station in the region; likewise, a grade B cool summer event was defined as $-1.0\sigma < \overline{T}_{6-8} \leqslant -0.4\sigma$. Both Changchun and Harbin were chosen as typical representatives of the temperature variation patterns in Northeast China (Lian et al., 1998). The summer of 1964, a cool summer event A, was chosen as a typical high-incidence example for diagnosis, and the summer of 1993, a cool summer event B, was chosen as a typical low-incidence example, based on the periodical variations of cold and warm events (high and low incidence of low temperature events) over the past 30—40 years recorded by two weather stations at which surface temperature observations were made for at least the prior 100 years (Lian et al., 1998; Shen et al., 2011).

The 500 hPa height field harmonic analysis (Chou, 1985) was applied to analyze the low-frequency variations of super-longwave perturbation that contributed to the blocking-high and cold-vortex activities in Northeast China (Lian et al., 2010). The band-pass filter used by Li (1990) was applied to calculate the 30—60-day low-frequency oscillations in the 500 hPa vorticity field. The Rossby wave energy dispersion process was analyzed using the approach developed by Bueh et al. (2007). Meanwhile, singular value decomposition (SVD) was employed for related analyses (Demmel et al., 1990).

The polar vortex activity area index was calculated based on the formula and parameters given by Liu (1986). The target Pacific region covers an area from 150°E—120°W. The Pacific-sector polar vortex area on the 500 hPa isobaric surface starts north of the contour bearing the signature of a polar vortex boundary to the pole:

$$S = \int_{\varphi}^{\frac{\pi}{2}}\int_{\lambda_1}^{\lambda_2} R^2 \cos\varphi \, d\lambda \, d\varphi = R^2(1-\sin\varphi)(\lambda_2-\lambda_1) \tag{1}$$

which can be calculated per 10° longitude in a deferential format. Here, R stands for the Earth's radius at 6378 km, the units of S are 10^5 km, φ represents latitude, λ represents longitude. Based on different regions, the indices of polar vortex area S were divided into five categories as follows:

when $\lambda_1=0°$, $\lambda_2=360°$, and S is the index of Northern Hemisphere polar vortex area;

$\lambda_1=60°$, $\lambda_2=150°$, and S is the index of Asian polar vortex area;

$\lambda_1=240°$, $\lambda_2=330°$, and S is the index of North America polar vortex area;

$\lambda_1=330°$, $\lambda_2=420°$, (i. e., region covering from 30°W—60°E, in an anticlockwise direction), and S is the index of Atlantic and European polar vortex area;

$\lambda_1=150°$, $\lambda_2=240°$, and S is the index of the Pacific polar vortex (PPV) area.

The definition area and indexes of the Northeast Cold Vortex (NCV) and the NPO can be found in the work of Zhao et al. (1992) and Liu et al. (2003); and the definition area of the central and western

Asian blocking highs can be found in the work of Liu et al. (2012). The westerly drift zone was confined to an area from 150°E—150°W to 30°—46°N.

The NCV was defined as follows: over the region (35°—60°N, 115°E—145°W), there is at least one closed contour on 500 hPa geopotential height, accompanied a cold center or cold trough, and it maintains this state for at least three days within the defined domain. The blocking high was defined as follows: There is a center of high pressure on 500 hPa geopotential height, and the position of the center moves <2.5 times the meridian distance each day; moreover, the high-pressure center is maintained for at least three days.

The index of NPO_L (L was the abbreviation for the family name of "LIAN Yi") was defined as follows:

$$NPO_L = P_{S1} - P_{S2}, \tag{2}$$

P_{S1}, P_{S2} means the area average of standardized sea surface pressure over the domain of (25°—40°N, 140°E—150°W) and (50°—65°N, 140°E—150°W).

The westerly drift index W was defined as follows: W = Asst where Asst is the area average of standardized SST over the domain (30°—46°N, 150°E—150°W).

The NAO index (NAOI) was defined as the difference in the normalized monthly sea level pressure (SLP) regionally zonal-averaged over the North Atlantic sector from 30°E—80°W to 35°—65°N (Li et al., 2003).

The statistical significance of a composite difference was assessed using the Student's test. The 95% confidence level was considered significant, unless otherwise stated.

3 Connections between the PPV and NPO seasonal evolution

3.1 Weak or strong anomalies for the spring PPV area

Four typical weak and four typical strong years were selected separately from the spring PPV area index series. Strong years (1984, 1993, 1996, and 2005) reported an averaged anomaly of $\geqslant 1.5\times10^6$ km^2, whereas the weak years (1948, 1964, 1967, and 2002) reported an averaged anomaly of $\leqslant 2.0\times10^6$ km^2.

Fig. 1a shows that the 500 hPa height-anomaly differences between four strong minus four weak years in the preceding winter formed in a pattern that features positive anomalies above the Aleutian Islands and negative anomalies on the southern flank of the North Pacific sector, a typical NPO-like negative phase. However, the 500 hPa height-anomaly differences (four years) depicted an entirely opposite picture against the preceding winter (Fig. 1b); they ended up in a north-south "−, +" distribution, suggesting a strong North PPV area, favoring an NPO-like positive phase.

The weak and strong PPV areas in the spring underwent a composite 1000 hPa wind anomalies dominated by the cyclonic circulation stretching from the Aleutian Islands in the center to the entire area north of 35°N over the North Pacific in a significant positive NPO phase, which in turn brought a significant westerly belt between 25°—40°N (Fig. 2a). The spring season weak and strong PPV areas weakened somewhat at the 1000 hPa wind fields in the summer, mainly dominated by an anticyclonic circulation stretching from west of the Aleutian Islands to the Sea of Okhotsk. Meanwhile, the north eastern and eastern parts of China were dominated by a strong northerly flow, which featured a weak summer monsoon but strong northerly cold air (Fig. 2b).

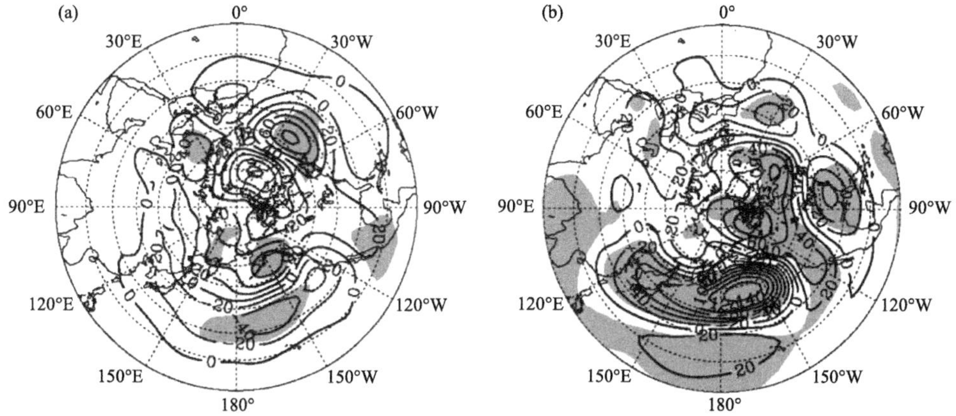

Fig. 1 Differences in 500 hPa height anomalies between four strong and four weak years: (a) the preceding winter, and (b) the following spring. Dotted lines represent the negative values, and shaded area indicates height anomalies beyond the 95% confidence level. Contour intervals are 20 gpm

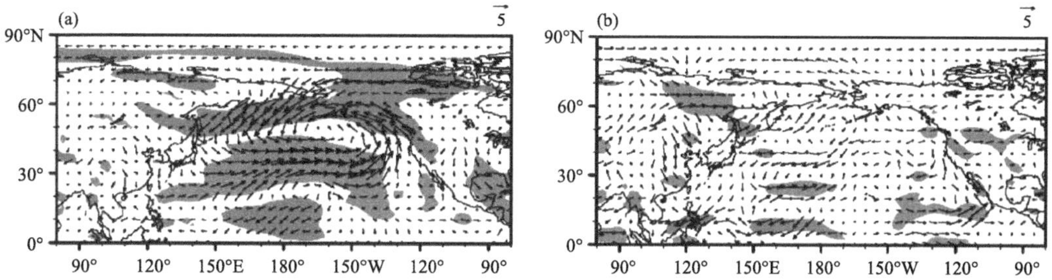

Fig. 2 Composite patterns of 1000 hPa wind anomaly differences between four strong and four weak years. Shaded area indicates the region of wind anomalies beyond the 95% confidence level. (a) the following spring, (b) summer (units: m s^{-1})

3.2 Main mode configuration of spring NPO and North Pacific SST

The spatial and temporal distribution of spring North Pacific SST and sea level pressure SVD1 are shown in Fig. 3. The mode accounted for 61% of the squared covariance and total contribution, with a field correlation coefficient of 0.74. These distributions indicate that the North Pacific SST is configured with the central and eastern equatorial Pacific La Niña phase and the warm westerly drift zone phase in the northwest, in line with the north-to-south SLP "+, −" wave train, or with the major coupled mode of negative NPO phases. Likewise, the North Pacific SST is configured with the central and eastern equatorial Pacific El Niño phase and the cold westerly drift zone phase in the northwest, in line with the north-to-south SLP "+, −" wave train, or with the major coupled mode of positive NPO phases. As far as the time coefficient trends are concerned, the two fields mostly ran apposite in phase distribution except the 1960s. The North Pacific SLP in the north, with the Aleutian Islands at the center, has been dominated by a low-pressure system since 1976—1977, when the Aleutian low suddenly strengthened in the winter. On the southern flank (a subtropical region), the SLP was mainly under the influence of high-value pressure systems. As a result, NPO entered a decadal variation period dominated mainly by a positive phase. Correspondingly, North Pacific sea surface temperature anomaly (SSTA) has been dominated by the cold northwest drift zone phase in the north since 1976, with the central and eastern equatorial

Pacific under an El Niño phase, which agreed with the Pacific Decadal Oscillation (PDO) warm phase analysis made by Mantua et al. (1997).

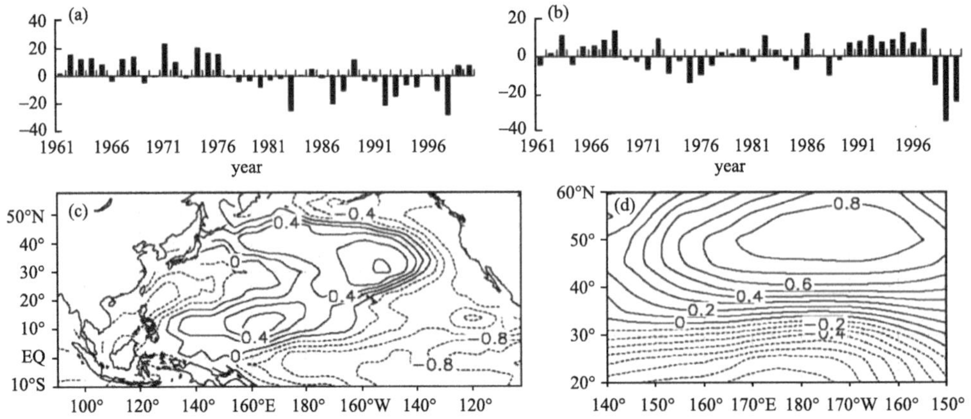

Fig. 3　Distribution of spring North Pacific SST (left) and sea-level pressure (right) in the same season and associated time coefficients: (a) and (b) show time coefficients, and (c) and (d) show distributions

For five typical years with deepened spring Pacific Aleutian lows (1980, 1998, 1991, 1981, and 1987) and weakened spring Pacific Aleutian lows (1989, 1964, 1976, 1972, and 1967), based on SVD1 time coefficient, the real strength of Aleutian low and NPO index were chosen. Then composite SST and wind anomalies were analyzed. Our results are presented in the next section.

Fig. 4a depicts the composite sea surface wind and temperature anomalies for five individual years with deepened North Pacific Aleutian lows. The deepened northern Aleutian low bred out an equatorial northwesterly flow, or even a westerly flow over east of 155°E, which weakened the Walker circulation. The deepened Aleutian low in the spring also led to the convergence of prevailing westerly jets in the eastern Pacific westerly drift zone (35°—45°N) in the south and the northwesterly north 45°N to the west, forcing the westerly drift zone SSTA to remain in a cold phase. Five individual years with weakened northern Pacific Aleutian lows (Fig. 4b) had a composite sea surface wind and temperature anomaly running just opposite in distribution, with easterly jet prevailed over the westerly drift zone, and hence a warm SSTA phase, and the southeasterly over the equatorial zone east of 155°E, was favorable for a strengthened easterly under the Walker circulation. In Fig. 4c, a cold SSTA phase in the westerly drift zone and a warm SSTA phase in the central-eastern equatorial Pacific ocean can be clearly seen. Our analysis confirms the rationality of the aforementioned coupled atmosphere-ocean SVD1 mode, as well as the dynamic role played by sea-level pressure in defining SSTA fields.

In this section, a composite analysis and the SVD method were employed to reveal the major mode configuration of the PPV, NPO, and northern Pacific SST (the westerly drift zone, ENSO phase), in an attempt to sort out the strong and weak spring Pacific sector polar vortex area that contributes to positive and negative NPO phase anomalies at the mid-troposphere and in SLP. Meanwhile, positive or negative NPO phase anomalies often agreed with the first coupled mode that enjoys a stable westerly drift zone SST cold and warm phase and a similarly stable El Niño or La Niña phase (i. e., SVD1 accounted for 60% of the squared covariance and total contribution). This conclusion was further confirmed by the composite sea surface wind and temperature anomalies in a typical year with deepened and weakened spring Pacific Aleutian low.

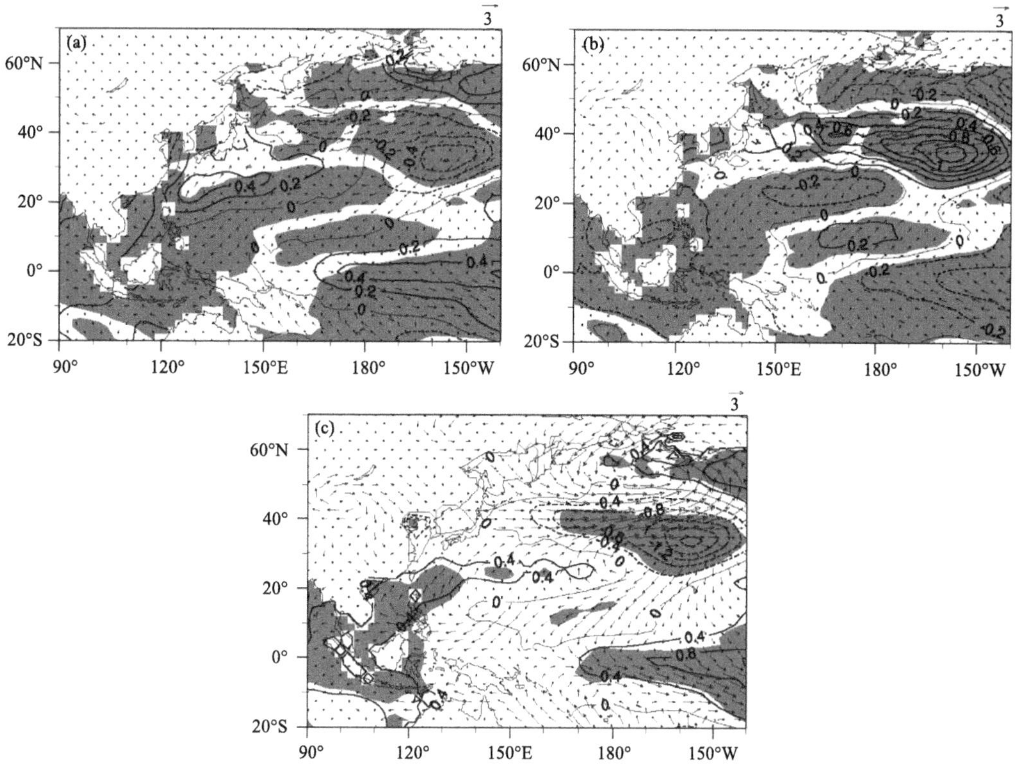

Fig. 4 Composite sea surface wind and temperature anomalies in a typical year with deepened and weakened spring Pacific Aleutian low: (a) deepened, (b) weakened, and (c) deepened minus weakened. Vectors: winds, unit: m s^{-1}; contours: SST anomalies. Dashed lines indicate negative values, shaded area indicate the region of wind anomalies beyond the 95% confidence level. Contour intervals are 0.2, and units are ℃

4 Seasonal evolution of two early signals and associated atmospheric low-frequency variations

We analyzed the basic mode configuration of atmosphere-ocean systems in the early stage and the associated seasonal evolution for cool summer years based on two cities (Changchun and Harbin) that reported significant unusually cool summers in two typical years featuring either a high (1964) or low (1993) incidence of cool summers on a decadal background. The diagnosis of the seasonal evolution of two different early signals and their associated mid-troposphere harmonic waves and low-frequency vorticity oscillation enhanced our knowledge of the low-frequency physical processes that characterizes the major mode configuration of the PPV, NPO, and North Pacific SSTA (the westerly drift zone, ENSO phase), and of their possible impacts on the summer middle and high latitudes atmospheric circulation over Asia.

4.1 Atmosphere-ocean system configuration in the early phase of cool summers in Northeast China

In Table 1, a large difference between two individual years (1964 and 1993) in terms of the basic atmosphere-ocean system configuration in the early phase and associated seasonal variations can be seen. In the preceding winter, the PPV area index, NPO, and NAO basically remained in opposite phases. Meanwhile, the westerly drift zone and the central and eastern equatorial Pacific SST agreed with one another in phase, though became opposite in the spring. This can be explained by the fact that the preceding winter of 1964 and the following spring also reported the opposite phase of NPO, NAO, westerly drift zone, and central and eastern equatorial Pacific index.

Table 1 Seasonal evolution of indexes in 1964 and 1993

Project	Year	Last winter	Spring	Summer
PPV area index	1964	−1.80	−20.60	17.20
(anomaly)	1993	4.30	24.70	9.20
NPO index	1964	0.67	−0.30	0.44
	1993	0.10	0.10	−1.19
NAO index	1964	−0.76	0.90	−0.62
	1993	1.41	−0.10	0.52
ENSO	1964	0.81	−0.40	−0.67
	1993	0.28	0.84	0.47
Westerly drift index	1964	−0.36	0.64	0.40
	1993	−0.29	−1.18	−1.90

4.2 Seasonal evolution of atmosphere-ocean system

4.2.1 500 hPa height anomalies

The 500 hPa height anomalies are given in Fig. 5, in the preceding winter of 1964 underwent a typical strong negative phase over the North Atlantic Ocean (a southbound "+, −" wave train from Iceland), and a positive NPO-like phase above the North Pacific (a southbound "−, +" wave train from the Aleutian Islands) can be seen. In the spring, NAO remained in the negative phase, though significantly weak, with the North Pacific turning into a significant negative phase (a southbound "−, +" wave train from the south). Positive anomalies started to occupy the entire middle and high latitudes across the North Pacific, with the strongest center over the Northern Hemisphere. In the summer, NAO continued at its weak negative status. However, it was very difficult to determine the NPO phase status, positive or negative, at middle and high latitudes, as the southbound "−, +" wave train has moved to Northeast Asia from the south. A positive 500 hPa height anomaly stretched from the Laptev Sea to northern Kamchatka Peninsula via the East Siberian Sea from the south. Negative anomalies covered the mid-latitudes in Asia and their southern neighbors, placing the southern part of the Sea of Okhotsk, Northeast China, and southern Yakutsk under the influence of the same negative anomalies. Meanwhile, a weak positive anomaly (Figs. 5a−c) prevailed over the Ural Mountains in the north. In the preceding winter, and in the spring and summer of 1993, both NAO and NPO registered a phase evolution opposite that of 1964. In the preceding winter, NAO was in a positive phase enjoying a southeast bound "+, −" wave train over Iceland from the northwest, while NPO remained in an atypical negative phase. In the spring, NAO was atypically positive in phase, exhibiting a southwest bound "−, +" wave train from the northeast, with NPO in a strong positive phase. The negative anomalies above the Aleutian Islands swept across the entire middle and high latitudes in the North Pacific, with its central value at its lowest over the Northern Hemisphere. In the summer, both NAO and NPO were positioned in a significantly weakened negative phase, with strong positive anomalies running from the Asian Ural Mountains in the north to Novaya Zemlya, suggesting that the blocking was particularly strong in western Asia, with a relatively weak positive anomaly blocking over a stretch from Lake Baikal to the Sea of Okhotsk. As a result, the Eurasian mid-latitudes were prevailed upon by negative anomalies, and a vast region from Northeast China to 30°N and the northwestern Pacific were dominated by 20 gpm negative anomalies (Figs. 5d—f).

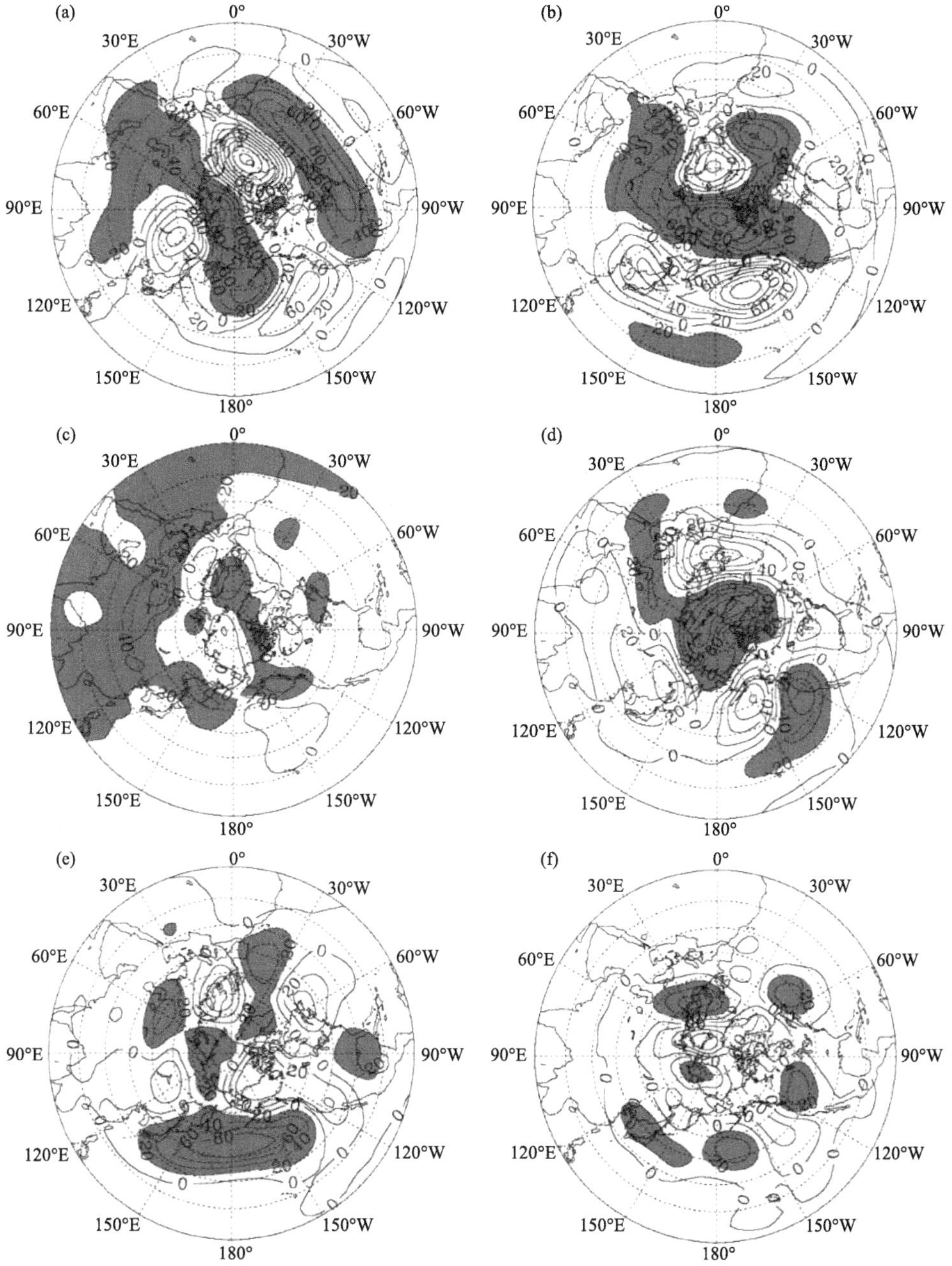

Fig. 5　Seasonally averaged 500 hPa height anomalies: (a) the preceding winter, (b) the spring, and (c) the summer of 1964; and (d) the preceding winter, (e) the spring, and (f) the summer of 1993. Shaded areas are for ≤−20. Contour intervals are 20, and units are gpm

Apparently, Northeast Asian mid-latitudes were dominated by significant negative anomalies, despite the fact that 1964 and 1993 had somewhat different atmosphere-ocean mode configurations and seasonal evolutions in the early stages, and that the summer of 1964 was prevailed upon by Northeast Asian high-latitude blocking and the summer of 1993 a western Asian mid-and high-latitude blocking.

4.2.2 SST anomalies

Fig. 6 depicts the seasonally averaged SST anomalies over the North Pacific. In 1964 the SSTA over the North Pacific westerly drift zone changed from the eastbound "−, +" wave train featuring a vague warm or cold phase in the preceding winter into an abruptly enhanced spring warm phase that sustained and further enhanced into the summer. In the preceding winter, the central and eastern equatorial Pacific SSTA remained in a pattern that featured a weak warm phase for the central Pacific and a weak cold phase for the eastern Pacific. The pattern was transformed into a consistent strong La Niña phase in the spring that continued into the summer. In the preceding winter of 1993, the North Pacific westerly drift zone SSTA was warm in the northwest and cold in the southeast. The SSTA underwent a cold phase in the spring and a noticeable cold phase in the summer. Meanwhile, the central and eastern equatorial Pacific SSTA changed from a weak warm phase in the preceding winter to a strong El Niño phase in the spring, proceeding into a noticeably weakened warm phase in the summer.

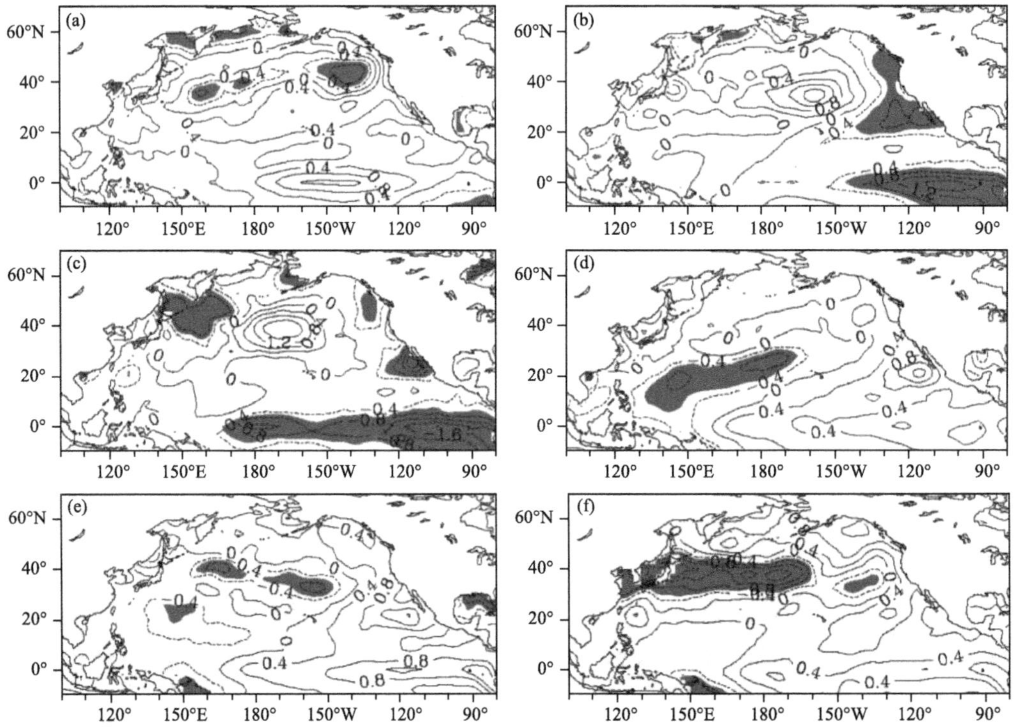

Fig. 6　Seasonally averaged SST anomalies in the Northern Hemisphere: SSTA in (a) the preceding winter, (b) the spring, and (c) the summer of 1964; and SSTA in (d) the preceding winter, (e) the spring, and (f) the summer of 1993 (Dashed lines are for negative values, and shaded areas are for SSTA ≤ −0.6. Contour intervals are 0.2, and units are ℃)

4.3 Harmonic waves and LFOs

4.3.1 500 hPa ultra-longwave and longwave activities

Fig. 7 shows the 500 hPa ultra-longwave perturbation along 45°N from April to July. During the period April—May 1964, the Northwest Pacific mostly underwent a mobile longwave trough and ridge, rather than a stable ultra-longwave trough above its central and western parts. The situation changed starting at the end of May, when a relatively stable ultra-longwave disturbance trough appeared, retreating westward and wandering over an area west of 120°E in June—July (Fig. 7a, thick arrows). During the period April—July 1993, except the first half of June, the slow eastbound movement of the upstream ultra-longwave disturbance confined the western boundary of the trough influence east of 130°E, though the trough made a rapid westbound advance in the second half of June, dominating the areas near 120°E. As a result, the central and western parts of the northwestern Pacific (Fig. 7b, 120°E—150°W, thick arrows) was solidly occupied by a stable ultra-longwave disturbance trough.

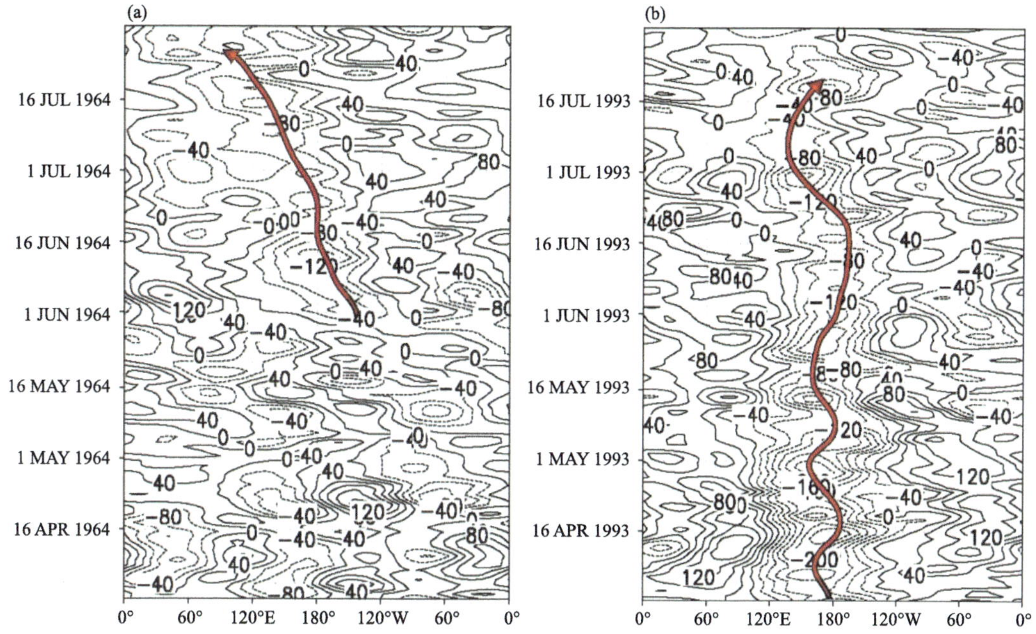

Fig. 7 500 hPa ultra-longwave perturbation along 45°N from April to July. Thick arrows represent the propagation of the low frequency: (a) 1964 and (b) 1993 (gpm)

Fig. 8 shows the 500 hPa ultra-longwave perturbation along 65°N from April to July. In late May 1964, an ultra-longwave disturbance ridge that originated from the central and eastern Pacific started to scale back to an area near 120°E. The same movement occurred twice in June—July (Fig. 8a, thick arrows). In mid-June 1993, a noticeable ultra-longwave ridge that originated from the central and eastern Pacific retreated to an area near 120°E (Fig. 8b, thick arrows).

4.3.2 500 hPa vorticity and associated 30-60-day LFO

Fig. 9 illustrates the time-longitude sections of 30-60-day vorticity at 500 hPa along 45°N from April to August. In the first half of June, the northeast cold vortex activity area (45°N, 125°E) saw the eastbound low-frequency vorticity originated from 100°E in 1964 (Fig. 9a, thick arrows) and the westbound low-frequency vorticity stemmed from 160°E in 1993 (Fig. 9b, thick arrows). In the same manner, Fig. 10, where the time-latitude sections of 30-60-day vorticity at 500 hPa along 125°E from April to August were given, shows that in most of June, a northbound positive low-frequency vorticity that originated from 15°—25°N appeared in 1964 (Fig. 10a, thick arrows), and a southbound negative low-frequency vorticity that originated north of 50°N turned up in 1993 (Fig. 10b, thick arrows). This above analysis indicates that 1964 and 1993 registered a vorticity propagation in reverses direction, regardless of longitudinal or zonal orientation.

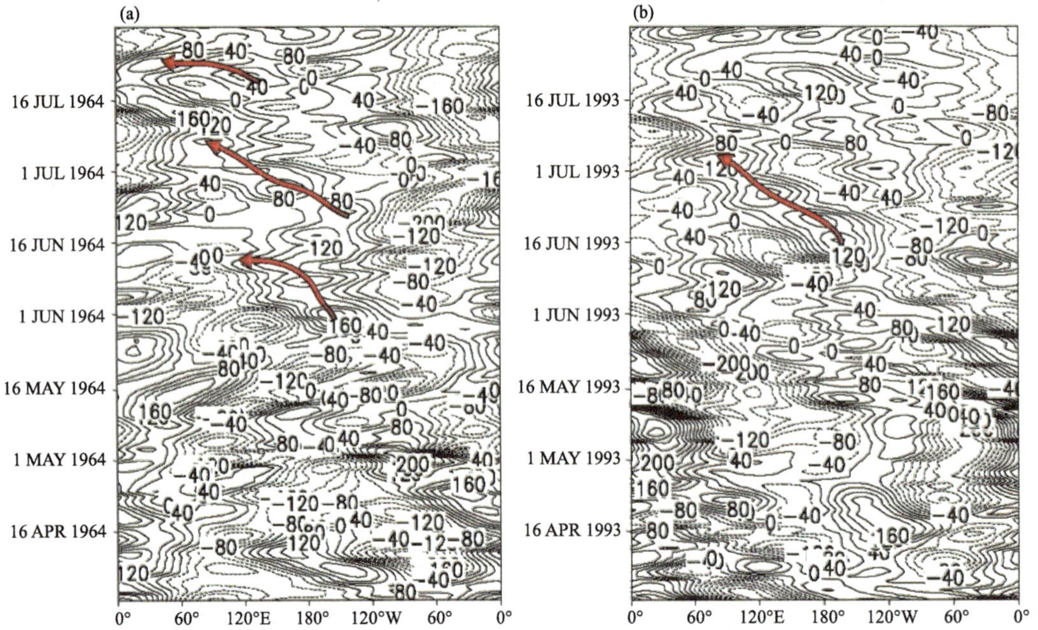

Fig. 8　500 hPa ultra-longwave perturbation along 65°N from April to July.
Thick arrows represent the propagation of the low frequency: (a) 1964, and (b) 1993 (gpm)

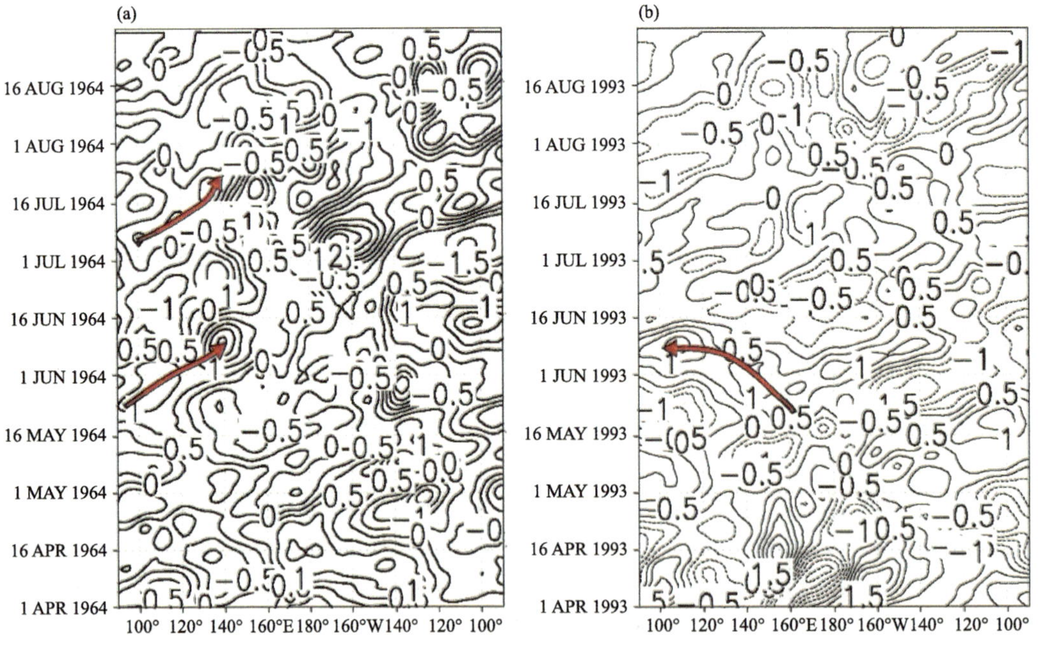

Fig. 9　Time-longitude sections of 500 hPa 30-60-day vorticity along 45°N from April to August ($10^{-5} s^{-2}$).
Thick arrows represent the propagation of the low frequency: (a) 1964 and (b) 1993

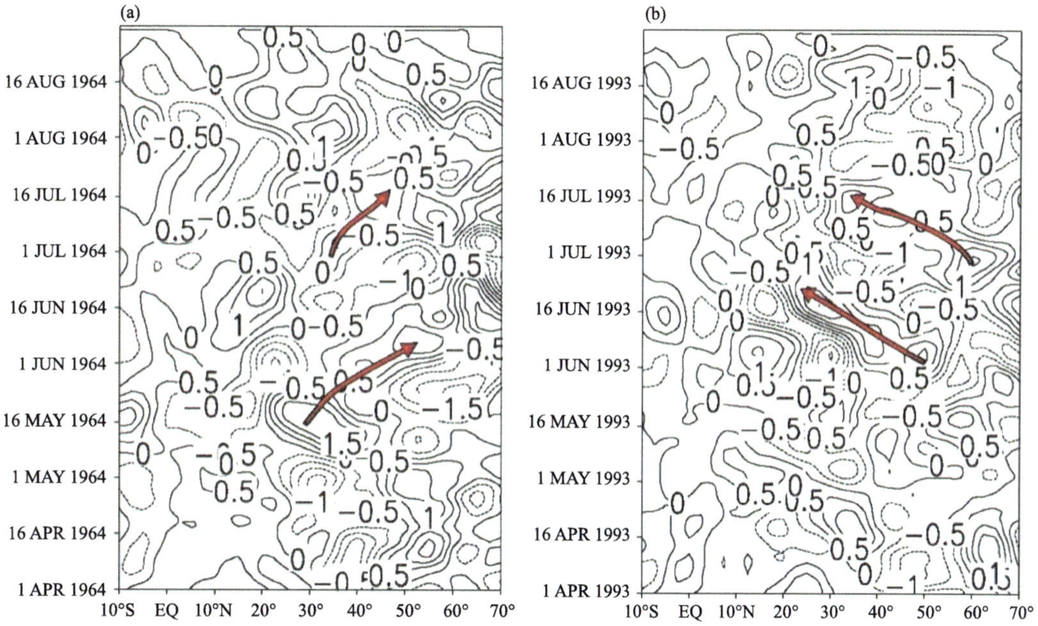

Fig. 10　Time-latitude sections of 500 hPa 30-60-day vorticity oscillation along 125°E from April to August ($10^{-5}s^{-2}$). Thick arrows represent the propagation of the low frequency: (a) 1964 and (b) 1993

5　Rossby waves and abnormal summer dipole

The preceding two sections discuss the relationship between a strong or a weak PPV in the spring and a positive or a negative NPO phase, as well as associated North Pacific SSTA (cold or warm westerly drift zone and ENSO phase). The configuration resulted in a strong PPV in the spring of 1993 and a weak PPV in the spring of 1964. As a typical case, both of them contributed to the unusually cool summer events in Northeast China. The configuration also caused a 30—60-day LFO to the 500 hPa height anomaly field, ultra-longwave, longwave and vorticity, and defined the seasonal and monthly evolution pattern of SST anomalies. To further understand the northeast cold vortex, a system having a direct bearing on the summer low-temperature events in Northeast China, and the sustained abnormal activities of the Asia blocking high (dipole), a major system that has an impact on summer low-temperature events, this section discusses the Rossby wave energy dispersion process reported in April—June of 1964 and 1993, to establish a connection between the dipole and early summer circulation anomalies and to sort out their possible wave sources (Bueh et al., 2007).

Fig. 11 shows the 500 hPa circulation anomalies and the corresponding wave action in April—June of 1964 and 1993. To compare the circulation anomalies at different latitudes, the figure also shows geopotential height anomalies in the form of stream functions (i. e., geopotential height anomalies multiplied by $\sin 45°/\sin\varphi$, where φ represents latitude). Comparisons among Figs. 11a—f indicate that a "—, +, —" height anomaly wave train appeared over an area stretching from the 500 hPa polar area to the North Pacific, and from East Asia to North America, in the April of 1964, and a "+, —, +" height anomaly wave train appeared over this region in 1993. The wave sources were basically positioned over the North Pacific region in 1993 (Fig. 11d). In 1964, the wave sources were positioned mainly over East Asia, with the strongest wave source near the western coast of North America (in Fig. 11a, the light gray area represents the largest combined area). No large variations of height anomaly wave train occurred in the May

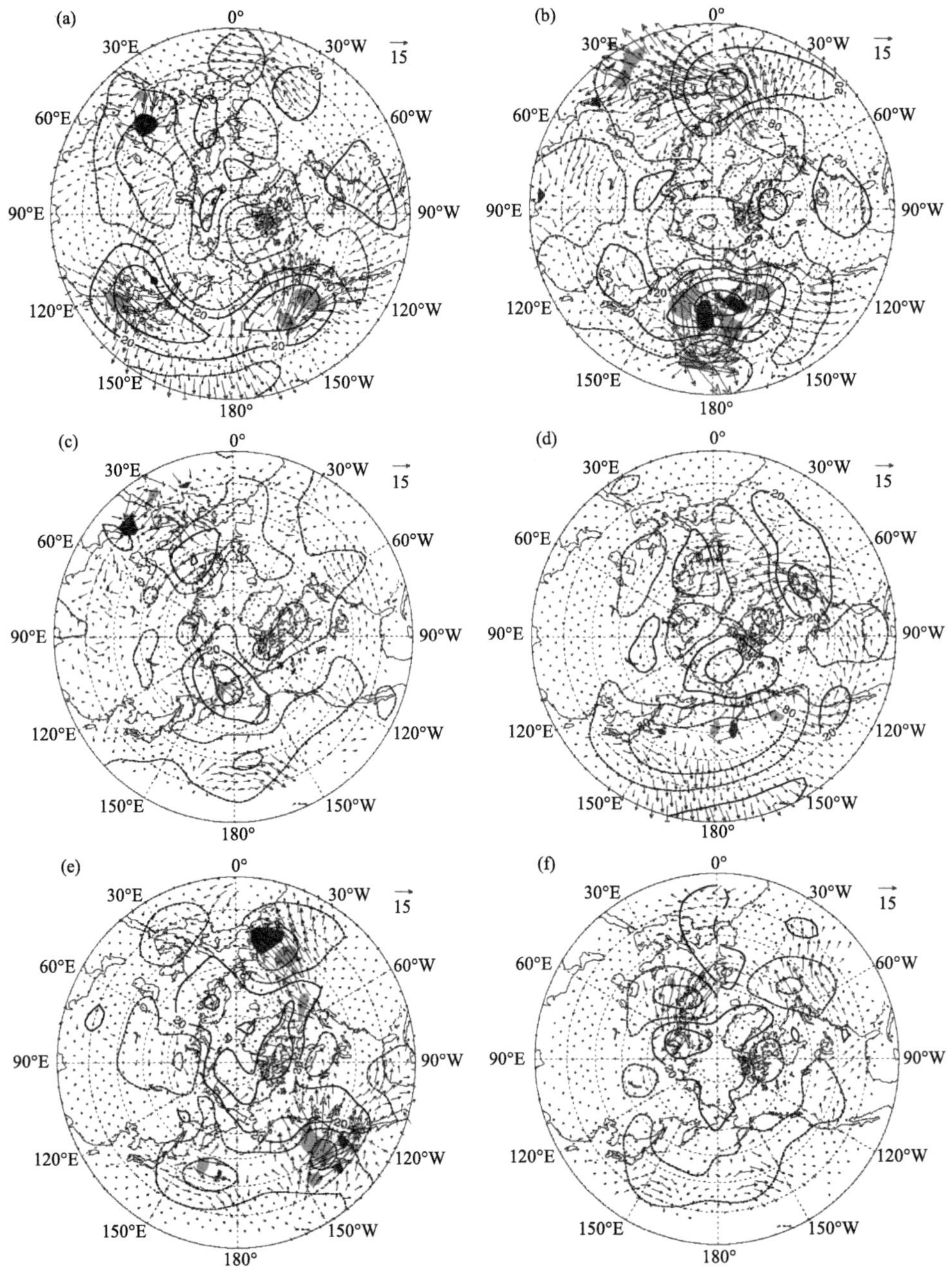

Fig. 11　500 hPa geopotential height anomalies (gpm, contours) from April to June and associated wave action flux (arrows). Contours are set at ± 20, ± 80, and ± 160 gpm, respectively. Dashed lines represent negative values. Light shading indicates wave action flux divergence that is $>2\times10^{-5}$ m s^{-2} area. Dark shading indicates wave action flux convergence that is $<-2\times10^{-5}$ m s^{-2} area: (a) April, (b) May, and (c) June of 1964; and (d) April, (e) May, and (f) June of 1993

months of both 1964 and 1993 in the polar sector and in the North Pacific. However, in 1964, the polar sector registered a negative anomaly, with AO in a typical positive phase. Meanwhile, the Aleutian low weakened noticeably again, allowing the positive anomalies to sweep across the North Pacific with a center in the south of the Aleutian Islands. The positive anomalies enhanced rapidly, with the central value reaching 160 gpm, and they became the strongest wave source and sink at middle and high latitudes over the Northern Hemisphere (Fig. 11b). The wave action was basically northerly and southerly in distribution, either pointing to the equator above the northwestern Pacific or moving toward the pole over the northeastern Pacific. In May 1993, the positive polar anomalies left AO in a typical negative phase. At that moment, both wave source and sink displayed a large variation in distribution, mainly over the western coast of North America, or near Great Britain east of the North Atlantic (Fig. 11e). In June 1964, the strongest positive anomaly zone in the Northern Hemisphere was found in an area stretching from the Aleutian Islands to the north of the Kamchatka Peninsula, accompanied by wave sources. The positive anomaly zone was apparently the result of the northwestward retreat of the positive anomalies across the North Pacific with a center in the south of the Aleutian Islands in May 1964 (the Rossby wave action flux in early, mid, and late June 1964 makes a further evidence, though not shown in the figure). In the south of the anomalies sat a vast negative anomaly zone, a typical dipole anomaly pattern in Northeast Asia (Fig. 11c). In June 1993, the most significant positive anomalies prevailed over an area in the north of the Ural Mountains, constituted the west blocking in Asia, which could mostly be the result of the eastbound shift of the positive anomalies that wandered above the Scandinavian Peninsula in May 1993 (Fig. 11f). The Rossby wave action flux in early, mid, and late June 1993 makes a further evidence to support the scenario (not shown in the figure). It can be seen from Tables 2 and 3 that the June and summer of 1964 was prevailed upon by the east blocking and abnormally increased northeast cold vortex days in Asia, which defined a typical Northeast Asia dipole configuration (Fig. 11c). However, in the June and summer of 1993, the Ural blocking high (west blocking) prevailed, and abnormally increased north-east cold vortex days, which constituted another Asia dipole configuration (Fig. 11f).

Table 2　Number of blocking high days in the June of 1964 and 1993 (units: d)

Year	Project	Easterly	In the middle	Westerly	Others
1964	June	16	0	0	14
	Early	8	0	0	0
	Mid	8	0	0	10
	Late	0	0	0	4
1993	June	12	12	12	13
	Early	3	12	6	7
	Mid	0	0	6	0
	Late	9	0	0	6

Table 3　Summer blocking high and abnormally increased northeast cold vortex days in 1964 and 1993 (units: d)

Year	Blocking high			NCV activities (anomaly)
	East	Central	West	
1964	+6.4	−12.5	−4.7	52(+13)
1993	−3.6	−11.5	+5.4	46(+7)

6　Decadal variation background of 1964—1993 spring polar vortex area index

Fig. 12 depicts the opposite variation trends of the PPV area index (green line) and the polar vortex index in other Northern Hemisphere regions in the spring seasons of 1948—2009. The Mann-Kendall (M-K) method was used to test the point mutation, which indicated that the former underwent an abrupt change in the late 1960s (Fig. 12b) that was noticeably different from the one reported by other regions. The PPV saw a noticeable decline in the spring of 1964, while the same indicator saw a noticeable boost in the spring of 1993. In Fig. 12a, one can see that five polar vortex index curves ended up in a continued downward trend heading to 2009.

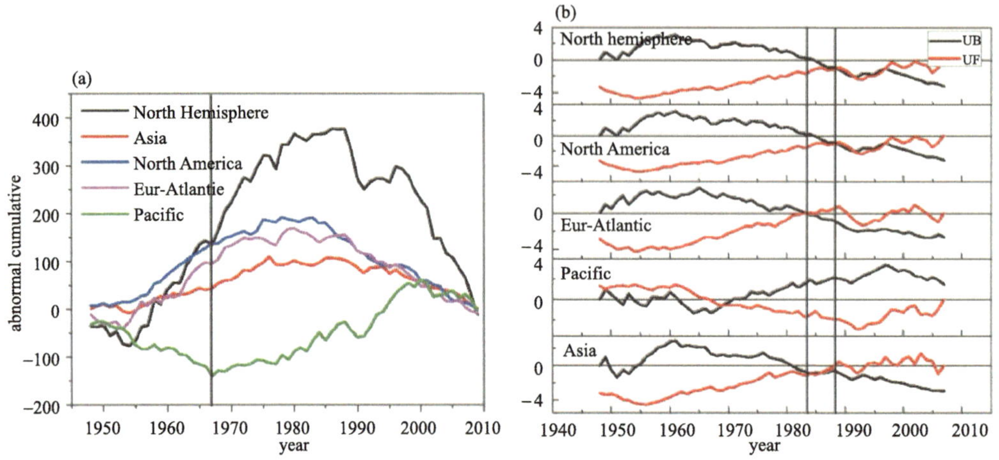

Fig. 12　(a) Combined spring polar vortex area index anomalies, and (b) M-K tests

7　Discussion and conclusion

(1) In this study, the composite anomalies and SVD method were employed to reveal the major mode configuration of the PPV, NPO, and North Pacific SSTA (westerly drift zone, ENSO phase), in an attempt to sort out the strong and weak spring PPV area that may contribute to positive and negative NPO phase anomalies at the mid-troposphere and in the sea level pressure fields. Meanwhile, positive or negative NPO phase anomalies often agreed with the first coupled mode that enjoys a stable cold or warm westerly drift zone SST phase and an equally stable El Niño or La Niña phase (e. g., SVD1 accounted for 60% of the squared covariance and total contribution). This conclusion was further confirmed by the combined sea surface wind and temperature anomalies in a typical year, with deepened or weakened spring Pacific Aleutian low.

(2) The year 1964, with a high cool summer incidence, and the year 1993, with a low cool summer incidence, were different only in basic atmospheric mode configuration, and their major North Pacific SST phases were similar. Compared with the preceding winter, the spring of 1964 included a reversed phase realignment of the PPV index, NPO/NAO, westerly drift zone, and the central and eastern equatorial Pacific SSTA. However, 1993 was consistent in both the preceding winter and the following spring. As a result, the opposite phases of major atmosphere-ocean system modes in the spring of two individual years would naturally produce a different evolution from the spring to the summer. For example, in the spring of 1964 a very weak PPV occurred, with NPO in a typical negative phase, the westerly drift zone in a strong warm phase, and the central and eastern equatorial Pacific SSTA in La Niña phase. In contrast, in the spring of 1993, the polar vortex was strong, and NPO occurred in a typical positive phase, with the westerly drift zone in a zero phase and the central and eastern equatorial Pacific SSTA in an El Niño phase. Entering the summer, the westerly drift zone was positioned in a noticeably cold phase, and the central and eastern equatorial Pacific SSTA was in a weakened El Niño phase. However, unusually cool summers occurred in both individual years, regardless of the difference between the two. The reversed realignment of basic atmosphere-ocean system modes from the preceding winter to the following spring, creating the "spring forecast threshold."

(3) In this study, diagnoses were conducted to understand the evolution patterns of mid-troposphere harmonic waves, low-frequency vorticity oscillations, and Rossby wave energy, to further reveal the fact that both 1964 and 1993 shared a low-frequency physical process that contributed to the mid-and high-latitude summer dipole circulation anomalies, even though the two individual years exhibited entirely different modes of configuration for spring signal, PPV, NPO, and North Pacific SSTA (westerly drift zone, ENSO phase). However, in the spring of 1964, a weak polar vortex and a typical negative NPO combined with the warm westerly drift zone and the La Niña-oriented central and eastern equatorial Pacific SSTA to create a dominant Okhotsk blocking high (eastern blocking) in the coupled atmosphere-ocean process from the spring to the summer. The northeast cold vortex acquired its positive vorticity mainly from the western path, and from the low-latitude transport as well. However, in the spring of 1993, the strong polar vortex area and positive NPO phase collaborated with the zero phase of westerly drift zone SSTA and the El Niño-oriented central and eastern equatorial Pacific SSTA to stage a Ural blocking high (western blocking), with the northeast cold vortex obtaining its positive vorticity mainly from the westbound retreating ultra-longwave disturbance trough (eastern path) in the central North Pacific, and from high-latitude transport as well.

(4) The PPV index exhibited an opposite variation trend against the polar vortex index in other Northern Hemisphere regions in the spring seasons of 1948—2009. The M-K method was employed to test the point mutation, which indicated that the former registered an abrupt change in the late 1960s that was noticeably different from the one reported by other regions. The PPV area noticeably declined in the spring of 1964, while the same indicator noticeably enhanced in the spring of 1993. The pattern revealed from a different perspective that both 1964 and 1993 underwent an unusually cool summer in Northeast China, regardless of the large difference between the two in early basic atmosphere-ocean mode configuration and in seasonal evolution. On one hand, the two individual years, either as a high incidence or as low incidence years, had different configurations of early signals and summer dipole, which could probably be explained by the difference in interdecadal variation. On the other hand, both 1964 and 1993 shared a low-frequency physical process that contributed to the mid-and high-latitude summer dipole circulation anomalies.

In this study, only the spring polar vortex area anomalies in four typical strong and four typical weak years, and the basic patterns of abnormal phase couplings between NPO and the North Pacific westerly drift zone and the central and eastern equatorial Pacific SSTA were analyzed. The years 1964 and 1993, where a significant cool summers occurred in Northeast China, were chosen to represent typical years of a weak or a strong PPV area corresponding to a high or a low cool summer incidence. The two individual years were simply opposite in phase anomalies, in terms of polar vortex area, NPO, the westerly drift zone, and the central and eastern equatorial Pacific SSTA. The significant difference, as such, probably represents the major footprints of interdecadal variations of summer low-temperature events in the northeast. Some other points, including early signals and atmosphere-ocean coupling processes in a typical low-or high-incidence year and their applicability to other individual years, require further study.

Acknowledgements. This work was jointly supported by the National Natural Science Foundation of China (Grant Nos. 41175083 and 41275096), and the Special Fund for Meteorological Scientific Research in the Public Interest (Grant Nos. GYHY201006020, GYHY201106016, and GYHY201106015).

References

BJERKNES J,1966. A possible response of the atmospheric Hadley circulation to equatorial anomalies of ocean temperature [J]. Tellus,18(4):820-829.

BJERKNES J,1969. Atmospheric teleconnection from the equatorial Pacific[J]. Mon Wea Rev,97(3):163-172.

BUEH C L,NAKAMURA H,2007. Scandinavian pattern and its climatic impacts[J]. Quart J Roy Meteor Soc,133(629), 2117-2131.

CHOU Y Y,1985. Medium Term Weather Forecast[M]. Beijing: Science Press:920. (in Chinese)

DEMMEL J,Kahan W,1990. Computing small singular values of bidiagonal matrices with guaranteed high relative accuracy [J]. SIAM Journal on Scientific and Statistical Computing,11(5):873-912.

HUANG R H,LI W J,1987. Influence of heat source anomaly over the western tropical Pacific on the sub-tropical high in East Asia[C]. Chengdu:Chinese Academy of Science:40-51.

HUANG R H,LI W J,1988. Heat source anomalies over the western tropical Pacific in the summer and its impacts on the East Asia subtropical high and associated physical mechanism[J]. Sci Atmos Sinica,12(9):106-107. (in Chinese)

HUANG R H,SUN F Y,1992. Impacts of the tropical western Pacific on the East Asian summer monsoon[J]. J Meteor Soc Japan,70(1):243-256.

KLEIN W H,1952. Some empirical characteristics of long waves on monthly mean charts[J]. Mon Wea Rev,80(1):203-219.

LI C Y,1989. El Niño events and temperature anomalies in eastern China[J]. J Trop Meteor,5(3):210-219. (in Chinese)

LI C Y,1990. Intraseasonal oscillation in the atmosphere[J]. Sci Atmos Sinica,14:32-45. (in Chinese)

LI J,WANG J,2003. A new North Atlantic Oscillation index and its variability[J]. Adv Atmos Sci,20(5):661-676.

LIAN Y,2007. Correlation between North Pacific oscillation and East Asian summer monsoon[J]. Sci Geographica Sinica,27 (S1):19-27. (in Chinese)

LIAN Y,AN G, 1998. The relationship among East Asia summer monsoon El Niño and low temperature in the Songliao Plains Northeast China[J]. Acta Meteor Sinica,56(6):724-735. (in Chinese)

LIAN Y,BUEH C L,XIE Z W,et al,2010. The anomalous cold vortex activity in Northeast China during the early summer and the low-frequency variability of the Northern Hemispheric atmosphere circulation[J]. Sci Atmos Sinica,34(2):429-439. (in Chinese)

LIANG F,TAO S Y,WEI J,et al,2011. Variation in summer rainfall in North China during the period 1956—2007 and links with atmospheric circulation[J]. Adv Atmos Sci,28(2):363-374,doi:10. 1007/s00376-010-9220-2.

LIU G,SHEN B Z,LIAN Y,et al,2012. The sorts of 500 hPa blocking high in Asia and it's relations to cold vortex and aestival low temperature in northeast of China[J]. Sci Geographica Sinica,32(10):1269-1274. (in Chinese)

LIU Z X,1986. Calculation of the intensity of the Northern hemisphere polar vortex area index and its association with temperature variations in China[J]. Meteor Monthly,2(S1):84-89. (in Chinese)

LIU Z X,LIAN Y,GAO Z T,et al,2002. Analyses of the Northern Hemisphere circulation characters during northeast cold vortex persistence[J]. Sci Atmos Sinica,26(3):361-372. (in Chinese)

LIU Z X,LIAN Y,SHEN B Z,et al,2003. Seasonal variation features of 500 hPa height in North Pacific oscillation region and its effect on precipitation in Northeast China[J]. J Appl Meteor Sci,14(5):553-561. (in Chinese)

MADDEN R D,JULIAN P R,1971. Detection of a 40—50 day oscillation in the zonal wind in the tropical pacific[J]. J Atmos Sci,28(5):702-708.

MADDEN R D,JULIAN P R,1972. Description of global scale circulation cells in the tropics with 40—50 day period[J]. J Atmos Sci,29(6):1109-1123.

MANTUA N J,HARE S R,ZHANG Y,et al,1997. A Pacific interdecadal climate oscillation with impacts on salmon production[J]. Bull Amer Meteor Soc,78(6):1069-1079.

NAMIAS J,1951. The great Pacific anticyclone of the winter 1948—1950:A case study in the evolution of climatic anomalies [J]. J Meteor,8(4):251-261.

NATIONAL CLIMATE CENTER,2010. 2009 Annual Report on Climate System Monitoring and Diagnosis[M]. Beijing: China Meteorological Press:176. (in Chinese)

PENG S Q,ZHANG J J,1998. The statistical diagnosis and numerical experiment of atmospheric teleconection patterns in Northern Hemisphere[J]. J Appl Meteor Sci,9(1):65-71. (in Chinese)

ROGERS G T,1981. The north-Pacific oscillation[J]. J Climate,1(1):68-83.

SHEN B Z,LIU S,LIAN Y,et al,2011. An investigation into 2009 summer low temperature in Northeast China and its association with prophase changes of the air-sea system[J]. Acta Meteor Sinica,69(2):320-333. (in Chinese)

SUN J Q,WANG H J,2006. Regional difference of summer air temperature anomalies in Northeast China and its relationship to atmospheric general circulation and sea surface temperature[J]. Chinese J Geophysics,49(3):662-671. (in Chinese)

WALLACE J M,GUTZLER D S,1981. Teleconnections in the geopotential height field during the Northern Hemisphere winter[J]. Mon Wea Rev,109(4):784-812.

YANG H,SUN S Q,2005a. The characteristics of longitudinal movement of the subtropical high in the Western Pacific in the pre-rainy season in China[J]. Adv Atmos Sci,22(3):392-400.

YANG Q,LIAN Y,HE J H,2005b. Predicting north-east cool summer using singular value decomposition method[J]. Meteor Monthly,31(1):31-35. (in Chinese)

ZHANG S Q,1990. Cold Temperature in Northeast China and its Association with El Niño[M]//National Science and Technology Commission. No. 5,China Science and Technology Blue Book. Beijing: China Science and Technology Press:308-310. (in Chinese)

ZHANG S Q,YU T J,LI F Y,et al,1985. Seasonal variation of Northern Hemisphere polar vortex area and associated intensity,and their association with the temperatures in Northeast China[J]. Sci Atmos Sinica,9(2):178-185. (in Chinese)

ZHAO Z G,LIAO Q S,1992. Winter North Pacific oscillation and summer rainfall in China[J]. Meteor Monthly,18(1):11-16. (in Chinese)

ZHENG W Z,NI Y Q,1999. Tropical and mid-latitude Pacific SSTA and its impacts on summer low temperature stress in East Asia[J]. J Appl Meteor Sci,10(4):394-401. (in Chinese)

ZHOU B T,WANG H J,CUI X,2008. Significant relationship between Hadley circulation and North Pacific Oscillation[J]. Chinese J Geophysics,51(4):999-1006. (in Chinese)

Numerical Experiments on the Impact of Spring North Pacific SSTA on NPO and Unusually Cool Summers in Northeast China*

LIAN Yi[1,2]　ZHAO Bin[3]　SHEN Baizhu[1,2]　LI Shangfeng[1,2]　LIU Gang[1,4]

(1 Laboratory of Research for Middle-High Latitude Circulation Systems and East Asian Monsoon, Changchun 130062; 2 Institute of Meteorological Sciences of Jilin Province, Changchun 130062; 3 National Meteorological Center, Beijing 100081; 4 Jilin Meteorological Science and Technology Service Center, Changchun 130062)

Abstract: A set of numerical experiments designed to analyze the oceanic forcing in spring show that the combined forcing of cold (warm) El Niño (La Niña) phases in the Niño 4 region and sea surface temperature anomalies (SSTA) in the westerly drifts region would result in abnormally enhanced NorthEast Cold Vortex (NECV) activities in early summer. In spring, the central equatorial Pacific El Niño phase and westerly drift SSTA forcing would lead to the retreat of non-adiabatic waves, inducing elliptic low-frequency anomalies of tropical air flows. This would enhance the anomalous cyclone-anticyclone-cyclone-anticyclone low-frequency wave train that propagates from the tropics to the extratropics and further to the mid-high latitudes, constituting a major physical mechanism that contributes to the early summer circulation anomalies in the subtropics and in the North Pacific mid-high latitudes. The central equatorial Pacific La Niña forcing in the spring would, on the one hand, induce teleconnection anomalies of high pressure from the Sea of Okhotsk to the Sea of Japan in early summer, and on the other hand indirectly trigger a positive low-frequency East Asia-Pacific teleconnection (EAP) wave train in the lower troposphere.

Keywords: atmospheric model; westerly drifts; Niño 4 SSTA; low-frequency variation; circulation pattern; cool summer; Northeast China

1　Introduction

The concept of the "local circulation pattern" was first introduced in the 1980s and 1990s to represent weather systems that repeatedly appear in a particular region in a sustained or quasi-stationary manner as a result of a remote response to low-frequency oscillation, and to elucidate the forming mechanism of each stage of the circulation pattern (Legras et al., 1985; Mo et al., 1988). Molteni (1993) pointed out that the presence of an extratropical atmospheric flow type does not need the support of SSTA forcing anomalies, though the frequency and stability of a circulation pattern has to be determined by the nature of boundary forcing, including the intensity and sustainability. A study conducted by Bueh et al. (1999)

* 本文原载于 *Advances in Atmospheric Sciences*, 2014, 31(6): 1305-1315

showed that tropical SSTA may turn a seemingly chaotic tropical atmosphere into a circulation pattern, resulting in partial confinement. Xu et al. (2001,2002) pointed out that blocking is not only an atmospheric nonlinear response to the local forcing, such as a heat source or terrain, but also a result of the nonlinear interactions between fluctuations and local sources (Lu et al. ,1998;Lu,2001).

In Northeast China, the influential local circulation patterns are the northeast cold vortex (NECV) and the Asia blocking high. The flow patterns change depending on the presence of internal or external tropical climate systems (Shen et al. ,2012;Lian et al. ,2013). Liu et al. (2003) revealed that the North Pacific Oscillation (NPO) region's negative 500 hPa height phase in the downstream of Northeast Asia in the winter results in enhanced cold vortex activity in the summer across Northeast China. The enhanced cold vortex activity also produces a climatic effect (He et al. ,2006;Liu et al. ,2012;Shen et al. , 2012). Lian et al. (2010) studied typical northeast cold vortex cases in early summer, and found that the sustained upstream Ural blocking enhances northeast cold vortex activity due to the combined effects of downstream Rossby wave energy dispersion and transient eddy forcing, resulting in precipitation anomalies across the northeastern region (Shen et al. , 2011). Numerical experiments have shown that the North Pacific mid-latitude SSTA in the summer of 1993 could have played a major role in shaping the abnormal southerly position of North Pacific subtropical highs in June and July (Gong et al. ,2006). The results, obtained from an improved high spectral model, indicated that the spatial distribution pattern of El Niño's early external thermal forcing created a desirable atmospheric circulation background for the frequent occurrence of Northeast Asian blocking in the summer (Cao et al. ,2006).

Lian et al. (1998) suggested that El Niño warming events have different OLR (outgoing long-wave radiation) meridional low-frequency oscillations close to 120°E, depending on whether they originate from the central equatorial or the eastern Pacific. In recent years, researchers have become increasingly interested in understanding the atmospheric response to the central equatorial Pacific SSTA forcing (Ashok et al. ,2007;Ratnam et al. ,2012;Zhou et al. ,2012). Lian et al. (2013) performed a diagnosis which found that the unusually cool summers of 1964 and 1993 in Northeast China were supported by an opposite phase configuration of the North Pacific polar vortex, NPO, and North Pacific SSTA in the spring. Though noticeably differing in type, they were equally able to trigger the occurrence of an unusually cool summer in Northeast China. Meanwhile, the northeastern region's early summer temperature has a significant positive correlation with the region's summer temperature as a whole (Li et al. ,2012).

In this paper, based on numerical experiments using the CMA 3. 1 model and experiments on the North Pacific westerly drifts and Niño 4 SSTA forcing, we investigate major circulation patterns' responses to SSTA forcing in an attempt to verify the diagnostic conclusion drawn by Lian et al. (2013). We also explore the physical mechanism that allows the low-frequency variation, a response of Niño 4 SSTA to the tropical atmospheric warming or cooling in the spring, to propagate to the extratropical region, and its possible impact on the different blocking patterns that could lead to an unusually cool early summer in Northeast China.

2 Data and method

The definition of 500 hPa NPO (25°—65°N, 140°E—150°W) and NPO_L index can be found in the literature (Lian et al. ,2013), while NECV is defined in the area (35°—60°N, 115°—145°E) where a closed cyclonic circulation at 500 hPa forms and maintains for more than 3 days (Liu et al. ,2003). Following the definition of the Okhotsk and Ural blocking highs in East Asia by Liu et al. (2012) (termed the "east" and "west" blockings, respectively, in this paper), the westerly drift index is defined as the normalized area-weighted SSTA in the domain (30°—46°N, 150°E—150°W) (Lian et al., 2013).

In this study, CAM 3.1, a 26-level model with a mixed vertical coordinate system developed by National Center for Atmospheric Research (NCAR), is applied. Based on Hadley Center monthly global SST, the standard SST data are taken from the International Comprehensive Ocean-Atmosphere Data Set (ICOADS) and a climatological dataset containing 12 monthly time samples. The monthly mean SSTA forcing is used to define the annual climatic cycle. The North Pacific SSTA variation is defined at a scale of 2 ℃. The tests are initiated from 1 March with NCAR analysis data. The time step is 1200 s, and the horizontal resolution is chosen as T42 (128×64 grid points for the globe). A long-term integration of 50 years is carried out and the last 45 years are selected for analysis.

We (Lian et al., 2013) have pointed out that North Pacific SST in spring and 500 hPa height SVD1 is about 61% of covariance square. The left field is the westerly drifts over North Pacific SST (30°—46°N, 150°E—150°W) and the spatial feature vector is "+". The Niño 3.4 spatial feature vector is "−". Matching with the right field 500 hPa height SVD1, in the southern part of NPO_L (25°—45°N, 140°E—150°W), the spatial feature vector is mainly "−" and the northern part of it (50°—65°N, 140°E—150°W) is "+". Therefore, it shows that when the basic mode represents the warm (cold) phase SSTA in the North Pacific and La Niña (El Niño) in Niño 3.4, the 500 hPa height over the North Pacific presents a negative (positive) phase NPO_L. In this paper, we study in detail the effects of spring NPO_L and cool early summer circulation in Northeast China by westerly drifts and Niño 4 SSTA forcing.

Based on the observed results, six numerical sensitivity experiments are designed to analyze the impacts of different combinations of westerly drifts and Niño 4 SSTA on northern hemispheric atmospheric circulations, especially their possible impacts on east/west blocking highs and NECV anomalies in Asia that could induce the occurrence of unusually cool summers in Northeast China. The sensitivity experiments are designed as follows:

(1) Experimental regions: region I is a westerly drift region that tilts eastwards (20°—50°N, 170°E—130°W); and region II is a Niño 4 region (5°N—5°S, 160°E—150°W) designed to show the effect of central equatorial Pacific SSTA forcing.

(2) Region configurations: region I, Niño 4, and region I+Niño 4 (see the seven configurations in Table 1).

Table 1 Numerical experiments and associated abbreviations

Experiment	Definition
Control	SST unchanged for 50 years
Experiment 1: SUM+2	Niño 4 SST increased by 2 ℃ and westerly drift SSTs decreased by 2 ℃ from March to May for 50 years
Experiment 2: SUM−2	Niño 4 SST decreased by 2 ℃ and westerly drift SSTs increased by 2 ℃ from March to May for 50 years
Experiment 3: $N+2$	Niño 4 SST increased by 2 ℃ from March to May for 50 years
Experiment 4: $N-2$	Niño 4 SST decreased by 2 ℃ from March to May for 50 years
Experiment 5: $X+2$	Westerly drift SSTs increased by 2 ℃ from March to May for 50 years
Experiment6: $X-2$	Westerly drift SSTs decreased by 2 ℃ from March to May for 50 years

3 The contribution of spring SSTA forcing to NPO phase patterns

Fig. 1a shows that when Niño 4 SSTA increases by 2 ℃ and westerly drift SSTA decrease by 2 ℃, it represents a typical positive NPO phase as a response to atmospheric circulations, where the 500 hPa height anomalies above the Aleutian Islands aligns with those above the subtropics in a north-south "－, ＋" pattern, constituting a planetary-scale disturbance wave train crossing the North Pacific, with a negative value center up to －12 dagpm. A positive value center in the south stretches from the subtropics to the equator and further to the southern hemisphere. On the contrary, when Niño 4 SSTA decreases by 2 ℃ and westerly drift SSTA increases by 2 ℃, a typical negative NPO phase is the response of atmospheric circulations. Consequently, the 500 hPa height anomalies become a strong planetary disturbance wave train crossing the North Pacific in a north-south "＋, －" manner from the Aleutians to the subtropics, with a positive value center up to 6 dagpm (Fig. 1b).

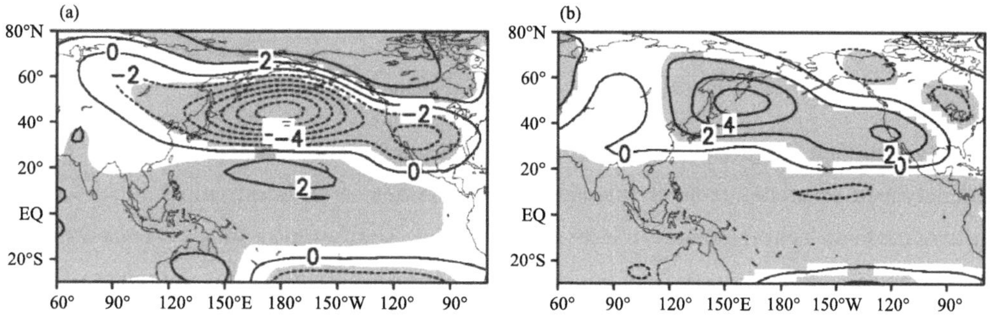

Fig. 1　500 hPa height anomalies of experiments (1 and 2) and the control, (a) experiment 1, (b) experiment 2. The shaded areas are the one that have passed the 95% confidence level, with the thin solid lines for the 500 hPa positive geopotential anomalies and the thin dashed lines for the 500 hPa negative geopotential anomalies in dagpm

The combination of Niño 4 and westerly drift SSTA shows that the North Pacific atmospheric circulations have a significant positive or negative mode as a response. Here, we also discuss the respective contribution made by Niño 4 and westerly drift SSTA to the mode of NPO. Fig. 2 presents the 500 hPa height anomalies of experiments 3 and 4 and the control with the presence of El Niño/La Niña SSTA forcing in the Niño4 region. Similarly, the 500 hPa height anomaly fields show two completely different wave train responses from the North Pacific to the North Pole: the 500 hPa height anomaly field has an opposite dipole response in the spring from "－, ＋" to "＋, －". As a result, the scope of the ≥95% confidence level not only covers most parts of the mid-high latitudes, but also symmetrically covers the entire tropical and subtropical Pacific regions (shaded areas in Figs. 2a and 2b), a typical positive or negative NPO phase, suggesting that Niño 4 SSTA may result in an NPO teleconnection, which agrees with the previous diagnosis of the positive NPO phase corresponding to the coupled SVD1 mode of El Niño (61% of the total sum of square covariance) (Lian et al., 2013). Previous studies have shown that most PNA-like teleconnections appear in the winter (Van et al., 1981; Van et al., 1981). However, the experiments performed in the current study show that a strong spring El Niño SSTA phase in the Niño 4 region results in a "＋, －, ＋" wave train from the North Pacific to the Bering Strait, and further to northwestern Canada, suggesting that the warm central equatorial Pacific SSTA forcing in the spring produces a PNA-like teleconnection as a response to atmospheric circulations.

The results derived of experiments 3 and 4 and the control are quite similar to those from experiments 1 and 2, albeit with a 1 dagpm difference (Figs. 2a, 2b and Figs. 1a, 1b), which fully agrees with the fact that El Niño's (La Niña's) positive (negative) NPO phase in the Niño 4 region is the most significant forcing source contributing to a positive or negative NPO phase.

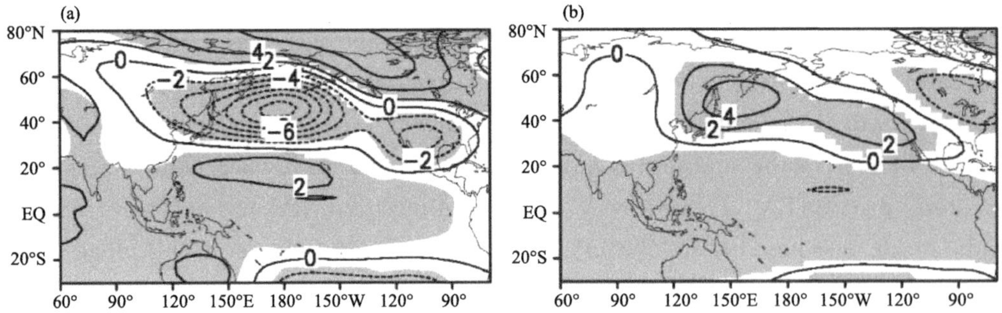

Fig. 2　500 hPa height anomalies of experiments (3 and 4) and the control,(a) experiment 3,(b) experiment 4. The shaded areas are the one that have passed the 95% confidence level,with the thin solid lines for the 500 hPa positive geopotential anomalies and the thin dashed lines for the 500 hPa negative geopotential anomalies in dagpm

Fig. 3 shows the 500 hPa height anomalies of experiment 6 and the control,where a planetary-scale negative anomaly belt stretches from Balkhash Lake to the Sea of Okhotsk and further to the Aleutians, with a negative anomaly center located in the south of the Aleutians,supported by a value center of < 2.5 dagpm. The whole of Northeast Asia and the Aleutians show significant anomalies (≥95%). There is a weak positive anomaly region above the mid-latitudes and the subtropics,with a north-south "－,＋" wave train above the North Pacific and the Aleutians,suggesting a typical positive NPO phase. The significant positive anomaly region (≥95%) stretches from eastern Siberia to the northern part of the Sea of Okhotsk and further to Alaska,constituting a north-south "＋,－" dipole circulation pattern together with the negative anomalies in the south of the Aleutians. In the west stands a typical Northeast Asian blocking made up of the Sea of Okhotsk high and NECV. Unfortunately,the pattern,as such,only appears in the spring rather than in early summer. In addition,the 500 hPa height anomalies of experiment 5 and the control,in which westerly drifts are given an increase of temperature by 2 ℃,failed to show the negative NPO phase (figure omitted).

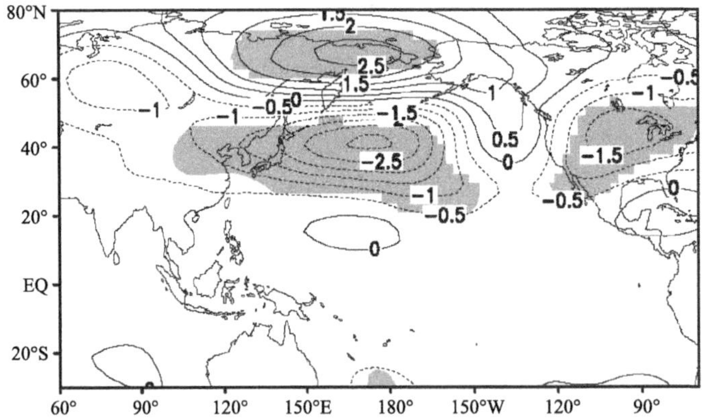

Fig. 3　500 hPa height anomalies of experiment 6 and the control. The shaded areas are the one that have passed the 95% confidence level,with the thin solid lines for the 500 hPa positive geopotential anomalies and the thin dashed lines for the 500 hPa negative geopotential anomalies in dagpm

The simulation results of Niño4 SSTA forcing are consistent with the atmospheric results from the combination of Niño4 and westerly drift SSTA forcing,except that the positive NPO phase forced out by El Niño has a North Pacific anomaly center that is smaller in magnitude compared with the combined anomalies,and that the circulation pattern derived from La Niña has an anomaly magnitude that is equivalent to the one from the combined anomalies. Cold SSTA contributes to westerly drift forcing. The magnitude of anomaly becomes larger when the model anomaly is added to the Niño4 SSTA forcing anomaly, with a result closer to the combined anomalies, suggesting a major contribution of Niño4 to a positive NPO phase. Cold westerly drift SSTA contributes to an enhanced positive NPO phase. Meanwhile, it is only Niño4 SSTA, rather than westerly drift SSTA, that contributes to a negative NPO phase.

4 Impact of spring SSTA forcing on early summer atmospheric circulation anomalies in Northeast Asia

As reported above, the impact of the combined forcing of Niño4 and westerly drift SSTA on NPO phases has been analyzed. In this section, we investigate the role played by a non-adiabatic NPO, arising from the joint effects of sea surface temperature and atmosphere (Chou, 1986), in inducing a retreating wave that could lead to early summer blocking abnormalities across Northeast Asia, and the physical process that would trigger an unusually cool early summer in Northeast China.

4.1 Combined forcing of warm Niño4 and cold westerly drifts encourages NECV activities in early summer

Fig. 4 presents the 500 hPa height anomalies of experiment 1 and the control. In March, it shows a negative anomaly center stretching from Northeast Asia to the Aleutians, with a central value of up to 8 dagpm. There is a negative anomaly center that covers some areas in the central and southern part of North America. Meanwhile, significant anomalies cover most parts of the region. There is an extensive positive anomaly corridor distributed from the East Siberian Sea in the north of the Aleutians to Alaska and further to northwestern Canada, with significant anomalies covering most parts of the region, conducive for Polar ridge activities. A weak height anomaly region can be seen over the central subtropical and

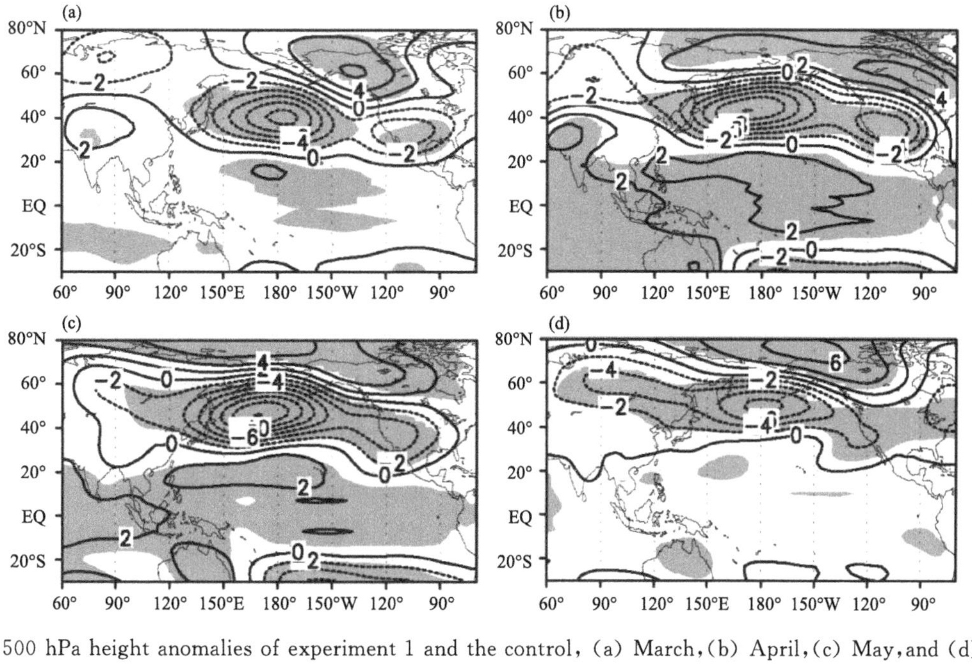

Fig. 4　500 hPa height anomalies of experiment 1 and the control, (a) March, (b) April, (c) May, and (d) June. The shaded areas are the one that have passed the 95% confidence level, with the thin solid lines for the 500 hPa positive geopotential anomalies, and the thin dashed lines for the 500 hPa negative geopotential anomalies in dagpm tropical Pacific in the south of the Aleutians, although an impressive part of the region passed the ≥95% confidence level (Fig. 4a). April—May shows a peak anomaly that runs north-south in a "+, −, +" manner, forcing out the westbound retreat of the negative anomaly center above the Aleutians by some 30° of longitude (Figs. 4b and 4c). In May, the largest negative anomaly reaches ≤−12 dagpm above the Aleutians, with a further expanded domain of ≥95% confidence (Figs. 4c; 4a). In June, the north-south "+, −" anomaly wave train is sustained over a vast region from the North Pole to the North Pacific high latitudes, albeit with a noticeably weakened strength. Whereas, the negative value region remains able to stretch westwards to Northeast Asia and even to central Asia, with a negative anomaly of up to −4 dagpm across Northeast Asia (Fig. 4d).

Experiment 1 shows that the combined forcing of warm Niño4 and cold westerly drifts would have a lasting impact on the presence of a positive NPO phase, and that the significant westbound retreat of negative Aleutian anomalies in the positive NPO phase would encourage early summer NECV activities in Northeast Asia, which is quite similar to the physical process that staged an NECV circulation pattern in the early summer of 1993 across Northeast Asia that led to an unusually cool summer in Northeast China (Lian et al., 2013).

4.2 Cold Niño4 SSTA and warm westerly drifts encourage positive Okhotsk high anomalies

Fig. 5 shows the 500 hPa height anomalies of experiment 2 and the control. It indicates that during the period of March—May, the 500 hPa height field responds to the broad symmetric negative anomaly belt, with the equator as the axis, in the subtropical regions of both the northern and southern hemispheres, with the largest negative anomaly at ≤−2 dagpm in May. The significant anomalies (≥95%) basically envelope the entire region, suggesting southerly and easterly subtropical high activities (Figs. 5a—c). In April—May, the positive anomaly center stretches into a planetary-scale belt from the Sea of Okhotsk to the west coast of the United States, with the central value in the south of the Sea of Okhotsk

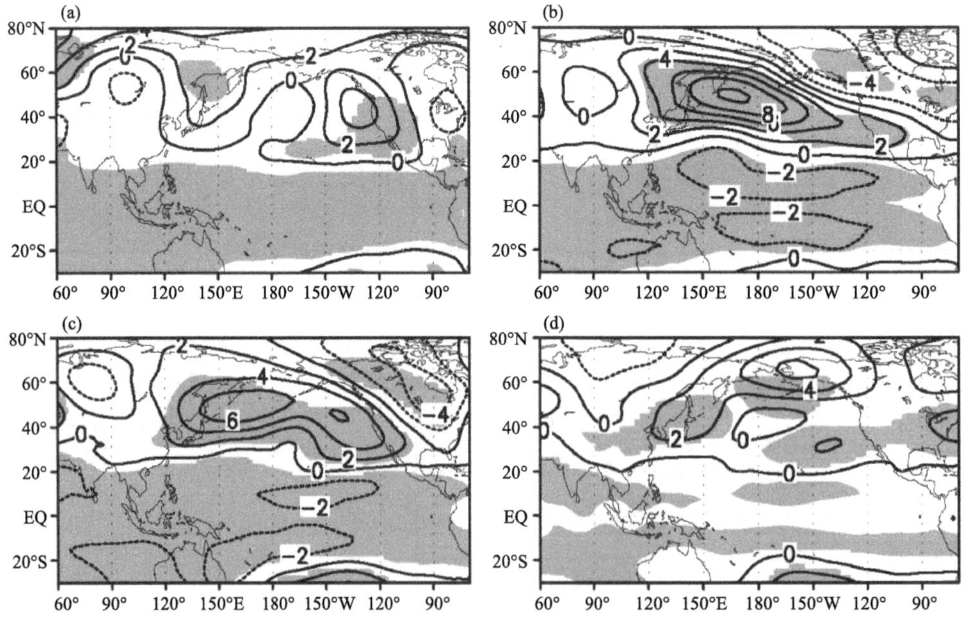

Fig. 5 500 hPa height anomalies of experiment 2 and the control, (a) March, (b) April, (c) May, and (d) June. The shaded areas are the one that have passed the 95% confidence level, with the thin solid lines for the 500 hPa positive geopotential anomalies, and the thin dashed lines for the 500 hPa negative geopotential anomalies in dagpm of up to 8 dagpm. Significant anomalies (≥95%) cover the entire region, supporting a stable and sustained negative NPO phase (Figs. 5b and 5c). In June, the negative NPO anomaly center becomes noticeably weakened, although remains over a region from the Sea of Okhotsk to the Sea of Japan (≥95%), constituting a "low in the west and high in the east" dipole circulation pattern together with the relatively weak negative anomalies that flank Lake Baikal in the west.

Experiment 2 produces a combined forcing of cold Niño4 and warm westerly drifts that encourage the formation of a negative NPO phase. Meanwhile, the stationary wave of positive anomalies in the negative NPO phase makes a significant contribution to the early summer positive anomalies of the Sea of Okhotsk, bearing some similarity to the eastern blocking that prevailed across Northeast Asia in the summer of 1964, where a severe unusually cool summer was reported in Northeast China (Lian et al., 2013).

4.3 Niño4 La Niña and El Niño encourage high anomalies in the south of Okhotsk and low anomalies in the northeast

Fig. 6 shows the 500 hPa height anomalies of June from experiments 3 and 4 featured with La Niña SSTA and El Niño SSTA forcing in the Niño4 region. Two completely different responses can be seen at 500 hPa in early summer. Under the thermal forcing of El Niño (+2 ℃), significant early summer negative anomalies appear as a response over the mid-high latitudes, from the central Pacific subtropics to the Aleutians, Alaska, and further to the North Atlantic in the north, in a "+, −, +" wave train pattern, with the latter two (−, +) being stronger, suggesting that El Niño encourages low-pressure activities in East Asia, the Aleutians, and the central and northern parts of the United States, with the ≥95% confidence level covering the entire region. Additionally, there is a weak positive anomaly region above the Ural Mountains, albeit at a poor confidence level. It produces a "high in the west and low in the east" pattern similar to the circulation pattern that prevailed in the summer of 1964, where an unusually severe cool summer occurred in Northeast China (Fig. 6a) (Lian et al., 2013). The thermal forcing of cold Niño4 SSTA (La Niña −2 ℃) results in a significant positive anomaly in the south of the Sea of Okhotsk, including Northeast China, suggesting that cold Niño4 SSTA is desirable for the maintenance and development of southerly eastern blocking. Meanwhile, there is a relatively weak negative anomaly region in the west of Lake Baikal. There is also a relatively strong positive anomaly belt from the East Siberian Sea to Alaska. The positive anomaly belt from the Northeast Pacific to the central and eastern parts of the United States is as strong as the one in the south of the Sea of Okhotsk, with a ≥95% confidence level covering the entire region (Fig. 6b).

The results from experiments 3 and 4 (Figs. 6a and 6b) are very similar to those from experiments 1

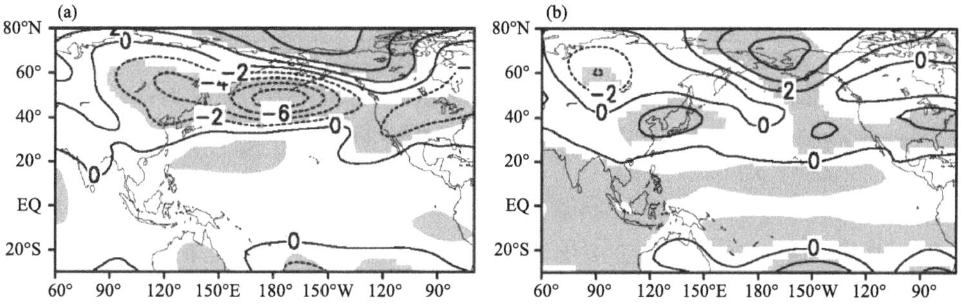

Fig. 6 500 hPa height anomalies of experiments (3 and 4) and the control in June, (a) experiment 3, (b) experiment 4. The shaded areas are the one that have passed the 95% confidence level, with the thin solid lines for the 500 hPa positive geopotential anomalies and the thin dashed lines for the 500 hPa negative geopotential anomalies in dagpm and 2 (Figs. 4d and 5d), suggesting that El Niño or La Niña forcing in the Niño4 region plays a dominant role in shaping the major North Pacific SSTA phases.

4.4 Weak impact of cold/warm westerly drift forcing on East Asian blocking

Fig. 7 shows the 500 hPa height anomalies of experiment 6 and the control with a decreased (by 2 ℃) westerly drift SSTA from March—May. The results show that a planetary-scale 500 hPa positive anomaly belt stretches from eastern Siberia to Alaska throughout March—May with a confidence level up to ≥95%. Meanwhile, there is a planetary-scale negative anomaly belt in East Asia from the Sea of Japan to the Aleutians, constituting a "+, −" wave train from the polar region to Northeast Asia and further to the Aleutians, supported by ≥95% confidence. In addition, there is an unstable negative anomaly region above the North American continent (Figs. 7a—c). In June, the 500 hPa height anomaly field, despite a "+, −" wave train remaining over the polar region and the Pacific Ocean, becomes noticeably weakened and shrinks eastwards, compared with the strongest "+, −" wave train appearing in May from the polar region to Northeast Asia and further to the Aleutians, suggesting that the cold westerly drift forcing in the spring would have an impact on NECV anomalies that would be strong in May but weak in early summer. In early summer, it only has an effect on the "+, −" wave train from the polar region to the North Pacific. The 500 hPa height anomalies of experiment 5 and the control, with a westerly drift SSTA increased by 2 ℃ from March to May, show that there is neither a stationary anomaly wave train nor the support of ≥95% confidence (figure omitted).

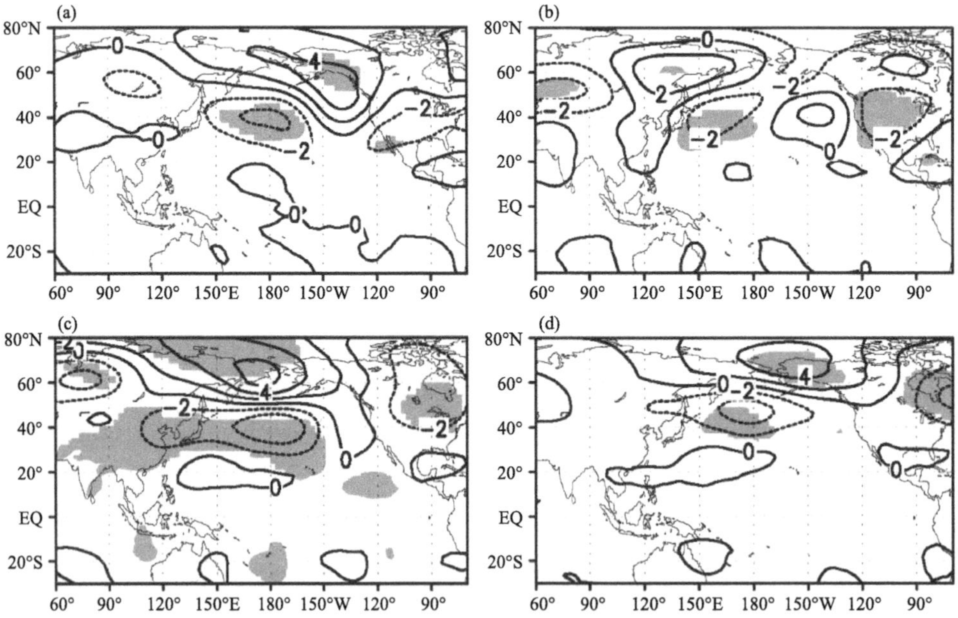

Fig. 7 500 hPa height anomalies of experiment 6 and the control, (a) March, (b) April, (c) May, and (d) June. The shaded areas are the one that have passed the 95% confidence level, with the thin solid lines for the 500 hPa positive geopotential anomalies and the thin dashed lines for the 500 hPa negative geopotential anomalies in dagpm

5 Physical mechanism underpinning the Niño4 forcing and low-frequency atmospheric response

In this section we analyze the 850 hPa tropical and extratropical low-frequency variation pattern induced by sustained Niño4 SSTA forcing (+2 ℃ or −2 ℃) from March to June, and explain the possible physical mechanisms that produce the numerical experiment results described in the previous two sections through the effect of energy dispersion. In doing so we hope to facilitate a comparison of the results with those from other major national and international numerical experiment studies.

5.1 Elliptic extratropical low-frequency variation pattern induced by El Niño makes the "C-A-C-A" low-frequency wave train more significant

Fig. 8 shows the evolution of 850 hPa flow fields of experiment 3 and the control from March to June. In March, it is under the influence of strong cold surges sweeping over China's eastern coasts and the west Pacific warm pool around the Philippines (see Fig. 8a, in which strong northwesterly flow anomalies can be seen spreading from the west of Lake Baikal to China's east coast and further to the central Pacific, before forming a closed anticyclonic flow anomaly field in the north of the Philippines, as shown in the north of the black box). The 850 hPa Niño4 flow anomaly field in the central tropical equator sits in a symmetric parabolic low-frequency variation pattern with the equator as the axis (see the square area in Fig. 8a that tracks the flow anomaly variations in other months). With an SSTA increased by 2 ℃ in April, a literally closed cyclonic low-frequency circulation pattern appears in the north of the central equatorial Pacific Niño4 region. Meanwhile, the subtropical area at 20°N quickly forms a closed anticyclonic circulation pattern, which enhances the cyclonic circulation that has been sustained above the Aleutians since March (compare Figs. 8a and 8b). In May, a significant cyclonic flow pattern appears in the northern part of the central equatorial Pacific Niño4 region, with an enhanced anticyclonic flow pattern in the northwest of the subtropics. As a result, the cyclonic flows above the Aleutians prevail over the North Pacific mid-high latitudes, with their westbound retreating center joining the cyclonic flows in the west of Lake Baikal. The evolution leaves no intact anticyclonic circulation above the Sea of Okhotsk. In the spring NPO sits in a positive phase, overseeing the northeast-southwest presence of cyclone (C) and anticyclonic (A) over the North Pacific. The 850 hPa flow anomaly field of experiment 3 and the

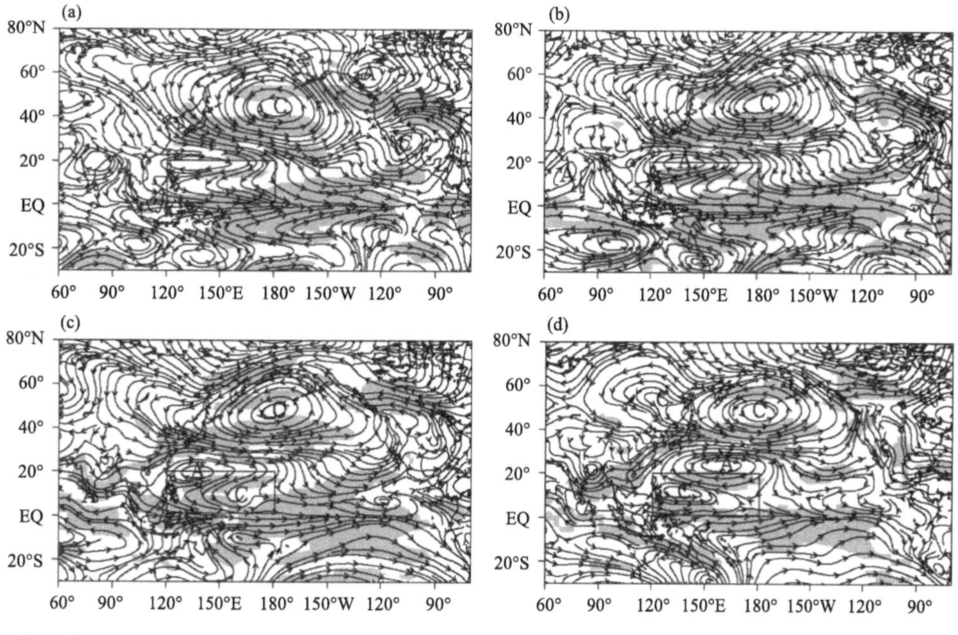

Fig. 8 850 hPa flow anomalies of experiment 3 and the control from March to June, (a) March, (b) April, (c) May, and (d) June. The shaded areas are the one that have passed the 95% confidence level, with the thin solid lines for the 850 hPa flow lines

control in June shows that an intact elliptical low-frequency cyclonic circulation pattern appears above the tropical ocean east of the Philippines (the black box in Fig. 8d). Meanwhile, the northeast bound subtropical anticyclone moves northwards by 5° of latitude and eastwards by 30° of longitude (compare Figs. 8c and 8d), before traveling through the powerful Aleutian low-frequency cycle and the Arctic anticyclone in northern Alaska, forming a "C-A-C-A" low-frequency wave train (Fig. 8d). In addition, the westbound expansion of the Aleutian low-frequency cyclone "C" results in another cyclonic circulation center ("C") over Northeast Asia, encouraging NECV activities. Both the easterly and westerly flow anomalies in the southern and northern flanks of each wave train center reach the ≥95% confidence level (Fig. 8).

The evolution of 850 hPa flow anomalies in Experiment 3 and the control from March to June show that, under the strong influence of non-adiabatic and regional divergence flows stemming from the mid-latitudes, the tropical region would see a parabolic low-frequency variation pattern (Fig. 8a). When the warm tropical SSTA disturbance exceeds the influence of non-adiabatic and regional divergence flows in April—May, there would be a full elliptical low-frequency cyclonic circulation variation pattern over the central and west part of the tropical Pacific. This, under the effect of energy dispersion, would reinforce the southwest-northeast C-A-C-A low-frequency variation wave train that travels from the tropics to the subtropics, and further to the mid-high latitudes (Fig. 8d). This explains the physical mechanism that allows the warm spring Niño4 SSTA forcing to encourage the presence of positive NPO anomalies, which would in turn significantly weaken the Okhotsk blocking high (east blocking) in early summer.

5.2 La Niña contributes to a positive low-frequency EAP wave train

Fig. 9 shows the 850 hPa flow anomalies of experiment 4 and the control from March to June. It indicates a symmetric parabolic low-frequency variation pattern, with the equator as the axis, during that period (the square area in Fig. 9a), without sending waves to the extratropical high latitudes of the

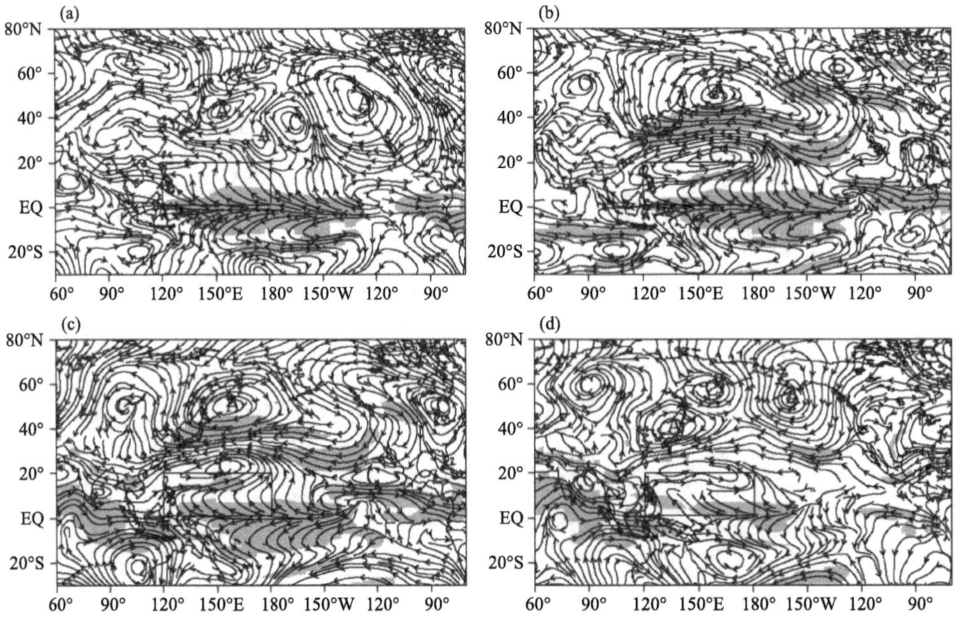

Fig. 9 850 hPa flow anomalies of experiment 4 and the control from March to June, (a) March, (b) April, (c) May, and (d) June. The shaded areas are the one that have passed the 95% confidence level, with the thin solid lines for the 850 hPa flow lines

Northern Hemisphere. This constitutes a substantive difference from the elliptic pattern derived from experiment 3, in which the La Niña phase in the Niño4 region shows a parabola, dual curve and straight trajectory without tropical non-adiabatic heating (Xu et al., 2002). In June, a "C-A-C-A-C" wave train forms along Asian east coasts, traveling across the western Pacific warm pool, China's east coast, Japan's Honshu Island, the Aleutians, and America, becoming a typical positive EAP phase, which is obviously associated with the indirect thermal forcing of the western Pacific warm pool in the Niño4 region with an SSTA decreased by 2 ℃ (Nitta, 1987; Huang et al., 1987). The westerly and easterly anomalies above the western Pacific warm pool and the subtropical region register a confidence level of ≥95% (Fig. 9d).

6 Summary and discussion

The combined results from the numerical sensitivity experiments (1～4) conducted in this study provide further evidence that two different modal forcing of El Niño (La Niña) and cold (warm) westerly drifts in the Niño4 region would have a positive (negative) NPO phase as an atmospheric response over the North Pacific in spring. Hereon, SST forcing over the Niño4 region presents a major contribution to the NPO positive (negative) phase modal the westerly drift SSTA forcing shows obviously asymmetry in response to mid-high latitude circulation, and only cold SST forcing contributes to strengthening the NPO positive phase model; while the NPO negative phase modal only depends on the Niño4 region SST, regardless of the westerly drift. The Niño4 region SST forcing in the North Pacific is far more important than SSTA of the westerly drift from the perspective of ocean forcing, this provides further evidence of a sea-air coupled mode over the North Pacific in the spring. It also confirms that these two major coupled modes can serve as precursor signals for the commencement of an unusually cool summer in Northeast China (Lian et al., 2013).

The results of experiment 1 and the control, featuring a warm Niño4 and cold westerly drifts, show that this combination would lead to a significant westbound retreat of the non-adiabatic positive NPO phase, rendering a significant contribution to the enhanced early summer NECV activities at 500 hPa. This is similar to the major configurations of the North Pacific SSTA pattern and NECV circulation pattern that prevailed in Northeast Asia in the early summer of 1993, and that encouraged the occurrence of an unusually cool early summer in Northeast China (Lian et al., 2013). Experiment 2, designed with a combination of cold Niño4 and warm westerly drifts, shows that the non-adiabatic positive NPO phase anomaly center, jointly formed by SSTA and the atmosphere in the spring, would allow the stationary wave to make a significant contribution to the positive anomalies in the region from the Sea of Okhotsk to the Sea of Japan. This bears a resemblance to the major North Pacific SSTA configurations in the spring of 1964, and to the eastern blocking that also prevailed over Northeast Asia in the summer of that same year (Lian et al., 2013).

Experiments 3 and 4 and the control produced a range of results that are very similar to those derived from the combination of experiments 1 and 2, suggesting that El Niño/La Niña forcing in the Niño4 region may play a dominant role in shaping the major North Pacific SSTA modes, although the impact on the Ural High anomalies in early summer would be insignificant. The circulation pattern forced out by La Niña would have a magnitude that matches the combined phase anomalies. Meanwhile, experiments 5 and 6 and the control produced a result showing that only cold westerly drift forcings would have a significant impact. The anomaly would become larger in magnitude, if the mode anomaly is added to the Niño4 SSTA forcing anomaly, making it closer to the combined anomalies.

A reasonable explanation for the above-mentioned results is found in the 850 hPa flow anomalies stemming from experiments 3 and 4 and the control from March to June. When the warm central equatorial Pacific Niño4 SSTA disturbance exceeds the non-adiabatic influence from the mid-latitudes, there would be a full elliptical low-frequency cyclonic variation in the region, reinforcing the C-A-C-A low-frequency variation wave train ($\geqslant 95\%$). The significant westbound retreat of "C" above the Aleutians would result in a new cyclonic circulation center, "C", encouraging the early summer NECV activities in the northeast. The cold central equatorial Pacific SSTA forcing would see a parabolic or linear low-frequency variation pattern. Although not directly triggering the northward propagation of the low-frequency wave train, the pattern, as such, would indirectly trigger a positive EAP wave train in early summer (June), due to the fact that the cold central equatorial Pacific phase would remain an obvious heat source compared with the western Pacific warm pool. Such a wave train pattern would encourage cyclonic circulations above the Sea of Okhotsk, albeit with a weak strength at a poor confidence level. As a result, the central equatorial Pacific La Niña forcing would encourage abnormal high pressure activities in the south of Okhotsk in early summer, which supports the theoretical results on tropical low-frequency oscillation trajectories using the WKBJ method (Xu et al., 2001, 2002). It confirms the conclusion drawn in the study by Lim et al. (1983), and reveals the low-frequency variation of different phase trajectories, including cold and warm central equatorial Pacific SSTA forcing, which would in turn help determine if there is a low-frequency variation wave train between the tropical, subtropical and mid-high latitudes.

In Northeast China, the early summer temperature has a strong correlation with the summer temperature as a whole. In this context, the classified extraction and integration of the early signals that would foresee an unusually cool early summer plays a pivotal role in the short-term prediction of the early summer temperatures of Northeast China.

The two spring precursor signals that foretell an unusually cool summer in Northeast China would have an opposite sea-air configuration in the North Pacific. This study has presented a reasonable explanation through a range of numerical experiments: in the spring, the different combinations of North Pacific westerly drifts and Niño4 SSTA forcing would result in a different positive or negative NPO phase. The different sea-air coupling configurations would make a noticeable difference to the early summer blocking anomalies prevailing in Northeast Asia. The paper has also discussed the low-frequency variation process triggered by the central equatorial Pacific SSTA thermal forcing, and the physical mechanisms underpinning the mid-high latitude circulation anomalies. Another major mechanism that may also have an impact on the occurrence of unusually cool summers in Northeast China, or the Ural blocking, may involve forcings derived from other oceans, which requires further investigation in a separate paper.

Acknowledgements. This study was supported by a National Natural Science Foundation project approved under Grant Nos. 41175083, 41275096 and 41305091, and a China Meteorological Administration special public welfare research funds registered under Grant Nos. GYHY201006020, GYHY 201106016, and GYHY201106015.

References

ASHOK K, BEHERA S K, RAO S A, et al, 2007. El Niño Modoki and its possible teleconnection [J]. J Geophys Res, 112 (C11), doi: 10.1029/2006JC003798.

BUEH C L, JI L R, 1999. The ocean-atmosphere coupled regimes and east Asian winter monsoon (EAWM) activity[J]. Adv Atmos Sci, 16(1): 91-106, doi: 10.1007/s00376-999-0006-3.

CAO J, HUANG R H, TAO Y, 2006. A mechanism for interannual variations of Northeast Asian Blocking High in summer

[J]. Chinese J Geophys,49(1):37-44. (in Chinese)

CHOU J F,1986. Long Range Numerical Weather Forecast[M]. Beijing:China Meteorological Press:329. (in Chinese)

GONG Y F ,HE J H,DUAN T Y,et al,2006. Numerical experiment on the influences of minus-ssta over mid-latitude northern Pacific on the subtropical anticyclone[J]. J Trop Meteor,22(4):386-392. (in Chinese)

HE J H,WU Z W,HONG Z J,et al,2006. Climate effects of northeast cold vortex and its impacts on plum rains[J]. Chinese Sci Bull,5(23):2803-2809. (in Chinese)

HUANG R H,LI W J,1987. Influence of the heat source anomaly over the western tropical Pacific on the subtropical high over East Asia[C]. Chengdu:International Conference on the General Circulation of East Asia.

LEGRAS B,GHIL M,1985. Persistent anomalies,blocking and variations in atmospheric predictability [J]. J Atmos Sci,42 (5):433-471.

LI S F,LIAN Y,CHEN S B,et al,2012. Distribution of extreme cool events over Northeast China in early summer and the related dynamical processes[J]. Sci Geographica Sinica,32(6):752-758. (in Chinese)

LIAN Y,AN G,1998. The relationship among East Asia summer monsoon,El Niño and low temperature in Songliao Plains Northeast China[J]. Acta Meteor Sinica,56(6):724-735. (in Chinese)

LIAN Y,BUEH C L,XIE Z W,et al,2010. The anomalous cold vortex activity in Northeast China during the early summer and the low-frequency variability of the northern hemispheric atmosphere circulation[J]. Chinese J Atmos Sci,34(2):429-439. (in Chinese)

LIAN Y,SHEN B Z,LI S F,et al,2013. Impacts of polar vortex,NPO,and SST configurations on unusually cool summers in Northeast China. Part I:Analysis and diagnosis[J]. Adv Atmos Sci,30(1):193-209,doi:10. 1007/s00376-012-1258-x.

LIM H,CHANG C P,1983. Dynamics of teleconnections and Walker circulations forced by equatorial heating[J]. J Atmos Sci,40(8):1897-1915.

LIU G,SHEN B Z,LIAN Y,et al,2012. The sorts of 500 hPa blocking high in Asia and it's relations to cold vortex and aestival low temperature in Northeast of China[J]. Sci Geographica Sinica,32(10):1269-1274. (in Chinese)

LIU Z X,LIAN Y,SHEN B Z,et al,2003. Variation features of 500 hPa height in north pacific oscillation region and its effect on precipitation in Northeast China[J]. J Appl Meteor Sci,14(5):553-561. (in Chinese)

LU R Y,2001. Eddies during the blocking maintenance over the Northeastern Asia in summer[J]. Chinese J Atmos Sci,25 (3):289-302. (in Chinese)

LU R Y,HUANG R H,1998. Influence of East Asia/Pacific teleconnection pattern on the interannual variations of the blocking highs over the Northeastern Asia in summer[J]. Chinese J Atmos Sci,22(5):727-734. (in Chinese)

MO K C,GHIL M,1988. Cluster analysis of multiple planetary flow regimes[J]. J Geophys Res,93(D9):10927-10951.

MOLTENI,1993. Research activities in atmospheric and oceanic modelling[R]. Geneva:CAS/JSC Working Group Numerical Experimentation.

NITTA T,1987. Convective activities in the tropical western Pacific and their impact on the Northern Hemisphere summer circulation[J]. J Meteor Soc Japan,65(3):373-390.

RATNAM J V,BEHERA S K,MASUMOTO Y,et al,2012. Anomalous climatic conditions associated with the El Niño Modoki during boreal winter of 2009[J]. Climate Dyn,39(1-2):227-238.

SHEN B Z,LIN Z D,LU R Y,et al,2011. Circulation anomalies associated with interannual variation of early-and late-summer precipitation in Northeast China[J]. Sci China (Earth Sci),54(7):1095-1104.

SHEN B Z,LIU S,LIAN Y,et al,2012. The 2009 summer low temperature in Northeast China and its association with prophase changes of the air-sea system[J]. Acta Meteor Sinica,26(4):320-333.

VAN L H,MADDEN R A,1981a. The southern oscillation,Part I:Global associations with pressure and temperature in northern winter[J]. Mon Wea Rev,109(6):1150-1162.

VAN L H,ROGERS J C,1981b. The southern oscillation,Part Ⅱ:Associations with changes in the middle troposphere in the northern winter[J]. Mon Wea Rev,109(6):1163-1168.

XU X D,ZHOU M Y, CHEN J Y,et al,2001. Comprehensive physical picture of dynamic and thermal structure in Qinghai-Tibet Plateau[J]. Sci China (Ser D),31(5):428-440. (in Chinese)

XU X D,GAO S T,2002. Dynamics of External Forcing and Wave-Current Interaction[M]. Beijing:China Ocean Press:291.

(in Chinese)
ZHOU B T, XIA D D, 2012. Interdecadal change of the connection between winter North Pacific Oscillation and summer precipitation in the Huaihe River valley[J]. Sci China (Earth Sci), 55(12): 2049-2057.

Mechanisms for the Formation of Northeast China Cold Vortex and Its Activities and Impacts: An Overview*

LIAN Yi[1,2]　SHEN Baizhu[1,2]　LI Shangfeng[1,2]　LIU Gang[1,2]　YANG Xu[1,2]

(1 Laboratory of Research for Middle-High Latitude Circulation Systems and East Asian Monsoon, Changchun 130062; 2 Institute of Meteorological Sciences of Jilin Province, Changchun 130062)

Abstract: In the mid 20th century, great efforts were made to investigate the formation process of high-latitude cold vortex, which is regarded as a major weather system in the atmospheric circulation. In the late 1970s, Chinese researchers noticed that the Northeast China cold vortex (NECV) is an active and frequently occurring weather system over Northeast Asia, which is generated under specific conditions of topography and land-sea thermal contrast on the local and regional scales. Thereby, the NECV study was broadened to include synoptic situations, mesoscale and dynamic features, the heavy rain process, etc. Since the 21st century, in the context of the global warming, more attention has been paid to studies of the mechanisms that cause the NECV variations during spring and early summer as well as the climatic impacts of the NECV system. Note that the NECV activity, frequent or not, not only affects local temperature and precipitation anomalies, but also regulates the amount of precipitation over northern China, the Huai River Basin, and the middle and lower reaches of Yangtze River. The NECV influence can even reach the Guangdong—Guangxi region. However, compared to the achievements for the blocking system study, theoretical studies with regard to the NECV system are still insuffcient. Research activities regarding the mechanisms for the NECV formation, particularly theoretical studies using linear or weak nonlinear methods need to be strengthened in the future. Meanwhile, great efforts should be made to deepen our understanding of the relations of the NECV system to the oceanic thermal forcing, the low-frequency atmospheric variations over mid-high latitudes, and the global warming.

Keywords: Northeast China cold vortex; blocking; global warming; higland low-frequency variations; oceanic forcing; weak nonlinear method

1 Introduction

In the late 1940s, when analyzing the strong fluctuations in the westerlies, some overseas researchers found that closed low-or high-pressure systems often formed in the westerly flows in the mid and upper troposphere. These low-and high-pressure systems may have a prolonged stay in a specific region and

* 本文原载于 *Acta Meteorologica Sinica*, 2016, 30(6): 881-896

have profound impacts on local and regional weather (Crocker et al. ,1947;Schwerdtfeger,1948;Palmén, 1949). Palmén et al. (1949) defined the cutoff high in the north as a warm system, and the cutoff low in the south as a cold system. This concept was later demonstrated by using a schematic diagram. A well-known Chinese researcher, Professor Y. B. Xie, proposed a classic four phase model to depict the high-latitude cold vortex cutoff process over North America (Hsieh,1949). Since then, researchers started to pay more attention to the high-latitude cutoff low and the associated blocking pattern in the Northern Hemisphere.

The first theoretical interpretation of the formation of high-latitude cold cutoff low can be traced back to a cyclonic wave model and an instability theory developed by the Norwegian School (Bjerknes et al. ,1922). With the advantages of the advancement of upper air observing techniques, Rossby employed a vorticity equation to examine the evolution of mid-troposphere disturbance. Sutcliffe (1947) developed a baroclinic vorticity equation to interpret the development of cyclone. Petterssen (1955) made a similar attempt. Both Charney (1947) and Eady (1949) introduced a baroclinic instability theory, followed by a barotropic instability theory proposed by Kuo (1949). These theories offered important insights into the development of large-scale weather systems, including the high-latitude cutoff low. During the 1960s—1980s, Chen (1963a,1963b,1980,1987) developed a thermal wind adaptation theory based on decomposition analysis. Ye et al. (1965) highly praised this work, and stated that "when the discussion of geostrophic adaptation process remains largely a theoretical one, Chen et al. (1963a,1963b,1980,1987) have applied the theory to weather analysis, and the results are encouraging".

The introduction of nonlinear theory enhanced the blocking studies, and the theoretical study of cutoff low benefited from advances in the blocking studies since the cutoff low is closely associated with the blocking. For example, Egger (1978) elaborated a theory focusing on the nonlinear interactions between the slow-moving free waves and the terrain forcing waves. Charney et al. (1979) proposed a multiple equilibrium theory, and Tung et al. (1979) developed a resonance theory. Meanwhile, several theories for local blocking were developed, including the "modons" theory proposed by Mcwilliams (1980) and the korteweg de-Vries (KdV) Rossby soliton theory proposed by Malguzzi et al. (1984). Chinese researchers developed an enveloped Rossby soliton theory (Luo et al. ,1989; Luo, 1991, 1999). Unfortunately, these cold vortex theories were mainly developed based on the blocking theory and independent judgment was missing in these studies.

In the late 1970s, Chinese scholars raised the concept of the NECV and started to investigate its formation mechanisms and impacts. The NECV became a major topic of the meteorological studies in China thanks to the advocate of Tao (1980), who believed that the NECV creates a circulation pattern that can eventually result in heavy rains or thunderstorms in the northeastern and northern parts of China.

In the 1980s, Chinese scientists studied various aspects of the NECV, including its definition, its mesoscale features relevant to heavy rains, its energy cycle, especially the dynamic mechanisms for its formation, etc. (Lian,1987;Zheng et al. ,1992;Sun,1998;Sun et al. ,2001).

Since the beginning of the 21st century, a significant warming trend is found in Northeast Asia, which makes Chinese researchers to pay more attention to the mechanisms behind the NECV variation from the spring to early summer, and the related climatic effects of the variation. An automated technique was introduced to identify the NECV event instead of using the subjective manual judgment method. On the one hand, the automated technique allows us to characterize the NECV activity in a quantitative way (Liu et al. ,2002;Miao et al. ,2006;Hu et al. ,2011;Yang et al. ,2012;Xie et al. ,2012); on the other hand, this method helps us to further understand the climatological aspects of the atmospheric circulations that would lead to a prolonged NECV event and its possible link with the precursor signals of North Pacific Oscillation (NPO) and Northern Annular Mode (NAM) (Sun et al. ,2000;Liu et al. ,2002;Wan et al. , 2013). Meanwhile, some researchers also started to investigate the climatic effects and low-frequency characteristics of the NECV anomalies (He et al. ,2006;Lian et al. ,2010;Liu et al. ,2012a;Bueh et al. , 2015).

Though NECV is a synoptic phenomenon, its activity, frequent or not, not only affects the local temperature and precipitation anomalies, but also regulates the amount of precipitation over the northern part of China, the Huai River basin, and the middle and lower reaches of Yangtze River. Its influence can even reach the Guangdong—Guangxi region. By far we are still facing a range of unanswered questions. Under the influence of global warming, what mutations will the NECV activities have as a major Northern Hemispheric blocking pattern? How to reveal the relationship between the NECV activity that has a high-frequency variation and the prolonged low-frequency anomalies? How to parallel the theoretical study of the NECV to the study of blocking? The mechanism study for the NECV formation and associated abnormal activities not only involves in the regional circulation pattern of Northeast Asia, but also makes an important component of the study of Northern Hemispheric circulation anomalies. The current paper intends to review the progresses achieved in the study of the high-latitude cold vortex cutoff, along with some suggestions on the selected scientific issues and methodologies to be studied and developed in the future.

2 Dynamic and thermodynamics for the NECV formation and maintenance

2.1 Terrain and land-sea distribution

In the past, researchers preferred to use the barotropic vorticity equation to interpret the impacts of mid-latitude terrains on the physical processes of large-scale circulation and movement with a focus on the formation of blocking high. For example, Egger (1978) developed a theory that emphasizes the nonlinear interaction between slow-moving free waves and terrain forcing waves. The decomposition analysis proposed by Chen (1980) not only revealed the role played by the dynamic action of advection in the development of super-long waves in the atmosphere, but also high-lighted the effects of the non-uniform distribution of advection. He believed that when the relative vorticity becomes increasingly weak downstream in the direction of thermal winds, the development can produce a thermal wind vorticity in the airflow field that is larger than the thermal wind vorticity in the temperature field. Consequently, a strong ascending movement would occur in the process of adaptation. When the long waves in the atmosphere hit the Da Hinggan Mountains area in Northeast China, local terrains will exert impacts on the pressure field, the shear flow field, and the temperature field, since a large part of the plain is located at the leeside of the mountains to the east. The pattern, as such, is favorable for the development of cyclonic vorticity over the leeside slope. Meanwhile, the low-pressure trough will move faster than the temperature trough. On the other hand, the westward retreat of long waves in the atmosphere leaves the Northeast Plains to be located on the leeward slope of the Xiao Hinggan Mountains and the Changbai Mountains, which inhibits the development of high-pressure ridges but promotes the development of super-long wave troughs (Chen, 1980, 1987). Zheng et al. (1992) further elaborated the mechanism that allows the NECV to prevail in the northeastern part of China based on Chen's thermal wind adaptation theory. Using a simplified large-scale vorticity equation, they proved that the local variation of isobaric vorticity is mainly the result of the combined effects of vorticity advection, thermal wind vorticity advection, and terrains. Lian et al. (2014) demonstrated in a recent numerical experiment study that the steady retreat of the vortex waves above the Aleutian Islands in early summer is one of the sources that can trigger low-frequency NECV activities, which echoes the argument made by Chen (1987) that the westbound retreat of super-long waves enhances the terrain effects in the development of high-latitude cyclones.

As early as in the 1950s, Pettersen (1950, 1955) analyzed spatial distribution statistics of global cyclone activities and showed that in addition to the terrain effects, the continent is warm in the summer. In this context, temperature intercomparison should not focus on the area outside the boundary of the icy land and water body, but should focus on the coastal region in the north of the continent and between the warm and cold ocean currents in high latitudes. In the north, vorticity mainly feeds the warm current. As a result, the low-pressure zone tends to sustain over the northern region of the continent in the summer. This argument also applies to mid latitudes. He believed that the final stage of a cyclone development is often associated with the high-latitude cold vortex cutoff, which indicates that the region of high-latitude cold vortex activities has a close relation with the distribution of land and sea. Lian et al. (2014) demonstrated in a recent numerical experiment that the steady retreat of low-latitude vortex above the Aleutians creates a source that triggers NECV activities in the spring, which is similar to the result of Chen (1987) that the westward retreat of long waves enhances the terrain effect and promotes the development of high-latitude cyclones. In addition, in warm season, the land-sea thermal contrast in the coastal region of East Asia is most obvious, easily resulting in the weakening of the westerlies and backward movement of the planetary waves. The semi-permanent presence of the Aleutian low activity center over North Pacific behaves as a steady low-frequency disturbance source for the formation of the NECV in early summer (Lian et al., 2013, 2014).

2.2 Atmospheric thermodynamics

Petterssen (1950, 1955) singled out three possible processes that may lead to the formation of high-latitude cutoff low, i. e., a low-pressure system that develops first on the ground and later is cut off to become a cold vortex in mid latitudes, or a vortex that develops first in high latitudes, or a vortex that develops both on the ground and in high latitudes.

A vortex that develops first in high latitudes is often associated with a barotropic process, which can be explained by the theory of the Modons, i. e., the nonlinearity is very important as β becomes smaller in high latitudes. Meanwhile, both the background and the baroclinic westerlies are relatively weak over the ocean, which facilitates the formation of a dipole blocking circulation (vortex) over the two oceans (Mcwilliams, 1980). Chao et al. (1980) tried to understand the blocking high and cutoff low in the atmosphere based on the Rossby solitary waves. Haines et al. (1987) reconsidered the role of synoptic waves, as the large-scale vortex forcing that they produced can trigger the resonant excitation effect on Modons. Since the Modons are accurate and non-dispersive interpretations of strong nonlinear Rossby waves, it is difficult to find a reason for the prolonged two-week maintenance of the dipole blocking pattern. Subsequent studies also showed that the transient large-scale vortex forcing caused by synoptic waves contributes to the maintenance of the circulation anomaly with a dipole structure (Guan, 1993). A number of diagnostic case studies also found that cold vortex cutoff can start from an upper trough with strong vorticity, or appears in an area with strong cyclonic shear on the left side of the zonal jet axis. Apparently, temperature advection can impose significant impacts on the high-latitude cold vortex cutoff (Palmén, 1951). The background circulation of high-latitude cold vortex cutoff could be branched out in the initial field, which was thoroughly discussed by Rex (1950,1951).

In the case in which a vortex develops from a low-pressure system on the ground before being cut off to become a cold vortex in mid latitudes, or a vortex develops both on the ground and in high latitudes, the process is more often than not associated with a baroclinic process. According to the baroclinic instability theory associated with small perturbations, certain medium wavelengths would encourage the growth of disturbances. For the unstable wavelength, the growth rate is proportional to the vertical shear. As a result, the mid troposphere would become a non-divergent layer, while substantial divergence and convergence would occur in the upper and lower layers, respectively. The strength of divergence, as such, is determined by the baroclinicity of the troposphere. One can use a vorticity equation to explain the positive divergence-free vorticity advection that is associated with the development of sea-level vorticity. The divergence zone in the upper troposphere corresponds to strong ascending movements and favors the development of cyclones in low latitudes (Arnason, 1963). Palmén (1951) elaborated that a cold or a warm advection is not necessarily sufficient to explain the temperature variations that accompany the cutoff process, as the vertical motion may also play a role in the process. Sometimes the core temperature would be dropping during a certain period when the cutoff cold vortex is developing, suggesting that ascending motions occur in the center of the vortex. This is consistent with the cross-section chart presented by Shen et al. (2012), which showed that the formation of NECV activities caused unusually cool summers in Northeast China. Fig. 1 depicts the four temperature anomaly cases along 120°E in 1969, 1971, 1972, and 1976. From Fig. 1a, compared to the mean airflow, the northerly flow originating from adjacent areas of the Northern Hemispheric polar region joins the airflow from areas near the equator, leading to a forceful ascending motion over 50°—52°N. Significant convergence can be found in the upper troposphere within 600—200 hPa with a deep cold air layer within 850—250 hPa and a cold center with temperature below −1.5 ℃ within 500—400 hPa that swings back and forth between 40°and 60°N. A warm tropopause that is associated with the latent heat release from the condensation caused by ascending motions in the convective cold vortex can be found in the upper level within 200—100 hPa. Fig. 1b indicates an opposite story. Using the NCEP reanalysis data, the FY-2C data, Cloud Sat data, and real-time observations, Liu et al. (2012b) analyzed the thermodynamic structure, the three-dimensional airflow field, and the cloud evolution of an NECV event that appeared during 27—30 June 2009. The results showed that the matured cold vortex had a cold core structure throughout the entire troposphere with a cold center in the upper troposphere and a warm core in the lower stratosphere. Based on a simplified large-scale atmospheric motion equation, Chen (1987) pointed out that the mid and high latitudes were the region most prone to long and super-long waves on the scale of $L > Lc$. In this context, the vorticity advection in the equation can be decomposed into several components for analysis. One of these components represents the transformation from a barotropic vorticity advection to a baroclinic one, which would always occur when temperature gradients induced a thermal wind vorticity in the turbulent temperature field. The readjustments of the non-thermal winds stemming from the development are favorable for the formation of cyclonic vortex, which in turn would induce a cold vortex cut off. In their analysis of the vertical structure of a heavy rain event caused by the NECV, Qiao et al. (2007) found that the NECV's initial cyclonic vorticity first appeared in the mid troposphere, and then stretched to the lower or upper levels. In the development stage, the cyclonic vorticity enhanced most rapidly in the upper troposphere and then steadily spread to the mid and lower levels, triggering the development of cyclones on the ground. This finding coincides with the second possible formation process of high-latitude cold vortex cutoff elaborated by Petterssen (1955). In general, the two baroclinic development processes of the

NECV show that the ascending motion in the atmosphere induced by positive vorticity advection must be accompanied by cooling in the mid troposphere, which is a favorable condition for the cutoff of the upper cold vortex. The convective condensation heating released by the ascending motion heats the tropopause, which helps maintain the balance of the atmospheric motion in both the horizontal and vertical directions.

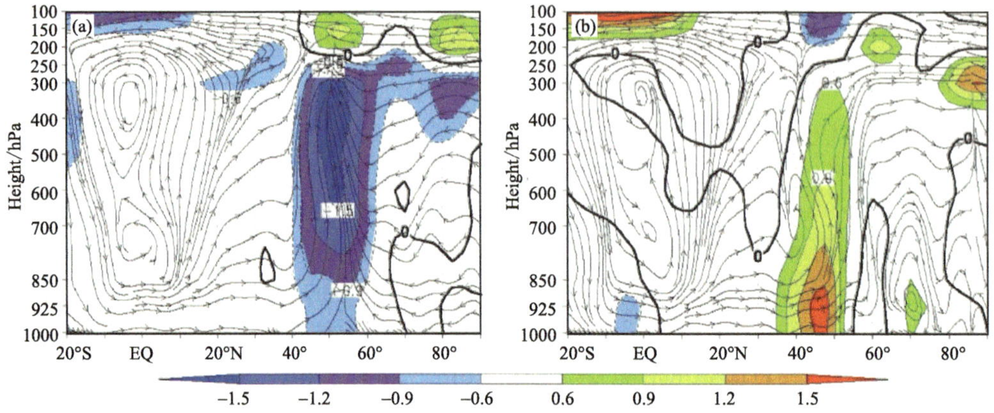

Fig. 1 Zonal vertical cross-sections of (w, v) streamlines at 120°E and cold and warm temperature anomalies in summer in Northeast China. (a) The composite cold temperature anomaly for the events that occurred in 1969, 1971, 1972, and 1976. (b) The composite warm temperature anomaly for the summers of 1982, 1988, 1994, and 1997. Dark solid lines represent speed of the (w, v) vectors, light solid lines show positive temperature anomalies, and dashed lines show negative temperature anomalies (Shen et al., 2012)

Taking into account the fact that a blocking high often lasts longer than the duration of a high-latitude cold vortex cutoff, and that the former has more significant impacts on abnormal weather events in the Northern Hemisphere, both domestic and overseas researchers started to pay more attention to the formation, maintenance, and collapse mechanisms for the blocking high (Berggren et al., 1949; Elliott et al., 1949; Yeh, 1949; Rex, 1950). As a matter of fact, a blocking high is often accompanied with a high-latitude cold vortex cutoff. The theoretical advancement of nonlinear dynamics spurred the study of blocking dynamics in the 1980s. Mcwilliams (1980) proposed the concept of Modons (dipole), while Malguzzi et al. (1984) introduced the KdV Rossby soliton. Chinese researchers developed the theory of the enveloped Rossby soliton (Luo et al., 1989; Luo, 1991, 1999). Meanwhile, researchers worldwide started to shift their attention to the formation process of the dipole pattern featured by the coexistence of the cold vortex and blocking high. Theoretically, the dipole pattern represents the state when the westerly jet split reaches maximum in the frontal zone. It is also the most important pattern in the NECV formation, and an important circulation pattern conducive to the extreme low-temperature events in northeastern China (Li et al., 2012). The Rossby soliton enveloped in the barotropic atmosphere has a dipole structure that is characterized by high pressure in the north and low pressure in the south. A dipole pattern may habitually last for 10 days. Normally, a baroclinically enveloped Rossby soliton is able to couple with baroclinic synoptic-scale waves. Baroclinic synoptic waves are supposed to enhance the amplitude of the soliton (Modons) and reduce the group and phase velocities, so as to form and maintain the dipole blocking. This is consistent with the results derived from the barotropic model (Berggren et al., 1949; Green, 1977; Shutts, 1983; Luo, 1999). The dipole blocking is more obvious in the vertical direction than in the horizontal direction. Luo (1999) reached a preliminary conclusion using a two-level baroclinic model that both the lower and upper levels are able to generate Modons, although the low-level Modons are stronger than those in the upper level. When the vertical shear becomes stronger between the lower and upper levels, the Modons generated by the enveloped Rossby solitons would move westward slowly and last shorter than normal, suggesting that the strong vertical shear inhibits the maintenance of the dipole blocking. In this context, a dipole blocking with a prolonged life span requires a weak vertical shear between the basic airflows in the lower and upper levels. This indicates from a different perspective that the dipole blocking is supposed to have a solid barotropic structure. During this period, the NECV studies were less independent as they were mostly conducted based on blocking high studies. In these studies, the dipole theory was simply associated with the cold vortex cutoff.

On the other hand, some progresses have been achieved on studies of the NECV energy budget. The cutoff low is a disturbance typically observed in the temperate zone. It plays an important role in the meridional and vertical exchanges of angular momentum and heat fluxes (Palmén, 1951). The exchange processes, although differed, determine the maintenance or collapse of a high-latitude cold vortex cut off. Sun (1998) analyzed a typical NECV process and showed that the energy distribution and transport pattern of ascending/descending airflows in the early stage of the NECV play an important role in the subsequent stages of the cold vortex development. NECV activities would weaken when the massive disturbance energy transfers to zonal kinetic energy, and would further decay and even disappear when the effective potential energy is converted to kinetic energy while the boundary flux effects gradually decrease. Based on an E-vector analysis and computation of monthly budget of quasigeostrophic potential vorticity, he also suggested that there exists a kinetic energy transfer from the transient eddy energy to the averaged airflow energy near the areas controlled by the East Asian blocking high and the NECV dipole subsystems. Meanwhile, there is an obvious convergence of E-vector, which results in a weakened westerly wind field but promotes the development of meridional circulation. The potential vorticity advection of time-averaged air flows is not conductive to the formation of blocking high/cold vortex dipole subsystems. Instead, it would widen the radial distance between the high-and low-pressure centers of the dipole subsystems. The transport of potential vorticity by the transient disturbance facilitates the prolonged local presence of high potential vorticity of cold vortex and low potential vorticity of blocking high, preventing their downstream transports (Sun, 1997).

Bueh et al. (2013) summarized the dynamic features related to the NECV formation and maintenance from the perspective of transient forcing and low-frequency waves. They pointed out that the quasi-zonal low-frequency Rossby waves that were associated with the Eurasian blocking circulations would directly transport their energy to the northeastern region. Meanwhile, the Rossby waves stemmed from the subtropical circulation anomalies would propagate in the meridional direction and reach Northeast China, leading to a circulation pattern favorable for the formation of the NECV. The transient disturbances from the upstream of Northeast China cooperate with the downstream circulation anomalies, creating an effective transient eddy forcing mechanism. In addition, the negative phase of the western Pacific (WP) teleconnection would result in a background downstream circulation favorable for the NECV activities. Furthermore, Xie et al. (2015) also showed that the abnormal NECV activities were closely associated with the Rossby wave breaking on the synoptic scale. The cyclonic and anticyclonic wave breaking processes on the synoptic scale are an important reason why all types of northeast cold vortex circulation can maintain persistently and cause consecutive meteorological disasters.

3 Correlation between NECV anomalies and atmospheric circulation and marine forcing

3.1 Statistical features of NECV activities

Zheng et al. (1992) indicated that the frequency of the NECV activity could reach 45% during April—October. The peak number of NECV days occurred in June, followed by that in July, August, and May. An automatic identification technique developed by Liu et al. (2015) showed that the average days of NECV activity were 14, 18, 18, and 14 days, respectively (64 days in total) for the months during May—August of 1948—2012, which accounted for 52% of the total days. A typical cold vortex process can last for 3~6 days. Xie et al. (2012) examined the temporal and spatial distribution of the NECV activity in the summers (May—August) of 1965—2007 and the associated background circulation patterns over the same period. The results showed that the NECV activities were mainly confined to the north of 45°N in May, and expanded to the south of 40°N in June before they retreated to the north month by month. The NECV activity frequency reached the peak around June 6. The number of NECV days declined during the rainy season. The end of the rainy season marked a further reduction in the NECV incidence. The frequency of NECV incidence displayed a quasi-2-year oscillation pattern. Hu et al. (2011) highlighted the fact that the correlation coefficient between NECV activities in the winter and East Asian winter monsoon intensity is significant at the 99% confidence level.

3.2 NECV with regional temperature and precipitation anomalies

As early as in the 1980s, Tao (1980) pointed out that the NECV was a circulation pattern that can result in heavy rainfall or thunderstorms in the northeastern and northern parts of China. Chinese researchers have paid more attention to the climatological aspect of the NECV activity since the end of the 20th century. Sun et al. (2000, 2001) examined the linkage between the prolonged NECV activities and atmospheric circulation anomalies, suggesting that the prolonged NECV activities are a key element not only responsible for the occurrence of unusually cool summer in Northeast China, but also for the precipitation and concentrated heavy rainfall in the region. Lian et al. (2003) and Shen et al. (2004) divided the summer precipitation over the northeastern region into two phases, i. e., the cold vortex rainy season (June—early July) and the subtropical summer rainy season. Shen et al. (2011) also summarized the main circulation patterns associated with the two rainy seasons. He et al. (2006) elaborated that the NECV makes a major weather system over mid-high latitudes of East Asia. Note that the prolonged NECV activities can generate a significant climatological effect, regardless of its synoptic scale. During the rainy season, the correlation between the NECV activity and rainfall indicates that enhanced NECV activities correspond to abnormally abundant rainfall, and vice versa. Miao et al. (2006) found that the intensity of the NECV activity in the pre-Meiyu rainy season is significantly correlated with the presence of rainfall in South China.

3.3 Circulation anomalies

Liu et al. (2002) conducted a composite analysis of the Northern Hemispheric 500 hPa geopotential height anomaly field during May and June—August in an effort to sort out the years that had more or fewer typical NECV activities. Their results showed that the prolonged NECV activities are not only closely associated with the Northern Hemispheric circulation anomalies in the pre-NECV, simultaneous-NECV, and post-NECV periods, but also constitute an important part of the anomalies. The linkage between the NECV and circulation anomalies is reflected in the correlation between the NECV and the teleconnection patterns of atmospheric circulations. Liu et al. (2012a) summarized the anomalous circulation patterns that support the formation and maintenance of NECV activities, pointing out that when the low-frequency Eurasian (EU) and East Asia/Pacific (EAP) teleconnection wave trains meet in East Asia, the interaction between the wave trains can eventually affect the moving direction and speed of the NECV system. The negative WP teleconnection phase is a downstream background circulation pattern that supports the NECV activity. Its prevalence can make an important contribution to the recurrence and prolonged stay of the NECV circulations (Bueh et al., 2013; Li et al., 2016; Xie et al., 2015). Meanwhile, it also contributes to the north-to-south movement of the massive NECV belt, which corresponds to the south-north swing of the 28 m s^{-1} wind speed axis in the 200 hPa flow field. This phenomenon suggests that enhanced NECV activities would force the southern jet stream to shift northward over high latitudes (Zheng et al., 1992). The southern jet stream branch is more often than not a result of the jet stream split over East Asia (Sun et al., 2000).

Zheng et al. (1992) summarized two major patterns that may shape up the high-latitude NECV situation due to the cutoff process. One is the Yakutsk blocking. The stability of Yakutsk blocking high in the east contributes to the cutoff of the cold and low trough movement in the northeastern region, facilitating the formation of NECV. The other is a blocking pattern made up of the high pressure in the north of Baikal and the cold vortex in the northeast. Lian et al. (2010) found that abnormal cold vortex activities in the summer correspond to the circulation anomalies near the upstream of the Ural Mountains and downstream of Japan. The concerted work of the downstream Rossby wave energy propagation and the local transient eddy forcing creates a blocking circulation pattern near the Ural Mountains, which is closely associated with abnormal cold vortex activities over northeastern China. On the one hand, it drives cold air to move from the polar region to the northeastern region of China; on the other hand, it enhances NECV activities through the propagation of Rossby wave energy. Basically opposite circulation patterns occurred in the years with reduced cold vortex anomalies. Wan et al. (2013) studied the impact of stratospheric NAM anomalies on NECV activities. Their composite analysis of different stratospheric NAM phase anomalies indicated that negative stratospheric NAM anomalies would contribute to the increased occurrence, prolonged maintenance, and enhanced strength of NECV activities, which have a closer tie to stratospheric circulation anomalies. Meanwhile, negative stratospheric NAM anomalies would enhance the activities of planetary waves, which in turn facilitate and enhance the development of NECV activities. Research results have demonstrated the importance of using stratospheric NAM anomalies to predict possible NECV activities in the winter. Li et al. (2015) also revealed the mass exchange between the stratosphere and troposphere caused by the cutoff low over Northeast Asia. Xie et al. (2012) classified the NECV pattern into four categories based on its high-pressure ridge position in the upstream: Yenisei pattern, Baikal pattern, Ural pattern, and Yakutsk-Okhotsk pattern. The WP teleconnection pattern creates a background circulation for NECV activities over the same period. A negative phase would encourage the formation of NECV activities. In addition, the NECV activities are closely associated with the Pacific-North America (PNA) pattern during May—June, and with the North Atlantic Oscillation (NAO) in August.

Shen et al. (2012) made a successful attempt to predict abnormal NECV activities in the early summer of 2009. Based on the composite analysis, they singled out the six-month, seasonal, and monthly precursor signals for NECV activities during June—July. In the preceding winter, North Pacific Oscillation (NPO) would remain in significant negative phase until the following spring. It is also correlated with the change of super-long wave disturbance into a steady wave at 500 hPa over North Pacific during April—May. In this context, the NPO phase anomalies not only behave as an important precursor signal for summer temperature variations in the northeastern region, but also become a stable planetary-scale component that can be extracted from the atmosphere.

Chinese researchers also paid sufficient attention to the extraction of precursor signals from dynamic atmospheric processes for predicting abnormal NECV activities in the summer. Harmonic and low-frequency oscillation analysis showed that during May—June of the years with abnormally more (fewer) NECV days, the northeastern region often had a positive (negative) vorticity phase, corresponding to the westward retreat (eastward advancement) of the super-long wave trough over North Pacific. The standing long-wave pattern over North Pacific—North America—Atlantic worked with the propagation of Rossby wave energy, facilitating the prolonged stay of anomalous circulations near the Ural Mountains. In May, the anomalous circulation near the Ural Mountains was a significant precursory signal for the possible NECV activities in early summer (Lian et al., 2010) and possible position and duration of precipitation in the eastern part of China as well. Apparently, the development status of EU and EAP low-frequency wave trains at the initial stage may determine the subsequent evolution of precipitation in different parts of eastern China (Liu et al., 2012a).

3.4 External forcing

Shen et al. (2004) took the period of June and early July as a cold-vortex rainy season. They conducted an SVD analysis using the NCEP/NCAR reanalysis data of 500 hPa geopotential height field and the snow depth data in the Tibetan Plateau as well as the daily summer rainfall data collected at 69 stations in northeastern China. The results showed that the abnormally affluent presence of snow cover over the Tibetan Plateau is associated with the abnormally affluent precipitation in the cold-vortex rainy season across the northeastern part of China. The plateau's snow cover field demonstrated a "+ −" pattern from north to south during their study period. When the positive or negative center is located above the major mountains of the plateau, the Sanjiang plains in the northeastern region and Jilin Province would become abnormally wet. When the snow cover field in the central and eastern parts of the plateau turns positive, the cold-vortex rainy season in the northeastern region would have a reversed "− +" pattern of precipitation distribution from east to west. Chen et al. (2013) indirectly discussed the impact of NECV activities based on the analysis of the impact of Eurasian mid-high latitudes snowmelt on the summer low-temperature anomalies in the northeastern region. They indicated that when the Eurasian mid-high latitudes have a massive springtime snowmelt in the west but little snowmelt in the east, the subsequent summer would have an abnormally affluent snow cover in the east. In the summer, snowmelt means heat absorption. The soil moisture that increases later would jointly result in an abnormally low summer temperature in the region. As the geopotential height field declines, the Eurasian mid-high latitudes are under the control of a deepened longwave trough in the east at 500 hPa, with enhanced northerly jets in the rear of the trough. Consequently, the cold air from the polar region can easily reach the East Asian mid-high latitudes, resulting in a low-temperature summer in the northeastern part of China.

David et al. (1995) demonstrated that El Niño events could enhance westerlies over the mid-high latitudes. As a result, the Aleutian low would become stronger than normal. Li (1989) performed a diagnostic analysis of the July—September 500 hPa geopotential height anomalies across East Asia and western Pacific in the El Niño years from 1951 to 1979, and found that there was a negative anomaly center above Northeast China in the summer. This result suggested that the low pressure trough activities over this region enhanced in the summer of El Niño years, which would in turn lead to an abnormally low-temperature summer in the region. Miao et al. (2006) conducted an analysis and showed that the abnormally strong NECV activities in the pre-flood season corresponded to the mature or developing stage of a La Niña event, while the abnormally weak NECV activities in the pre-flood season corresponded to the mature or developing stage of an El Niño event.

How to judge the association between the opposite ENSO phases and NECV activities? We start from the cyclonic vorticity source of low-latitude troughs above the northeastern part of China. Chou (1986) offered a theoretical explanation, stating that this is a retreating wave. He studied on an algebraic wave equation using a simplified quasi-geostrophic adiabatic model. When the wavelength is longer than 3000—4000 km, the wave will have a moving speed of 1000—2000 km month^{-1}. The westward slow-moving wave, as such, represents a semi-permanent atmospheric activity center on the ocean that corresponds to the marine activities. The diagnostic analysis of Liu et al. (2012b) indicated that the NECV activities during June—August are correlated with the semi-permanent atmospheric activity center in North Pacific. Looking at the NPO-like 500 hPa height anomaly field, it is found that a strong Aleutian low often corresponds to a positive NPO phase in the preceding winter and weak NECV activities during June—August. In the opposite situation when the Aleutian low is weak, NECV activities are strong. Shen et al. (2008) selected 10 typical cases that represent strong, weak, and normal summer NECV activities for a harmonic analysis. The results showed that the strength of the NECV activity is quite different among these cases, which highly depends on the position of the long waves, the wave axis tilting direction to West Pacific or to East Pacific, and the westward retreat or eastward advance of the waves.

Lian et al. (2013) employed an SVD method to reveal the presence of a major coupling mode between the semi-permanent atmospheric activity center and the marine forcing over North Pacific. The combined effect of the Pacific polar vortex, the NPO and North Pacific (westerly drifts and ENSO phases) determines the size (large or small) of the Pacific polar vortex in the spring, which would in turn determine the positive or negative NPO anomaly phase in the mid troposphere and in the sea surface pressure field. A positive or negative NPO phase anomaly can be tightly coupled with a cold or warm westerly drift phase as well as with El Niño or La Niña, resulting in up to 60% contribution made by SVD1 to the covariance square. Lian et al. (2014) conducted an SST forcing related numerical experiment using the CAM3. 1 model. The results indicated that the combined effects of El Niño and cold SST phase in the westerly drift region in the spring would be conducive to an enhanced NECV anomaly in early summer (Experiment 3 and associated controls, Fig. 2). As shown in the 850 hPa stream function anomaly field from March to June derived from differences between results of Experiment 3 and the control experiment, the low-frequency cyclone "C" that originally stayed above the Aleutian Islands in April expanded westward, leading to the formation of a cyclonic circulation center above Northeast Asia, which then encouraged abnormal NECV activities. In spring, the equatorial Pacific El Niño phase would induce elliptic low-frequency variations in the tropical airflow field, which contributes to and enhances the "C-A-C-A" low-frequency wave train in the tropical-subtropical-high latitudes. It creates an important physical mechanism that supports the North Pacific subtropical and high-latitude circulation anomalies in early summer with a 95% confidence level (Fig. 2). In spring, the equatorial Pacific La Niña phase forcing would trigger an abnormal high pressure tele-response in early summer over the Sea of Okhotsk and the Sea of Japan on the one hand, and indirectly induce a positive EAP low-frequency wave train in the lower troposphere on the other hand (figure omitted).

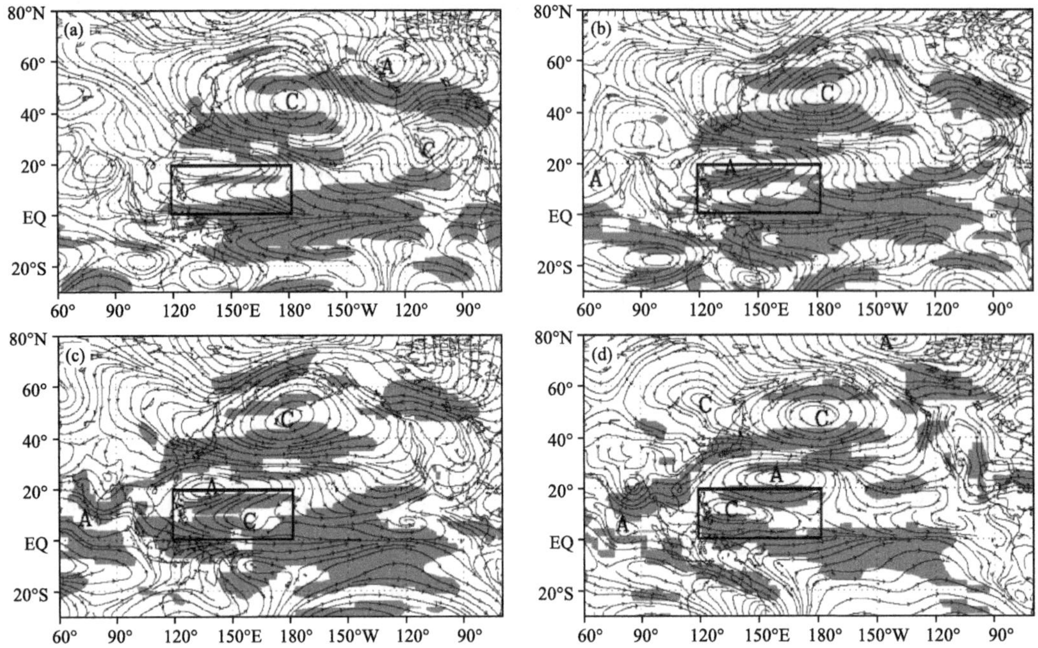

Fig. 2　Differences in 850 hPa airflow between Experiment 3 and the control experiment for the period of March—June. Gray shaded areas indicate the difference significant at the 95% confidence level and thin solid lines represent 850 hPa streamlines. (a) March, (b) April, (c) May, and (d) June (Lian et al., 2014)

(The 850 hPa Niño4 flow anomaly field in the central tropical equator sits in a symmetric parabolic low-frequency variation pattern with the axis as the square area)

4　Thinking points

The following scientific issues are raised for future studies:

(1) Greater efforts shall be made to parallel the study of the NECV formation and occurrence frequency to the study of Northern Hemispheric blocking high. As far as the blocking high studies are concerned, there are three major theories using both linear and nonlinear methodologies. One is the blocking theory in the global context represented abroad by Egger (1978), Charney et al. (1979), Tung et al. (1979), and domestically by Zhu et al. (1991); another is a blocking theory in the local context represented abroad by Mcwalliams' Modons (dipole), and Malguzzi and Malanotte-Rizzoli (KdV Rossby soliton theory); the third is an enveloped Rossby soliton theory proposed by Luo (1999). The NECV studies have made great progresses in the dipole theory that goes with blocking highs. Chinese researchers not only have studied the weather processes related to cold vortex, but also investigated the climatic effects stemmed from such processes (Bueh et al., 2015). These studies have made a great contribution to the theoretical study of high-latitude cold vortex cutoff. Unfortunately, substantive deficiency still exists in the NECV studies compared to the studies for blocking highs. The studies of high-latitude cold vortex cutoff have not yet resulted in a solid theoretical system, since these studies were mainly conducted as part of the blocking high study. Chen's thermal wind adaptation theory provides a reasonable explanation for the thermodynamics involved in the cold vortex cutoff. Bueh et al. (2011, 2015) introduced a theory and an identification process to explain the formation of baroclinic ridges and troughs,

which are favorable for prolonged wintertime low-temperature events in China. This theory presents a new configuration model that is different from the Modons for the study of cold vortex and transverse trough. Both the NECV and other high-latitude cold vortex activities are synoptic as far as the scale is concerned, which makes them quite different from the blocking highs that often have a lifespan up to 7 days. In the future, efforts have to be made to address a variety of related issues, including the dynamic and thermodynamic patterns of multiple NECV activities, the configurations between the NECV activities and Northern Hemispheric blockings, and mutations of planetary-scale flow patterns as well as the linkages between internal atmospheric dynamics and external forcings, among others. Answering these questions will help us to better understand the diversified formation processes of high-latitude cold vortex cutoff and associated maintenance mechanisms, and further enrich the connotations of the NECV theory.

(2) The studies of changes in the NECV activity in the context of global warming should be enhanced. The variability of temperature, an indicator sensitive to climate change, may result in changes in the atmospheric circulations and major circulation systems. The cooling effect occurred during the 1960s—1980s across Europe and North America altered atmospheric circulation patterns, resulting in changed precipitation distributions in North Africa (Rotstayn et al., 2002). During the 1960s—1970s, both northeastern China and the Northern Hemispheric high latitudes experienced increased incidence of unusually cool summer. In the summer, northeastern China may have 42 NECV days on average. Since 1993, or in the 1990s and 2010s, the same part of China experienced summers with sustained high temperature. As a result, the number of NECV days dropped to 34 (Liu et al., 2015). It can be found that under the influence of global warming, the number of NECV days has significantly reduced in northeastern China. What else the change could be? Would the NECV formation mechanism also be changed to some extent because of the global warming? More attention has to be paid to address these scientific questions.

(3) Mechanism studies for the formation of extreme events shall take advantage of the combined strengths of both linear and nonlinear methodologies, especially that of weak nonlinear methodology. The enhanced study of the formation mechanism for the Northern Hemisphere blocking highs embodies an evolution of linear methodologies with constantly added nonlinear elements. The same is also applied to the study of high-latitude cold vortex cutoff and NECV activities. Theoretical studies would be advanced with constant methodological improvements. The study needs to examine the relationship between NECV activities and extreme weather and climate events in individual seasons. For example, it is necessary to understand how the tilted trough that causes extensive low-temperature events in the winter can become an NECV. The question also needs to be addressed why some prolonged early summer NECV activities are able to induce extreme low-temperature events while other similarly prolonged NECV activities cannot. Something can be learned from the nonlinear methodologies, especially from the weak nonlinear methodologies. It is worth reading the book "Mathematic and Physic Methodologies Applied in Atmospheric Sciences" written by Huang et al. (2005). The book discusses the association between the formation of one cold vortex event and the increased incidence of multiple cold vortex events. Specifically, it talks about how to systematically link the fast varying variables with the relatively stable and slow varying variables in the chaotic atmosphere. A Chinese researcher Xu (1992) pioneered in this area. He simplified the tropical atmospheric motion equation using the WKBJ (Wentzel-Kramers-Brillouin-Jeffreys) process to study the impact of coeffcient matrix on tropical atmospheric low-frequency oscillations. The theoretical model result indicates that when the tropical region is free from adiabatic heating and

divergence, the thermal disturbance would always lead to periodic atmospheric oscillation in the tropics or an elliptic trajectory phase. Otherwise, the phase trajectory would be parabolic, hyperbolic, or linear, without the periodic low-frequency oscillation (Xu et al., 2002). Lian et al. (2013) employed the CAM3.1 model to justify Xu's theory and promoted its applications. They concluded that the low-frequency wave train forced by the El Niño phase has an important impact on facilitating NECV activities in early summer. Guan (1993) also pioneered in the study of the atmospheric circulations utilizing nonlinear methodology. He employed a barotropic model that is based on the quasi-geostrophic evolution hypothesis to reveal, from a theoretical perspective, the restoring force imposed on atmospheric particles in the Rossby wave process and a variety of possible flow patterns and particle trajectories. He provided a reasonable explanation for the presence of the blocking cutoff. Generally speaking, the nonlinear methodology makes a worthy tool in studies of the NECV anomalies. It may give us some unexpected results. In addition, we need to know which quadrant of the cold vortex is responsible for the severe convective weather events induced by NECV activities, including heavy rain, severe hail, thunderstorms, and high winds, and what formation mechanisms are behind. Enhanced studies are necessary in order to answer these questions. At present stage, it is still difficult for NWP models to predict severe convective weather systems caused by the NECV activities.

(4) More efforts should be made to study the low-frequency oscillation of the NECV activity and the precursor signals of its anomaly. Ding (2007) pointed out that northeastern China is a rendezvous of low-frequency waves stemmed from diversified sources, from which new disturbances would be generated and propagated northwards to the Pacific. Sun (1997) revealed the impact of a low-frequency oscillation and transient disturbance event that lasted for 10—20 days in East Asia on the maintenance of NECV activities through the 1986 case study. Lian et al. (2010) made an analysis of two typical years to examine the ties and possible mechanisms between significant early summer NECV anomalies and low-frequency oscillations in the Northern Hemispheric circulation. In the year with significant NECV anomalies, the concerted effects of downstream Rossby wave propagation and transient eddy forcing created a blocking circulation near the Ural Mountains. A low-frequency oscillation study showed that during May—June of the year with significantly more (less) cold vortex anomalies, the northeastern region was in a prevailing positive (negative) vorticity phase, corresponding to the westward retreat (eastward advance) of longwave troughs over North Pacific. As the longwave becomes stationary across North Pacific—North America—Atlantic, it facilitates the propagation of Rossby wave energy and is favorable for the prolonged maintenance of anomalous circulations near the Ural Mountains. Unfortunately, the conclusions in the above-mentioned studies were mainly based on individual cases. Using the data from 1979 to 2009, Liu et al. (2012a) made a detailed analysis of the meridional low-frequency oscillation stemmed from the northwest of the westerlies and from the tropical EAP based on an NECV index (NECVI). Significant progresses were made in this study, which was designed for 10—30-day low-frequency oscillations that were split into 8 different phases. Lian et al. (2013) revealed a coupled oscillation between NPO, westerly drifts and central/eastern equatorial Pacific SSTA, and investigated the impact of resulted non-diabatic retreating waves on NECV activities. The study, however, did not discuss the superimposition and phasing of eastward advance and westward retreat of low-frequency oscillations stemmed from the polar and tropical regions. Consequently, the physical process of the rampant NECV activity period (May—September) and associated early evolution is not fully unfolded, which would compromise the scientific judgment of the ties between internal atmospheric dynamic process and external forcing. Furthermore, such evolution could be differed from year to year, which makes it hard to extract a precursor signal of the

NECV anomalies. This is an important issue that needs further studies.

This paper mainly discussed the current status of the NECV formation and maintenance studies, with little room spared for an automated NECV identification and the methodologies applied in projecting climatological anomalies. The authors also proposed some scientific issues that are worth further investigation. On the one hand, more efforts in the NECV study will further raise the academic level of the NECV study and improve the weather forecast and short-term climate prediction; on the other hand, results of the NECV study will help us better understand the impact of mid-and high-latitude circulation systems on the distribution of precipitation across East Asia during the period from the spring to early summer. Overall the NECV study is an important scientific issue that cannot be ignored.

References

ÁRNASON G,1963. The stability of nongeostrophic perturbations in a baroclinic zonal flow[J]. Tellus,15(3):205-229.

BERGGREN R,BOLIN B,ROSSBY C G,1949. An aerological study of zonal motion,its perturbations and breakdown[J]. Tellus,1(1):14-37.

BJERKNES J,SOLBERG H,1922. Life cycle of cyclones and the polar front theory of atmospheric circulation[J]. Geopys Publ,3(1):1-18.

BUEH C,XIE Z W,2013. Northeastern China cold vortex circulation and its dynamical features[J]. Adv Meteor Sci Technol, 3(1):34-39. (in Chinese)

BUEH C,FU X Y,XIE Z W,2011. Large-scale circulation features typical of wintertime extensive and persistent low temperature events in China[J]. Atmos Oceanic Sci Lett,4(4):235-241.

BUEH C,PENG J B,XIE Z W,et al,2015. A Study of Winter Time and Persistent Low Temperature Events in China[M]. Beijing:China Meteorological Press:230. (in Chinese)

CHAO J P,HUANG R X,1980. The Cnoidal waves of rotating barotropic atmosphere[J]. Sci Sinica,23:1266-1277.

CHARNEY J G,1947. The dynamics of long waves in a baroclinic westerly current[J]. J Meteor,4(5):135-162.

CHARNEY J G,DEVORE J G,1979. Multiple flow equilibria in the atmosphere and blocking[J]. J Atmos Sci,36(7):1205-1216.

CHEN C,1963a. On the formation and the destruction of thermal wind in a simple baroclinic atmosphere[J]. Acta Meteor Sinica,33(1):51-63. (in Chinese)

CHEN C,1963b. On the formation and the destruction of thermal wind in a simple baroclinic atmosphere[J]. Acta Meteor Sinica,33(2):153-162. (in Chinese)

CHEN H S,QI D,XU B,2013. Influence of snow melt anomaly over the mid-high latitudes of the Eurasian continent on summer low temperatures in northeastern China[J]. Chinese J Atmos Sci,37(6):1337-1347. (in Chinese)

CHEN Q S,1980. Instable development of terrain long and super-long waves[J]. Sci Sinica,10(6):576-585. (in Chinese)

CHEN Q S,1987. Dynamics of Synoptic and Supper Synoptic Scale Systems[M]. Beijing:Science Press:166. (in Chinese)

CHOU J F,1986. Long-term Numerical Weather Prediction[M]. Beijing:China Meteorological Press:32. (in Chinses)

CROCKER A M,GODSON W L,PENNER C M,1947. Frontal contour charts[J]. J Meteor,4(3):95-99.

DAVID L,ANDERSON T,1995. The tropical ocean global atmosphere programme[J]. Contemporary Physics,36(4):245-265.

DING Y H,2007. Preface[J]. Scientia Geographica Sinica,27(S1),1. (in Chinese).

EADY E T,1949. Long waves and cyclone waves[J]. Tellus,1:33-52.

EGGER J,1978. Dynamics of blocking highs[J]. J Atmos Sci,35(10):1788-1801.

ELLIOTT R D,SMITH T B,1949. A study of effects of large blocking high on the general circulation of the Northern Hemispheric westerlies[J]. J Atmos Sci,6(2):67-85.

GREEN J S A,1977. The weather during July 1976:Some dynamical considerations of the drought[J]. Weather,32(4):120-126.

GUAN Z Y,1993. A further study on the characteristics of nonlinear Rossby waves[J]. J Trop Meteor,9(2):160-168. (in

Chinese)
HAINES K,MARSHALL J,1987. Eddy-forced coherent structures as a prototype of atmospheric blocking[J]. Quart J Roy Meteor Soc,113(476):681-704.
HE J H,WU Z W,JIANG Z H,et al,2006. Climate effects of NECV and its impact on the rainy season[J]. Chinese Sci Bull,51(23):2803-2809. (in Chinese)
HSIEH Y P,1949. An investigation of a selected cold vortex over North America[J]. J Meteor,6(6):401-410.
HU K X,LU R Y,WANG D H,2011. Cold vortex over Northeast China and its climate effect[J]. Chinese J Atmos Sci,35(1):179-191. (in Chinese)
HUANG S X,WU R S,2005. Mathematics and Physics in Atmospheric Science Question[M]. Beijing:China Meteorological Press:540. (in Chinese)
KUO H L,1949. Dynamic instability of two-dimensional nondivergent flow in a barotropic atmosphere[J]. J Meteor,6(2):105-122.
LI C Y,1989. El Niño Event and the temperature anomalies in eastern China[J]. J Trop Meteor,5(3):210-219 (in Chinese).
LI D,BIAN J C,FAN Q J,2015. A deep stratospheric intrusion associated with an intense cutoff low event over East Asia [J]. Sci China (Earth Sci),58(1):116-128.
LI S F,LIAN Y,CHEN S B,et al,2012. Distribution of extreme cool events over Northeast China in early summer and the related dynamical processes[J]. Sci Geographica Sinica,32(6):752-758. (in Chinese)
LI S F,JIANG D B,LIAN Y,et al,2016. Interdecadal variations of cold air activities in Northeast China during springtime [J]. J Meteor Res,30(5):645-661,doi:10. 1007/s13351-016-5912-6.
LIAN Y,1987. Analysis of a Heavy Rain Process Induced by NECV (Task Force for Northern Heavy Rains). Torrential Rain Digest[M]. Beijing:China Meteorological Press:223. (in Chinese)
LIAN Y,SHEN B Z,GAO Z T,et al,2003. The study of the onset criterion and the date of East Asian summer monsoon in Northeast China and its main characteristic analysis[J]. Acta Meteor Sinica,61(5):548-558. (in Chinese)
LIAN Y,BUHE C,XIE Z W,et al,2010. The anomalous Cold Vortex activity in Northeast China during the early summer and the low-frequency variability of the Northern Hemispheric atmosphere circulation[J]. Chinese J Atmos Sci,34(2):429-439. (in Chinese)
LIAN Y,SHEN B Z,LI S F,et al,2013. Impacts of polar vortex,NPO,and SST configurations on unusually cool summers in Northeast China. Part I:Analysis and diagnosis[J]. Adv Atmos Sci,30(1):193-209.
LIAN Y,ZHAO B,SHEN B Z,et al,2014. Numerical experiments on the impact of spring North Pacific SSTA on NPO and unusually cool summers in Northeast China[J]. Adv Atmos Sci,31(6):1305-1315.
LIU Z X,LIAN Y,GAO Z T,et al,2002. Analyses of the Northern Hemisphere circulation characters during Northeast Cold Vortex persistence[J]. Chinese J Atmos Sci,26(3):361-372. (in Chinese)
LIU H M,WEN M,HE J H,et al,2012a. Characteristics of the Northeast Cold Vortex at intraseasonal time scale and its impact[J]. Chinese J Atmos Sci,36(5):959-973. (in Chinese)
LIU Y,WANG D H,ZHANG Z F,et al,2012b. A comprehensive analysis of the structure of a Northeast China cold vortex and its characteristics of evolution[J]. Acta Meteor Sinica,70(3):354-370. (in Chinese)
LIU G,LIAN Y,YAN P C,et al,2015. The objective recognition and classification of Northeast Cold Vortex and the Northern Hemisphere atmospheric circulation characters in May to August[J]. Sci Geographica Sinica,35(8):1042-1050. (in Chinese)
LUO D H,1991. Nonlinear Schrodinger equation in the rotational barotropic atmosphere and atmospheric blocking[J]. Acta Meteor Sinica,5:587-597.
LUO D H,1999. Near-resonant topographically forced envelope Rossby solitons in a barotropic flow[J]. Geophys Astrophys Fluid Dyn,90:161-188.
LUO D H,JI L R,1990. A theory of blocking formation in the atmosphere[J]. Sci China (Series B),33(3):323-333.
MALGUZZI P,MALANOTTE RIZZOLI P,1984. Nonlinear stationary rossby waves on nonuniform zonal winds and atmospheric blocking. Part I:The analytical theory[J]. J Atmos Sci,41(17):2620-2628.
MCWILLIAMS J C,1980. An application of equivalent modons to atmospheric blocking[J]. Dyn Atmos Oceans,5(1):43-66.

MIAO C S, WU Z W, HE J H, et al, 2006. The anomalous features of the northeast cold vortex during the first flood period in the last 50 years and its correlation with rainfall in South China[J]. Chinese J Atmos Sci, 30(6): 1249-1256. (in Chinese)

PALMÉN E, 1949. On the origin and structure of high-level cyclones south of the maximum westerlies[J]. Tellus, 1(1): 22-31.

PALMÉN E, 1951. The aerology of extratropical disturbances. Compendium of Meteorology[C]. T. F. Malone, Ed., Amer. Meteor. Soc., Boston: 599-622.

PALMÉN E, NAGLER K M, 1949. The formation and structure of a large-scale disturbance in the westerlies[J]. J Meteor, 6 (4): 228-242.

PETTERSSEN S, 1950. Some aspects of the general circulation of the atmosphere[C]. Cent Porc Roy Meteor Soc. London: Royal Meteorological Society: 120-155.

PETTERSSEN S, 1955. A general survey of factors influencing development at sea level[J]. J Meteor, 12(1): 36-42.

QIAO F X, ZHAO S X, SUN J H, 2007. Study of the vorticity and moisture budget of a Northeast Vortex producing heavy rainfall[J]. Clim Environ Res, 12(3): 397-412. (in Chinese)

REX D F, 1950. Blocking action in the middle troposphere and its effect upon regional climate. I: An aerological study of blocking action[J]. Tellus, 2: 196-211.

REX D F, 1951. The effect of Atlantic blocking action upon European climate[J]. Tellus, 3: 100-112.

ROTSTAYN L D, LOHMANN U, 2002. Tropical rainfall trends and the indirect aerosol effect[J]. J Clim, 15(15): 2103-2116.

SCHWERDTFEGER W, 1948. Untersuchungen Äuber den Aufbau von Fronten und Kaltlufttropfen[J]. Ber Deut Wetterdienstes US Zone, 3: 1-35.

SHEN B Z, LIAN Y, 2004. The preliminary analysis on the relationship between snow cover over the Tibetan Plateau and summer Cold Vortex precipitation in Northeast China[J]. J Jilin Univ (Earth Sci), 34(S1): 112-118. (in Chinese)

SHEN B Z, LI S F, LIAN Y, 2008. The harmonic analysis of Cold Vortex of Northeast China in early summer[J]. J Jilin Univ (Earth Sci), 38(S1): 204-208. (in Chinese)

SHEN B Z, LIN Z D, LU R Y, et al, 2011. Circulation anomalies associated with interannual variation of early and late summer precipitation in Northeast China[J]. Sci China (Earth Sci), 54(7): 1095-1104.

SHEN B Z, LIU S, LIAN Y, et al, 2012. The 2009 summer low temperature in Northeast China and its association with prophase changes of the air-sea system[J]. Acta Meteor Sinica, 26(4): 438-453.

SHUTTS G J, 1983. The propagation of eddies in different jetstreams: Eddy vorticity forcing of "blocking" flow fields[J]. Quart J Roy Meteor Soc, 109(462): 737-761.

SUN L, 1997. A study of the persistence activity of Northeast Cold Vortex in China[J]. Chinese J Atmos Sci, 21(3): 297-307. (in Chinese)

SUN L, 1998. A energetics study of a northeast cold vortex[J]. Acta Meteor Sinica, 56(3): 349-361. (in Chinese)

SUN L, AN G, 2001. A diagnostic study of northeast cold vortex heavy rain over the Songhuajiang-Nenjiang River basin in the summer of 1998[J]. Chinese J Atmos Sci, 25(3): 342-354. (in Chinese)

SUN L, AN G, LIAN Y, et al, 2000. A study of the persistent activity of Northeast Cold Vortex in summer and its genral circulation anomaly characteristics[J]. Acta Meteor Sicica, 58(6): 704-714. (in Chinese)

SUTCLIFFE R C, 1947. A contribution to the problem of development[J]. Quart J Roy Meteor Soc, 73: 370-383.

TAO S Y, 1980. Rainstorm in China[M]. Beijing: Science Press: 225. (in Chinese)

TUNG K K, LINDZEN R S, 1979. A theory of stationary long waves. Part I: A simple theory of blocking[J]. Mon Wea Rev, 107(6): 714-734.

WAN X M, FU Z T, HU Y Y, 2013. The influence of stratospheric anomalies on the Northeast Cold Vortex[J]. Acta Scientiarum Naturalium Universitatis Pekinensis, 49(3): 417-425. (in Chinese)

XIE Z W, BUEH C, 2012. Low frequency characteristics of Northeast China Cold Vortex and its background circulation pattern[J]. Acta Meteor Sinica, 70(4): 704-716. (in Chinese)

XIE Z W, BUEH C, 2015. Different types of cold vortex circulations over Northeast China and their weather impacts[J]. Mon Wea Rev, 143(3): 845-863.

XIE Z W, BUEH C, 2017. Cold vortex events over Northeast China associated with the Yakutsk-Okhotsk blocking[J]. Int J Climatol, 37(1): 381-398, doi: 10.1002/joc.4711.

XU X D, 1992. Correlations between the phase diagram structures of periodic oscillations in the tropical atmosphere and thermal forcing from midlatitudes of the Southern and Northern Hemispheres[J]. Sci Atmos Sinica, 16(2): 147-157. (in Chinese)

XU X D, GAO S T, 2002. Dynamics of Exogenous Forcing and Wavecurrent Interaction[M]. Beijing: China Oceanography Press: 291. (in Chinese)

YANG H W, FENG G L, SHEN B Z, et al, 2012. The quantitative research on cold vortex in summer over Northeast China [J]. Chinese J Atmos Sci, 36(3): 487-494. (in Chinese)

YE D Z, LI M C, 1965. Adaptations in Atmospheric Motion[M]. Beijing: Science Press: 126. (in Chinese)

YEH T C, 1949. On energy dispersion in the atmosphere[J]. J Meteor, 6(1): 1-16.

ZHENG X Y, ZHANG T Z, BAI R H, 1992. Rainstorm in Northeastern China[M]. Beijing: China Meteorological Press: 219. (in Chinese)

ZHU B Z, JIN F F, LIU Z Y, 1991. An Introduction to the Nonlinear Dynamics of the Atmosphere and Ocean[M]. Beijing: Oceanography Press: 328. (in Chinese)

Circulation Anomalies Associated with Interannual Variation of Early-and Late-Summer Precipitation in Northeast China*

SHEN Baizhu[1,2] LIN Zhongda[3] LU Riyu[3] LIAN Yi[1,2]

(1 Institute of Meteorological Science of Jilin Province, Changchun 130062; 2 Laboratory of Research for Mid-High Latitude Circulation System and East Asian Monsoon, Changchun 130062; 3 Institute of Atmospheric Physics, Chinese Academy of Sciences, Beijing 100029)

Abstract: Summer rainfall is vital for crops in Northeast China. In this study, we investigated large-scale circulation anomalies related to monthly summer rainfall in Northeast China using European Center for Medium-Range Weather Forecast ERA-40 reanalysis data and monthly rainfall data from 79 stations in Northeast China. The results show that the interannual variation in rainfall over Northeast China is mainly dominated by a cold vortex in early summer (May—June) and by the East Asian summer monsoon in late summer (July—August). In early summer, corresponding to increased rainfall in Northeast China, an anomalous cyclonic anomaly tilted westward with height appears to the northwest and cold vortices occur frequently. In late summer, the rainfall anomaly is mainly controlled by a northward shift of the local East Asian jet stream in the upper troposphere and the northwest extension of the Western Pacific subtropical high (WPSH) in the lower troposphere. The enhanced southwesterly anomaly in the west of the WPSH transports more moisture into Northeast China and results in more rainfall. In addition, compared with that in July, the rainfall in Northeast China in August is also influenced by a mid-and high-latitude blocking high over Northeast Asia.

Keywords: summer rainfall; Northeast China; interannual variation; cold vortex; western Pacific subtropical high; East Asian jet stream

1 Introduction

Northeast China is a region with high potential for the development of agriculture, especially grain production. Summer is the main season for rainfall and the growth of crops. Therefore, the interannual variation of summer rainfall and its spatial distribution in Northeast China is important for agriculture and crop safety in China.

The seasonal march of the East Asian summer monsoon (EASM) is characterized by three stages in China. In May and June, the rainy belt is located to the south of the Yangtze River, and it shifts northward to the Yangtze—Huaihe Rivers valley in late June and North and Northeast China in August (Guo

* 本文原载于 *Science China (Earth Sciences)*, 2011, 54(7): 1095-1104

et al. ,1981;Ding,1992;Ding et al. ,2004). Moreover, Lian et al. (2003) analyzed the seasonal march of the EASM in Northeast China and revealed that the EASM onsets in late July and retreats southward in mid-August.

The rainfall in Northeast China also has significant interannual variability. Through empirical orthogonal function analysis of summer rainfall in Northeast China, Sun et al. (2000) identified that the first leading mode depicts coherent variation in the whole region of Northeast China, and the associated moisture mainly originates from the tropical South China Sea, the South Asian monsoon region and the subtropical western North Pacific(Sun et al. ,2007). Because of the onset of the EASM in Northeast China in late July, the southerly brings a large amount of moisture and starts the rainy season in Northeast China. Therefore, the interannual variation in the intensity of the EASM may affect the advance and delay of the rainy season in Northeast China and thus determine the amount of precipitation and the extent of flood or drought in Northeast China (Sun et al. ,2003).

In addition to the impact of the EASM, rainfall in the Northeast China is also influenced by mid-and high-latitude circulation. He et al. (2006) found that cold vortices occur more frequently in summer and the rainfall increases in Northeast China when the Northern Hemisphere annular mode is weak in February, and vice versa. Furthermore, Wu et al. (2008) suggested a close relationship between the Northeast China summer rainfall and the dipole anomaly of concurrent sea level pressure north of 70°N. When the Arctic vortex is centered near North America, the Northeast China summer rainfall is greater. Previous studies also indicate that the Northeast China summer rainfall is possibly related to snow cover on the Tibet Plateau in winter and spring (Shen et al. ,2004).

The above studies focused on either the seasonal evolution of summer rainfall in Northeast China, or interannual variability of the summer-mean rainfall in Northeast China. In fact, the rainfall has a crucial subseasonal feature with a rainy season in late summer (July—August). In this study, we ask whether the interannual variation in rainfall in early summer (May—June) is similar to that in late summer and what the difference is in the associated large-scale circulation anomalies. The answers will be useful in understanding the impact of the circulation system at different latitudes on the interannual variation of the Northeast China summer rainfall.

2 Data

Data of monthly rainfall accumulated from 753 stations in mainland China was provided by the Chinese Meteorological Data Center. In this study, we restrict our analysis of rainfall to the region of Northeast China, which has 79 stations (Fig. 1), during the period of 1961—2005. The monthly atmospheric data are taken from the European Center for Medium-Range Weather Forecasts (ECMWF) ERA-40 reanalysis dataset (Uppala et al. ,2005). The variables used in this study include zonal and meridional winds, vertical pressure velocity, and geopotential height. To calculate moisture transport and its divergence, four-times-per-day data of zonal, meridional winds, specific humidity and sea level pressure from ECMWF(Uppala et al. ,2005) are used, which follows the work of Chen et al. (2008). Both types of atmospheric data cover the period of 1961—2002. We also used National Centers for Environmental Prediction/National Center for Atmospheric Research reanalysis data and obtained similar results. Therefore, for brevity, only the results derived from ECWMF reanalysis data are presented in this study.

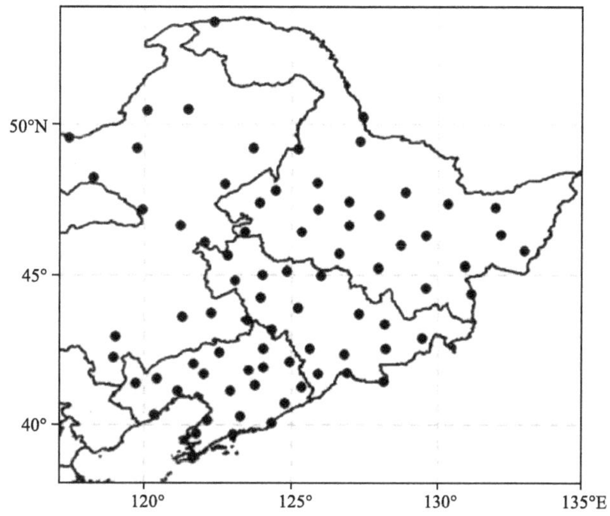

Fig. 1　The 79 stations in Northeast China

3　Climatology of summer rainfall in Northeast China

The spatial and temporal features of summer rainfall in Northeast China are analyzed in this section using rainfall data for 1961—2005 recorded at the 79 stations. The annual mean rainfall averaged over the 79 stations is 552 mm, which is mostly contributed by summer rainfall and then spring and autumn rainfall. The rainfall in winter is least. In summer, the rainfall is about 154 mm in July and 126 mm in August. Generally, rainfall during May—June—July—August (MJJA), which is crucial for the growth of crops in Northeast China and affects the output of grain, is about three quarters of the annual total rainfall. Therefore, in this study, the large-scale atmospheric circulation anomalies related to the interannual variation of summer rainfall in Northeast China in MJJA are investigated.

Fig. 2 shows the spatial distribution of the climatology of monthly rainfall in Northeast China from May to August. The summer rainfall has a decreasing trend from southeast to northwest in Northeast China and the rainfall in the southeast is twice that in the northwest. The irregular distribution is determined by the topography of Northeast China. The southerly from the tropics rises against the northeast—southwest Changbai Mountains, which are located in the east of Northeast China, and leads to rainfall over the windward slopes southeast of Northeast China. In addition, the mid-latitude westerly flow meets the northeast—southwest Da Hinggan Mountains in the northwest of Northeast China and results in rainfall to the west and less rainfall to the east (Figs. 2b and 2c). Generally, the rainfall in late summer is remarkably greater than that in early summer. In early summer, the rainfall ranges 20～80 mm in May and 60～110 mm in June. The rainfall is twice that in July and August, ranging 100～300 mm.

4　Interannual variation of summer rainfall in Northeast China

The monthly rainfall in Northeast China in summer has strong interannual variability. Fig. 3 shows the time series of the monthly rainfall averaged over the 79 stations in Northeast China in MJJA. In May, the maximum rainfall is about 80 mm in 2005, twice the climatological mean, and the mean is about 20 mm in 2003, half the climatological mean. The year-to-year variation of rainfall is also significant in

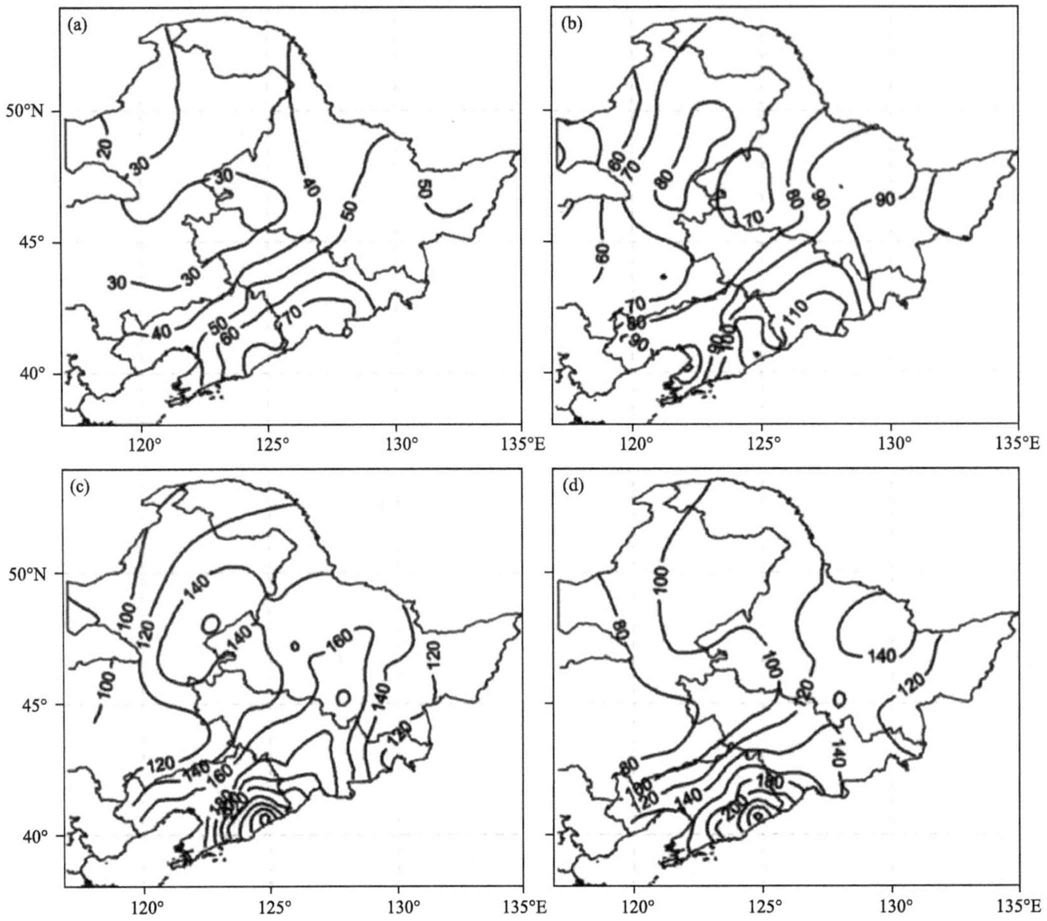

Fig. 2　Monthly rainfall in Northeast China. Unit: mm

(a) May, (b) June, (c) July, (d) August

June, July and August. The standard deviations of the averaged rainfall in Northeast China are 14, 21, 34 and 40 mm in May, June, July and August, respectively. Note that during the 45 years from 1961 to 2005, the monthly rainfalls in Northeast China in MJJA do not have a significant trend.

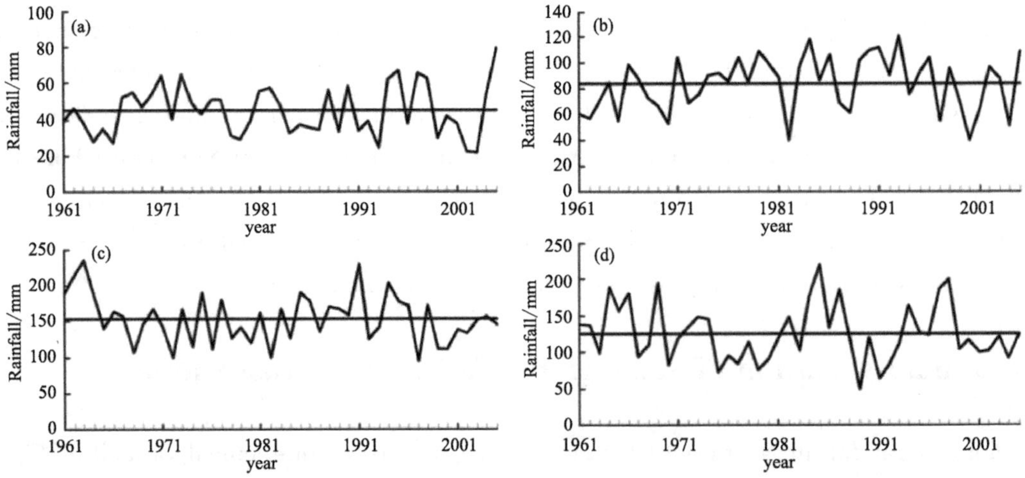

Fig. 3　Time series of monthly rainfall averaged over the 79 stations in Northeast China from 1961 to 2005

(a) May, (b) June, (c) July, (d) August

We now ask whether the rainfall averaged over the 79 stations represents the coherent variation of rainfall over the entire Northeast China. Fig. 4 shows the spatial distribution of the simultaneous correlation between monthly rainfall averaged over the 79 stations and the individual rainfalls of the 79 stations in Northeast China. The correlation coefficients are positive over the whole of Northeast China. That is, the rainfall averaged over the 79 stations well depicts the coherent interannual variation in rainfall in Northeast China.

On the other hand, there is a difference in the spatial distribution of correlation coefficients between early summer (Figs. 4a and 4b) and late summer (Figs. 4c and 4d). In early summer, the positive correlation is significant over the whole Northeast China, with the maximum correlation exceeding 0.6 in central and western regions. In late summer, there is significant positive correlation in the southeast, with the maximum exceeding 0.7. Moreover, the location of the maximum shifts northward from Liaoning Province in July to Jilin and Heilongjiang Provinces in August. The northward shift is probably related to the northward march of the EASM from July to August in Northeast China. The difference in the spatial distribution of correlation coefficients between early summer and late summer indicates possibly different circulation anomalies responsible for the interannual variation of rainfall in Northeast China in early summer and late summer. In the following section, the lower-and upper-tropospheric circulation anomalies associated with the interannual variation of monthly rainfall in Northeast China in MJJA are investigated.

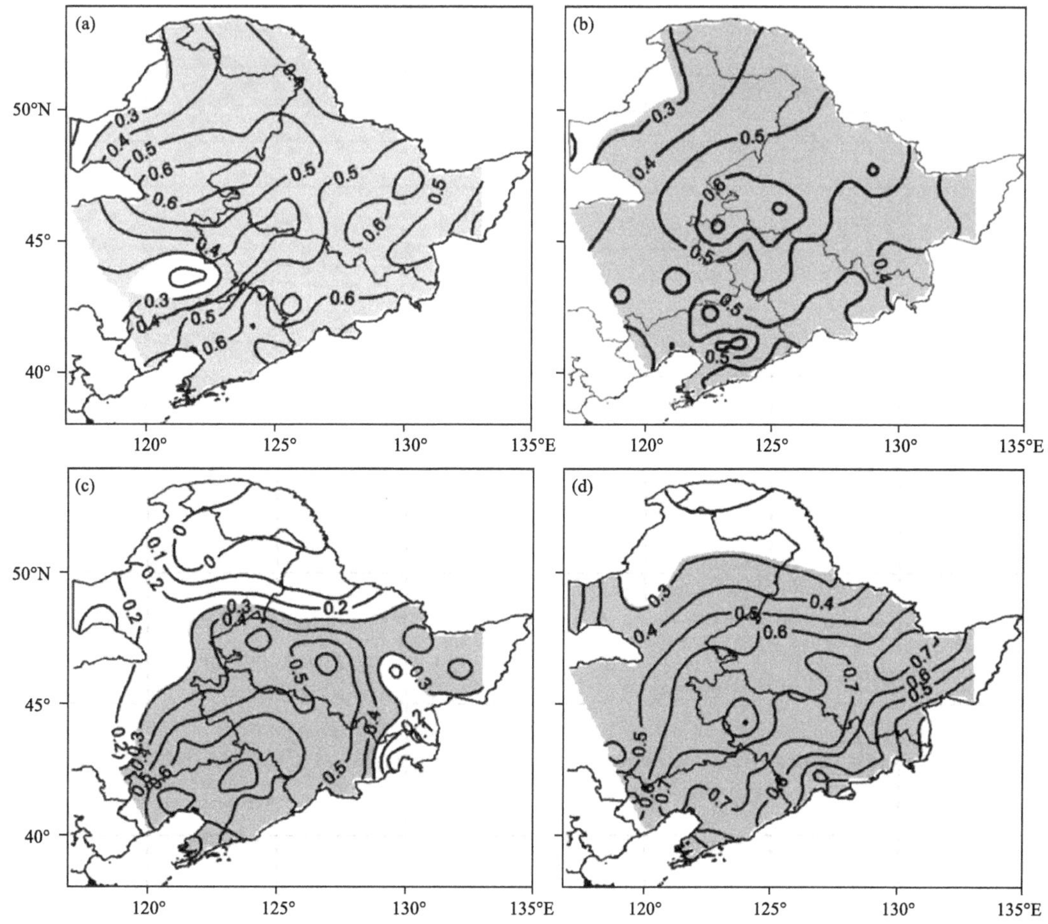

Fig. 4　Correlation coefficients of monthly rainfall with the time series of simultaneous rainfall averaged over the 79 stations in Northeast China. Shading illustrates significance at the 95% confidence level

(a) May, (b) June, (c) July, (d) August

5 Circulation anomalies associated with monthly rainfall in Northeast China in MJJA

5.1 Lower-and mid-tropospheric circulation anomalies

In early summer, the enhanced rainfall in Northeast China is correlated with an in situ anomalous low at the level of 500 hPa (Figs. 5a and 5b). The southwesterly anomaly to the east of the anomalous low (Figs. 6a and 6b) converges the moisture transported from the immediate Yellow Sea, Bohai Sea and Japan Sea over Northeast China (Figs. 7a and 7b) and results in enhanced rainfall. Note that the ridge of the western Pacific subtropical high (WPSH) is still located between 15°N and 25°N (Figs. 5a and 5b) and does not directly link to the cyclonic anomaly in Northeast China.

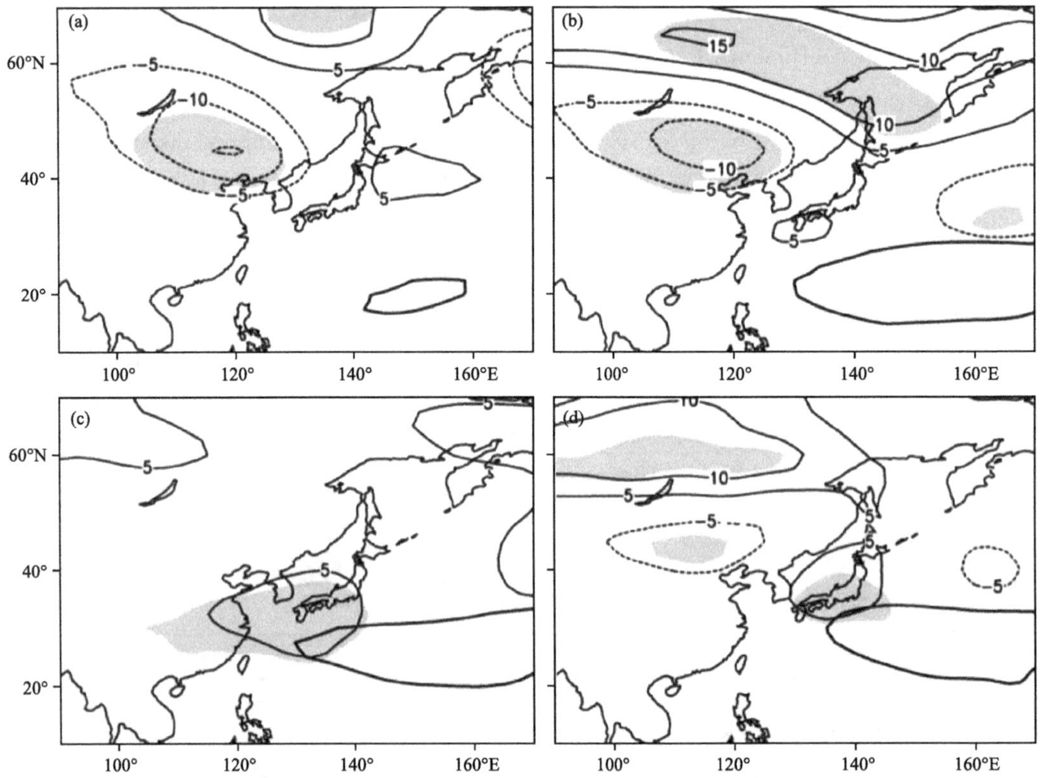

Fig. 5　Monthly geopotential height anomalies at 500 hPa (unit: gpm) regressed upon the simultaneous standardized time series of rainfall averaged over the 79 stations. Shading illustrates significance at the 95% confidence level, and the thick solid line denotes the contour of 5880 gpm at 500 hPa

(a) May, (b) June, (c) July, (d) August

In late summer, the interannual variation in rainfall in Northeast China is related to an anomalous high over Japan (Figs. 5c and 5d). The anomalous high resides in the northwest of the WPSH and leads to the westward and northward advance of the WPSH. The advance of the WPSH then favors northern transport of moisture in a southwesterly direction to the west of the WPSH into Northeast China (Figs. 7c and 7d). Compared with the north edge of moisture transport and its center of convergence in the southeast of Northeast China in July (Fig. 7c), the southerly anomaly marches more northward into Jilin and Heilongjiang Provinces (Fig. 6d) and induces the northward advance of moisture transport and its convergence center (Fig. 7d). This northward advance in August is consistent with the northward shift

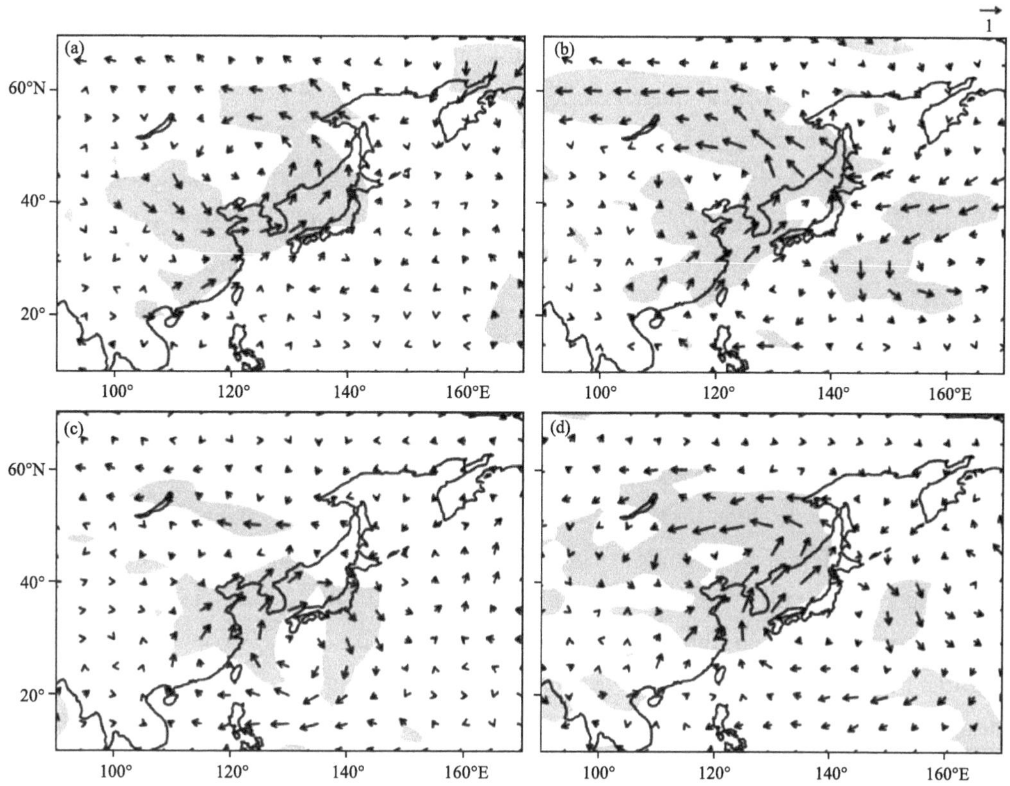

Fig. 6　Same as Fig. 5 but for wind at 850 hPa. Unit: m s^{-1}

(a) May, (b) June, (c) July, (d) August

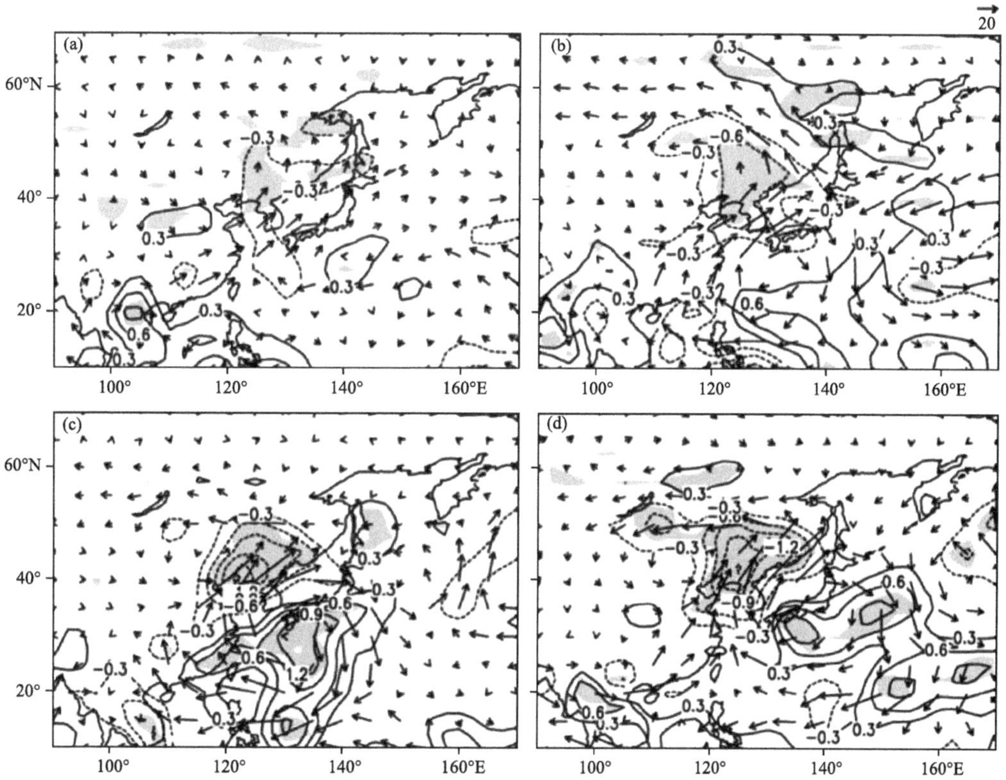

Fig. 7　Same as Fig. 5 but for vertically integrated moisture transport from the surface to 100 hPa (unit: kg/(m $s)^{-1}$) and its divergence (unit: mm d^{-1}). Shading illustrates significance of divergence at the 95% confidence level

(a) May, (b) June, (c) July, (d) August

of the location of the maximal correlation coefficient (Fig. 4d) of rainfall with the time series of the rainfall averaged over the 79 stations in Northeast China. On the other hand, there is also a significant cyclonic anomaly in Northeast China and an anticyclonic anomaly to the north of Lake Baikal in August (Fig. 5d). The cyclonic anomaly is related to the eastward propagating Rossby wave along the Asian westerly jet stream in the upper troposphere, which will be discussed in the next section. The anticyclonic anomaly possibly results from the mid-and high-latitude blocking high in Northeast Asia.

We also make composites of circulation considering a possible nonlinear effect. The years with rainfall anomalies one standard deviation larger (less) than that in climatological monthly rainfall in Northeast China, are selected for May, June, July and August, respectively. During 1961—2002, the numbers of years with enhanced rainfall are seven, seven, seven, and eight and those for suppressed rainfall are seven, eight, eight, and seven in May, June, July and August, respectively.

Fig. 8 shows the composite geopotential height in years with enhanced rainfall (left) and suppressed rainfall (right) in Northeast China in May, June, July and August, respectively. In early summer, the geopotential anomalies in Northeast China are basically linear, with a significant cyclonic anomaly in the years of enhanced rainfall and a significant anticyclonic anomaly in the years with suppressed rainfall. In late summer, an anomalous high appears in the northwest of the WPSH when the rainfall increases in Northeast China and an anomalous low appears in the northwest of the WPSH when the rainfall decreases. In addition, an anomalous low in the years with increased rainfall contrasts an anomalous high in the years with decreased rainfall. In August (Figs. 8d and 8h), besides the anomalies in the northwest of the WPSH and in Northeast China there is an anomalous high (low) in the years with enhanced (suppressed) rainfall in Northeast Asia, which is consistent with that indicated in Fig. 5d. That is, the interannual variation of rainfall in Northeast China in August is the result of the EASM and the high-latitude circulation. This is a possible reason why the anomaly in August is not significant, especially in the western North Pacific and East Asia. We composite the difference in geopotential height anomalies in the years with enhanced rainfall from those in the years with suppressed rainfall in Northeast China (figures are not shown). The composite difference is similar to that in Fig. 5. We also make composites of the 850 hPa wind and vertical integral moisture transport and its divergence and find that the anomalies are mostly linear (figures are not shown).

5.2 Upper-tropospheric circulation anomalies

Corresponding to enhanced rainfall in Northeast China in early summer, a cyclonic anomaly appears to the northwest near Lake Baikal in the upper troposphere (Figs. 9a and 9b). The cyclonic anomaly exhibits a westward tilt with height; the center of the cyclonic anomaly is located over Lake Baikal at 200 hPa (Figs. 9a and 9b) and east of it at 850 hPa (Figs. 6a and 6b). The westward-tilting cyclonic anomaly suggests atmospheric baroclinic instability (Holton, 1992) and favors anomalous upward flow and formation of the cold vortex to the east over Northeast China. Statistical analysis conducted by Hu et al. (2010) on the cold vortex in Northeast China showed that the vortex mainly originates near Lake Baikal, whereas the meridional cyclonic shear to the north of the East Asian upper-tropospheric jet stream (EAJS) favors the formation of a cold vortex close to the location of the cyclonic anomaly in Figs. 9a and 9b. Moreover, we calculate the correlation between the numbers of days for a cold vortex in early summer from the work of Hu et al. (2010) and simultaneous rainfall in Northeast China. The correlation coefficient is 0.43 during 1961—2002, which is significant at the 99% confidence level (Fig. 10a). The interannual variation in rainfall in Northeast China in early summer is related to the cold vortex in Northeast

China.

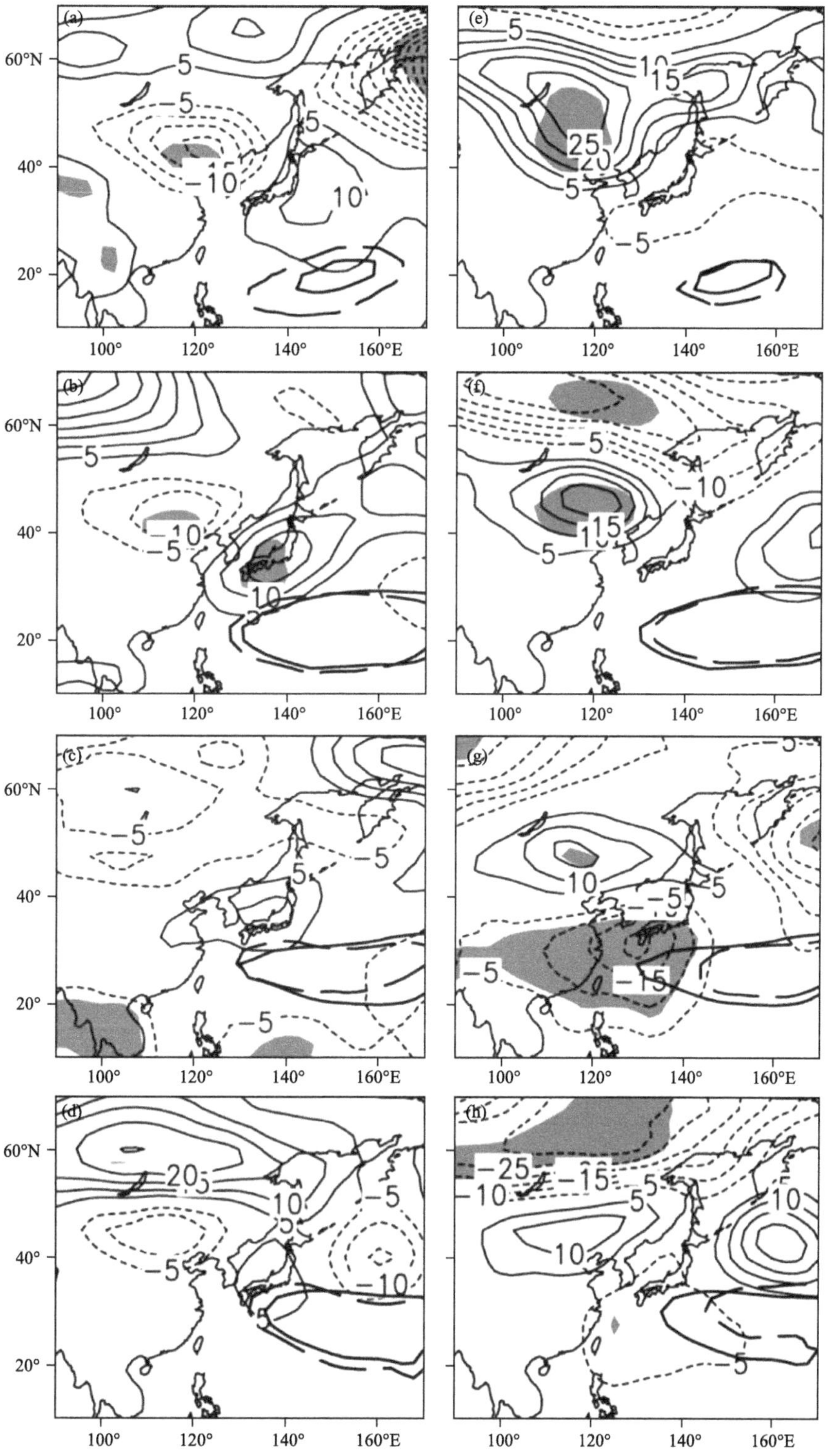

Fig. 8 Composite 500 hPa geopotential height and its anomalies in years with enhanced rainfall in Northeast China (left) and suppressed rainfall (right). Shading illustrates significance at the 90% confidence level, the thick solid and dashed lines denotes the monthly geopotential height contour of 5880 gpm at 500 hPa for the climatology and the composite result, respectively (unit: gpm)

(a) (e) May, (b) (f) June, (c) (g) July, (d) (h) August

In late summer, the enhanced rainfall in Northeast China is associated with an anticyclonic anomaly in the upper troposphere over East Asia. The anticyclonic anomaly consists of an easterly anomaly to the south of the axis of the EAJS and a westerly anomaly to the north, leading to a northward shift of the EAJS (Lin et al., 2005). The cyclonic anomaly near Lake Baikal in early summer (Figs. 9a and 9b)

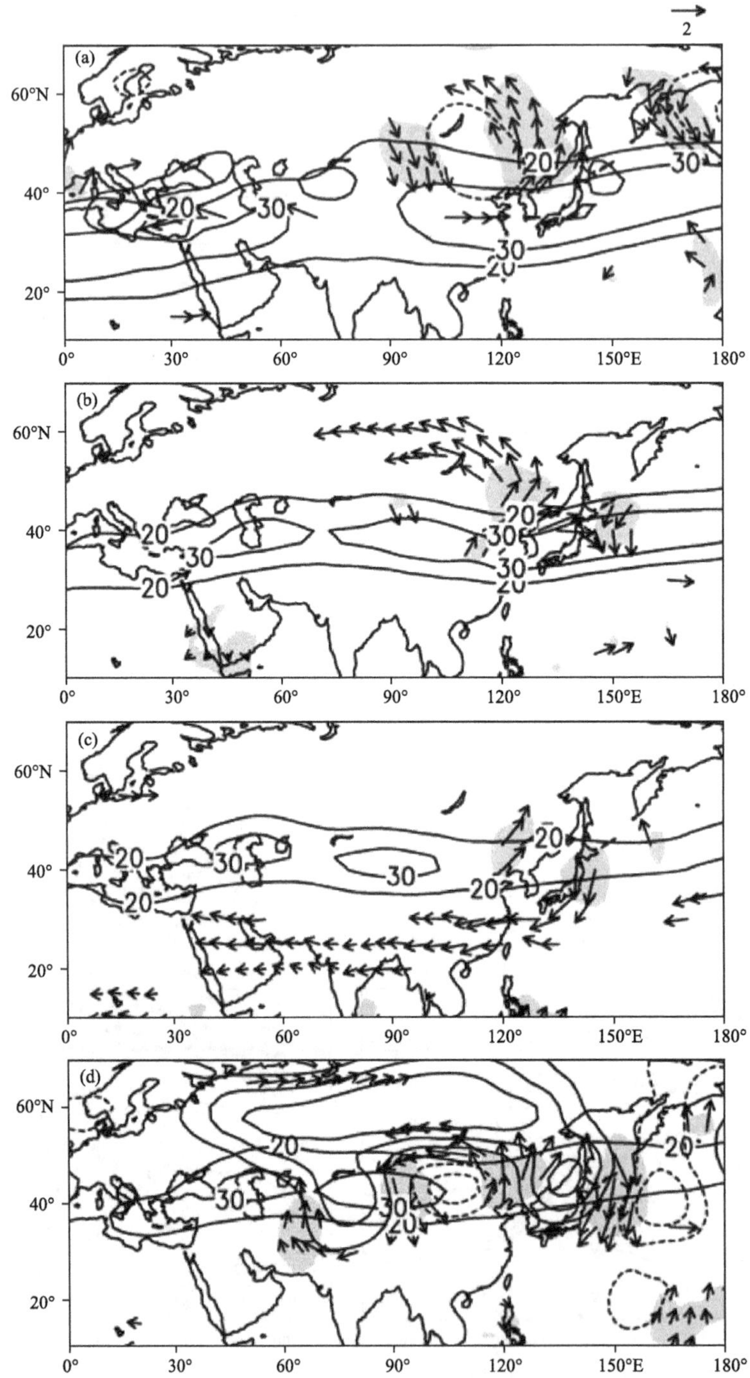

Fig. 9　Same as that in Fig. 5 but for monthly horizontal wind anomalies (vector, unit: m s^{-1}) and geopotential height anomalies in August (contour, the interval is 5 gpm and the zero contour is omitted) at 200 hPa. Shading illustrates significance for meridional wind at the 95% confidence level. The solid lines denote the location of the climatological westerly jet stream with 200 hPa zonal wind exceeding 20 m s^{-1}

(a) May, (b) June, (c) July, (d) August

disappears in July; i. e. , the effect of the cold vortex on rainfall in Northeast China is negligible. In August, although there is a cyclonic anomaly to the west of Northeast China, it resides on the EAJS at about 40°N, more southward than the cyclonic anomaly in early summer and the most frequent location of cold vortices at 50°N. The coefficient of the correlation between the numbers of days of the cold vortex in late summer from Hu et al. (2010) and simultaneous rainfall in Northeast China is only 0. 14 (Fig. 10b). That is, in late summer, the impact of the cold vortex on the interannual variation of rainfall in Northeast China is not significant.

The anticyclonic anomaly over East Asia seems to be associated with circulation anomalies over the Eurasian continent in late summer (Figs. 9c and 9d). In July, the anomaly links to the easterly anomaly extending from South China westward through India to the Arabian Peninsula south of the axis of the Asian westerly jet stream. However, the mechanism is not clear. In August, the anti-cyclonic anomalies over East Asia and Central Asia together with a cyclonic anomaly to the south of Lake Baikal form a wave train along the Asian westerly jet stream.

Previous studies (Lu et al. , 2002; Enomoto et al. , 2003; Enomoto, 2004; Ding et al. , 2005; Liao et al. , 2006; Yang et al. , 2008) have identified an eastward-propagating Rossby wave along the Asian westerly jet stream in the upper troposphere. Lu et al. (2002) revealed a significant teleconnection pattern from North Africa to East Asia and proposed the remote effect of heating anomalies over India and the Atlantic on the EASM through this teleconnection pattern at about 40°N. This teleconnection is named the "Silk Road" pattern (Enomoto et al. , 2003). Bueh et al. (2008) analyzed the evolution of the East Asia-Pacific pattern during the period of Meiyu. They suggested that the locking and interaction between mid-and high-latitude eastward-propagating Rossby wave over the Eurasian continent and northward-propagating Rossby wave along the East Asia coast leads to intraseasonal variation of Meiyu. In August, the Rossby wave propagates from 60°E eastward to the east of Japan along the Asian westerly jet stream and results in cyclonic anomaly to the south of Lake Baikal, as shown in Fig. 9d.

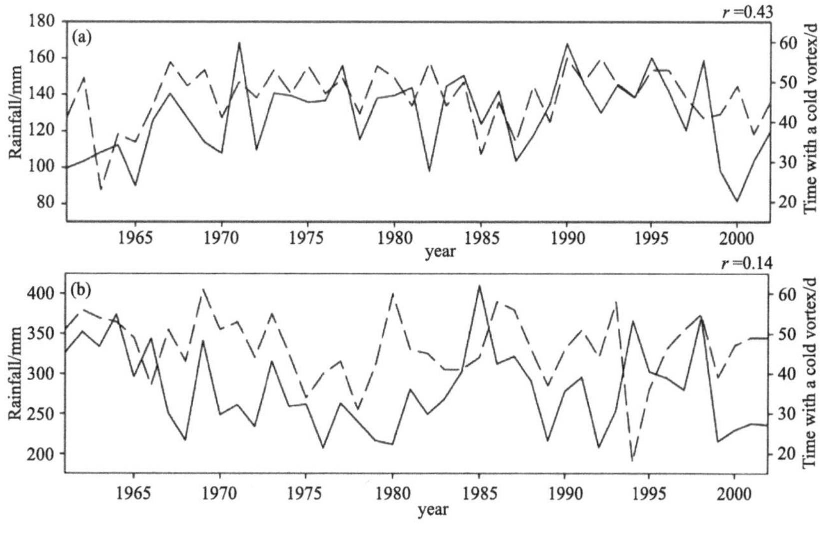

Fig. 10 Time series of rainfall (solid line) and the number of days with a cold vortex (dashed line) in Northeast China in early summer (a) and late summer (b) from 1961 to 2002. The coefficients of correlation between them are 0. 43 in early summer and 0. 14 in late summer

In summary, in early summer, enhanced rainfall in Northeast China mainly results from in situ frequent cold vortices. In late summer, enhanced rainfall in Northeast China is dominated by the northward shift of the EAJS in the upper troposphere and the westward and northward advance of the WPSH in the lower troposphere, which is possibly affected by low-and mid-latitude circulation anomalies over the Asian continent upstream of East Asia.

6 Summary and discussion

This study analyzes the interannual variation of monthly rainfall in Northeast China in summer using monthly rainfall data from 79 stations across Northeast China and the ECMWF ERA-40 reanalysis dataset. The interannual variation of rainfall in Northeast China is remarkably different between early summer (May—June) and late summer (July—August). In early summer, over the whole region of Northeast China the rainfall significantly correlates with the time series of rainfall averaged over the 79 stations over the central and western regions of Northeast China, while there is significant correlation in the southeast of Northeast China in late summer. Moreover, there is a significant difference in circulation anomalies associated with rainfall in Northeast China between early summer and late summer.

In early summer, the interannual variation of rainfall in Northeast China is related to an in situ cold vortex. When the vortices occur frequently in Northeast China, a cyclonic anomaly appears to the west with a westward tilt with height. The westward-tilting cyclonic anomaly favors anomalous upward flow to the east over Northeast China, and thus leads to enhanced rainfall.

In late summer, the enhanced rainfall in Northeast China is dominated by a northward shift of the EAJS in the upper troposphere and the westward and northward advance of the WPSH in the lower troposphere. Corresponding to the westward and northward advance of the WPSH, the enhanced southwesterly transports more moisture from the tropics into Northeast China and results in more rainfall. In addition, compared with that in July, the impact of the WPSH is a little weaker and the impact of the mid-and high-latitude blocking high in Northeast Asia is significantly enhanced in August.

It is interesting that the rainfall in Northeast China in late summer is also associated with an easternly anomaly from southern China westward through to India and the Arabian Peninsula in July, and an eastward-propagating Rossby wave along the Asian westerly jet stream originating at 60°E in August. The results indicate the possible role of upstream low-and mid-latitude circulation anomalies over the Asian continent in rainfall over Northeast China in late summer. The associated mechanism is not clear and requires further study.

Acknowledgements. We thank reviewers for their valuable comments and suggestions. We also thank Dr Hu K. X. for providing data on cold vortices in Northeast China. This work was supported by National Technology Support Project (Grant Nos. 2009BAC51B04, 2007BAC29B01), Key Knowledge Innovation Programs of the Chinese Academy of Sciences (Grant No. KZCX2-YW-220), National Natural Science Foundation of China (Grant Nos. 40575047 and 40705036) and the New Technology Projects of China Meteorological Administration (Grant No. CMATG2009MS01).

References

BUEH C L, SHI N, JI L R, et al, 2008. Features of the EAP event on the medium-range evolution process and the mid-and high-latitude Rossby wave activities[J]. Chinese Sci Bull, 53(4): 610-623.

CHEN J L, HUANG R H, 2008. Interannual and interdecadal variations of moisture transport by Asian summer monsoon and their association with droughts and floods in China [J]. Chinese J Geophys, 51(2): 352-359. (in Chinese)

DING Q H, WANG B, 2005. Circumglobal teleconnection in the northern hemisphere summer[J]. J Clim, 18(17): 3483-3505.

DING Y H,1992. Summer monsoon rainfalls in China[J]. J Meteor Soc Japan,70(1B):373-396.

DING Y H,LI C Y,HE J H,et al,2004. South China Sea monsoon experiment (SCSMEX) and the East Asian monsoon[J]. Acta Meteor Sinica,62(5):561-586. (in Chinese)

ENOMOTO T,2004. Interannual variability of the Bonin high associated with the propagation of Rossby waves along the Asian Jet[J]. J Meteor Soc Japan,82(4):1019-1034.

ENOMOTO T,HOSKINS B J,MATSUDA Y,2003. The formation mechanism of the Bonin high in August[J]. Quart J Roy Meteor Soc,129(587):157-178.

GUO Q Y,WANG J Q,1981. Interannual variations of rain spell during predominant summer monsoon over China for recent thirty years[J]. Acta Geographica Sinica,36(2):187-195. (in Chinese)

HE J H,WU Z W,QI L,et al,2006. Relationship among the Northern Hemisphere annual mode,the northeast cold vortex and the summer rainfall in Northeast China[J]. J Meteor Environ,22(1):1-5. (in Chinese)

HOLTON J R,1992. An Introduction to Dynamic Meteorology 3rd ed[M]. California:Academic Press.

HU K X,LU R Y,WANG D X,2010. Seasonal climatology of cutoff lows and associated precipitation patterns over Northeast China[J]. Meteor Atmos Phys,106(1-2):37-48.

LIAN Y,SHEN B Z,GAO Z T,et al,2003. The study of the onset criterion and the date of East Asian summer monsoon in Northeast China and its main characteristic analysis[J]. Acta Meteor Sinica,61(5):548-558. (in Chinese)

LIAO Q H,TAO S Y,WANG H J,2006. Internal dynamics related to anomalies of seasonal evolution of summer circulations in East Asia during July—August[J]. Chinese J Geophys,49(1):28-36. (in Chinese)

LIN Z D,LU R Y,2005. Interannual meridional displacement of the East Asian upper-tropospheric jet stream in summer[J]. Adv Atmos Sci,22(2):199-211.

LU R Y,OH J H,KIM B J,2002. A teleconnection pattern in upper-level meridional wind over the North Africa and Eurasian continent in summer[J]. Tellus,54(1):44-55.

SHEN B Z,LIAN Y,2004. The preliminary analysis on the relationship between snow cover over the Tibetan Plateau and summer cold vortex precipitation in northeast China[J]. J Jilin Univ (Earth Sci),34 (Spec Ed):112-118. (in Chinese)

SUN L,AN G,TANG X L,2003. Relationship between the northeast Asian summer south wind anomaly and the precipitation in Northeast China[J]. Chinese J Atmos Sci,27(3):425-434. (in Chinese)

SUN L,AN G,DING L,et al,2000. A climatic analysis of summer precipitation features and anomaly in Northeast China[J]. Acta Meteor Sinica,58(1):16-19. (in Chinese)

SUN L,SHEN B Z,GAO Z T,et al,2007. The impacts of moisture transport of East Asian monsoon on summer precipitation in Northeast China[J]. Adv Atmos Sci,24(4):606-618.

UPPALA S M,ALLBERG K,SIMMONS P W,et al,2005. The ERA-40 reanalysis[J]. Quart J Roy Meteor Soc,131(612):2961-3012.

WU B Y,ZHANG R H,ROSANNE D A,2008. Arctic dipole anomaly and summer rainfall in Northeast China[J]. Chinese Sci Bull,53(14):2222-2229.

YANG L M,ZHANG Q Y,2008. Climate features of summer Asia subtropical westerly jet stream[J]. Clim Environ Res,13(1):10-20. (in Chinese)

The 2009 Summer Low Temperature in Northeast China and Its Association with Prophase Changes of the Air-Sea System*

SHEN Baizhu[1,2]　LIU Shi[1,2]　LIAN Yi[1,2]　FENG Guolin[3]
LI Shangfeng[1,2]　GONG Zhiqiang[3]

(1 Laboratory of Research for Middle-High Latitude Circulation Systems and East Asian Monsoon, Changchun 130062; 2 Institute of Meteorological Science of Jilin Province, Changchun 130062; 3 National Climate Center, Beijing 100081)

Abstract: Under the background of global warming, summer (JJA) low temperature events in Northeast China had not occurred for about 15 years since 1994, but one such event took place in 2009. By using the NCEP/NCAR reanalysis data, the 100 years station temperature data at Harbin and Changchun, and the Hadley Center sea surface temperature (SST) data, this paper intends to reveal the cause, circulation background, and influencing factors of this event. Analysis of both horizontal and vertical circulations of a low-value system over Northeast China in summer 2009 during the low temperature event shows that anomalous activities of the Northeast China cold vortex (NECV) played the most direct role. A decadal cooling trend of −0.8 ℃/10 years over 1999—2008 at Changchun and Harbin was found, which is obviously out-of-phase with the linear warming trend (0.2 ℃/10 years) over 1961—2000 for Northeast China in response to the global warming. The previous winter North Pacific polar vortex (NPPV) area index, significantly positively related to the observed summer temperatures of Harbin and Changchun, was also in a significantly declining tendency. These provide favorable decadal backgrounds for the 2009 low temperature event. Different from the average anomaly field of 500 hPa height for summer 1994—2008 in Northeast China, in the summer of 2009, the Arctic Oscillation (AO) showed a strong negative phase distribution, and significant negative height anomalies dominated Northeast Asia, Aleutian Islands, and North Atlantic. Furthermore, the negative phase of North Pacific Oscillation (NPO) in the winter of 2008 was obviously strong, and it maintained in the spring of 2009. Meanwhile, the SSTA in the equatorial eastern-central Pacific Ocean in the winter of 2008 showed a La Niña phase, but the strength of the La Niña weakened obviously in the spring of 2009. The abnormally strong activities of NECV in June and July of 2009 were related to the disturbances of stationary waves that replaced the original ultra-long waves over the North Pacific region in April and May 2009. The singular value decomposition (SVD) and harmonic analysis results suggest that the anomalous phase of NPO is an important precursor for summer temperature variations over Northeast China, and also a stable planetary-scale component that can be extracted from the atmospheric circulation in addition to the chaotic components on the synoptic scale.

* 本文原载于 *Acta Meteorologica Sinica*, 2012, 26(4): 438-453

Keywords: global warming; summer low temperature in Northeast China; decadal variability; Northeast China cold vortex; polar vortex; North Pacific Oscillation

1 Introduction

Low temperature cold injury events frequently occur in summer during the crop growth period (May—September) in Northeast China from the 1950s to 1970s. Different from low temperature/freezing damage, low temperature/cold injury usually refers to crop injury resulted from the low temperature above 0 ℃ (sometimes even about 20 ℃) in its growth phase (Zhang, 1990; Wang, 1995). Low temperature/cold injury events in the summers of 1969, 1971, 1972, 1976, and 1979 resulted in about 20% reduction of the annual total grain output of Jilin Province, especially the rice as high as 30%~50% (The General Office of the People's Government Of Jilin Province, 2002). This has been attracted much attention by the Chinese researchers, and some investigations have been performed since the end of the 1970s. It was found that the Northeast China summer low temperature (NESLT) events are correlated with the negative temperature anomaly in the mid-high latitudes of Northern Hemisphere (excluding the area north of 60°N), and they exhibit planetary-scale characteristics (Jilin Province Meteorological Observatory, 1980; Zhang et al., 1982, 1983). Temperature changes at Harbin in 1881—1983 showed distinctive phases (Wang, 1995). During the cold phase of NESLT, the occurrence of El Niño adds a negative effect on the temperature, i. e., making it much lower than normal (Zhang, 1990). Occurrences of cool or hot summer in Northeast China were closely related to major large-scale circulation patterns in the Northern Hemisphere (Northeast China Summer Low Temperature Research Collaboration Team, 1979; Jilin Province Meteorological Observatory, 1980). Typical case analysis by Fu (1980) concluded that when the snow cover in the Northern Hemisphere and the sea ice in the Atlantic Ocean increase in previous winter and spring, the summer temperature in Northeast China would be lower than normal. Zhang et al. (1985) and Liu (1986) investigated the association between NESLT and monthly variations of polar vortex. They defined the area and intensity indices of polar vortex, which reflect the interaction between polar vortex and mid-latitude circulation, and found that the polar vortex area index in the domain from Asia to the central Pacific was negatively correlated with temperature in Northeast China and most areas of China and the polar vortex intensity index was also negatively correlated with temperature in Northeast China.

Chinese researchers have conducted many studies to reveal the prophase influential signals of the air-sea system related to summer climate in Northeast China. Much attention has been paid to atmospheric semi-permanent activity centers that bear a barotropic structure with seasonal changes (Chou, 1986, 2007). Liu et al. (2002, 2003) and Lian et al. (2007) showed that there was a remarkable teleconnection between the oscillations of both the Aleutian low and the North Pacific Oscillation (NPO) in the subtropical region south of the Aleutian islands and the summer Northeast China cold vortex (NECV) and precipitation over eastern China. The intensity of summer NECV was positively correlated with the Northern Hemisphere Annular Mode (NAM) in February (He et al., 2006). Some decadal characteristics of NESLT events have been found (International CLIVAR Project Office, 1997): from the 1950s to 1970s, cool summers in Northeast China frequently occurred corresponding to the El Niño events in this period; however, since the 1980s, the El Niño events were often accompanied by warm or less cool summers (Wang et al., 1997; Lian et al., 1998; An et al., 1998; Zheng et al., 1999; Sun et al., 2006). This indicates that interannual changes of NESLT are constrained by the longer term (10 years or more) climate changes (Ding et al. ,2003;Ding,2007). How the decadal variability of climate affects the interannual variability of NESLT needs further investigations. Some consistency between the interdecadal and interannual variations of rainy-season precipitation in North China has been reported (Lu,2003).

Global climate has entered the warmest period since the 1990s (IPCC,2007),and Northeast China is one of the regions that exhibit a remarkable regional warming in response to the global warming (National Assessment Report of Climate Change Editorial Committee,2007). The frequency of cool summer has significantly reduced since the 1980s;in particular,no large-scale summer low temperature event happened in Northeast China from 1994 to 2008 (Chen et al. ,2004;Yang et al. ,2005). However,a significant summer low temperature event occurred in Jilin and Heilongjiang Provinces in 2009. This raises a question:Is it an accidental event in the current warm phase or an early signal for the bulk NESLT events similar to those during the 1950s—1970s? At present,the conclusion of continuous global warming based on climate model simulations under the various IPCC SERS scenarios was obtained without considering the impact of natural forcing (Zhao et al. ,2005,2008). As far as adaptation to the global warming is concerned,carrying out appropriate research on natural variability of temperature is helpful (Ye et al. ,2008). Northeast China has concentrated thermophilic rice farming in higher latitudes such as Heilongjiang Province. If the climate in Northeast China enters a relatively cool phase,this would pose a major threat to the grain production there,as Northeast China is a major grain production base with annual total grain yield accounting for 1/6 of China's total grain output. It is necessary to perform a case study on the summer low temperature in 2009,and to promote the studies on the decadal variability of NESLTs.

This paper diagnoses the circulation characteristics during the summer low temperature event in 2009,aiming to reveal the relationship between early-stage changes of the air-sea system and the summer low temperature,and the decadal background favorable for the NESLTs. The present study also tries to identify planetary-scale stable atmospheric components,in addition to the synoptic-scale chaotic components (Chou,1986,1997;Feng et al. ,2006;Ren et al. ,2010;Lin et al. ,2010;Zhi et al. ,2010). Based on multiple spatial/temporal-scale analysis,causes of the 2009 summer low temperature in Northeast China are to be revealed.

2 Data and methods

The data used in this paper include daily temperature observations at Changchun Station from 1909 to 2008 and at Harbin Station from 1881 to 2008,the reanalysis data of NCEP/NCAR from 1948 to 2009 and that of ECMWF from 1958 to 2002,and the SST data of Hadley Center from 1881 to 2005. The temporal filtering of long period ($\geqslant$ 9 years;decadal-scale component) and short period ($<$ 9 years;interannual-scale component) was performed on the time series of the above data,and the correlations between the interannual and interdecadal frequency spectra were then computed.

Based on the standard deviation (σ) of observed temperature series for about 100 years at Changchun and Harbin stations,an NESLT event is defined as follows:At any station in Northeast China,if $\overline{T}_{6-8} \leqslant -1.0\sigma$(subscript 6—8 denotes June—August,same below),a severe cool summer (A event) occurs;if $-1.0\sigma < \overline{T}_{6-8} \leqslant -0.4\sigma$,a cool summer (B event) occurs;if $-0.4\sigma < \overline{T}_{6-8} \leqslant 0.3\sigma$,a normal summer occurs;if $0.3\sigma < \overline{T}_{6-8} \leqslant 0.8\sigma$,a warm summer (C event) occurs;while $\overline{T}_{6-8} > 0.8\sigma$ denotes a severe warm summer (D event).

The temperature analysis by Lian et al. (1998) showed that Changchun and Harbin stations can be taken as stations representative of the regional climate of Northeast China.

Table 1 shows that the frequencies of A events (levels 1—3) and B events (level 4) in summer were 14% and 23% at Changchun and 15% and 20% at Harbin, respectively. Therefore, A events occur less frequently than B events.

Based on Table 1, four (three) years of 1969, 1971, 1972, and 1976 (1954, 1957, and 1993) were chosen as the A (B) event years for Northeast China (represented by Changchun and Harbin) and seven (four) years of 1952, 1955, 1982, 1988, 1994, 1997, and 2000 (1982, 1988, 1994, and 1997) as the typical summer high temperature years. Then, composite wind and temperature anomaly fields were plotted. The vertical velocity was calculated by following Ding (1989).

Table 1　Classification of summer average temperature based on the standard deviation (σ) of observed temperation at Changchun and Harbin in the period 1909—2008 and corresponding occurrence percentages

Level	1	2	3	4	5	6	7	8	9
Threshold/σ	$\leqslant -2.0$	$(-2.0,-1.4]$	$(-1.4,-1.0]$	$(-1.0,-4.0]$	$(-0.4,0.3]$	$(0.3,0.8]$	$(0.8,1.4]$	$(1.4,2.2]$	>2.2
Changchun(Freq, %)	2	2	10	23	28	18	8	6	3
Harbin(Freq, %)	2	5	8	20	29	18	9	6	3

In this paper, we adopted the definitions of NECV and NPO given in Liu et al. (2003), and we followed Liu (1986) for computation of relevant polar vortex indices and selection of the domain of the Pacific polar vortex.

3　The 2009 NESLT case

3.1　500 hPa height anomaly in summer 2009

Fig. 1a shows the summer (JJA) average 500 hPa height anomaly field for 2009. It can be seen that anomalous disturbances in the "two-trough-one-ridge" pattern appeared over Asia, i. e. , there was a stable negative disturbance south of 55°N in Northeast Asia with its center in coastal Russia, and a positive anomaly region (a persistent active area of blocking highs) from the Lake Baikal to the Okhotsk Sea. The two anomaly regions with opposite signs formed a stable dipole pattern over Northeast Asia (including Northeast China), and sustained for two months. This was also to certain extent related to the stable activities of the low vortices in the semi-permanent active center over the Aleutian Islands.

Fig. 1b displays summer average 500 hPa height anomaly field during 1994—2008. The Northern Hemispheric general circulation anomalies in Fig. 1a had signs almost opposite to those in Fig. 1b. In summer 2009, there was a strong positive anomaly over the polar region (indicating that AO was in a significant negative phase), and there were distinctive negative anomalies over Northeast Asia, Aleutian Islands, and North Atlantic.

3.2　Average temperature anomalies over China in summer 2009

It can be seen from the summer average temperature anomalies over China in 2009 (Fig. omitted) that a wide range of low temperatures happened in Jilin Province and the central eastern part of Heilongjiang Province, where negative temperature anomalies ranging from -0.5 to -1.0 ℃. Because the main body of the cold air mass resided in the eastern part of Jilin Province and the standard deviation of summer average temperature at Changchun was -0.49σ ($<-0.4\sigma$), the 2009 NESLT case reached the standard of a B event (cool summer).

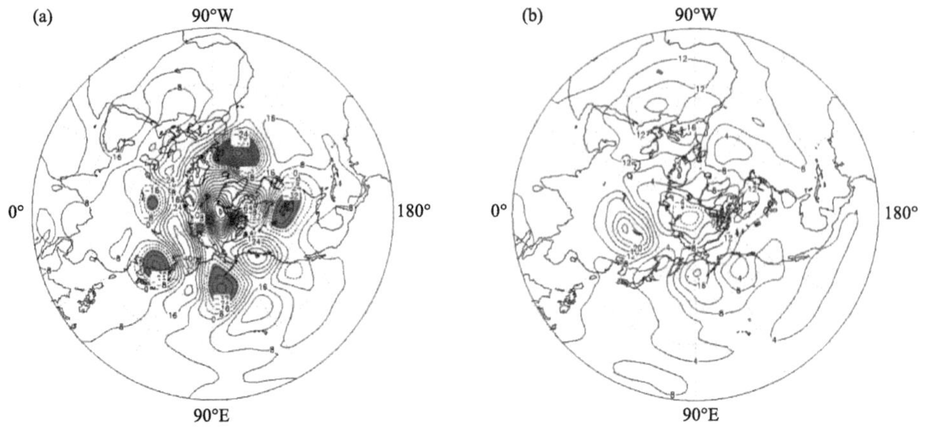

Fig. 1　Summer average 500 hPa geopotential height anomaly (gpm) patterns for (a) 2009 and (b) 1994—2008. Shadings denote areas with anomaly values less than −20. The contour interval is 4

4　Composite analysis

4.1　NECV activities

Table 2 shows the anomaly of NECV days in association with temperature deviations at Changchun and Harbin. The number of NECV days was more than normal in 8 years out of the 11 A and B event years at Harbin and Changchun (the summer temperature of 2009 did not meet the threshold of A or B event at Harbin), normal in 1 year, and slightly less in 2 years (1976 and 1979). Even in 1976 and 1979, the NECV days for June only were also slightly more than normal (table omitted). In the 5 typical summer high temperature years of 1982, 1988, 1994, 1997, and 2000, the NECV days were 10～15 days less than normal (table omitted). Therefore, it is inferred that the positive anomaly of NECV days is correlated with NESLT, and vice versa, and the NECV is the system that directly influences the NESLT.

Table 2　Standard deviations of summer temperature at Changchun and Harbin and NECV days in NESLT years

Year	1957	1964	1969	1971	1972	1976	1979	1983	1986	1993	2009
Anomaly of NECV days	+29	+13	+21	+4	0	−3	−7	+6	+8	+7	+13
Changchun/σT	−1.848	−1.390	−2.076	−1.172	−1.274	−1.527	−0.878	−1.155	−0.970	−0.865	−0.490
Harbin/σT	−0.654	−1.109	−1.488	−1.298	−1.602	−0.578	−0.388	−1.867	−0.426	−1.222	−0.274

4.2　500 hPa height anomaly composites

Figs. 2a and 2b show the summer 500 hPa height anomaly composites over 7 A events and summer high temperature years of Northeast China, respectively. It is seen that over the Asian mid-high latitudes, Aleutian Islands, and the northwest of North America, there were two hemispheric-scale wide-range negative and positive anomaly bands, with two low value centers over Northeast China and the Aleutian Islands, respectively. In the NESLT years, there were two summer low value activity centers in the Northern Hemisphere, one of them is the NECV center, where the active NECVs directly result in NESLT events. It is inferred that summer surface temperature over North Pacific under the active center of the Aleutian low might also be lower (Fig. 2a).

Fig. 3 exhibits spatial distributions of the correlation coefficient of the summer average temperature at Changchun and Harbin with Northern Hemispheric summer 500 hPa heights. It can be seen that the high correlation belt with values ≥0.3 (significant at the 95% confidence level) spreads zonally from central Asia to central North Pacific, with two large value centers over Northeast China and the Aleutian Islands, respectively. The maximum correlation value was over 0.7 (significant at the 99% confidence level) over the first center. The correlation coefficient pattern over Asia and North Pacific (Fig. 3a) is similar to the 500 hPa height anomaly pattern over the same region (Fig. 2a), indicating that the summer temperature was positively correlated with the 500 hPa height anomaly in Northeast China, that is to say, the NESLT corresponded to the negative summer 500 hPa height anomalies in Northeast China, and vice versa.

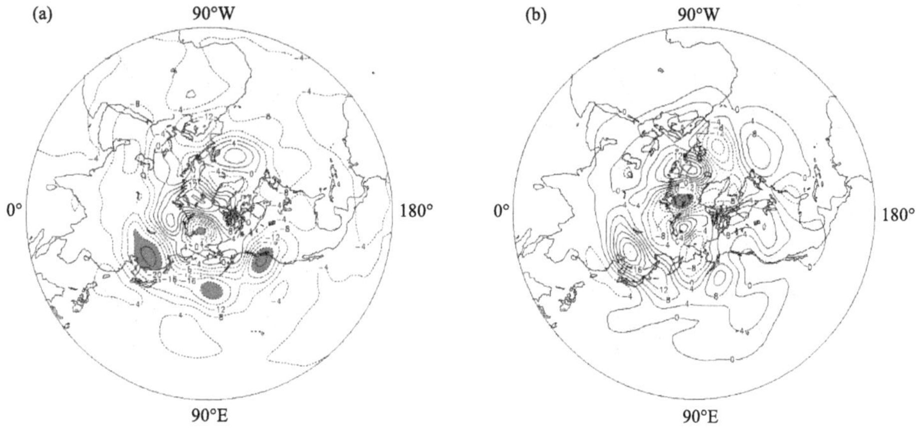

Fig. 2　Summer 500 hPa height anomaly (gpm) composites over A event years (a) and typical summer high temperature years (b) of Northeast China. Shadings illustrate the value of anomaly less than −20; The contour interval is 4

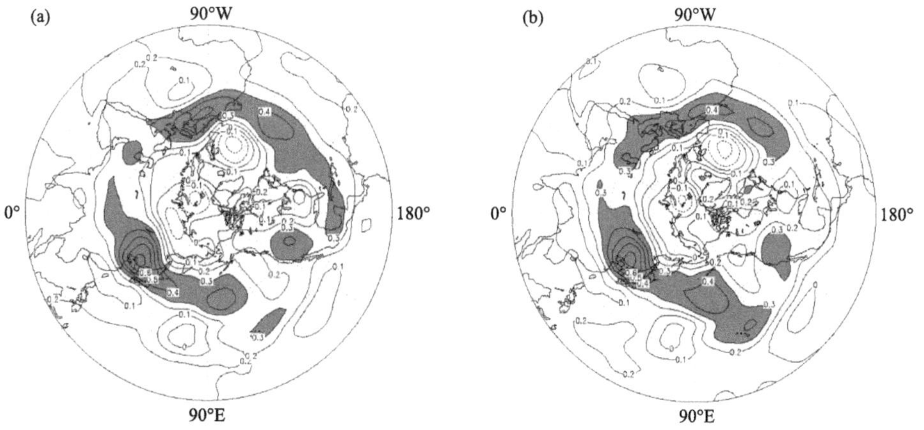

Fig. 3　Simultaneous dependence patterns of the summer temperature of Changchun (a) and Harbin (b) with North Hemispheric 500 hPa geopotential height. The significant correlations at the 95% confidence level are shaded; The contour interval is 0.1

4.3　Meridional circulation

Figs. 4a and 4b are the cross-sections of vertical and meridional winds and temperature anomaly for A event years and high temperature years of Northeast China along 120°E, respectively. It is seen from Fig. 4a that the southward air flow from the polar region of Northern Hemisphere met the northward flow originated from the equatorial region over 50°—52°N, then ascended, and obviously converged in the mid-upper troposphere of 600—200 hPa. Correspondingly, a thick cold air layer at 850—250 hPa with a core cold air layer at 500—400 hPa, where the negative anomaly center was −1.5 ℃, controlled the whole Northeast China, and above the thick cold air layer was a warm tropopause with a warm core of +0.9 ℃ at 150 hPa. However, in the summer high temperature years (Fig. 4b), the polar cold air could reach the vicinity of 75°N only at the tropopause, but the warm air originated from the equatorial region could advance northward within the planetary layer to the north of 50°N, indicating that the warm air at low levels was very strong. Over 40°—50°N, there was a warm core of 1.5 ℃, with its warm tongue extending upward to the mid-upper troposphere, and the relevant flow also ascended and advanced towards the vicinity of the polar region. The above reveals that the cold core of NESLT lay in the vicinity of 500 hPa in the middle troposphere, so we examine the evolution of 500 hPa circulation systems for the diagnosis and prediction of the NESLT events.

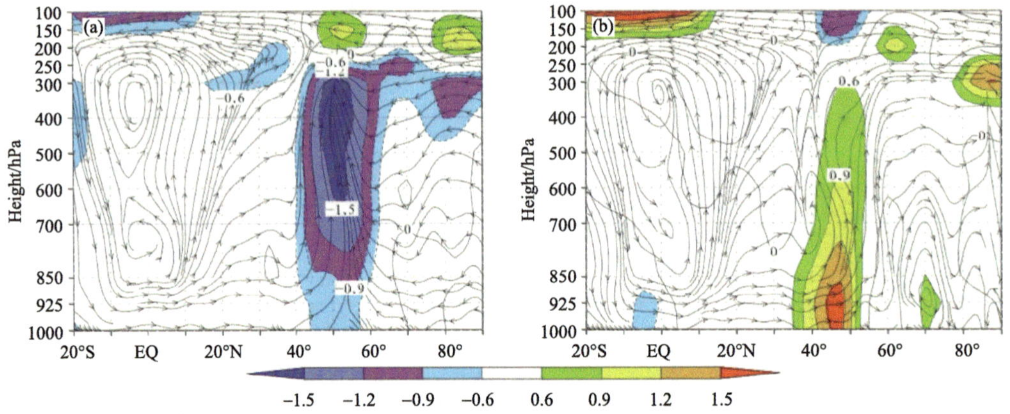

Fig. 4 Composite latitude-height cross-sections of wind and temperature anomaly (℃) over (a) A event years (1969, 1971, 1972, and 1976) and (b) typical summer high temperature years (1982, 1988, 1994, and 1997) of Northeast China along 120°E. Thick solid line with arrows denotes streamline of vertical and meridional winds; thin solid line with color shaded area denotes positive temperature anomalies; thin dashed line with color shaded area denotes negative temperature anomalies. The multi-year mean was calculated for 1958—1997

5 Decadal variability

5.1 Summer average temperature at Changchun and Harbin

Summer average temperature anomaly series at Changchun (Fig. 5a) and Harbin (Fig. 5b) clearly show that the temperature has exhibited a rising trend since the 1920s, and no obvious negative anomaly has occurred since 1994. However, in the summer of 2009, negative anomalies of −0.34 ℃ and −0.24 ℃ occurred at Changchun and Harbin stations, respectively.

Analysis of the decadal component (≥9 years) of summer average temperature anomaly series (Figs. 5c and 5d) shows that the summer temperature of Harbin varies with an about 80 years dominant period, e. g., a cold phase in 1881—1913, a warm phase in 1914—1954, a cold phase again in 1955—1992, and a warm phase in 1992—2008 (Fig. 5d). The dominant period for the summer temperature of Changchun is about 70 years (Fig. 5c), e. g., a cold phase before 1918, a warm phase in 1919—1953, a cold phase again in 1954—1991, and a warm phase in 1991—2008. Most of the cold and warm phases of the two series corresponded and overlapped. Especially, from the late 20th century to the early 21th century, after the positive temperature anomaly has reached the peak value among the 100 year temperature observations, the temperature series at the two stations both underwent a declining phase, with a declining rate of about 0. 7—0. 8 ℃/10 year, that even exceeded the linear warming rate of 0. 2 ℃/10 year during 1961—2000 (Yan, 2007).

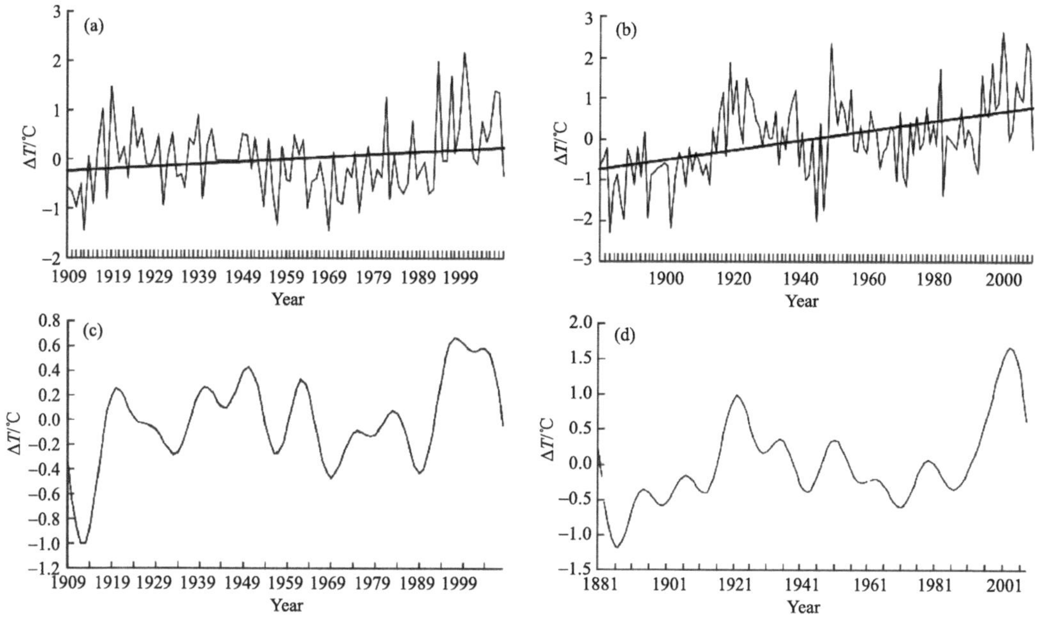

Fig. 5　(a, b) Summer average temperature anomaly series (thin solid line: temperature anomaly; thick solid line: linear trend) and (c, d) their interdecadal variations at (a, c) Changchun and (b, d) Harbin, respectively

Power spectrum analysis of summer average temperature series of Changchun (Fig. 6a) and Harbin stations (Fig. 6b) indicates that the temperature series of Changchun has 66 year and 33 year periods significant at the 95% confidence level (Fig. 6a); while the temperature series of Harbin has 82 year, 3 year, and 2. 3 year significant periods (Fig. 6b). This further confirms that the cold and warm phase alteration of about 30—40 year is significant at the 95% confidence level.

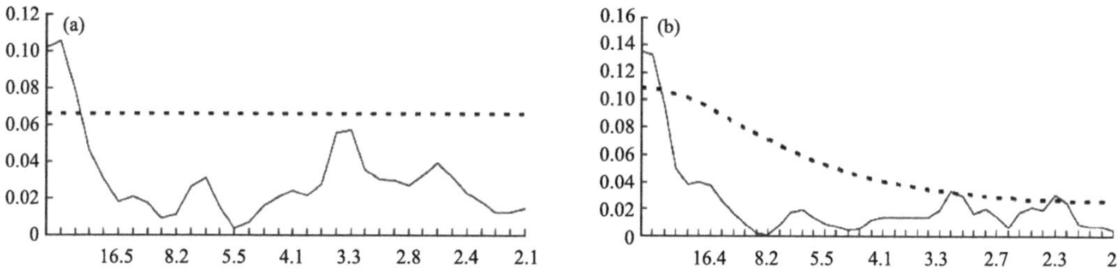

Fig. 6　Power spectra of observed yearly summer temperatures at (a) Changchun and (b) Harbin. Solid line: power spectrum value; dotted line: threshold line for significance at the 95% confidence level

5. 2　Decadal variability of NPPV

The time series of North Pacific polar vortex (NPPV) area index for previous winter (relative to the summer of the indicated year; Fig. 7a) shows that after a period of rapid rising (1958—1963), the area index is low (1964—1992), and its variation is similar to that of the Changchun summer average temperature (1959—1992) (Fig. 5c). The NPPV area index for spring (Fig. 7b) is basically similar to that for previous winter (Fig. 7a), but no second peak appears after 1988, thereby indicating that the NPPV area index has strong persistence from the previous winter to spring. This persistent anomaly on the half-year scale might provide a precursor signal for short-term prediction of the anomalous East Asian summer climate.

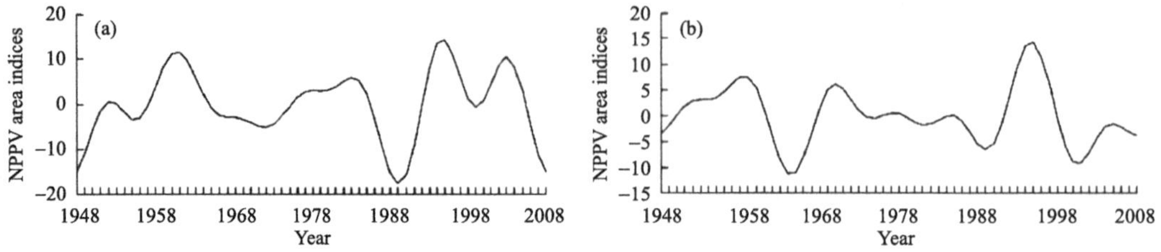

Fig. 7　Decadal variations of NPPV area indices for previous winter (a) and spring (b)

5.3　Relations of Changchun and Harbin summer temperatures to decadal variations of NPPV and North Pacific SSTs

Table 3 shows that the correlations between summer average temperatures of Changchun and Harbin and the previous winter and spring SSTs of the west wind drift region and Nino 3+4 region did not pass their significance tests of the 95% confidence level, but those for summer and autumn SSTs (simultaneous or lagging) were significant at the 95% confidence level. After decomposing the summer average temperature time series of the two stations into interannual and decadal component series, it is found that the decadal components of Changchun and Harbin summer temperatures are positively correlated with the previous winter NPPV area index at the 99% and 95% confidence level, respectively (Table 4); and that the interannual component of Changchun summer temperature is negatively correlated with the summer and autumn NPPV area indices, as well as previous winter SST of the west wind drift region, significant at the 95% confidence level. It is also known from Table 4 that the decadal component of Changchun summer temperature is positively correlated with those of all four season SSTs in the west wind drift region and those of the previous winter SST in the Nino 3+4 region, at the 95% confidence level; and the interdecadal component of Harbin summer temperature is also significantly correlated with that of the summer SST in the west wind drift region at the 95% confidence level. This indicates that the previous winter NPPV area index provides an important precursor signal for changes in both Changchun and Harbin summer temperatures on decadal scale; and changes in the previous winter SST in the west wind drift region are also important prophase signals for summer temperature changes at Changchun Station, and changes in the SST of Nino 3+4 region in previous winter also to some extent affect Changchun summer temperature. Besides, the SST anomaly over east-central equatorial Pacific in winter 2008 presented a La Niña pattern, which obviously weakened in the spring of 2009 (figure omitted).

Table 3　Correlation coeffcients between North Pacific SST and the summer average temperatures of Changchun and Harbin

area	SST location	Previous winter	Spring	Summer	Autumn
Changchun	West wind drift region*	−0.07	0.01	0.26	0.35
	Nino 3+4	−0.01	−0.01	−0.02	−0.04
Harbin	West wind drift region*	−0.02	0.08	0.33	0.34
	Nino 3+4	−0.03	−0.06	−0.05	−0.04

* Correlation coefficient for Changchun and Harbin were calculated based on the data of 1909—2005 and 1881—2005, respectively.

Table 4 Correlation coefficient of the interannual and decadal components of the seasonal average temperatures of Changchun and Harbin with the NPPV area index and North Pacific SSTs in four seasons

area		NPPV area index*				North Pacific SST**							
						West wind drift region				Nino 3+4			
		Previous winter	Spring	Summer	Autumn	Previous winter	Spring	Summer	Autumn	Previous winter	Spring	Summer	Autumn
Changchun	Interannual component	0.08	0.05	−0.28	−0.36	−0.21	−0.14	−0.12	−0.02	0.21	0.15	0.08	0.10
	Decadal component	0.56	−0.08	−0.16	0.23	0.24	0.23	0.40	0.53	−0.15	−0.17	−0.07	−0.05
Harbin	Interannual component	0.06	0.02	−0.14	−0.16	0.04	0.04	−0.02	−0.02	0.17	0.08	−0.07	−0.02
	Decadal component	0.26	−0.09	−0.08	0.07	−0.11	0.02	0.25	0.07	0.08	0.10	0.09	0.21

* Correlation coefficient for Changchun and Harbin were both calculated based on the data of 1948—2008;

** Correlation coefficient for Changchun and Harbin were calculated based on the data of 1909—2005 and 1881—2005, respectively. Values in bold exceed the 95% confidence level.

6 Teleconnection between Northeast China summer temperature and NPO

6.1 Half-year-scale teleconnection

Figs. 8a and 8b give the simultaneous correlation maps of the SVD1 of the previous winter 500 hPa geopotential height field in the NPO region and the summer average temperature field of 80 stations in Northeast China, respectively. The two maps both show an approximately south—north "+ −" phase distribution of correlation wave train; wherein a negative correlation region lies around the Aleutian Islands with its center value ≤−0.8 in the north of the islands, and a positive correlation center with its value ≥0.6 lies in the eastern Pacific near 65°N (Fig. 8a). Correspondingly, the summer temperature showed consistent positive values over Northeast China, with correlation coefficient mostly from 0.7 to

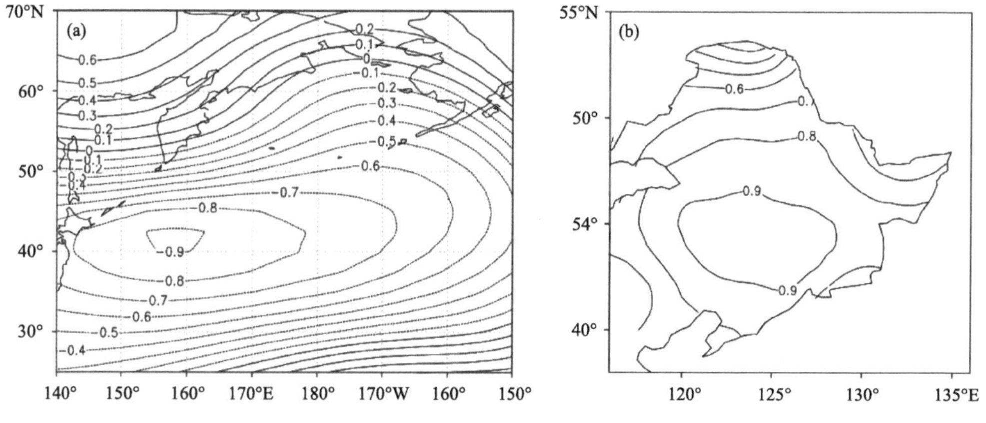

Fig. 8 Simultaneous correlation maps of the SVD1 of (a) the 500 hPa geopotential height field of the NPO region (25°—70°N, 140°E—150°W) in previous winter and (b) the summer average temperature field of 80 stations in Northeast China

0.9 (Fig. 8b); the half-year scale teleconnection of the two fields is significant at the 99% confidence level, and the contribution of SVD1 accounts for 75% (a very high weight) of the total sum of squared covariances.

The homogeneous correlation maps of the SVD2 (Figs. 9a and 9b) show a pattern opposite to that of the SVD1 (Figs. 8a and 8b). That is to say, over the Aleutian Islands and the Kamchatka Peninsula area, there is a large positive correlation region with its center values above 0.8; and over the smaller area south of 45°N, there is a negative correlation zone with its center value ≤−0.8 (Fig. 9a). Correspondingly, Northeast China is dominated mostly by negative correlations, especially over Jilin and Heilongjiang provinces, and the correlation coefficients range from −0.25 to −0.50 with most of them passing the 95% confidence level (Fig. 9b). The correlation flow for the SVD2 is the same as the SVD1, i. e., if there is a north—south "+−" (or "−+") phase distribution in the NPO region in the previous winter, the summer temperature in Northeast China will be lower (or higher). The variance contribution of the SVD2 accounts for 12% of the total covariance, and the contributions of the SVD1 and SVD2 account for 87% of the total covariance, which sufficiently indicates that the NPO phase distribution in previous winter is an important and strong precursor signal for the summer temperature of Northeast China. The anomalous phase of NPO might be a planetary-scale stable component that can be extracted from the atmospheric general circulation.

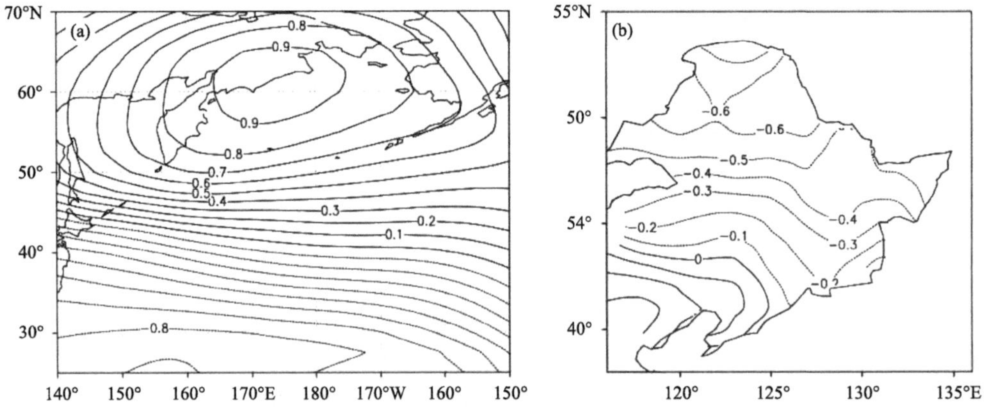

Fig. 9　As in Fig. 9, but for SVD2

6.2 Similar NPO situations between 2009 and 1972

It can be seen from Figs. 10a and 10d that the 500 hPa height anomaly patterns of the previous winters of 1972 and 2009 are very similar, i. e., the Northern Hemispheric 500 hPa circulation basically showed a two-wave disturbance distribution, with the strongest positive anomaly center both over the sea south of the Aleutian Islands, and in the NPO region where there appeared a typical negative phase of the north—south "+−" anomaly distribution of NPO. The positive anomaly center with its center value being 140 gpm in 2009 is slightly stronger than its counterpart in 1972, but the two-wave disturbance pattern in 1972 is clearer than that in 2009. Furthermore, there appeared also a stronger negative phase of North Atlantic Oscillation (NAO) over the Atlantic Ocean, relative to that in 1972.

Figs. 10b and 10e show that the spring 500 hPa height anomaly patterns of 1972 and 2009 are also very similar: The negative phase disturbances of NPO still maintained over the mid-high latitudes of North Pacific, except that the positive anomaly center over the sea south of the Aleutian Islands was

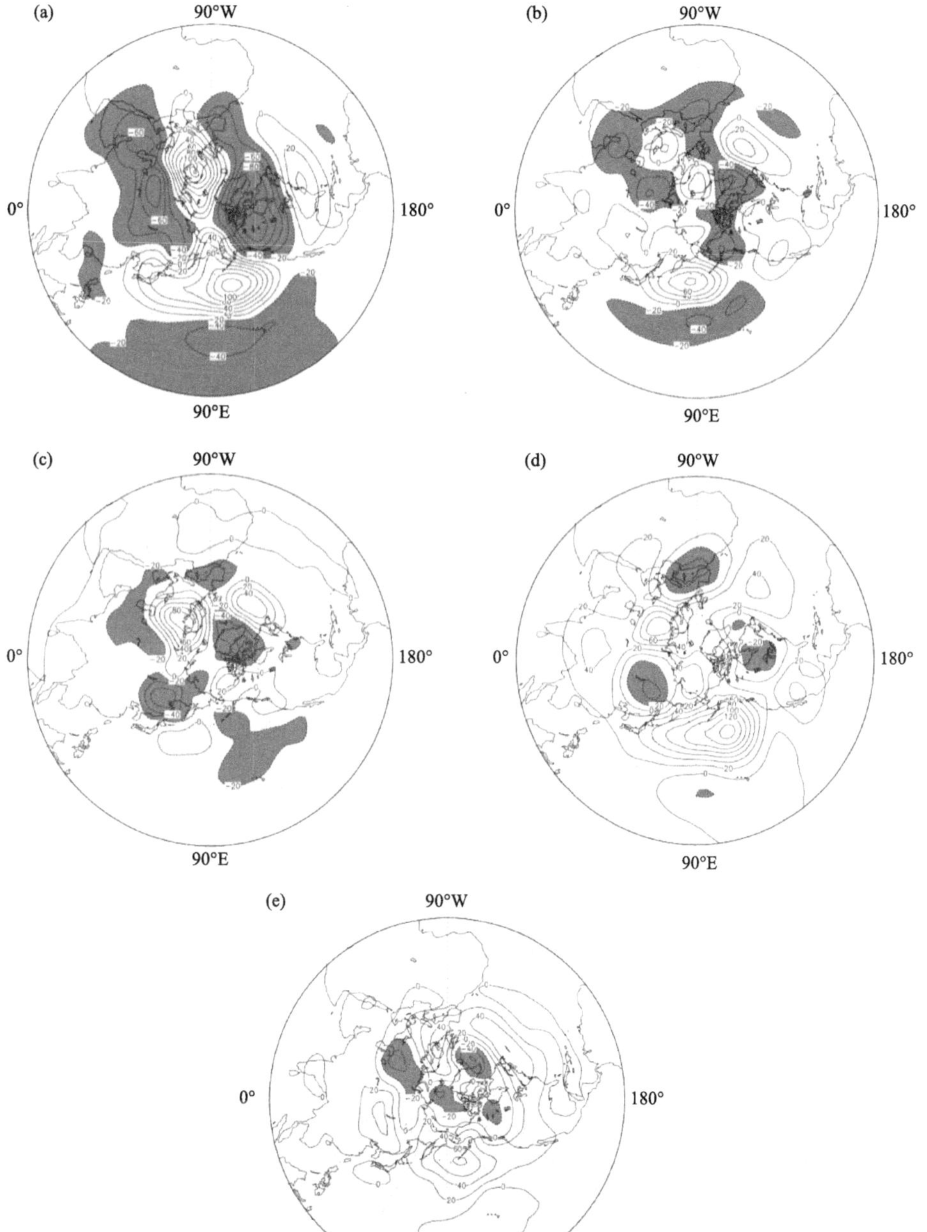

Fig. 10　500 hPa geopotential height anomaly fields (gpm) in previous winter of (a) 1972 and (d) 2009, spring of (b) 1972 and (e) 2009, and (c) summer of 1972. Solid (dashed) line denotes positive (negative) anomaly; contour interval is 20; shadings indicate values less than or equal to −20

obviously weakened relative to that in the previous winter, but it was still the strongest spring disturbance center over the Northern Hemisphere. Therefore, the negative phases of NPO in the previous winter and spring of 1972 and 2009 were very strong and sustained for a half-year, which must have impacted the summer Northern Hemispheric 500 hPa circulation. In summer 1972, A event occurred in Northeast China (Table 2) with stronger negative temperature anomalies (Fig. 10c), but in summer 2009, the negative 500 hPa height anomaly center moved to the coastal region of Russia, east of its normal position, and the NESLT mainly happened in the central eastern parts of Jilin and Heilongjiang provinces. The persistent negative phase of NPO over the previous winter and spring of 1972 and 2009 both resulted in the occurrence of NESLT events, again indicating that the anomalous phase of NPO is a stable planetary-scale signal worthy of notice.

6.3 Harmonic analysis of the preceding 500 hPa height field for anomalous NECV activities

Fig. 11 shows the ultra-long wave disturbance distribution of 500 hPa height along 45°N during April—May 2009. The ultra-long wave (wave numbers 1—3) trough always stayed over 120°—180°E especially the North Pacific region. It obviously shifted eastward only in early May and then distinctively retreated westward again in mid May. This is a typical precursor for anomalously strong activities of summer NECV (Shen et al., 2008), indicating the transformation of the North Pacific ultra-long wave disturbance trough along 45°N into a stationary wave disturbance trough, which is obviously related to the persistent north—south "+ −" anomaly wave train of the negative phase of NPO over the North Pacific region in spring, and is the major character of prophase circulation evolution or a precursor for the anomalous June NECV activities of 2009. The NECV days in June—July 2009 were 18 days more than normal, accounting for a percentage anomaly of 54%.

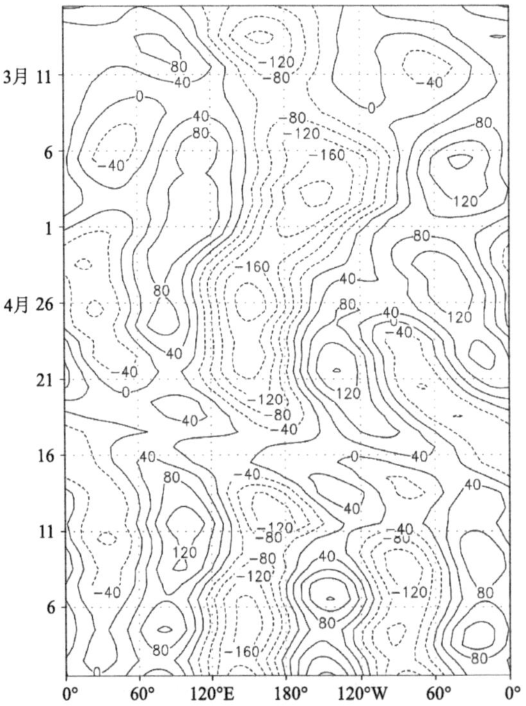

Fig. 11　Time-longitude cross-section showing the ultra-long wave disturbances of the 500 hPa geopotential height (gpm) along 45°N. Solid (dashed) line denotes positive (negative) value

7 Conclusions and discussion

A summer low temperature event occurred in the central-northern part of Northeast China in 2009. The temperature anomaly at Changchun Station reached -0.49σ, characterizing a B event according to the NESLT classification given in this paper.

It is found that a strong NECV directly resulted in the NESLT in central-eastern Northeast China in summer 2009. Summer average temperatures of Harbin and Changchun were significantly correlated with concurrent 500 hPa height over Northeast China, with correlation coefficients of more than 0.7, significant at the 0.01 confidence level. Composite analysis reveals that the meridional circulation for typical NESLT years is featured with a deep cold air layer between 850 and 250 hPa over the whole Northeast China. The horizontal and vertical circulation reflect anomalously strong NECV activities directly responsible for NESLT events. During June—July 2009, 18 more NECV days appeared, accounting for a percentage anomaly of 54%; and the negative 500 hPa height anomaly directly resulted in the NESLT over the coastal region of Russia, east of its normal position.

The decadal components (≥9 years) of observed 100 year summer average temperature series at Harbin and Changchun exhibited a distinctive declining trend in the past decade, which provides an important decadal background for the NESLT of Jilin and Heilongjiang provinces in 2009. The power spectrum analyses of the temperature series of the two stations uncover that the series had a significant 30—40-year cold/warm phase alteration and showed a remarkable declining trend in the period from the end of the 20th century to the beginning of the 21st century, with the declining rate of 0.8 ℃/10 year over 1999—2008, far greater than the linear warm rate of 0.2 ℃/10-year of Northeast China in response to the global warming over 1961—2000. The summer circulation anomaly over the mid-high latitudes of Northern Hemisphere in 2009 is almost opposite in sign (positive/negative) to that in 1994—2008 when no NESLT event happened. Meanwhile, in 2009, significant negative phase of AO and negative height anomalies over Northeast Asia, Aleutian Islands, and the North Atlantic occurred.

The decadal variation of previous winter NPPV activities is an important background condition for summer temperature change in Northeast China. The previous winter NPPV area index, which was significantly positively correlated with the summer temperatures of Changchun and Harbin, also showed a decadal declining trend. It is also known that the SST in the west wind drift region of North Pacific also affects the summer temperature in Northeast China, with its effect next secondary to that of the NPPV, and the SST in the east-central tropical Pacific has some impact as well.

The phase of NPO is an important precursor for summer temperature changes in Northeast China. This is also a planetary-scale stable atmospheric component in addition to the synoptic-scale chaotic component. The SVD1 and SVD2 of previous winter 500 hPa height fields in the NPO region and summer temperature fields at the 80 stations of Northeast China both reveal that the positive/negative phase of NPO was significantly correlated with the positive/negative anomaly of summer temperatures, with the covariance contributions of the SVD1 and SVD2 to the total covariance reaching 87%. In the previous winter of 2009, the negative phase of NPO was the strongest since the mid 1980s and it sustained into the spring of 2009. The evolution of NPO was very similar to that in 1972, when a severe NESLT event occurred in Northeast China. Therefore, occurrence of the 2009 NESLT event may hint that summer temperature in Northeast China might have entered a low temperature phase similar to that in the 1970s.

Harmonic analysis of April—May 500 hPa height along 45°N over North Pacific suggests that the ultra-long wave activities appearing as a form of stationary disturbances serve as a precursor for the anomalously strong activities of NECV in June—July 2009. The ultra-long wave disturbance trough over North Pacific was a stationary wave disturbance, which was related to the persistent negative phase of the north—south "+−" anomaly wave train of NPO, and formed a prophase mid-high latitude circulation background in favor of the June—July unusually active NECVs of 2009.

This paper analyzed the decadal variation of the 100-year temperature observations at Changchun and Harbin stations and the NPPV activities. The results showed a declining trend of summer temperature in the recent decade. On the annual and seasonal scale, the previous winter and spring precursor signals of the air-sea system over North Pacific favorable for the occurrence of NESLT and NECV are identified. On the intraseasonal and monthly scale, the lower frequency signal of June—July 500 hPa height in favor of NECV is uncovered by the power spectrum analysis. In early June 2009, we used the above data and methods, with reference to the 10—16 days 500 hPa height forecast (in which abnormally more frequent NECV activities were found), issued a summer low temperature warning. This is a successful effort on the prediction of NESLT.

Overall, this paper investigated the large-scale circulation features of the 2009 NESLT event. Meanwhile, composite analysis was performed to identify the decadal background favorable for the occurrence of NESLT, the relationships between NESLT/NECV activity and prophase changes in the air-sea system over the North Pacific region, and how to predict the NESLT event of 2009 using multi-spatiotemporal scale predictors. However, the physical mechanism of the NESLT and the air-sea system association is not clear. Is the NESLT event of 2009 only an isolated case, or an early warning signal for a new phase of massive occurrences of such events? Further studies are needed to address those issues.

Acknowledgments. The authors would like to thank Prof. Ding Yihui for his valuable advice; thanks also go to the two anonymous reviewers for their valuable comments. Contributions from Gao Zongting, Chen Changsheng, Guo Chunming, Su lixin, Tu Gang, Sui Bo, et al. of the Institute of Meteorological Science of Jilin Province were also appreciated.

References

AN G, LIAN Y, WANG Q, 1998. The wavelet analysis of temperature variations of crop growth period[J]. Acta Meteor Sinica, 56(4): 458-464. (in Chinese)

CHEN L X, ZHOU X J, LI W L, 2004. Characters of the climate changes and its formation mechanism in China in last 80 years[J]. Acta Meteor Sinica, 62(5): 634-646. (in Chinese)

CHOU J F, 1986. Long Range Numerical Weather Forecast[M]. Beijing: China Meteorological Press: 329. (in Chinese)

CHOU J F, 1997. Advances in non-linearity and complexity in atmospheric sciences[J]. Bull Chinese Acad Sci, 12(5): 325-329. (in Chinese)

CHOU J F, 2007. An innovation road to numerical weather prediction: From the initial value problem to the inverse problem [J]. Acta Meteor Sinica, 65(5): 673-682. (in Chinese)

DING Y H, 1989. Diagnosis of Weather Dynamics[M]. Beijing: Science Press: 292. (in Chinese)

DING Y H, 2007. Preface[J]. Scientia Geographica Sinica, 27(S1): 1. (in Chinese)

DING Y H, ZHANG J, XU Y, et al., 2003. Evolution of Climate System and Its Prediction[M]. Beijing: China Meteorological Press: 137. (in Chinese)

FENG G L, DONG W J, GONG Z Q, et al, 2006. Nonlinear Spatiotemporal Distribution Theory and Methods of Observations [M]. Beijing: China Meteorological Press: 227. (in Chinese)

FU C B, 1980. Relationships between the Northern Hemispheric winter and spring snow area and summer temperature in Northeast China[J]. Acta Atmos Sinica, 38(2): 167-192. (in Chinese)

HE J H, WU Z W, QI L, 2006. Relationships of Northern Hemispheric annual mode and Northeast China cold vortex with summer precipitation in East Asia and China[J]. Meteor Environ J, 22(1): 1-5. (in Chinese)

INTERNATIONAL CLIVAR PROJECT OFFICE, 1997. A Research Programme on Climate Variability and Prediction for the 21st Century. CLIVAR, World Climate Research Programme[M]. Bundsstr: Max-Plank-Institute für Meteorologie: 48.

IPCC, 2007. Summary for Policymakers of the Synthesis Report of the IPCC Fourth Assessment Report[M]. Cambridge: Cambridge University Press: 996.

JILIN PROVINCE METEOROLOGICAL OBSERVATORY, 1980. Analysis of the long-term weather processes causing summer (May—September) continuous low temperature[J]. Sci Atmos Sinica, 4(1): 54-67. (in Chinese)

LIAN Y, AN G,1998. The relationship among East Asian summer monsoon,El Niño and low temperature in Songliao Plain in Northeast China[J]. Acta Meteor Sinica,56(5):724-735. (in Chinese)

LIAN Y,GAO Z T,SHEN B Z,et al,2007. Climate change and its impacts on grain production in Jilin Province[J]. Adv Clim Change Res,3(1):46-49. (in Chinese)

LIN M J,FAN K,WANG H J,2010. An investigation into the climatic characteristics of vertical shear of zonal wind in the western North Pacific[J]. Acta Atmos Sinica,68(3):309-314. (in Chinese)

LIU Z X,1986. Northern Hemispheric polar vortex strength index and its relation with temperature changes in China[J]. Meteor Monthly,12 (Suppl II):84-89.

LIU Z X,LIAN Y,GAO Z T,et al,2002. Analyses of the Northern Hemisphere circulation characters during northeast cold vortex persistence[J]. Chinese J Atmos Sci,26(3):361-372.

LIU Z X,LIAN Y,SHEN B Z,et al,2003. Seasonal variation features of 500 hPa height in the North Pacific Oscillation region and its effect on precipitation in Northeast China[J]. Chinese J Appl Meteor,14(5):553-561. (in Chinese)

LU R Y,2003. Linear relation between interdecadal and interannual changes in summer rainfall in North China[J]. Chinese Sci Bull,48(7):718-722. (in Chinese)

NATIONAL ASSESSMENT ROPORT OF CLIMATE CHANGE EDITORIAL COMMITTEE, 2007. China's Climate Change Assessment Report[M]. Beijing:Science Press:422. (in Chinese)

NORTHEAST CHINA SUMMER LOW TEMPERATURE RESEARCH COLLABORATION TEAM,1979. A preliminary research on cool/hot summer in the Northeast China region[J]. Acta Atmos Sinica,37(3):44-58.

REN R C,XIANG C Y,2010. Temporal and spatial connections of the stratospheric polar vortex oscillation to the ENSO tropical SST anomalies[J]. Acta Atmos Sinica,68(3):285-295. (in Chinese)

SHEN B Z,LI S F,LIAN Y,2008. The harmonic analysis of cold vortex of Northeast China in early summer[J]. J Jilin Univ (Earth Scie),38(Suppl):204-208. (in Chinese)

SUN J Q,WANG H J,2006. Regional differences of summer air temperature anomalies in Northeast China and its relationship to atmospheric general circulation and sea surface temperatures[J]. Acta Geographica Sinica,49(3):662-671. (in Chinese)

THE GENERAL OFFICE OF THE PEOPLE'S GOVERNMENT OF JILIN PROVINCE,2002. Jilin Yearbook[M]. Changchun:Jilin Yearbook Press:240. (in Chinese)

WANG J F,WU G X,1997. Evolution and characteristic of the persistent cold summer in Northeast China[J]. Chinese J Atmos Sci,21(5):523-532. (in Chinese)

WANG S Y,1995. Studies on Crops Cool Injury[M]. Beijing:China Meteorological Press:246. (in Chinese)

YAN M H,2007. Instrumental Climate Change in Northeast China and Its Regional Differences[M]. Beijing:Science Press: 37-38. (in Chinese)

YANG Q,LIAN Y,HE J H,2005. Prediction of summer temperature with SVD in Northeast China[J]. Meteor Monthly,31 (3):31-35. (in Chinese)

YE D Z,YAN Z W,2008. On orderly adaptation to global warming[J]. Acta Meteor Sinica,66(6):855-856. (in Chinese)

ZHANG M L,FU C B,WANG M R,et al,1982. A study of global surface temperature field in the 1970s. Part I:The characteristics of global temperature and summer cold disaster in the Northeast China in the 1970s[J]. Scienita Atmos Sinica,6 (3):229-235. (in Chinese)

ZHANG M L,FU C B,WANG M R,et al,1983. A preliminary research on global surface temperature in the 1970s (III)[J]. Sci Atmos Sinica,7(1):23-31. (in Chinese)

ZHANG S Q,1990. Relationships between Northeast China summer low temperature and El Niño events[M]//No. 5-Climates/National Science and Technology Committee. Blue Book of Science and Technology. Beijing:Science and Technology Literature Press:367. (in Chinese)

ZHANG S Q,YU T J,LI F Y,et al,1985. Seasonal changes in Northern Hemispheric polar vortex area and intensity and their association with temperature in Northeast China[J]. Sci Atmos Sinica,9:178-185.

ZHAO Z C,LUO Y,JIANG Y,2008. Challenge on Prediction of East Asian Winter Monsoon as a Global Warming[C]. Changchun:ISSCCE:66.

ZHAO Z C,WANG S W,XU Y,et al,2005. Attribution of the 20th century climate warming in China[J]. Climate Environ Res,10(4):808-817.

ZHENG W Z,NI Y Q,1999. Diagnostic study on the impact of sea surface temperature anomalies over tropical and mid-latitude Pacific on summer low temperature/cool damage in Northeast China[J]. Quart J Appl Meteor,10(4):394-401. (in Chinese)

ZHI R,FENG G L,ZHOU L,et al,2010. Temporal and spatial correlation analysis of global temperature data based on the correlation matrix theory[J]. Acta Mereor Sinica,68(4):501-513.

Time-Frequency Characteristics of Regional Climate over Northeast China and Their Relationships with Atmospheric Circulation Patterns*

LIU Shi[1] YANG Song[2] LIAN Yi[1] ZHENG Dawai[3] WEN Min[4] TU Gang[1]
SHEN Baizhu[1] GAO Zongting[1] WANG Donghai[4]

(1 Institute of Meteorological Sciences of Jilin Province, Changchun, Changchun 130062; 2 NOAA/Climate Prediction Center, Camp Springs, Maryland; 3 Shanghai Astronomical Observatory, Chinese Academy of Sciences, Shanghai 200030; 4 Chinese Academy of Meteorological Sciences, Beijing 100081)

Abstract: The time-frequency characteristics of the variations of temperature and precipitation over the city of Changchun in northeast China and their associations with large-scale atmospheric and oceanic conditions are analyzed. It is found that the variations of the regional climate are characterized by strong semiannual signals. For precipitation, the amplitude of semiannual signal is about half of that of the annual cycle. The relationships of the Changchun temperature and precipitation with local winds and large-scale patterns of atmospheric circulation and sea surface temperature are also strongest on annual and semiannual time scales. These strong semiannual signals are potentially helpful for improving the prediction of the regional climate.

On the annual time scale, the northeast China climate is affected by both the thermal contrast between the Asian continent and the tropical Indo-Pacific Oceans and that between the continent and the extratropical North Pacific. These effects are manifested by the cyclonic (anticyclonic) pattern over the Asian continent (North Pacific) and the strong southerly flow over East Asia and northwestern Pacific associated with increases in temperature and precipitation. On the semiannual time scale, the northeast China climate is mainly related to the large-scale circulation pattern centered over the North Pacific, with its western portion over northeast China, North and South Korea, and Japan. While temperature signals are related to extratropical atmospheric process more apparently, both extratropical and tropical influences are seen in the semiannual variation of precipitation.

There exist strong relationships between Changchun temperature and precipitation and the North Pacific Oscillation (NPO) in the frequency band up to 7 months. Temperature increases and precipitation decreases when NPO is positive. The relationships were weak before 1980 but became stronger afterward, associated with the strengthening of the East Asian trough.

1 Introduction

Northeast China (NEC) is the major corn field of China. The variability in corn production, which

* 本文原载于 *Journal of Climate*, 2010, 23(18): 4965-4972

affects the total agricultural production of the country significantly, is influenced strongly by the conditions of local temperature and precipitation. Because of the mutual influences by tropical and extratropical atmospheric systems, the climate over NEC is characterized by many complex features, and improved understanding of these features is important not only for climate prediction operations but also for studies of hydrology, ecology, and agriculture.

Previous studies have revealed many features of variations of NEC climate in station observations, reanalysis data, and numerical models. Many studies have focused on the synoptic and seasonal-to-interannual features of the regional climate and their relations with tropical and high-latitude influences (e. g., Lian et al., 1998; Liu et al., 2001). Recent investigations have also been conducted to understand the decadal changes, regime shift, and long-term trends of changes in temperature and precipitation (Zhang et al., 2005; Sun et al., 2006a, 2006b; Yang et al., 2008). It is found that in the past century, there exists a positive trend in temperature, especially in winter, and a negative trend in total precipitation, especially in fall. However, as the regional climate becomes warmer, extreme precipitation events exhibit an increasing tendency.

Previous studies have also analyzed the variations of atmospheric circulation systems and their relationships with regional temperature and precipitation (e. g., Lian et al., 1998; Sun et al., 2002). These studies attempt not only to understand the physical mechanisms responsible for variations of the regional climate but also to improve the prediction skill of the climate. Li et al. (2005) examined the relationships of NEC precipitation with El Niño-Southern Oscillation (ENSO), the North Pacific Oscillation (NPO), the Arctic Oscillation, the North Atlantic Oscillation, and the Pacific decadal oscillation. They found that NPO was correlated with the NEC precipitation most significantly, although multiple factors should be considered for predicting the regional climate. Overall, the effect of ENSO on extratropical NEC climate is weak, and the skill of predicting the regional climate by numerical models is usually low (Wang et al., 2008) because of various reasons: model deficiencies, strong irregular features of the interannual signals and thus low predictability, and others. Nevertheless, further investigations into the characteristics of NEC climate and its relationship with large-scale climate control are important for understanding the mechanisms responsible for the variability of the regional climate, improving the prediction of the climate, and mitigating the economic loss due to meteorological disasters.

In this study, we conduct an analysis of the detailed features of NEC climate variations in both time and frequency domains. We attempt to understand how the dominant climate features on different time scales and during different periods of time behave and are accompanied by different atmospheric circulation patterns. For example, even though the NEC climate is closely related to NPO, as seen in Li et al. (2005), does this relationship vary with time and change with time scales? It is important to answer the question because the various features of different time scales are often attributed by different causes. It is expected that the study will also provide useful information for improving the prediction of the regional climate over northeast China.

More specifically, we apply several advanced analysis tools to investigate the variations of temperature and precipitation over the city of Changchun and their relationships with large-scale circulation and sea-surface temperature (SST) patterns. We reveal the characteristics of temperature and precipitation variations in time and frequency domains, focusing on the signals of dominant time scales, and assess the relationships of the dominant climate signals with large-scale circulation and SST patterns. For example, we explore the time-frequency characteristics of relationships between NPO and temperature/precipitation, which have not been explored before.

Since this study focuses on the detailed features of the variations of regional climate represented by time series of temperature, precipitation, and others, we mainly analyze the data over Changchun, instead of the area averages over all of NEC, for several reasons. First, values averaged over large spatial domain may smooth out some detailed features that would be otherwise revealed by individual station data. Second, Changchun is one of the few meteorological stations in NEC that have long records of meteorological observations. Third, the precipitation over NEC is often characterized by relatively homogeneous features in signifying droughts and floods, and the mean temperature over Changchun, located near the center of the Songliao Plains, represents well the warm and cold conditions over most of NEC (Ding, 1980; Sun et al., 2000, 2006c; also see the discussion in section 2).

The rest of this paper is organized as follows. In section 2, we describe the basic features of datasets and analysis methods. In section 3, we depict the detailed features, including the time-frequency characteristics of Changchun temperature and precipitation. We discuss the relationships of the temperature and precipitation with local and large-scale circulation and SST features in section 4 and with NPO in section 5. A summary of the results obtained is provided in section 6.

2 Data and analysis methods

The main data used in this study are the daily precipitation and surface air temperature over Changchun of NEC. These data cover the period since 1909, with missing records during 1943—1948. The data of Changchun station also include daily surface winds and sea level pressure over Changchun since 1980. The temperature and precipitation over 79 NEC stations are also used in this study. Among the 79 stations, there is only one station with a location change in 1957. Quality control of the datasets because of a systematic change in instrumentation and others has been conducted to meet the requirement of meteorological operations by the China Meteorological Administration.

Fig. 1 shows the patterns of correlation of seasonal surface temperature between Changchun and other NEC stations for spring (March—May), summer (June—August), autumn (September—November), and winter (December—February). It can be seen that the variability of temperature over Changchun is significantly related to the variations of temperature over all of NEC year round. The correlation patterns show uniform features over a large domain, with the strongest relationship over the Songliao Plains (around 42.4°—48.7°N, 118.7°—128°E). More specifically, the coefficients of correlation between Changchun temperature and NEC-averaged temperature for 1961—2002 are 0.94, 0.98, 0.96, and 0.98 for winter, spring, summer, and autumn, respectively. There is also a strong relationship in seasonal precipitation (Fig. 2), especially in winter. As for temperature, the strongest relationship is found in the correlation between Changchun precipitation and the precipitation over the Songliao Plains. The coefficients of correlation between Changchun precipitation and NEC-averaged precipitation for 1961—2002 are 0.67, 0.65, 0.68, and 0.72 for winter, spring, summer, and autumn, respectively, and all these numbers significantly exceed the 99% confidence level. However, compared to the correlation of temperature, the correlation of precipitation is weaker and exhibits more spatially inhomogeneous features. In particular, a weak relationship appears between Changchun precipitation and the precipitation over northern NEC in summer (Fig. 2b).

The datasets analyzed in this study also include the National Centers for Environmental Prediction-National Center for Atmospheric Research (NCEP-NCAR) reanalysis (Kalnay et al., 1996) and the National Oceanic and Atmospheric Administration (NOAA) extended reconstructed sea-surface temperature

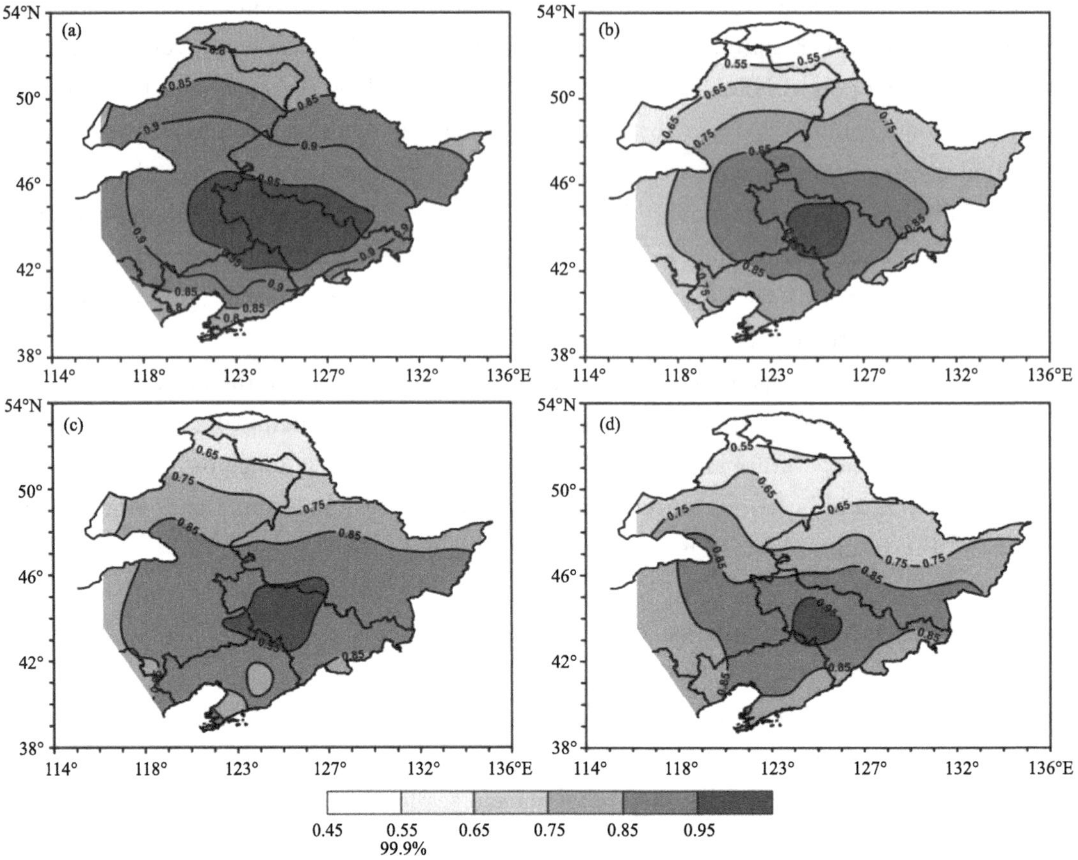

Fig. 1 Correlation of surface temperature between Changchun station and other stations over NEC (1961—2002) for (a) spring, (b) summer, (c) autumn, and (d) winter

(Smith et al., 2003). The NPO index applied in the analysis is defined as the normalized difference in sea level pressure between 25°—40°N, 130°—170°E and 50°—65°N, 130°—170°E (also see Wallace et al., 1981).

The study employs a number of statistical tools, which include wavelet analysis, least squares analysis, and methods of correlation, coherence, and regression analyses. Many features of these tools are discussed appropriately in the following sections [see Yang et al. (2007) for more a detailed description].

3 Time-frequency characteristics of temperature and precipitation variations

Fig. 3 shows the climatological means of monthly temperature and precipitation over Changchun. For 1949—2006, January is the coldest (−15.6 ℃) and driest (0.14 mm d^{-1}) month and July is the hottest (23.1 ℃) and wettest (5.56 mm d^{-1}) month of the year. Extremely high temperature was observed in July 1999 (25.3 ℃) and extremely low temperature in January 1970 (−20.4 ℃). Extremely heavy precipitation occurred in July 1986, which is 14.3 mm d^{-1}. For 1909—1942, January (July) is also the coldest (hottest) and driest (wettest) month of the year. The temperature and precipitation are −16.8 ℃ and 0.18 mm d^{-1} for January and 23.4 ℃ and 5.60 mm d^{-1} for July, respectively. Fig. 4 further shows the distributions of probability density function of temperature and precipitation over Changchun for 1649—2006, in January, the local temperature is mostly in the range for −16.8—−14.4 ℃, followed by a range for −19.2—−16.8 ℃ (Fig. 4a). In July, it is mostly in the range for 21.7—22.7 ℃, followed by

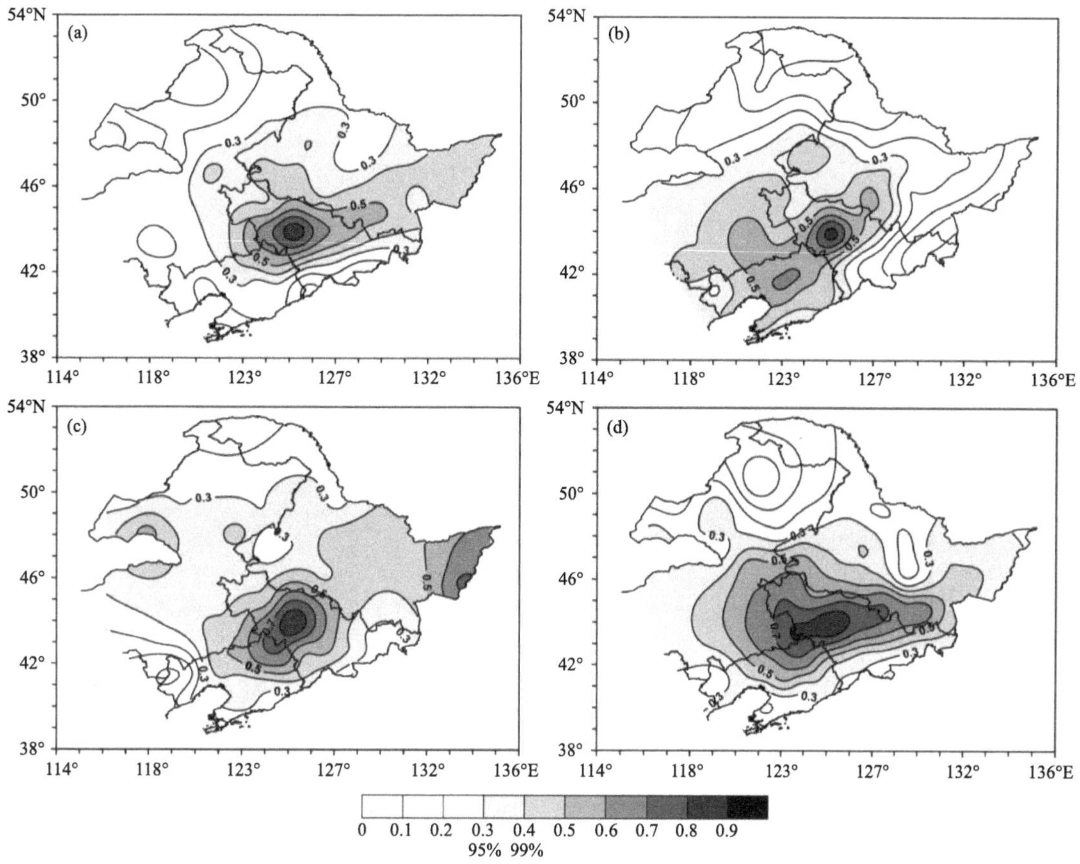

Fig. 2　As in Fig. 1, but for precipitation correlation

a range for 22.7—23.7 ℃ (Fig. 4b). For 1909—1942, the temperature has the largest density in the range of −19.2—−16.8 ℃ for January and the range of 22.7—23.7 ℃ for July. For both 1909—1942 and 1949—2006, the density of precipitation over Changchun is the largest in the range of 0—0.1 mm d^{-1} in January and 5.3—7.8 mm d^{-1} (followed by 2.7—5.3 mm d^{-1}) for July (Figs. 4c and 4d). It should be pointed out that the data of 1909—1942 may be less reliable compared to those of 1949—2006 and thus only the latter period is mainly analyzed in this study.

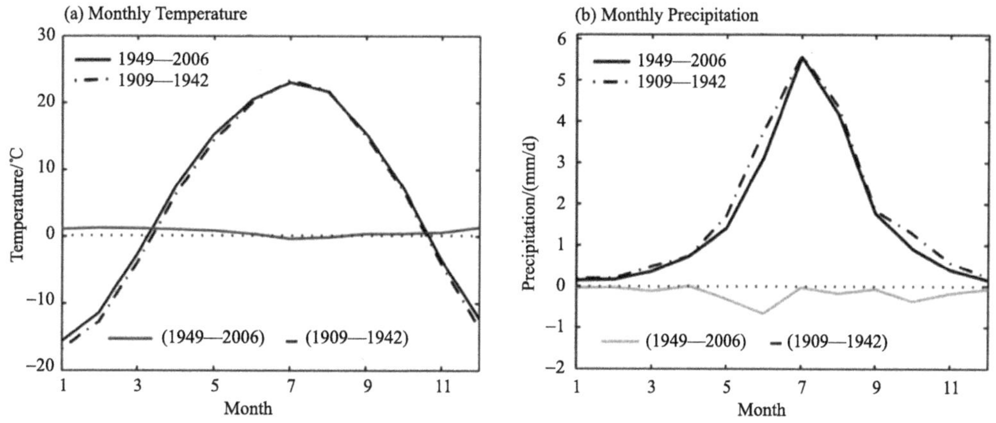

Fig. 3　Climatological means of monthly (a) temperature (℃) and (b) precipitation (mm d^{-1}) over Changchun station for 1909—1942 and 1949—2006. The differences in temperature and precipitation between the two periods are also shown

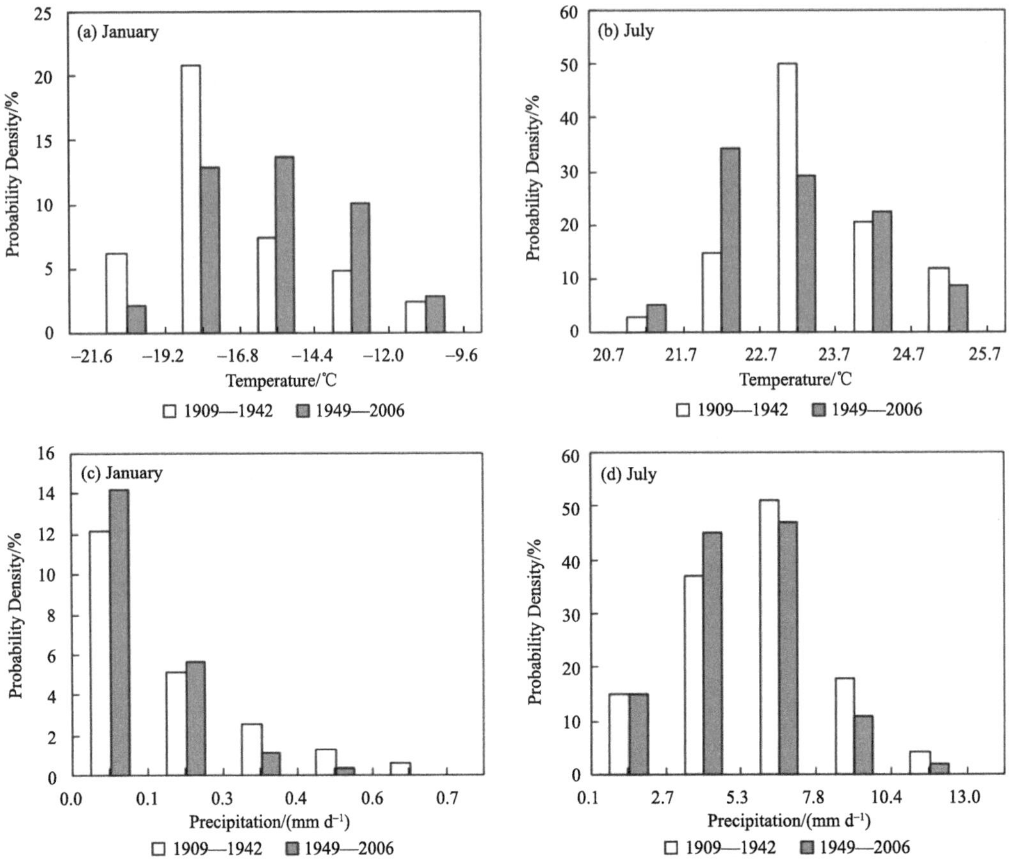

Fig. 4 Distributions of probability density function for temperature and precipitation over Changchun, for both January and July

We apply wavelet transform (Morlet et al. 1982) and other related analyses to reveal the time-frequency characteristics of the variations of temperature and precipitation. Since a distorted edge effect of wavelet spectrum may occur in the transform, especially in the lower frequency bands, as in Yang et al. (2007), who analyzed the time-frequency characteristics of the U. S. Great Plains precipitation, we apply the leap-step time series analysis (LSTSA) model developed by Zheng et al. (2000) to improve the data information of end points. The LSTSA is a nonlinear model. It decomposes a time series into deterministic and stochastic components, and the stochastic component is further characterized by several stochastic models, each of which is valid within a subdomain of the time series.

Fig. 5 shows the estimated time-frequency spectrum from the wavelet transform for temperature. Unsurprisingly, the variability of temperature is absolutely dominated by the annual cycle. Signals on intraseasonal, interannual, and interdecadal time scales can also be observed, despite their smaller mean power. The "first guess" provided by Fig. 5, largely qualitative, offers great efficiency to our next analysis in which the least squares method of Householder transforms (Powell et al., 1969; also see Yang et al., 2007), a linear regression problem, to quantitatively determine the mean magnitude and phase of the strong oscillating signals of temperature variations. We focus on various time scales (see rows 3～7 in Table 1): semiannual, annual, quasi-biennial, and interannual (about 4 and 7.5 years). Since the periods of interannual oscillations are relatively unstable and drifts in frequency may occur in the spectral estimate, we determine their mean values by a method of trial and error in the process of least squares computations. We identify and determine the optimal mean periods of spectral signals by adjusting the periodic values step by step. In the process, the amplitudes and phases of the spectral signals are also estimated and the uncertainties in the period of the signals are measured by the standard deviations of phases.

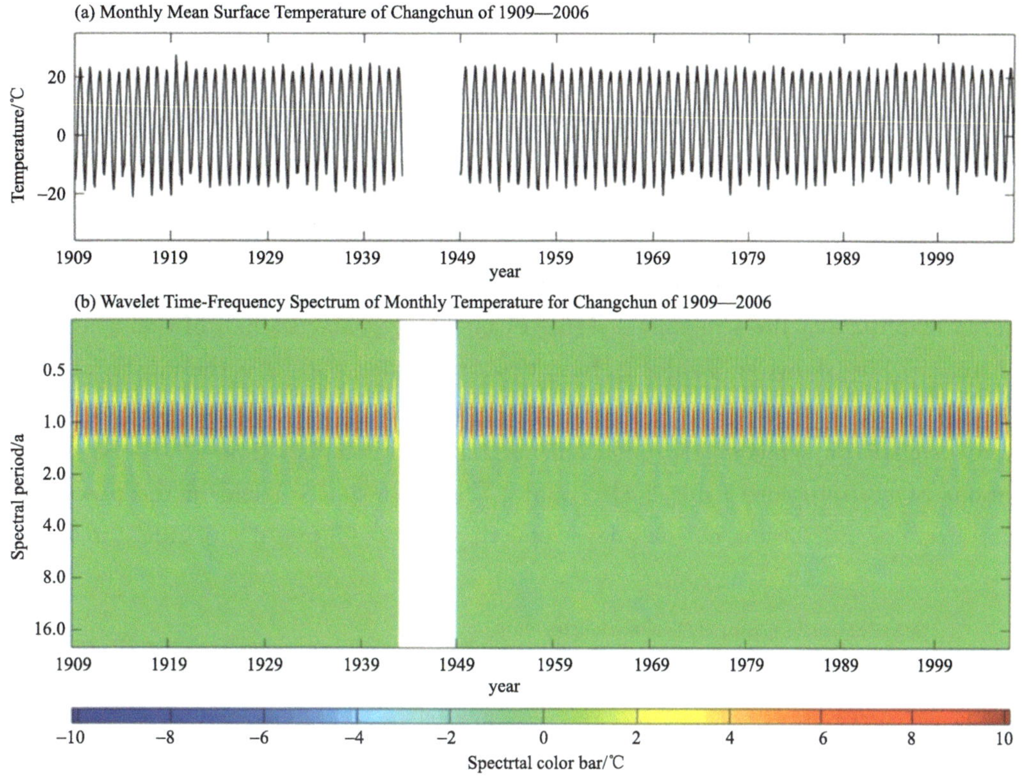

Fig. 5　(a) Monthly mean surface temperature (℃) of Changchun of 1909—2006. (b) wavelet time-frequency spectrum of monthly temperature for Changchun of 1909—2006. The red and blue colors represent the largest positive and negative amplitudes of wavelet spectra, respectively. They coordinate represents the periodic time scales of the time-frequency spectra

As seen in Table 1, the mean amplitude is 19.1 ℃ for the annual variation and 1.9 ℃ for the semi-annual signal. These two time scales have the most stable phase (see the smallest standard deviation of 0.01 in column 6). The table also shows clear signals on quasi-biennial and interannual time scales. Row

Table 1　Parameter estimations of the most apparent signals calculated from the Changchun temperature for 1909—1942 and 1949—2006. The estimated phases are relative to the epoch of January 1909 for data span 1909—1942 and January 1949 for 1949—2006

Temperature in 1909—1942			Temperature in 1949—2006		
Period/a	Amplitude/℃	Phase/a	Period/a	Amplitude/℃	Phase/a
6.90	0.27±0.12	−0.17±0.50	7.47	0.33±0.09	3.20±0.34
4.39	0.25±0.12	−0.60±0.34	3.94	0.38±0.09	−0.50±0.16
1.76	0.46±0.12	0.69±0.08	1.84	0.30±0.09	−0.10±0.09
1.00	19.71±0.12	−0.29±0.01	1.00	19.06±0.09	−0.28±0.01
0.50	1.73±0.12	−0.15±0.01	0.50	1.86±0.09	−0.15±0.01
Linear rate=0.026±0.009 ℃a^{-1}			Linear rate=0.034±0.004 ℃a^{-1}		
RMS=1.77 ℃(14.08 ℃,87%)			RMS=1.76 ℃(13.65 ℃,87%)		

9 indicates that, on average, these oscillatory signals explain 87% of the total variation of the temperature, which is obtained as the ratio of the mean temperature (13.65 ℃) minus root-mean-square (RMS) (1.76 ℃) to the mean. The RMS is calculated from the residual series after the five oscillating signals, and the constant and linear trend terms are removed from the original temperature data. In addition, there exists a positive trend in the temperature, which is 0.26 ℃ decade^{-1} in 1909—1942 but increases to 0.34 ℃ decade^{-1} in 1949—2006 (see column 9). The results discussed earlier are for 1949—2006; however, similar features can also be found for 1909—1942.

We also apply the same analysis to the precipitation over Changchun. Fig. 6 shows that the variability of precipitation is also characterized by a strong annual cycle, although it is not as dominant as the annual cycle of temperature. As in the temperature field, signals of precipitation variations on intraseasonal, interannual, and interdecadal time scales, especially on the intraseasonal time scale, are also apparent. The mean amplitude of the semiannual signal (1 mm d^{-1}) is about a half of that of the annual variation (2.13 mm d^{-1}), as shown in Table 2. These two time scales also have the most stable phase. Compared to temperature, the biennial oscillation of precipitation has a shorter time scale of 1.53 a. Overall, the oscillatory signals on semiannual, annual, biennial, and interannual time scales explain 40% of the total variations of precipitation. Table 2 also shows a small negative trend in the precipitation, which decreases at a rate of -0.03 mm d^{-1} decade^{-1} compared to -0.08 mm d^{-1} decade^{-1} in 1909—1942.

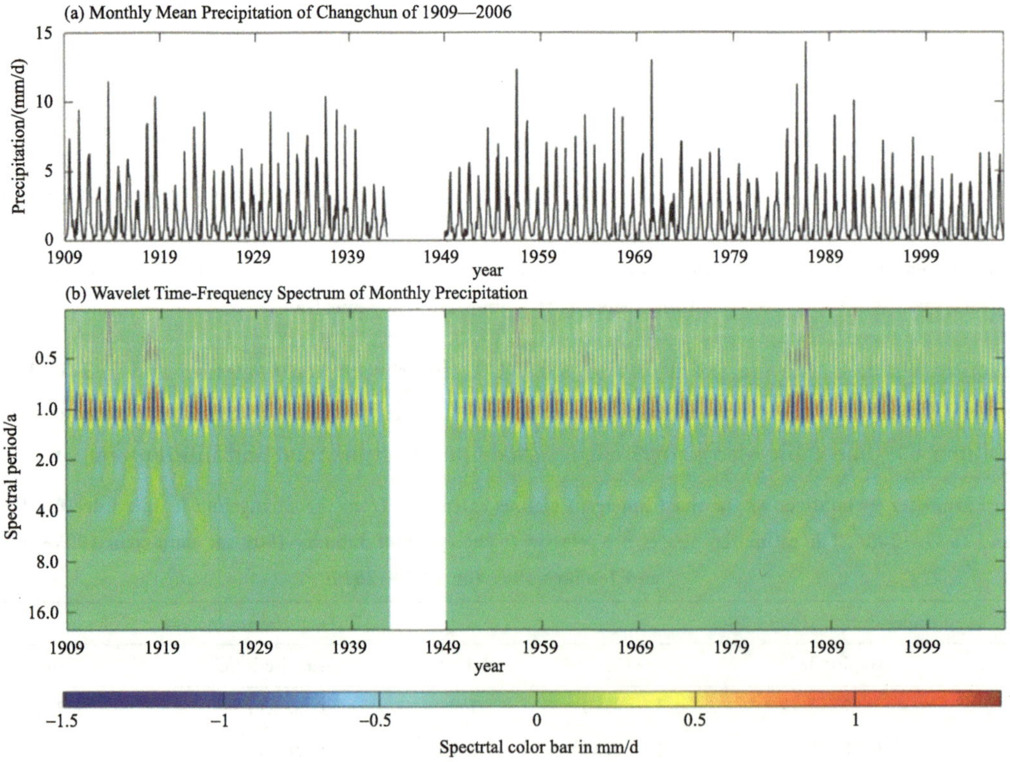

Fig. 6　As in Fig. 5, but for monthly-mean precipitation (mm d^{-1}) over Changchun

Table 2　As in Table 1, but for Changchun precipitation

Precipitation in 1909—1942			Precipitation in 1949—2006		
Period/a	Amplitude/(mm d^{-1})	Phase/a	Period/a	Amplitude/(mm d^{-1})	Phase/a
6.80	0.19±0.09	−3.06±0.51	7.24	0.09±0.07	−0.78±0.86
3.70	0.18±0.09	−1.34±0.29	3.47	0.15±0.07	0.57±0.25
1.76	0.21±0.09	−0.23±0.12	1.53	0.13±0.07	−0.68±0.12
1.00	2.23±0.09	−0.30±0.01	1.00	2.13±0.07	−0.30±0.01
0.50	0.96±0.09	0.08±0.01	0.50	1.00±0.07	0.05±0.01
Linear rate = −0.008±0.006 mm $d^{-1}a^{-1}$			Linear rale = −0.003±0.003 mm $d^{-1}a^{-1}$		
RMS=1.29 mm d^{-1}(2.15 mm d^{-1}, 40%)			RMS=1.25 mm d^{-1}(2.08 mm d^{-1}, 40%)		

4　Relationships with local and large-scale circulation features

4.1　Relationships with local winds

We depict the relationships of temperature and precipitation with local winds by analyzing the cross-correlation function $\rho(\tau)$ in the time domain and the squared coherence spectrum $\gamma^2(f)$ in the frequency domain. As in Jenkins et al. (1968), we compute these values between the two time series as

$$\rho(\tau) = \sigma_{12}/(\sigma_{11}\sigma_{22})^{1/2} \tag{1}$$

$$R(f) = S_{12}(f)/[S_{11}(f)S_{22}(f)]^{1/2} \tag{2}$$

$$\gamma^2(f) = | R(f) |^2 \tag{3}$$

In Eq. (1), σ_{12} is the cross-covariance function of phase lag τ, and σ_{11} and σ_{22} are the variances of the two time series. In Eqs. (2) and (3), f is the frequency, $S_{12}(f)$ is the cross-power spectrum between the two time series, and $S_{11}(f)$ and $S_{22}(f)$ are the auto-power spectra of the two series, respectively. The multiwindow spectrum technique of Thomason (1982) is used in the calculations of the power spectrum, with the application of the Fourier transform. These analysis methods have also been applied by Ding et al. (2002) to analyze the variations of surface temperature over Hong Kong. Fig. 7 shows the cross correlation and cross coherence between temperature and the zonal and meridional winds. (The constant and linear trend terms have been removed to meet the principle of statistics for correlation and coherence calculations.) Figs. 7a and 7b reveal significant correlations between temperature and the winds. Temperature increases under the influences of easterly and southerly winds and decreases under the influences of westerly and northerly winds. These correlations significantly exceed the threshold value at the 99% confidence level determined by the Monte Carlo test, especially the relationship between temperature and zonal wind. The lag relationships also indicate that meridional wind leads the change in temperature, which is followed by a change in the zonal wind, likely related to the meridional temperature gradient. As indicated by Figs. 7c and 7d, both the zonal and meridional winds have the strongest relationships on both annual and semiannual time scales. Apparently, the relationships on these time scales significantly exceed the threshold value at the 99% confidence level. Particularly for the relationship between temperature and meridional wind (Fig. 7d), the value of coherence for the semiannual time scale is clearly larger than for the annual signal.

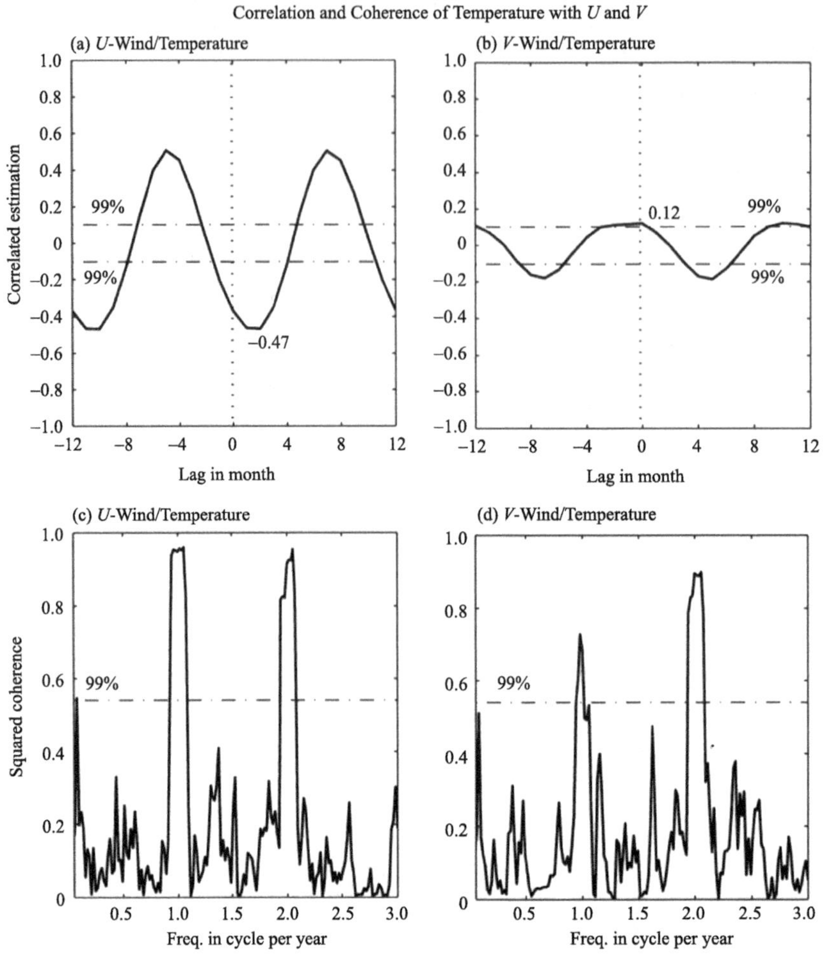

Fig. 7　Estimated (a),(b) cross correlation and (c),(d) squared coherence of Changchun temperature with zonal and meridional winds. In (a),(b),negative (positive) values shown in the x coordinate represent the correlations in which winds lead (lag) temperature. The dashed lines show the threshold values of significant test (assessed two-sided) at the 99% confidence level

A similar analysis is also applied to precipitation. Figs. 8a and 8b show that precipitation over Changchun increases when easterly and southerly winds prevail and decreases under the influences of westerly and northerly winds. The meridional wind is more significantly correlated with precipitation than with temperature. It leads to change in precipitation,indicating the importance of water vapor supply for precipitation over NEC. Like temperature,precipitation has significant relationships with the zonal and meridional winds on both annual and semiannual time scales (Figs. 8c and 8d). The strongest relationship is found on the annual time scale for precipitation and zonal wind but on the semiannual time scale for precipitation and meridional wind. Nevertheless,the relationships of winds with temperature and precipitation become weaker after the annual and semiannual time scales are removed.

4.2　Relationships with large-scale atmospheric and SST patterns

To examine the relationships of temperature and precipitation over Changchun with large-scale atmospheric and SST patterns,we exclusively analyze the features for the time scales at which the variations of temperature and precipitation exhibit large signals and have significant relationships with the local winds. That is,we focus only on the annual and semiannual time scales. First,we apply the multistage

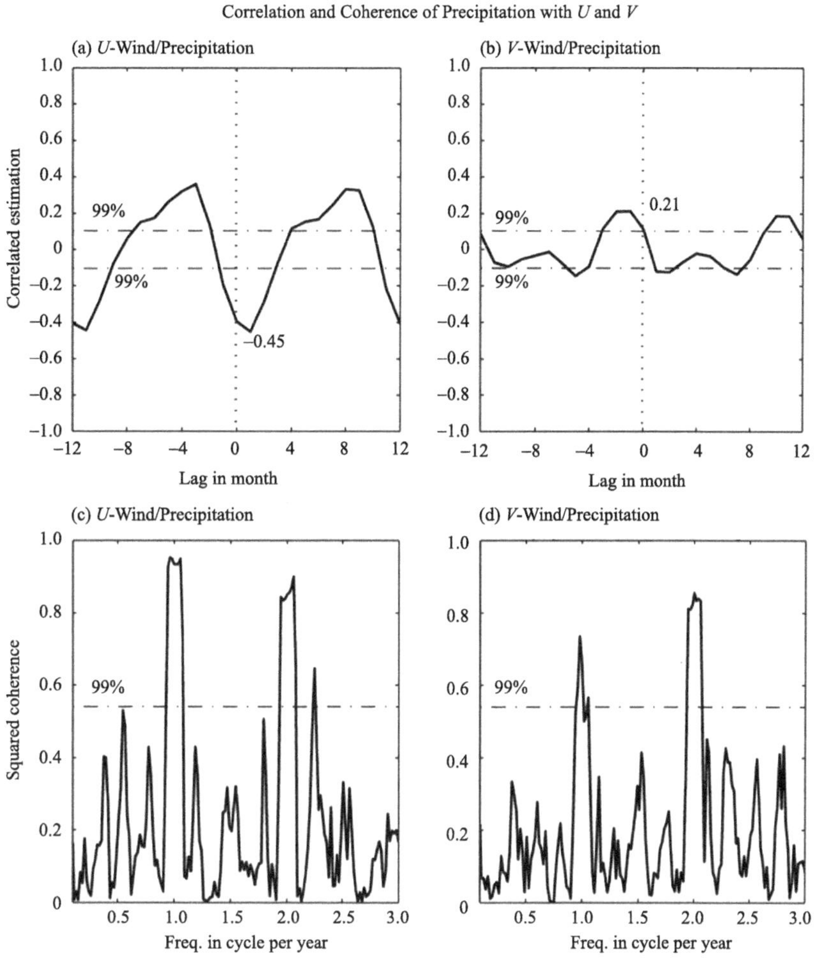

Fig. 8　As in Fig. 7, but for the correlation and coherence of precipitation with zonal and meridional winds

filter (MSF) developed by Zheng et al. (1986) to extract the signals in the windows of 10—14 months for the annual time scale and 5—7 months for the semiannual time scale. The theoretical formula for the frequency response function R of the MSF is

$$R = c[1 - A(f, e)^L]^M \tag{4}$$

In Eq. (4), c is a real constant, taken as 1 generally; L and M are positive integers, determined by the given bandwidth of truncated frequencies; and $A(f, e)$ is the frequency response in the smoothing method of Vondrák (1977):

$$A(f, e) = [1 + e^{-1}(2\pi f)^6]^{-1} \tag{5}$$

where f and e are the corresponding frequency component and the smoothing factor, respectively. The frequency response function of the MSF can be characterized by a small bandwidth of truncated frequency, so it can be treated as narrow-band filtering and improve signal resolutions, especially in separating the signals of close frequencies in data series. Thus, the MSF provides higher filtering resolutions of data series compared to the ordinary band pass filters, since it has a feature of more narrowly truncated frequency band in the frequency response function. In the computations, the LSTSA model is applied to improve the data information of end points to reduce the edge effects on the filtered output signals.

Then, we analyze the relationships of regional temperature and precipitation with 850 hPa circulation and SST patterns. Fig. 9a shows that, on the annual time scale, an increase in temperature is related to an anticyclonic pattern over the extratropical Pacific and a cyclonic pattern over Asia and the northern

Indian Ocean. Changchun is located between these two circulation patterns. An increase in temperature is also related to warming (cooling) over the Northern (Southern) Hemisphere and strong westerly flow over southern Asia, indicating a strong summer monsoon circulation. The figure also implies that in winter, the strong East Asian monsoon decreases the temperature over northeast China (Wang et al., 2010).

On the semiannual time scale (Fig. 9b), the temperature increases when Changchun is controlled by an anticyclonic pattern whose center is over the warm waters northeast of Japan. In the meantime, a cyclonic pattern appears over Southeast Asia and the subtropical western Pacific, centered over 20°N, 135°E and accompanied by low SST. Fig. 9b shows that westerly flow weakens over southern Asia and the northern Indian Ocean when the temperature increases over Changchun, suggesting the importance of extratropical atmospheric processes for affecting the variations of NEC temperature on semiannual time scale.

It can be seen from Fig. 10 that on the annual time scale, the large-scale circulation patterns associated

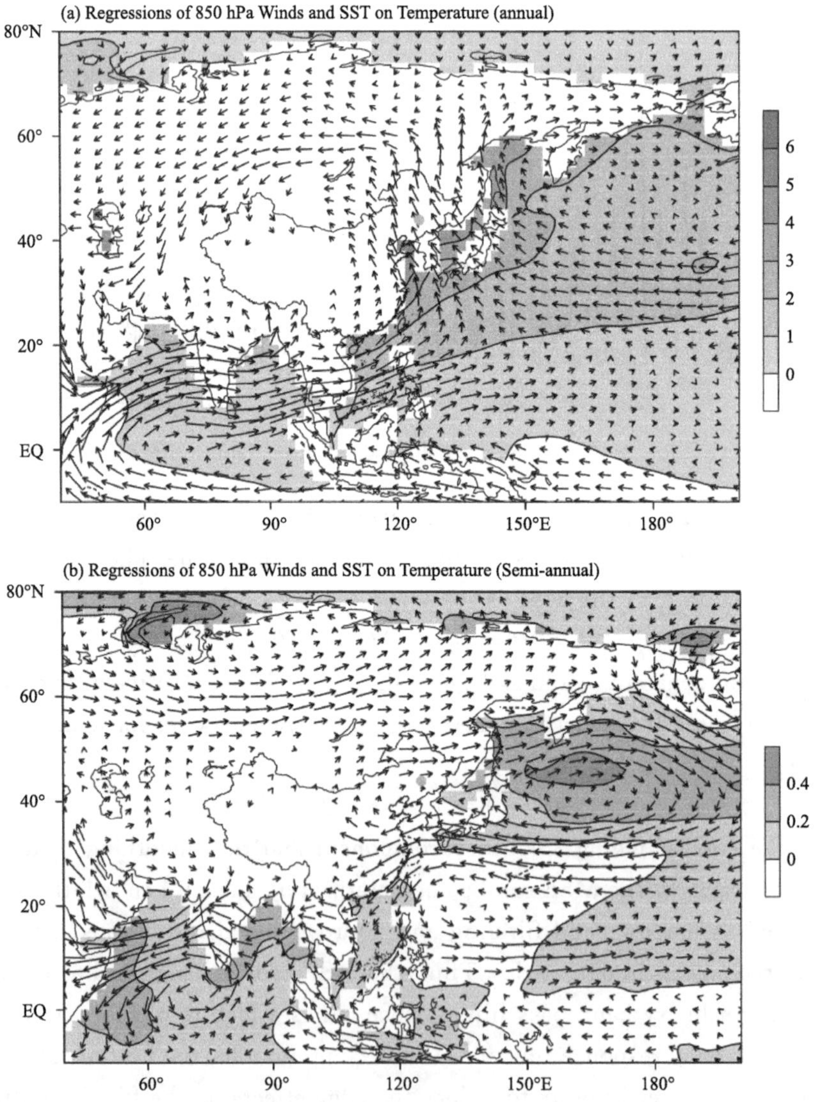

Fig. 9 (a) Regressions of 850 hPa winds (m s^{-1}; vectors) and SST (℃; shadings) against the annual signals of Changchun temperature for the period of 1949—2006; (b) As in (a), but for the semiannual signals of Changchun temperature

with an increase in precipitation (Fig. 10a) are similar to those associated with an increase in temperature (Fig. 9a). An increase in precipitation is linked to the intensification of southeasterly flow related to the anomalous anticyclonic over the extratropical North Pacific and to the anomalous cyclonic pattern over the Asian continent and the southwesterly flow over southern Asia and the northern Indian Ocean. An increase in precipitation is also linked to the apparent warming in the subtropical—extratropical western Pacific.

On the semiannual time scale, an increase in precipitation is more directly associated with the water vapor supply from the Sea of Japan and from the east (Fig. 10b). This moisture transportation is linked to an anomalous cyclonic pattern over the extratropical North Pacific, where SST decreases. Fig. 10b also suggests that the intensification of the large-scale Asian summer monsoon circulation increases the precipitation over NEC. Note also the apparent cooling in the Indian Ocean, driven by the strong monsoon flow.

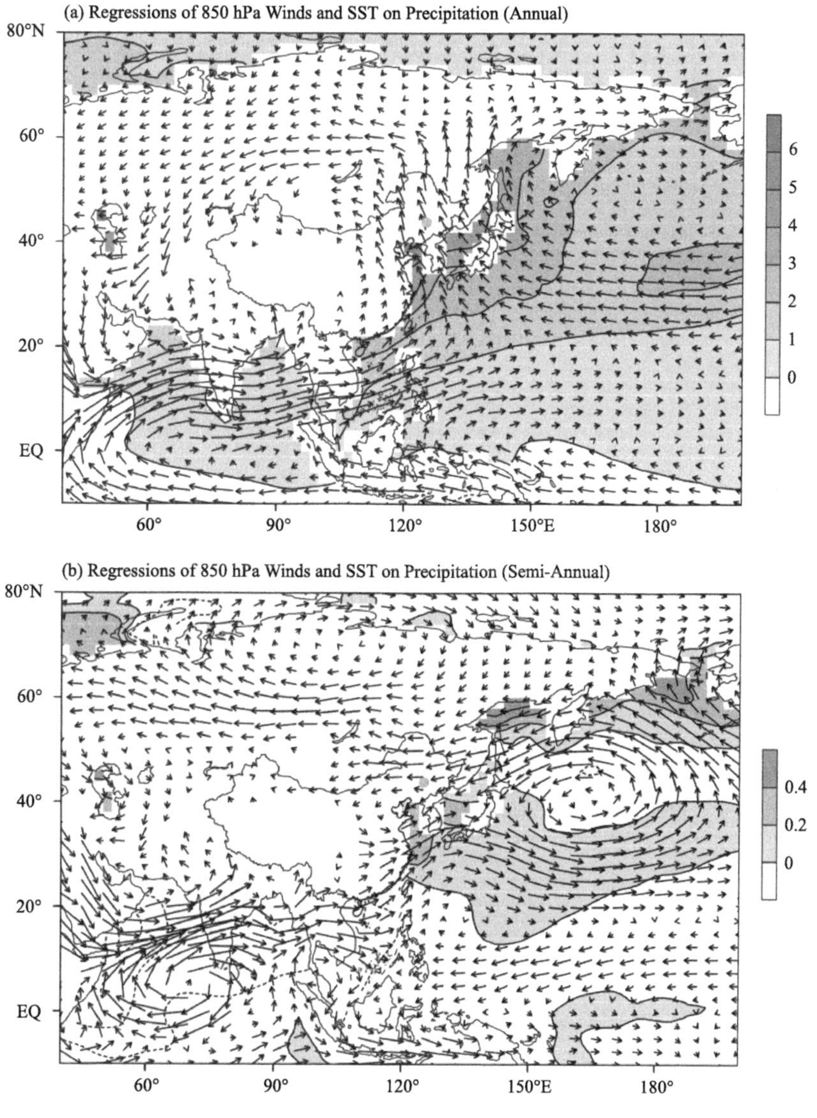

Fig. 10　As in Fig. 9, but for the regressions against Changchun precipitation

5 Relationships with NPO

Fig. 11 shows the time series of NPO and its estimated time-frequency spectra of wavelet transform. The variations of NPO at various frequencies are mostly unstable with time. For example, the spectrum of the annual time scale before the mid-1960s is larger than that afterward. The largest spectrum appears on the time scales of months. Least squares analysis indicates that, on average, the amplitude of NPO is largest and most stable on the semiannual time scale (see Table 3), followed by the annual time scale. The amplitude of the annual time scale is about a half of the amplitude of the semiannual time scale. A 9 a oscillation of NPO is also found. On this decadal time scale, the NPO is strongly related to the Pacific decadal oscillation.

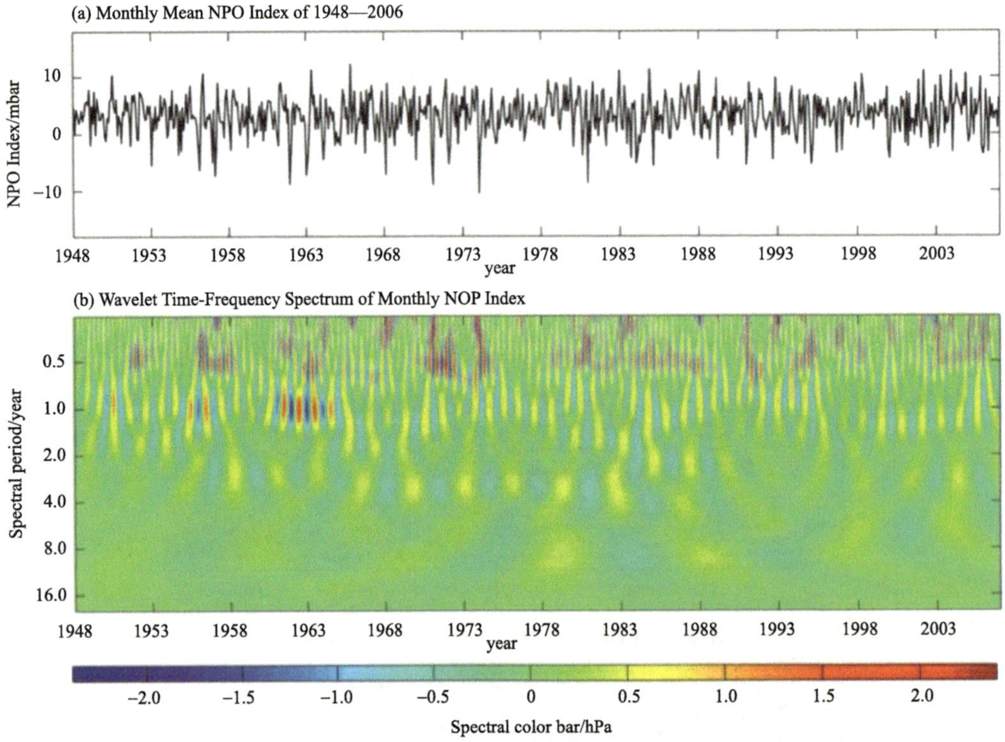

Fig. 11　(a) Index of monthly mean NPO; (b) Estimations of the time-frequency spectra of wavelet transform for NPO

Table 3　Parameter estimations of the most apparent signals calculated from NPO for 1948—2006. The estimated phase is relative to the epoch of January 1948

NPO for 1948—2006		
Period/a	Amplitude/hPa	Phase/a
9.28	0.51±0.17	−0.71±0.50
4.21	0.34±0.17	−0.10±0.34
1.82	0.22±0.17	0.44±0.23
1.00	0.76±0.17	−0.21±0.04
0.50	1.54±0.17	−0.20±0.01

Fig. 12 shows the cross correlation and cross coherence between the Changchun temperature and precipitation and the NPO index. In the computations, the constant and linear trend terms have been removed as required by the statistical analysis methods. The mean annual cycles have also been removed from the original data to focus on the features of other time scales. There is a highly strong relationship between the Changchun temperature and NPO. The temperature increases when NPO is positive and the correlation between the two time series significantly exceeds the 99% confidence level (Fig. 12a). Clearly, the largest coherence appears in the semiannual frequency band (Fig. 12c). There is also a significant relationship between the Changchun precipitation and NPO (Fig. 12b), although it is weaker than the temperature-NPO relationship. The correlation between the two, which indicates that precipitation decreases when NPO is positive, also significantly exceeds the 99% confidence level. The strongest precipitation-NPO coherence appears in the frequency band of 5—7 months, the semiannual time scale (Fig. 12d).

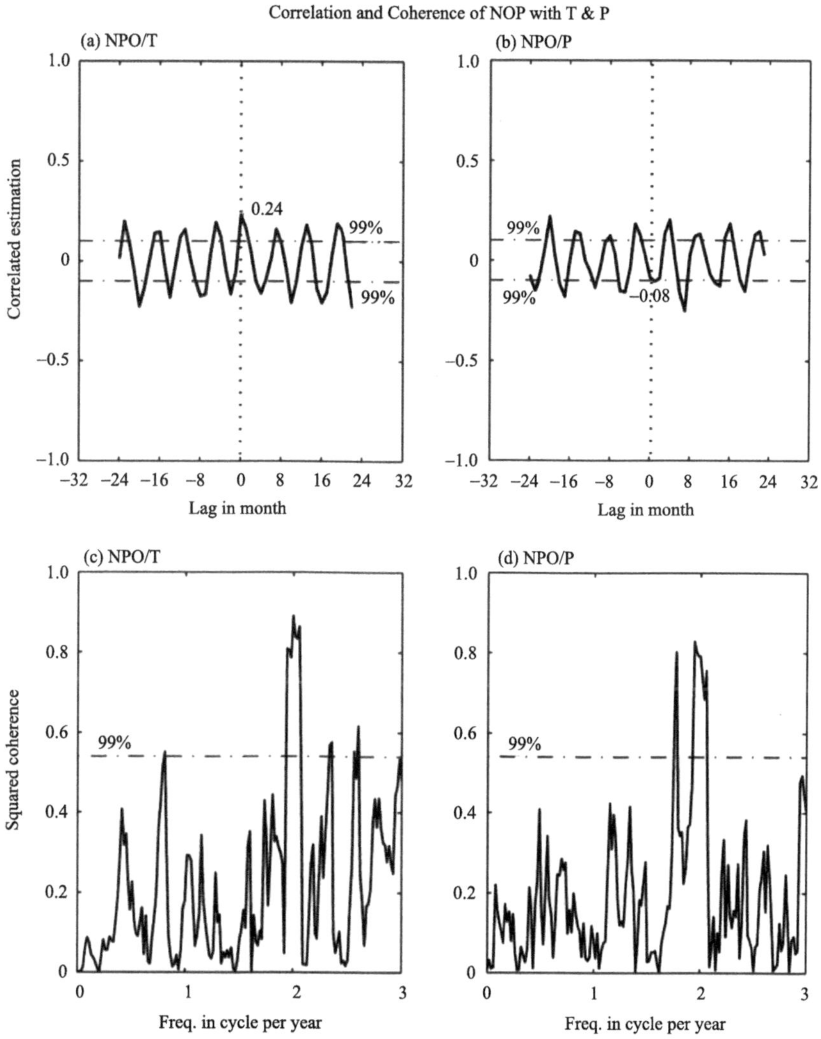

Fig. 12　Estimated (a),(b) cross correlation and (c),(d) squared coherence of Changchun temperature and precipitation with NPO. In (a),(b), negative (positive) values shown in the x coordinate represent the correlations in which NPO leads (lags) temperature or precipitation. The dashed lines show the threshold values of significant test (assessed two-sided) at the 99% confidence level. The mean annual cycles have been removed from the original data

It is known that the cross-coherence estimations shown in Fig. 12 only reveal the mean features of the entire time span, and that the estimations are unable to depict the stability and variability of these features in specific frequency bands with respect to the time process over the data span. Following Yang et al. (2007), we further explore the coherence features between NPO and Changchun temperature/precipitation in both time and frequency domains by applying a technique in which the coherence in time-frequency domains is derived by first estimating the values of a subseries of 10 a and then successively moving the data points for subsequent estimations. Fig. 13a shows that significant coherence between NPO and Changchun temperature appears mainly in the higher frequency bands. In particular, large coherence estimates appear in the frequency bands up to 7 months. They are unstable (varying with time) and have been large since the 1980s but small before that. This analysis thus manifests the advantage of the method in depicting the detailed features of the relationship between Changchun temperature and NPO.

A similar analysis for NPO and Changchun precipitation also reveals a strong relationship in the high-frequency hands and large (small) coherence since (before) the 1980s (Fig. 13b). The largest coherence appears in 1981—1992, mainly in the frequency band of 6—9 months. In addition, the large coherence is also seen on interannual time scales, including the biennial time scale from the late 1960s to the early 1970s.

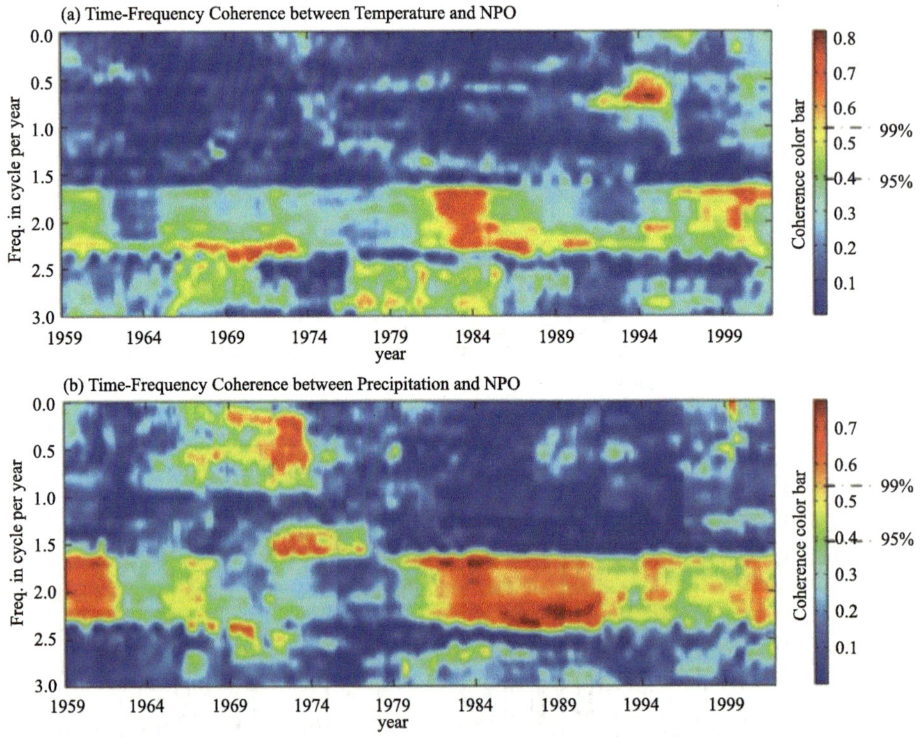

Fig. 13 (a) Estimations of coherence spectra between Changchun temperature and NPO shown in time-frequency domain; (b) Estimations of coherence spectra between Changchun precipitation and NPO. The threshold values of the significant test at the 95% and 99% confidence levels are given by the dashed-dotted lines in the color bar on the right-hand side. Note that the mean annual cycles have been removed from the original data and that a 5 a truncation occurs at each end of the figure

We further depict the large-scale atmospheric circulation and SST patterns associated with the temporally varying feature of the large coherence signals shown in Fig. 13. It can be seen from Fig. 14, which shows the mean patterns of 850 hPa winds and SST for 1968—1979 and their differences between 1981 and 1992 and 1968 and 1979, that atmospheric circulation and SST patterns exhibit many different features between the two periods [the selection of the two periods also addresses the potential problem of the NCEP-NCAR reanalysis, over Asia and prior to 1968, as demonstrated by Yang et al. (2002)]. As shown in the mean patterns for 1981—1992, NEC, including Changchun, is clearly affected by the East Asian trough (Fig. 14a). The East Asian monsoon circulation also influences the NEC climate. In the difference field (Fig. 14b), a cyclonic pattern appears over NEC, suggesting that the trough was stronger during 1981—1992 than during 1968—1979. Given the significant relationship between NPO and NEC climate and recalling the strong semiannual signals in temperature, precipitation, NPO, and their interrelationships, a strong semiannual component of the variation of the East Asian trough can be expected. The anomalous flow over East Asia indicates the stronger East Asian winter monsoon or the weaker summer monsoon in 1981—1992. It can also be seen from Fig. 14b that the easterly trade wind over the tropical Pacific and the westerlies over the Indian Ocean and tropical Asia was weaker in 1981—1992, associated with a general increase in SST. However, no apparent feature can be found in the change of SST over the extratropical northwestern Pacific Ocean.

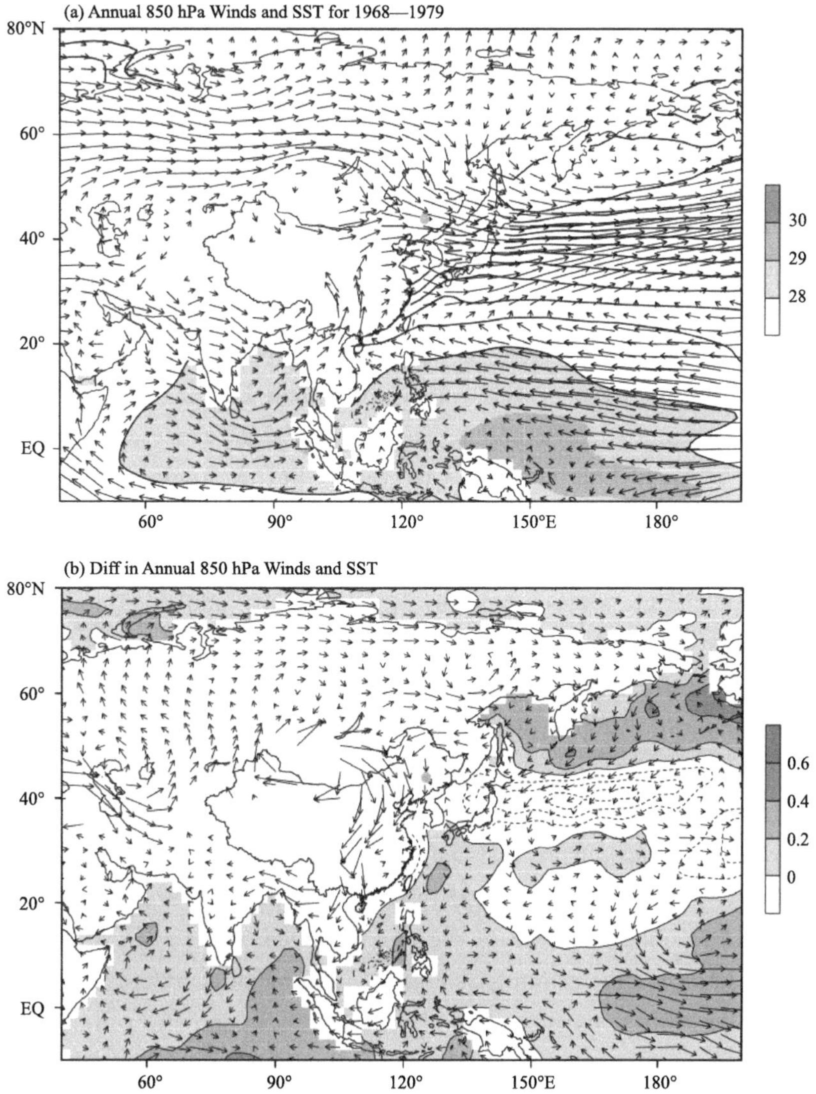

Fig. 14 (a) Mean patterns of annual SST (℃; shadings) and 850 hPa winds (m s^{-1}; vectors) for 1968—1979; (b) Differences in annual SST and 850 hPa winds (1981—1992 minus 1968—1979)

6 Summary and further discussion

In this study, we have applied several advanced analysis tools to depict the time-frequency characteristics of the variations of temperature and precipitation over the city of Changchun in northeast China and their associations with large-scale atmospheric and oceanic conditions. The analysis methods include wavelet transformation, least squares analysis, coherence analysis, and regression analysis. A multistage filter and a leap-step time series analysis technique have also been applied.

One of the most prominent features of the variations of Changchun temperature and precipitation is the strong semiannual signals. For example, the amplitude of the semiannual signal of precipitation is about half of that of the annual cycle. The relationships of Changchun temperature and precipitation with local winds and with large-scale atmospheric circulation and SST patterns on the semiannual time scale are also strong, only second to the strong relationships on the annual time scale.

On the annual time scale, the NEC climate is affected by both the thermal contrast between the Asian continent and tropical Indo Pacific Oceans and the difference in thermal condition between the continent and the extratropical North Pacific. Increases in temperature and precipitation are associated with cyclonic and anticyclonic patterns over the Asian continent and the North Pacific, respectively, and with strong southerly flow over East Asia and the northwestern Pacific. Clearly, a decrease (increase) in winter temperature is linked to the intensification (reduction) of the East Asian winter monsoon, and a decrease (increase) in summer precipitation is related to the weakening (strengthening) of the Asian summer monsoon circulation. On the semiannual time scale, the variability of NEC climate is mainly affected by the large-scale circulation pattern centered over the North Pacific, with the western portion over northeast China, North and South Korea, and Japan. Under the influence of the cyclonic (anticyclonic) pattern, precipitation over Changchun increases (decreases) but the regional temperature decreases (increases). While temperature may be mainly affected by the extratropical process, the variability of NEC precipitation is also related to the tropical Asian monsoon circulation on the semiannual time scale. On both annual and semiannual time scales, increases in temperature and precipitation are associated with warming in the subtropical extratropical northwestern Pacific Ocean.

Strong relationships occur between Changchun temperature/precipitation and NPO, especially in the frequency band up to 7 months. Temperature increases and precipitation decreases when NPO is positive. These relationships were weak before 1980; however, they become stronger afterward, associated with the strengthening of the East Asian trough.

It should be pointed out that our analysis is mainly for Changchun. Although we have shown that the temperature and precipitation are highly correlated with those over much of northeast China (see Figs. 1 and 2), we also carry out an analysis of the area-averaged temperature and precipitation over NEC (79 stations). The variations of Changchun temperature and precipitation are similar to those of NEC-averaged temperature and precipitation, consistent with the features shown in Figs. 1 and 2. For example, the variations of NEC temperature and precipitation are also characterized by the two strongest and most stable signals on annual and semiannual time scales. In particular, the amplitude of NEC temperature signal is 18.97 and 1.87 ℃ for the annual and semiannual time scales, respectively, comparable with 19.06 and 1.86 ℃ for the Changchun temperature. Similarly, the amplitude of NEC precipitation signal is 2.09 and 0.98 mm d^{-1} for the annual and semiannual time scales, comparable with 2.13 and 1.00 mm d^{-1} for the Changchun counterpart. Fig. 15 presents the cross correlation and cross coherence between the temperature and precipitation over NEC (79 stations) and the NPO index. It shows features similar to those presented in Fig. 12 for Changchun temperature and precipitation. Namely, there are highly strong relationships between NEC temperature and NPO (Fig. 15a) and between NEC precipitation and NPO (Fig. 15b). Temperature increases and precipitation decreases when NPO is positive. As shown in Fig. 12, the largest coherences also appear in the semiannual frequency bands (Figs. 15c and 15d), although significant coherences can also be found on intraseasonal time scales.

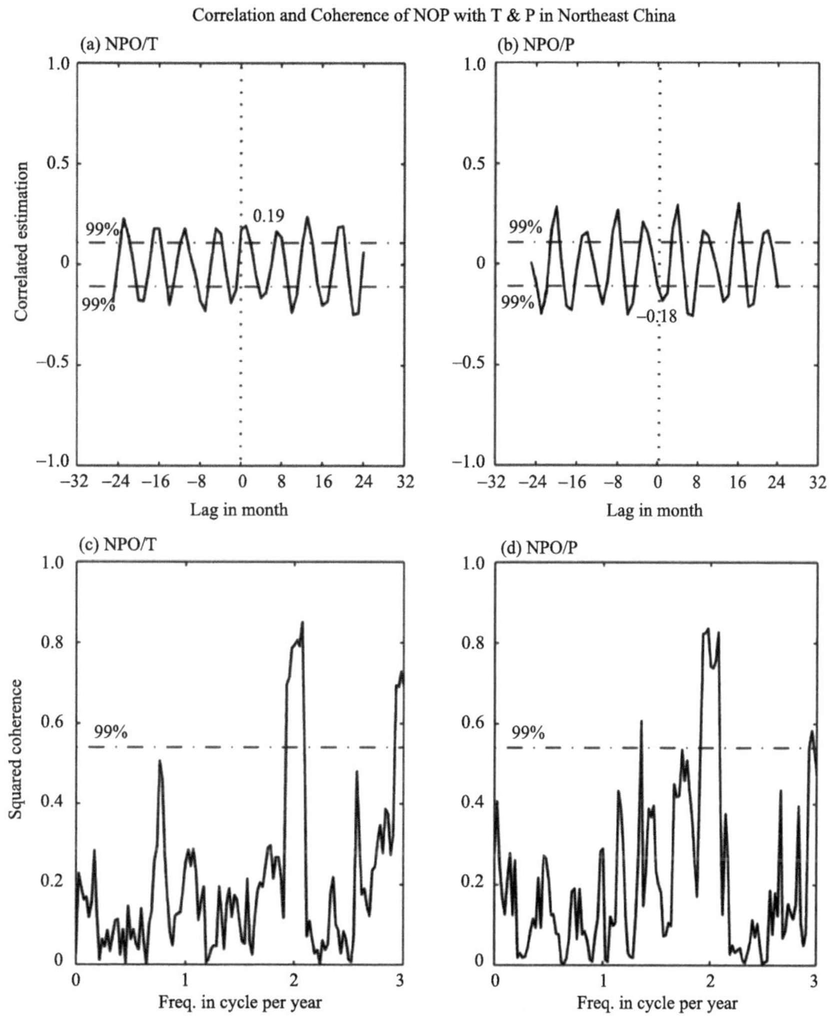

Fig. 15 Estimated (a), (b) cross correlation and (c), (d) squared coherence of temperature and precipitation over NEC (79 stations) with NPO. Values of negative (positive) lags shown in the x coordinate represent the correlations in which temperature or precipitation leads (lags) NPO. The dashed lines show the threshold values of significant test (assessed two-sided) at the 99% confidence level. The mean annual cycles have been removed from the original data

However, even though strong signals of semiannual time scale appear from the variations of NEC temperature/precipitation and NPO and the relationships between them, we should not claim that NPO exerts an influence on the regional climate at the particular time scale. It is possible that the variations of both the regional climate and NPO are associated with the variability of other climate patterns on the preferred time scale. The physical mechanisms responsible for these semiannual signals and the application of these signals in the prediction of the regional climate deserve further investigations. It should also be pointed out that in this study, we have depicted the features of NEC climate and its relationship with

NPO on a wide range of time scales with a focus on the semiannual time scale, instead of the interannual time scale that has been studied extensively. The features shown in this study should vary from season to season. For example, we have repeated the analysis of Fig. 12 using the seasonally stratified data for winter (December—February), spring (March—May), summer (June—August), and fall (September—November) and found that NPO is positively and significantly correlated with Changchun temperature in winter and summer. The correlation between NPO and Changchun seasonal-mean precipitation and the correlation between NPO and Changchun temperature in other seasons are insignificant. Nevertheless, the application of seasonally stratified data to some of the analyses, such as the technique shown in Fig. 13, requires much longer data records.

Acknowledgments. We are thankful for the helpful reviews by editor James Renwick and four anonymous reviewers. This study was partially supported by the Governor's Foundation of Jilin Province, China.

References

DING S C, 1980. The climatic analysis of low temperature in summer over the Northeast China and influence for agricultural product[J]. Acta Meteor Sinica, 38(3): 234-242. (in Chinese)

DING X L, ZHENG D W, YANG S, 2002. Variations of the surface temperature in Hong Kong during the last century[J]. Int J Clim, 22(6): 715-730.

JENKINS G M, WATTS D G, 1968. Spectral Analysis and Its Applications[M]. San Francisco: Holden-Day: 525.

KALNAY E, KANAMITSU M, KISTLER R, et al, 1996. The NCEP/NCAR 40-Year Reanalysis Project[J]. Bull Amer Meteor Soc, 77(3): 437-471.

LI Q, YANG S, KOUSKY V E, et al, 2005. Features of cross-Pacific climate shown in the variability of China and US precipitation[J]. Int J Clim, 25(13): 1675-1696.

LIAN Y, AN G, 1998. The relationship among East Asian summer monsoon, El Niño and low temperature in Songliao Plains Northeast China[J]. Acta Meteor Sinica, 56(6): 724-735. (in Chinese)

LIU S, WANG N, 2001. The impacts of antecedent ENSO event on air temperature over Northeast China in summer[J]. J Trop Meteor, 17(3): 314-319. (in Chinese)

MORLET J, ARENS G, FOURGEAU E, et al, 1982. Wave propagation and sampling theory—Part I: Complex signal and scattering in multilayered media[J]. Geophysics, 47(2): 203-221, doi: 10.1190/1.1441328.

POWELL M J D, REID J K, 1969. On applying Householder transformations to linear least squares problems[M]//A. J. H. Morrell, Ed. Mathematics, Software, Vol. 1, Information Processing 68: Proceedings of IFIP Congress 1968. Edinburgh North-Holland Publishing Company: 122-126.

SMITH T M, REYNOLDS R W, 2003. Extended reconstruction of global sea surface temperatures based on COADS data (1854-1997)[J]. J Clim, 16(10): 1495-1510.

SUN F H, YUAN J, LU S, 2006a. The change and test of climate in Northeast China over the last 100 years[J]. Clim Environ Res, 11(1): 101-108. (in Chinese)

SUN F H, YANG X Q, LU S, et al, 2006b. The contrast analysis on the average and extreme temperature trend in Northeast China[J]. J Sci Meteor Sinica, 26(2): 157-163. (in Chinese)

SUN J Q, WANG H J, 2006c. Regional difference of summer air temperature anomalies in Northeast China and its relationship to atmospheric general circulation and sea surface temperature[J]. Chinese J Geophys, 49(3): 662-671. (in Chinese)

SUN L, AN G, LIAN Y, et al, 2000. A study of the persistent activity of northeast cold vortex in summer and its general circulation anomaly characteristics[J]. Acta Meteor Sinica, 55(6): 70-82. (in Chinese)

SUN L, AN G, LIAN Y, et al, 2002. The unusual characteristics of general circulation in drought and flooding years of Northeast China[J]. Clim Environ Res, 7(1): 102-113. (in Chinese)

THOMASON D J, 1982. Spectrum estimation and harmonic analysis[J]. Proc IEEE, 70(9): 1055-1096.

VONDRÁK J, 1977. Problem of smoothing observational data Ⅱ [J]. Bull Astron Inst, 28(1): 83-93.

WALLACE J M, GUTZLER D S, 1981. Teleconnections in the geopotential field height during the Northern Hemisphere winter[J]. Mon Wea Rev, 109(4): 784-812.

WANG H J, SUN J Q, LANG X M, et al, 2008. Some new results in the research of the international climate variability and short-term climate prediction[J]. Chinese J Atmos Sci, 32(4): 806-814. (in Chinese)

WANG L, CHEN W, 2010. How well do existing indices measure the strength of the East Asian winter monsoon? [J] Adv Atmos Sci, 27(4): 855-875.

YANG S, LAU K M, KIM K M, 2002. Variations of the East Asian jet stream and Asian-Pacific-American winter climate anomalies[J]. J Clim, 15(3): 306-325.

YANG S, DING X, ZHENG D, et al, 2007. Depiction of the variations of Great Plains precipitation and its relationship with tropical central-eastern Pacific SST[J]. J Appl Meteor Clim, 46(2): 136-153.

YANG S Y, SUN F H, MA J Z, 2008. Evolvement of precipitation extremes in Northeast China on the background of climate warming[J]. Sci Geographica Sinica, 28(2): 224-228. (in Chinese)

ZHENG D W, DONG D N, 1986. Realization of narrow band filtering of the polar motion data with multistage filter[J]. Acta Agron Sinica, 27(4): 368-376.

ZHENG D, CHAO B F, ZHOU Y, et al, 2000. Improvement of edge effect of the wavelet time-frequency spectrum: Application to the length-of-day series[J]. J Geod, 74(2): 249-254.

ZHANG Y C, ZHANG L J, 2005. Precipitation and temperature probability characteristics in climatic and ecological transition zone of Northeast China in recent 50 years[J]. Sci Geographica Sinica, 25(5): 561-566. (in Chinese)

初夏东北冷涡活动异常与北半球环流低频变化*

廉毅[1] 布和朝鲁[2,3] 谢作威[2] 沈柏竹[1] 李尚锋[1]

(1 中高纬度环流系统与东亚季风研究开放实验室,长春 130062;2 中国科学院大气物理研究所国际气候与环境科学中心,北京 100029;3 中国科学院大气物理研究所大气科学和地球流体力学数值模拟国家重点实验室,北京 100029)

摘　要:在典型个例分析的基础上,揭示了初夏东北冷涡活动显著异常与北半球环流低频变化之间的联系及其可能机理。主要结果为:(1)初夏东北冷涡活动显著异常与上游乌拉尔山附近和下游日本附近的持续性异常环流相对应。冷涡异常多年,在向下游频散 Rossby 波能量和瞬变涡动强迫过程的共同作用下,乌拉尔山附近盛行阻塞型环流,它一方面使极地冷空气向东北地区移动,另一方面通过 Rossby 波能量的传播,使东北冷涡活动加强,冷涡异常少年的环流情况则基本相反。(2)谐波分析和低频振荡分析表明,在冷涡异常多(少)年的 5—6 月,东北地区正(负)涡度位相占优,这与北太平洋超长波槽西退(东进)相对应。长波在北太平洋—北美—大西洋地区呈驻波型,与 Rossby 波能量传播匹配,亦有利于乌拉尔山附近的异常环流得以长时间维持。(3)5 月,乌拉尔山附近的异常环流信号对初夏东北冷涡活动的显著异常具有前兆性意义。

关键词:东北冷涡;低频变化;谐波分析;Rossby 波

1　引言

初夏(6 月)的大气环流往往具有突变性,表现为东亚高空急流的突然向北推进及夏季风的开始暴发(叶笃正 等,1958)。因此,东亚夏季风的暴发和入梅等现象在这一时段一直备受关注。关于中高纬度大气环流的研究大多集中在东亚阻塞形势如何影响中国天气气候方面(汤懋苍,1957;陈汉耀,1957;毕慕莹等,1992;张庆云 等,1998),对与东亚阻塞高压相伴随的东北高空冷涡(切断低压)方面的研究则较少。

陶诗言(1980)指出,东北冷涡型是造成中国东北地区和华北北部暴雨或雷阵雨的环流形势。孙力等(1994)研究了东北冷涡的时空分布特征,并探讨了东北冷涡与东亚阻高的联系。孙力(1997,1998)从能量学的角度揭示了东北冷涡发生发展的天气过程,并讨论了大气低频振荡和瞬变扰动的位涡输送过程在东北冷涡活动中的作用。沈柏竹等(2008)对多个 6 月东北冷涡活动正常和异常年的环流进行谐波分析,并总结了沿 45°N 北半球超长波和长波活动的典型特征。布和朝鲁等(2008)发现,初夏东北低压的出现与中高纬度 Rossby 波活动密切相关。

近年来,引发暴雨的东北冷涡系统及其盛衰过程受到高度关注。乔枫雪等(2007)通过个例分析,再次揭示了东北冷涡暴雨过程与中高纬度亚洲阻塞形势的密切联系。张云等(2008)认为,东北冷涡在其衰退阶段也可引发暴雨。

尽管一次东北冷涡活动的时间尺度为天气(或中期天气)尺度,但持续和频繁的东北冷涡活动具有显著的"气候效应",不仅对东亚梅雨期降水有影响(何金海 等,2006a),还与前汛期华南降水量密切相关(苗春生 等,2006)。

由于东北冷涡具有显著的天气、气候效应,中国气象工作者一直在探索东北冷涡活动的短期气候预测

* 本文发表于《大气科学》,2010,34(2):429-439

方法。刘宗秀等(2002)揭示了前冬北太平洋涛动(NPO)与夏季东北冷涡持续活动的关联,并指出东北冷涡持续活动是北半球中高纬度大气环流异常持续与调整的重要表征。何金海等(2006b)指出,东北夏季冷涡活动强弱与前冬2月北半球环状模(NAM)呈正相关,与前期中国近海海温存在显著负相关。然而,目前东北冷涡的短期气候预测方法还很不完善,需要进行更深入的研究。

本文的基本思路是,从春季环流背景出发,揭示与初夏(6月)东北冷涡活动异常相关联的北半球低频环流的相关内部动力学特征(徐祥德 等,2002),一方面关注相伴出现的阻塞型环流(高守亭 等,1998),另一方面运用谐波分析方法探寻低频环流演变特征或低频信号的传播过程,旨在为初夏东北冷涡活动的短期气候预测寻找依据和前期信号。由此,我们选择了6月东北冷涡异常多的1991年和异常少的2007年作为典型个例,从超长波和长波活动贡献、30～60 d低频振荡、Rossby波能量频散以及瞬变涡动反馈强迫作用的角度,探讨初夏东北冷涡活动异常与北半球环流低频变化之间的内在关联。

2　资料和分析方法

选取1991年和2007年美国国家环境预报中心(National Centres for Enrivonmental Prediction, NCEP)逐日再分析资料,其水平分辨率为2.5°×2.5°,500 hPa高度平均场为1979—2007年的平均值。

本文采用500 hPa高度场的谐波分析(仇永炎,1985),取超长波为1、2、3波的合成,长波为4、5、6波的合成。因为6月东北冷涡活动的最大频数域在45°N附近(郑秀雅 等,1992),侧重分析了沿45°N纬圈的高度场演变特征。利用李崇银(1990)所用的带通滤波器,计算500 hPa涡度场的30～60 d低频振荡过程。

利用Takaya等(1997,2001)推导出的三维波作用通量,描述Rossby波能量频散过程,其中,多年(1979—2007年)平均500 hPa(或300 hPa)月(或旬)平均场作为基本场,与此对应的异常场则为准地转扰动场,位势高度的单位为gpm。本文根据准地转位涡方程计算瞬变涡动所导致的位势高度倾向场(Lau et al.,1984,1991;Bueh et al.,2007),并以此表征瞬变涡动反馈强迫场。

3　超长波和长波活动

3.1　冷涡活动异常与背景环流

图1给出东北冷涡活动异常多(比多年平均多7 d)的1991年和异常少(比多年平均少7 d)的2007年6月500 hPa高度距平场。根据孙力等(1994)所用东北冷涡的定义,计算得到了1991年和2007年6月冷涡活动天数以及多年平均(1965—2005年)冷涡活动天数(13 d)。6月北半球高纬度大气环流在1991年呈3波结构,而在2007年则为2波特征。2 a的波状环流位相分布差异也很大,在欧亚大陆上空尤为明显。1991年6月,乌拉尔山(60°E)附近呈现强大的正距平中心(+80 gpm),而在泰梅尔半岛和日本东南侧分别出现了强度达−60 gpm和+20 gpm的距平中心。乌拉尔山和泰梅尔半岛反气旋和气旋式环流相互匹配,形成了稳定的极地冷空气源。中国东北地区的上游和下游地区则均为高压控制,有利于在东北地区形成闭合低压(孙力 等,1994)。这些环流形势易于使极地冷空气南侵中国东北地区,形成东北冷涡环流形态。与此对应,东北地区的负距平达−20 gpm左右,东北冷涡活动天数多达20 d。2007年6月,在欧亚大陆上空的环流形势呈现与1991年基本相反的环流形势。乌拉尔山地区为强负距平中心(−80 gpm),贝加尔湖以东到日本东南侧为正距平区,特别是在东北地区有强达+60 gpm的正距平中心,东北冷涡活动天数仅为6 d。

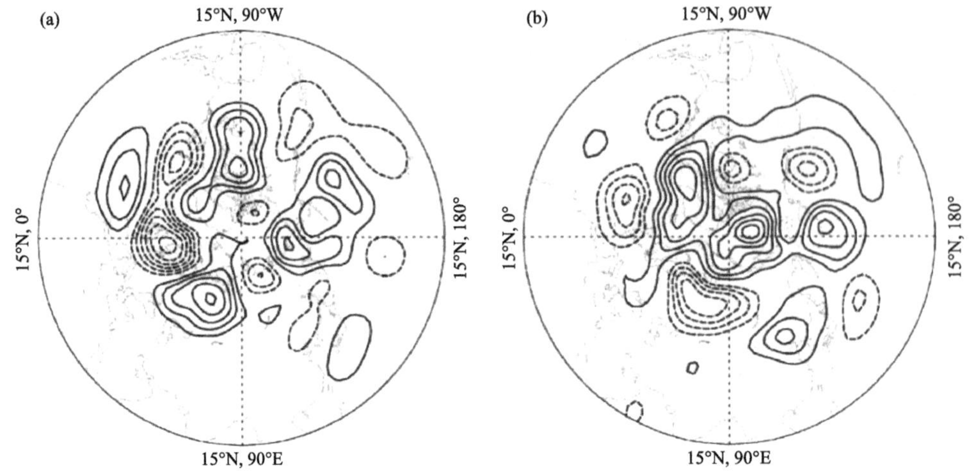

图 1　(a)1991 年和(b)2007 年 6 月 500 hPa 高度距平场(单位:gpm)。等值线间隔:20 gpm;实(虚) 线:正(负) 距平

3.2　5—6 月超长波活动

为了更清楚地勾画出 1991 年和 2007 年 6 月 500 hPa 异常环流的形成和演变特征,对 5—6 月高度场进行谐波分析,揭示超长波和长波活动的时空分布特征及方差贡献。超长波生命史一般为 10 d 左右(仇永炎,1985)。长波的生命史为 4～5 d(对中纬度而言),刚好与一次东北冷涡事件的时间尺度相当,长波槽脊究竟哪个占优亦将对冷涡活动区正涡度来源起着重要作用 (仇永炎,1985;徐祥德 等,2002)。

图 2 为 1991 年和 2007 年 5—6 月 500 hPa 高度场沿 45°N 超长波和长波的时间-经度剖面图。由图 2a 可见,在 6 月冷涡异常多的 1991 年,5—6 月 500 hPa 高度场沿 45°N 纬圈超长波活动的最显著特征为以北太平洋中西部上空(120°—180°E)的超长波槽为中心、2 波占优的环流形势。东北冷涡活动区以(120°E,45°N)为中心,在 5—6 月始终受北太平洋中西部超长波槽的影响,北太平洋超长波槽主体在 6 月中旬有一次明显的西退。北美大陆上空的超长波脊在 5—6 月变动较大,6 月上中旬,脊线基本稳定在北美西海岸。从图 2e 可见,1991 年 5—6 月超长波的方差贡献已达 40%。

从图 2b 可见,2007 年除了 5 月上旬和 6 月末以外的大部分时间,超长波槽中心位于北太平洋中部而且主体缓慢东进。这与 1991 年的情况(图 2a)形成鲜明对比。中亚超长波脊在 6 月初有一次明显的东移,到达东亚地区。与此对应,北太平洋超长波槽也向东移动。这表明东亚东北冷涡活动区在 2007 年 6 月基本处于超长波脊的控制之下。从图 2f 可见,2007 年 5—6 月超长波扰动的方差贡献总体上比 1991 年(图 2e)稍大,5—6 月平均可达 45%。

3.3　5—6 月长波活动

从图 2c 可见,1991 年 5—6 月 500 hPa 高度场沿 45°N 长波活动的主要特征为,在 5 月中旬至 6 月底,120°W 以东的地区长波自西向东呈"+－+"驻波特征,也就是北美大陆至北大西洋地区长波随时间近于无位相变化。总之,120°W 以东的北美至北大西洋的长波在整个 5—6 月份都显出驻波特征,6 月在 120°E 经线上长波槽位相占优,为东北冷涡活动提供了正涡度源。1991 年 5—6 月长波的方差贡献可达 40%(图 2e)。

与 1991 年相比,2007 年 5—6 月 500 hPa 高度场沿 45°N 的长波活动 (图 2d) 只是 5 月中旬和 6 月下旬在东亚至北美东岸呈现驻波特征。从 5 月下旬到 6 月中旬长波活动较弱,使整个 5—6 月份长波方差只有 30%左右 (图 2f)。6 月在 120°E 附近绝大部分时间为长波脊位相占优。

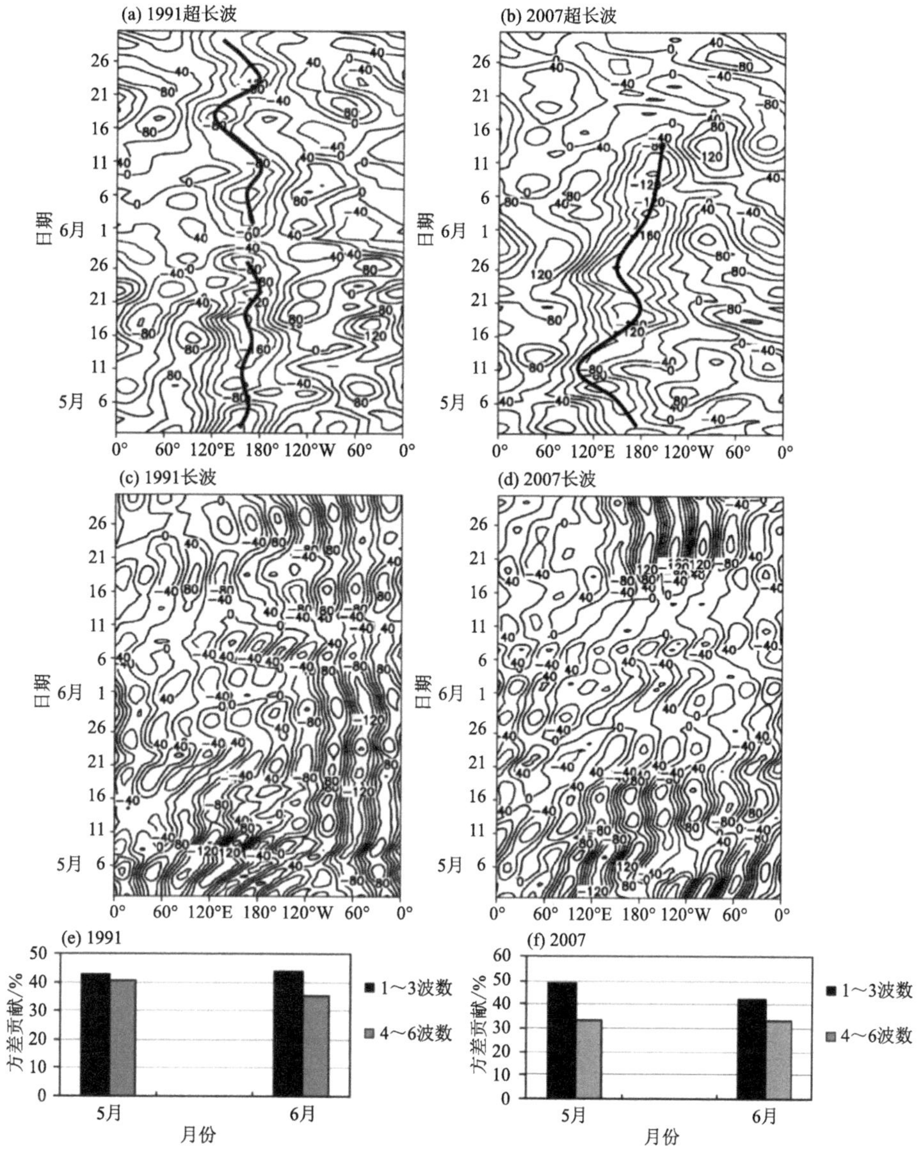

图 2　1991 年(a、c、e)和 2007 年(b、d、f)5—6 月 500 hPa 高度场(单位:gpm) 的 超长波(a、b)、长波(c、d)以及超长波和长波合成(e、f)方差贡献。等值线间隔:40 gpm;实(虚) 线:正(负) 距平

4　30～60 d 低频振荡

下面从 30～60 d 低频振荡的角度探讨相关环流的演变特征。30～60 d 低频振荡不仅是热带大气的特征频率,也是热带外大气运动的重要特点(李崇银,1990)。图 3 为 1991 年和 2007 年 5—6 月沿 45°N、120°E 的 500 hPa 涡度 30～60 d 低频振荡时间-经度(纬度) 剖面图。从图 3a 可见,1991 年 6 月上旬,来自 100°E 的低频涡度(向东传播)与来自 140°E 的低频涡度汇合于 120°E。其结果是在东北冷涡活动区(45°N,120°E 附近),1991 年 6 月间大部分时间为 30～60 d 的正涡度低频振荡位相占优。这一特征在图 2c 也十分明显。从图 3b 可见,在 120°E 的时间-纬度剖面上,一支来自 15°N 附近的低频正涡度自 6 月上旬向北传播,另一支来自 60°N 附近的低频正涡度自 5 月中旬向南传播,它们在 6 月中旬汇合于 40°N 附近,为东北冷涡活动区提供了低频正涡度源。

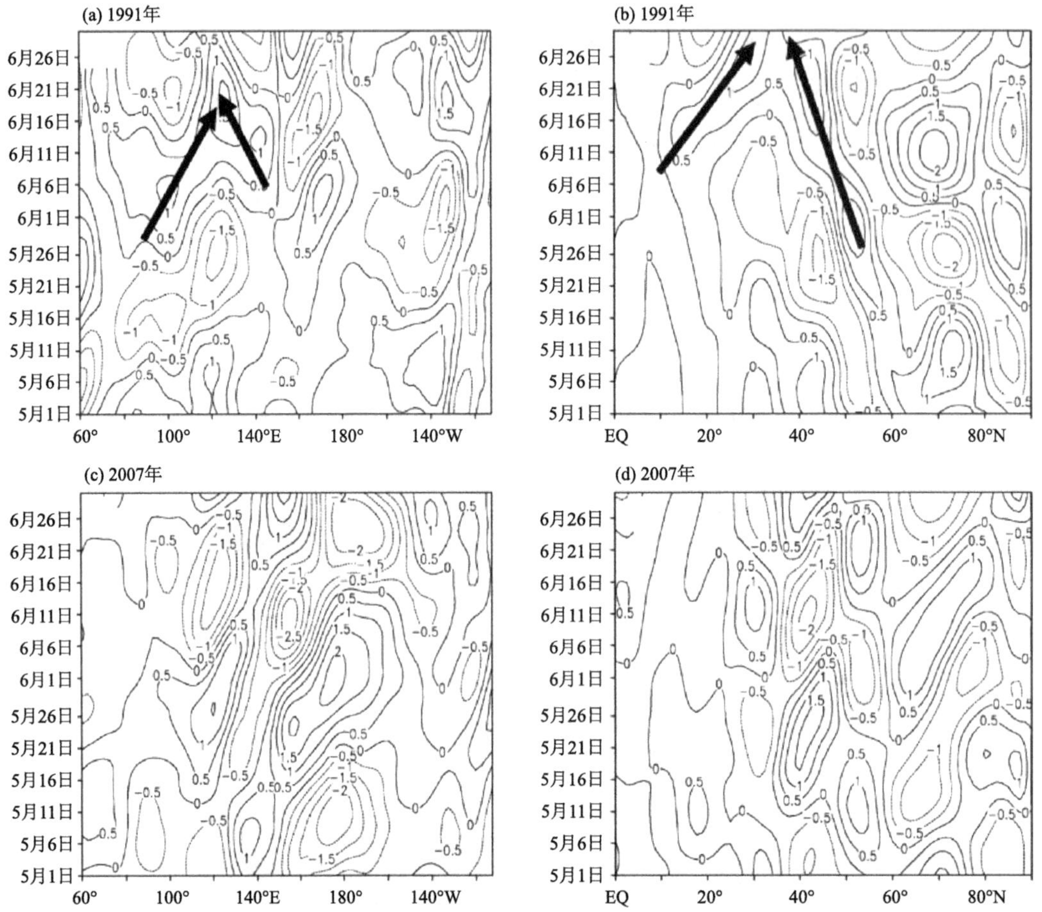

图 3　(a、b)1991 年和(c、d)2007 年 5—6 月 500 hPa 涡度的 30～60 d 低频振荡时间-经度(纬度)剖面图(单位:$10^{-5}s^{-2}$)。(a、c)沿 45°N;(b、d) 沿 120°E。等值线间隔:0.5。图中箭头表示低频振荡传播方向

图 3c、图 3d 清楚地反映出,2007 年低频涡度没有持续性的、跨月的向 (45°N,120°E) 传播的活动。在 6 月,(45°N,120°E)附近 30～60 d 振荡的负涡度位相占优。值得注意的是,自 5 月下旬 45°N 附近的低频正涡度向东传播,6 月在东北地区东侧形成了一个缓慢东移的正涡度区 (图 3c)。它们可能与较早建立的北太平洋正负低频涡度有关(图 3c)。从 6 月初开始,55°N 附近的负低频涡度向南传播到东北地区(图 3d)。另外,6 月上中旬东北地区南侧(30°N 附近)形成了一个低频正涡度区,几乎与东北地区的负涡度是同时的(图 3d)。

5　Rossby 波能量频散及瞬变涡动反馈强迫作用

上两节主要讨论了与东北冷涡活动有关的北半球环流异常及其演变特征。但上述分析无法辨识包含在北半球环流异常之间可能存在的因果关系。Rossby 波能量频散及瞬变涡动反馈强迫过程能够提供环流异常间的联系以及可能存在的波源(Bueh et al. ,2007),无疑有助于理解东北冷涡活动异常及其相关环流异常的成因机理。

图 4 给出 1991 年和 2007 年 4—5 月 500 hPa 位势高度异常场及其对应的波作用通量。为了便于不同纬度环流异常的对比,图中给出了流函数形式的位势高度异常(位势高度异常乘以 sin45°/sinφ,其中 φ 为纬度)。图 5 为 1991 年和 2007 年 4—6 月瞬变涡动强迫异常场,以 500 hPa 位势高度倾向异常给出。1991 年,乌拉尔山附近的阻塞型环流在 4 月已经形成,其中心距平达 150 gpm (图 4a),波作用通量在乌拉尔山地区大部显著辐散 (深阴影),说明这一地区成为波源区。由图 5a 可见,在乌拉尔山地区存在很强的

瞬变涡动强迫，可达 14 gpm/d，它是阻塞型环流形成的主要原因，而来自地中海地区的 Rossby 波能量注入也有利于该阻塞型环流的维持。5 月，随着局地瞬变涡动强迫的减弱（图 5b），乌拉尔山阻塞型环流有所减弱（图 4b），但仍维持＋80 gpm 左右的距平，这与波作用通量在乌拉尔山地区大部显著辐合有关（浅阴影）。5 月北美至北大西洋地区的驻波型长波扰动（图 2c）恰好与图 4b 的环流异常所对应。由此可见，Rossby 波能量传播与瞬变涡动强迫机制共同导致了 4—5 月乌拉尔山阻塞型环流的形成和维持。它在 6 月（图 1a）基本保持在 5 月的位置和强度，其中瞬变涡动强迫则有所加强（图 5c）。阿留申群岛附近 4—5 月存在气旋式异常环流，它在图 2a 中表现为驻波型超长波，它与上下游的 Rossby 波联系在 4 月比较明显（图 4a），但随着其强度的减弱，这种联系在 5 月变得不明显（图 4b）。

2007 年 4 月的环流与 1991 年明显不同。它在东北大西洋—欧亚大陆—北太平洋呈波状，其异常中心经向伸展，乌拉尔山及以西地区气旋式异常环流主导，而在贝加尔湖附近则为反气旋式异常环流。波能量的强迫和频散在这些地区比较明显（图 4c）。在西欧—地中海正高度异常区为波源区，它与瞬变涡动强迫正异常区一致（图 5d）。5 月（图 4d），乌拉尔山地区为气旋式异常环流（负距平），亚洲中纬度地区至东西伯利亚/白令海峡地区为反气旋式环流（正距平）。值得注意的是，在亚洲大陆上空，波作用通量不太明显，瞬变涡动强迫作用却变得十分明显。例如，中高纬地区的异常环流与瞬变涡动强迫分布非常一致（图 4d 和图 5e）。2007 年 5 月最显著的特点是乌拉尔山和东西伯利亚/白令海峡地区的负/正距平分布，它们与当年初夏的东北冷涡活动的异常少密切相关。

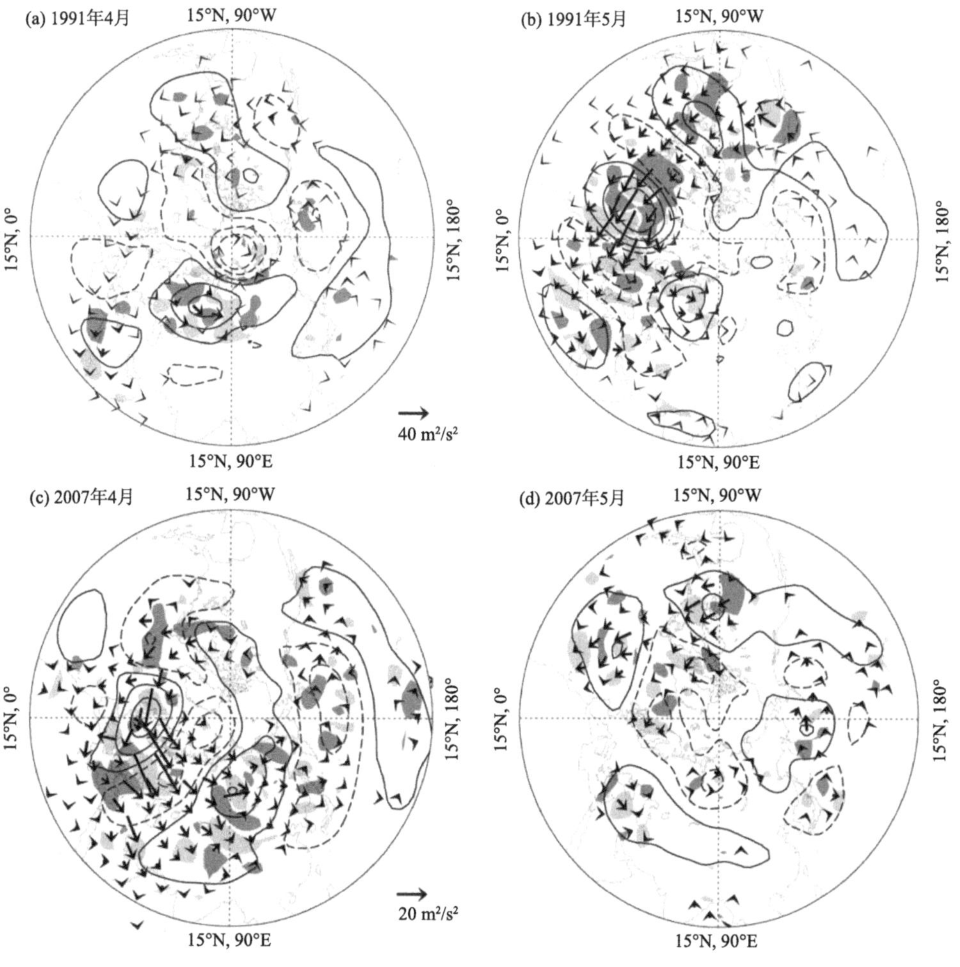

图 4　(a、b)1991 年和(c、d)2007 年 4—5 月 500 hPa 位势高度异常场(单位：gpm，等值线)及其对应的波作用通量(箭头)。等值线分别为±20、±80、±140 gpm，虚线表示负值，零线已略；深(浅)阴影表示波作用通量辐散(辐合)大于(小于)8×10^{-6}(-8×10^{-6}) m^2/s^2 的区域

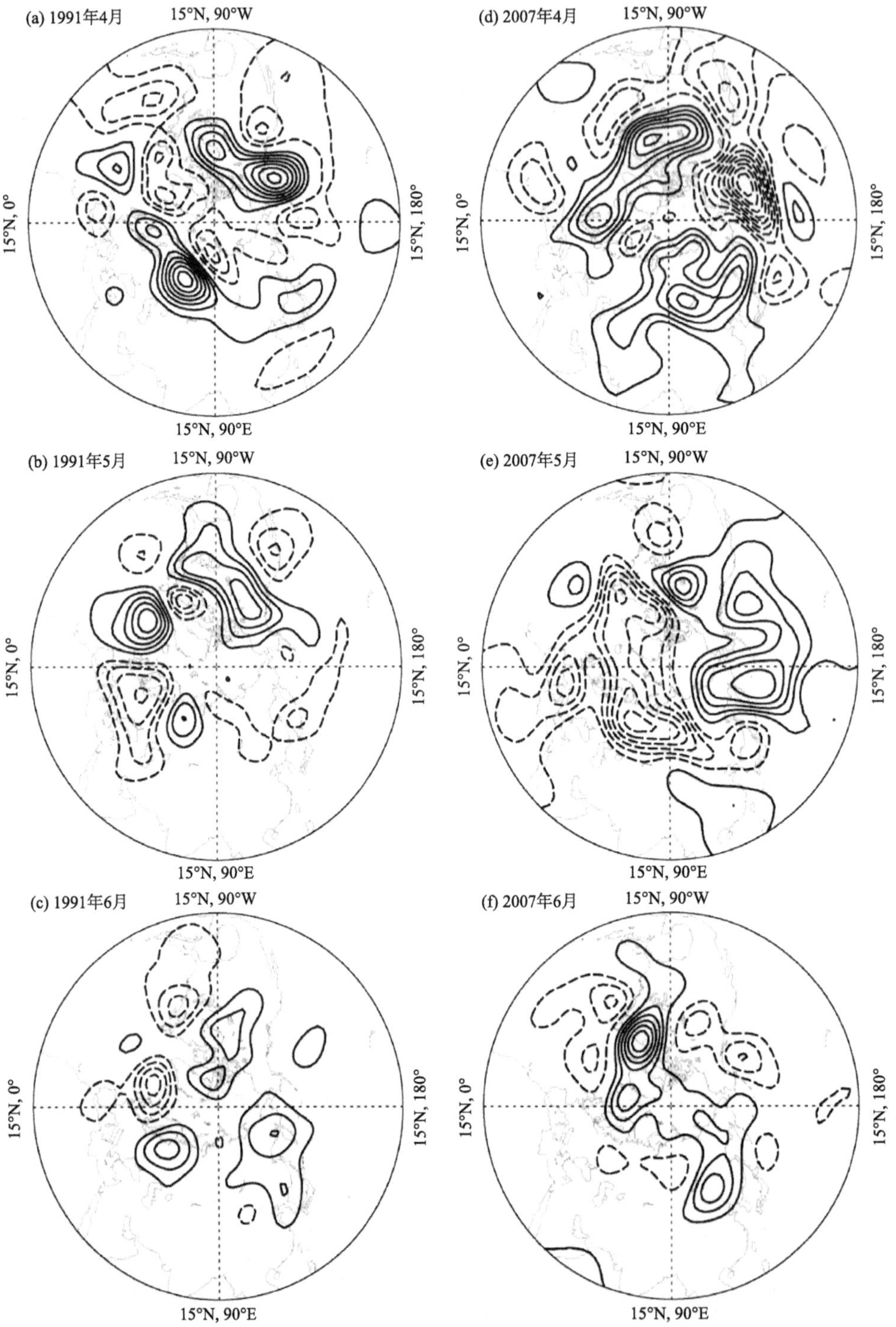

图 5　(a、b、c)1991 年和(d、e、f)2007 年 4—6 月 500 hPa 瞬变涡动强迫异常场
(单位:gpm/d,等值线)。等值线间隔为 2 gpm/d,虚线表示负值,零线已略

为了揭示初夏环流不同阶段的演变及其对应的波能量频散特征,在图 6 中绘制了 1991 年和 2007 年 6 月上、中、下旬的 500 hPa 位势高度异常场(流函数形式)及其对应的波作用通量。1991 年 6 月上旬(图 6a),北欧至里海北侧有－60 gpm 的负距平区,源自该地区的 Rossby 波向乌拉尔山地区传播,有利于乌拉尔山阻塞型环流的维持。在乌拉尔山区,6 月平均瞬变涡动强迫异常可达 6 gpm/d,它维持阻塞型环流,并成为波源(图 5c)。与 5 月的情况不同,这时起源于乌拉尔山阻塞型环流区的 Rossby 波向下游显著频散,一方面沿高纬度向雅库茨克地区频散,另一方面沿中纬度传播,分别在蒙古和环渤海地区形成了负、正

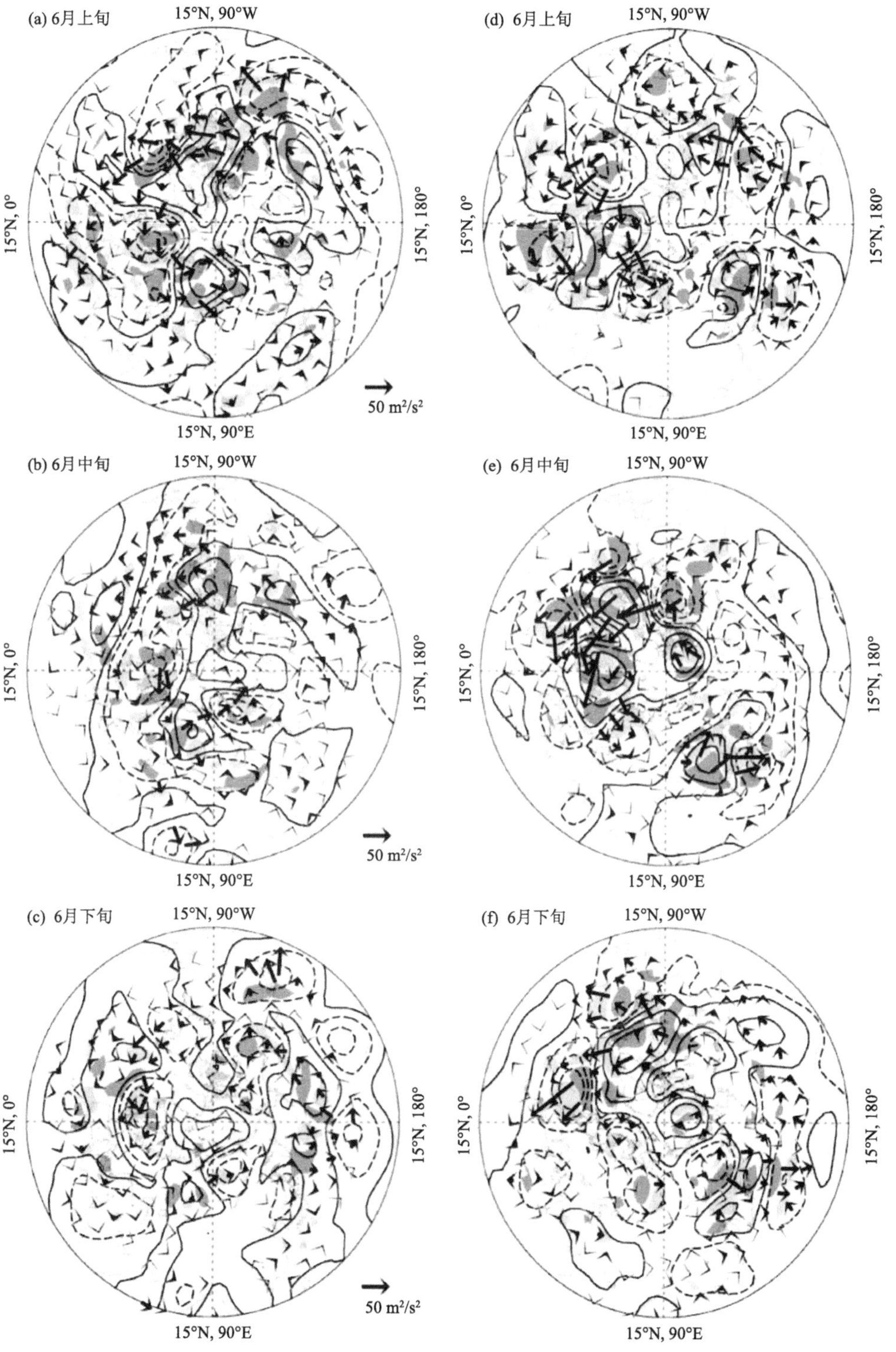

图 6　(a、b、c)1991 年和(d、e、f)2007 年 6 月上、中、下旬 500 hPa 位势高度异常场(单位:gpm,等值线)及其对应的波作用通量(箭头)。等值线分别为±20、±80、±160 gpm,虚线表示负值,零线已略;深(浅) 阴影表示波作用通量辐散(辐合) 大于(小于)2×10^{-5}(-2×10^{-5}) m^2/s^2 的区域

距平环流。这种环流配置比较有利于东北冷涡的形成,但冷涡的位置偏北,上旬的冷涡活动有 4 d。在 6 月中旬 (图 6b),基本保持上旬的环流形势,但贝加尔湖北侧负距平环流加强并向鄂霍次克海一侧伸展,这使冷涡位置偏东。蒙古的负距平环流也东伸到中国东北地区,这在图 3a 中表现为 30～60 d 的低频正涡度从西侧向东北地区传播。6 月中旬冷涡天数为 6 d。在 6 月下旬(图 6c),极涡凸向贝加尔湖方向,东

北地区由气旋式异常环流控制。这种现象正对应着由南北两侧低频正涡度向东北地区传播的特征(图3b)。冷涡活动区两侧的乌拉尔山和西北太平洋地区均有高压脊控制。这种环流配置极有利于东北地区闭合低压的形成,使整个6月下旬都有冷涡盛行。我们注意到,东北冷涡活动区的南侧6月的各旬都有正距平环流,对应的西太平洋副热带高压586线位于长江口附近。在300 hPa层,这一正距平环流由伊朗—青藏高原—中国东海的"+－+"波列所形成,Rossby波在亚洲副热带急流中的传播非常明显。由于青藏高原的存在,这一特征在500 hPa层不太明显。

2007年6月(图6d—f),乌拉尔山地区气旋式异常环流盛行,极地冷空气易于在乌拉尔山—亚洲北部—北太平洋带状区域活动,东北地区则为高压脊控制。乌拉尔山地区负距平环流与东北地区正距平环流之间没有明显的Rossby波链接。但是,东北地区为较强的波源区(深阴影区),它与局地瞬变涡动强迫异常分布所致,6月平均瞬变强迫异常可达6 gpm/d。乌拉尔山一带的冷空气活动及其对应的瞬变波活动可以强迫出持续性东北高压脊,这一机制最早由Green (1977) 在研究欧洲高温干旱时提出。自4月到6月,乌拉尔山地区一直由气旋式异常环流控制,在6月上中旬它主要由来自北欧/挪威海地区的Rossby波能量所维持(图5d—e)。极地冷空气长时间向乌拉尔山地区活动是东北冷涡异常少的一个重要原因。

6 结论和讨论

本文以初夏东北冷涡异常多的1991年和异常少的2007年为典型个例,利用谐波分析、30～60 d低频振荡分析、Rossby波能量频散以及瞬变涡动强迫过程分析相结合的方法,探讨了大气环流低频变化对初夏东北冷涡异常活动的影响,并得出如下结论。

(1)初夏东北冷涡活动显著异常与上游乌拉尔山附近和下游日本及以东(西北太平洋)的持续性异常环流所对应。冷涡异常多年(1991年),Rossby波能量向下游频散与瞬变扰动强迫致使乌拉尔山附近阻塞型环流盛行。该阻塞型环流一方面使极地冷空气向东北地区移动,另一方面通过Rossby波能量沿中纬度向东北地区传播,使东北冷涡活动加强。冷涡异常少年(2007年)的环流情况则基本相反。

(2)谐波分析和低频振荡分析表明,在冷涡异常多(少)年的5—6月正(负)涡度位相占优,这与北太平洋中西部的超长波槽西退(东进)所对应。在北太平洋—北美—大西洋地区呈驻波型长波活动有利于Rossby波能量的传播,亦有利于乌拉尔山附近的异常环流得以长时间维持。

(3)初夏东北冷涡活动显著异常的成因及信号至少可以追溯到当年的5月。冷涡活动区上游的乌拉尔山地区,其初夏环流特征均在5月环流上清晰可见。

本文通过两个异常个例分析,捕捉了关键性的低频异常强信号,为东北冷涡活动的月际尺度短期气候预测提供了有一定价值的依据,不仅着眼于冷涡活动区的上游地区,也要重视下游地区。

在6月份冷涡活动区的下游,即日本及以东(西北太平洋),不论在冷涡活动异常多还是异常少的年,均存在异常环流(图1)。它们都对应着一个减弱的局地西风气流,从而有利于波能量在冷涡活动区聚集。特别是1991年6月,在对流层上层(300 hPa) Rossby波在亚洲副热带急流中传播,形成了伊朗—青藏高原—中国东海的波列分布,使一个正高度异常区一直滞留于日本岛南侧至中国东海地区。由于篇幅关系,它与东北冷涡活动的具体关联,本文未能详加分析。另外,尽管在白令海峡和阿留申群岛一带有比较明显的异常环流,但它对东北冷涡活动的具体影响也尚不清楚。

参考文献

毕慕莹,丁一汇,1992.1980年夏季华北干旱时期东亚阻塞形势的位涡分析[J].应用气象学报,3(2):145-156.

布和朝鲁,施宁,纪立人,等,2008.梅雨期EAP事件的中期演变特征与中高纬Rossby波活动[J].科学通报,53(1):111-121.

陈汉耀,1957.1954年长江淮河流域洪水时期的环流特征[J].气象学报,28(1):1-12.

高守亭,朱文妹,董敏,1998.大气低频变异中的波流相互作用(阻塞形势)[J].气象学报,56(6):665-680.

何金海，吴志伟，江志红，等，2006a. 东北冷涡的“气候效应”及其对梅雨的影响[J]. 科学通报，51(23)：2803-2809.

何金海，吴志伟，祁莉，等，2006b. 北半球环状模和东北冷涡与中国东亚夏季降水关系分析 [J]. 气象与环境学报，22(1)：1-5.

李崇银，1990. 大气中的季节内振荡[J]. 大气科学，14(1)：32-45.

刘宗秀，廉毅，高枞亭，等，2002. 东北冷涡持续活动时期的北半球 500 hPa 环流特征分析[J]. 大气科学，26(3)：361-372.

苗春生，吴志伟，何金海，等，2006. 近 50 年东北冷涡异常特征及其与前汛期华南降水关系分析 [J]. 大气科学，30(6)：1249-1256.

乔枫雪，赵思雄，孙建华，2007. 一次引发暴雨的东北低涡的涡度和水汽收支分析[J]. 气候与环境研究，12(3)：397-412.

仇永炎，1985. 中期天气预报[M]. 北京：科学出版社：420.

沈柏竹，李尚锋，廉毅，2008. 初夏东北冷涡活动的谐波分析[J]. 吉林大学学报（地球科学版），38 (Z)：204-208.

孙力，1997. 东北冷涡持续活动的分析研究[J]. 大气科学，21(3)：297-307.

孙力，1998. 一次东北冷涡发展过程中的能量学研究[J]. 气象学报，56(3)：349-361.

孙力，郑秀雅，王琪，1994. 东北冷涡的时空分布特征及其与东亚大型环流系统之间的关系[J]. 应用气象学报，5(3)：297-303.

汤懋苍，1957. 亚洲东部的阻塞形势及其对天气气候的影响[J]. 气象学报，28(4)：282-293.

陶诗言，1980. 中国之暴雨[M]. 北京：科学出版社：225.

徐祥德，高守亭，2002. 外源强迫与波流作用动力学原理：2 版[M]. 北京：海洋出版社：291.

叶笃正，陶诗言，李麦村，1958. 在 6 月和 10 月大气环流的突变现象[J]. 气象学报，29(4)：249-263.

张庆云，陶诗言，1998. 亚洲中高纬度环流对东亚夏季降水的影响[J]. 气象学报，56(2)：199-211.

张云，雷恒池，钱贞成，2008. 一次东北冷涡衰退阶段暴雨成因分析[J]. 大气科学，32(3)：481-498.

郑秀雅，张延治，白人海，1992. 东北暴雨[M]. 北京：气象出版社：132.

BUEH C，NAKAMURA H，2007. Scandinavian pattern and its climatic impact[J]. Quart J Roy Meteor Soc，133(629)：2117-2131.

GREEN J S A，1977. The weather during July 1976：Some dynamical considerations of the drought[J]. Weather，32：120-128.

LAU N C，HOLOPAINEN E O，1984. Transient eddy forcing of the time-mean flow as identified by geopotential tendencies [J]. J Atmos Sci，41(3)：313-328.

LAU N C，NATH M J，1991. Variability of the baroclinic and barotropic transient eddy forcing associated with monthly changes in the mid-latitude storm tracks[J]. J Atmos Sci，48(24)：2589-2613.

TAKAYA K，NAKAMURA H，1997. A formulation of a wave-activity flux for stationary Rossby waves on a zonally varing basic flow [J]. Geophys Res Lett，24(23)：2985-2988.

TAKAYA K，NAKAMURA H，2001. A formulation of a phase-independent wave-activity flux for stationary and migratory quasigeostrophic eddies on zonally varying basic flow[J]. J Atmos Sci，58 (6)：608-627.

两种微物理方案对夏季东北亚阻塞过程影响的数值模拟结果分析*

杨旭[1,2]　钟中[2]　廉毅[1]　李彦良[3]　曲思邈[3]

(1 吉林省气象科学研究所长白山气象与气候变化吉林省重点实验室/中高纬度环流系统与东亚季风研究开放实验室，长春，130062；2 国防科技大学气象海洋学院，南京，211101；3 吉林省气象局，长春，130062)

摘　要：选用 NCEP/NCAR FNL 再分析资料，利用中尺度模式 WRF 对 2011 年 7 月 25 日至 8 月 3 日的一次东北亚阻塞高压过程进行数值模拟试验。通过数值对比试验，讨论有冰相物理过程和无冰相物理过程的两种微物理方案对此次阻塞高压过程的强度和维持时间模拟的差异，并简要阐述原因。结果表明：WRF 模式对阻塞高压过程有较好的模拟效果，有冰相物理过程的微物理方案模拟的阻高强度及维持时间较接近实况，整体阻塞闭合环流形势、高压中心值、高压中心位置以及高压中心上下游的低压槽强度均与实况接近；无冰相物理过程的微物理方案模拟阻塞高压在 500 hPa 和 200 hPa 平均位势高度场上没有闭合的高值中心出现，500 hPa 阻塞高压强度弱，持续时间短，200 hPa 则相反。进一步研究发现，这两种差异主要是由阻塞高压活动区域内的非绝热加热不同引起的，它对阻塞高压形成的强度和维持时间具有很好的指示意义。另外，阻塞高压附近区域的微物理潜热释放对阻塞高压过程的强度和维持时间也有一定影响。

关键词：数值模拟；阻塞高压；微物理方案；非绝热加热

1　引言

东北亚地区阻塞高压(本研究主要指鄂霍次克海附近的阻塞高压)的稳定维持在一定程度上会造成夏季中国东部地区的气候异常(毕幕莹 等，1992；吴国雄 等，1994)，它不仅是维持中国梅雨发生的大气环流背景，同时也是东亚夏季天气的重要影响系统之一(陶诗言 等，1958；张庆云 等，1998)，其发生频次与中国江淮地区降水多寡、江淮梅雨期持续时间及梅雨量都存在正反馈关系(陆日宇 等，1998；Wang，1992)。很多研究(赵汉光 等，1990；张存杰 等，2004；段廷扬 等，1997；徐祥德 等，1994)发现长江中下游洪涝灾害、盛夏华北地区的严重干旱、印度季风强弱、近海台风的路径等也与之有关。此外，东亚地区阻塞高压也是影响东北夏季低温、东北冷涡活动的重要的环流系统之一，超过 2/3 的东北冷涡与其发展变化有关(孙力 等，1994；李尚锋 等，2012)，对东北冷涡及气温都有一定的指示作用(刘刚 等，2012)。

目前，学者们对阻塞高压的研究已经从个例研究、统计分析、理论机理分析过渡到集合预报数值模拟分析，如果能利用中尺度模拟方法来模拟阻塞高压、通过敏感性试验尝试了解哪些因素能引起阻塞高压的强弱变化，对阻塞高压的数值模拟工作非常有意义。

20 世纪 40—50 年代，Elliott 等(1949)最先提出阻塞高压的定义并对北半球西风带阻塞活动做了系统的统计，研究发现中太平洋阻塞事件出现的数量是东北大西洋的两倍。之后诸多学者对阻塞高压的定义、活动频次及易发生区域、生命史的变化特征等进行了补充研究(Rex，1950a，1950b；Sanders，1953；Sumner，1954；Hartmann et al.，1980；Treidl et al.，1981；Dole，1986；周晓平，1957；叶笃正 等，1962；

* 本文原载于《气象学报》，2020，78(4)：593-607

White et al.,1975)。汤懋苍(1957)统计了1951—1956年各月500 hPa高度场100°—170°E区域阻塞形势及其对中国天气的影响,Okawa(1973)较早分析过东北亚阻塞形势中重要的地面系统——鄂霍次克海高压形成的机制,徐爱华等(1991)曾给出500 hPa鄂霍次克海高压的定义域,分析6月的7个典型年鄂霍次克海高压,发现5月阿拉斯加高压脊能否建立与西退是鄂霍次克海阻塞高压形成的关键因素,并与北太平洋海温距平场形势有一定的联系。

随着科学技术发展,数值预报和模拟能力得到进一步提升,Charney等(1979)用高截断谱分解方法讨论了强迫耗散正压模式的多平衡态,推测出阻塞高压是一类准稳定的低指数平衡态。随后朱正心等(1982)采用高截断谱分解方法,运用能反映超长波特点的斜压模式,从非线性和纬向不对称热力强迫作用角度出发,得出阻塞高压是一种在纬向不对称强迫源作用下稳定的超长波非线性平衡态,推断出只有当纬向加热差异超过一定的临界值才有可能出现阻塞高压的结论,侧面验证了Charney等(1979)的猜想。朱正心(1984)、朱正心等(1984,1985)开展了一系列基于准地转两层波谱模式的数值试验,并着重考虑了热力强迫作用、平流非线性作用以及实际地形情况,得出了纬向不对称热力强迫及平流非线性这两个物理因子共同作用下才会产生稳定维持的阻塞高压形势,温度平流在阻塞高压发展期间起重要作用,当热源强度增强时,阻塞高压中心强度增强,位置偏西,维持时间也略有延长的重要结论。罗哲贤(1989)运用强迫耗散KdV方程(Kortewegde Vries equation)实施数值积分方法,指出了阻塞流型形成的3类因子。周学鸣等(1995)采用郭晓岚-钱永甫p-σ坐标系5层原始方程球带模式,研究了乌拉尔山阻塞高压对亚洲季风环流系统及中国东部旱涝分布的影响。Lu等(1996)采用IAP-GCM模式模拟发现导致东北亚阻塞形势形成和维持的一个重要原因是海温异常。陈希等(2002)利用T63L9大气环流模式,详细地研究了阻塞高压位于不同地区对中国东部及西太平洋副热带高压地区的大气环流影响。魏东等(2006)利用NCAR CAM3模式模拟发现初夏日本东部附近海域负的海温异常对鄂霍次克海高压的形成有重要贡献。

近年来,集合预报迅速发展并广泛应用于阻塞高压的数值预报系统中,主要以检验和评估模式对阻塞高压的预报效果为主,Matsueda等(2009)利用各中心模式的集合预报资料检验了冬季阻塞形势的预报效果,发现模式对于欧洲—大西洋地区和太平洋地区的阻塞形势预报能力较高,而对于乌拉尔山地区阻塞形势的预报能力较低。潘婕等(2009)利用IAP T42L9全球大气环流谱模式对气候平均意义下的欧亚中高纬度环流由春入夏的演变规律进行模拟分析,模拟试验较成功地再现了北半球中高纬度的气候环流形势,只是槽、脊形势略强于实际结果。吴静等(2015)评估了耦合模式比较计划5(CMIP5)中17个气候模式对于20世纪北半球夏季气候态阻塞频数和振幅的模拟能力。此外,很多学者检验评估了NCEP全球谱模式、ECMWF集合预报系统、TIGGE延伸期天气预报模式、NCEP CFSv2气候模式对阻塞高压的模拟能力及检验阻塞高压的预报效果及技巧,得到了很多有意义的结果(Watson et al.,2002;Pelly et al.,2003;王毅 等,2014;Jia et al.,2014;周宁芳 等,2018)。

对于阻塞高压的数值模拟,有人利用多种数值模式做了很多研究、评估和检验工作,但鲜有学者运用WRF直接模拟阻塞高压,也甚少有学者在模拟阻塞高压的时候考虑有无冰相过程的微物理方案的选取对其持续时间及强度的影响。因为通常微物理方案主要用于降水、降雪等与天气现象相关的数值模拟,阻塞高压作为大尺度的环流背景,是否也对微物理方案中是否含有冰相物理过程有一定的敏感性和响应呢?

另外,Dole(1986)在研究北太平洋上空持续性阻塞高压生命周期的垂直剖面结构时曾指出,“由于持续性阻塞结构与受局地异常热源所强迫的扰动的垂直结构有明显区别,表明局地非绝热加热异常不太可能是它们(持续性阻塞高压)发展的最直接热源”。那么,紧邻亚洲大陆和北太平洋上空的东北亚阻塞高压,局地热源对其影响如何,也是本研究要进一步讨论的科学问题。

在2011年7月25日至8月3日的东北亚阻塞高压活动过程中,200 hPa、500 hPa的平均环流形势非常接近,高压中心上下游两侧处低压槽,形成了类似Ω的环流,在此次东北亚阻塞高压过程维持期间,中国出现东北—西南方向明显的干旱带,湖南、贵州、广西、重庆、云南等地干旱持续或发展,吉林东部、郑州—贵阳—昆明、长沙等地区甚至出现了不同程度大范围的重旱及特旱区域,给当地的农业生产及生活带来很大影响。

本研究利用 WRF 模式对此次阻塞高压过程进行模拟，用得到的高时空分辨模拟资料对比分析两种微物理方案的阻塞高压强度及维持时间的模拟结果的差异，并讨论有无冰相物理过程的条件下，非绝热加热对阻塞高压强度及维持时间的影响，以及微物理过程潜热释放（局地热源）对阻塞高压强度及维持是否有影响。

2　资料及方法

所用资料为美国 NCEP/NCAR 提供的 FNL 全球再分析资料，时间分辨率为 6 h，空间水平分辨率为 1°×1°，时间为 2011 年 7 月 25 日 00 时至 8 月 3 日 18 时（世界时，下同）。

大气的非绝热加热分布状况可用视热源表示（Yanai et al.，1973）

$$Q_1 = c_p\left[\frac{\partial T}{\partial t} + V \times \nabla T + \left(\frac{p}{p_0}\right)^k \times \omega \frac{\partial \theta}{\partial p}\right] = c_p(Q_{11} + Q_{12} + Q_{13}) \tag{1}$$

式中，$\omega = \mathrm{d}p/\mathrm{d}t$ 为 p 坐标系下的垂直速度，其余为常见的气象参数，Q_{11}、Q_{12}、Q_{13} 分别表示视热源的局地变化项、水平平流项、垂直输送项。

3　模式方案设计及模拟结果检验

3.1　模式方案设计

运用 WRF 模式，采取单层嵌套方案，模拟区域（图 1）中心为（54°N，142°E），格点数为 250×242，水平格距为 20 km，模式顶层气压取 50 hPa，积分步长为 60 s，地形数据采用的是分辨率为 30"的 MODIS 资料，输出时间间隔为 1 h，模拟时段为 2011 年 7 月 25 日 00 时至 8 月 3 日 18 时，设计两组数值试验，对照试验和敏感性试验选取的微物理方案分别为 WSM6-Class 方案（Hong et al.，2006a）和 Kessler 方案（Kessler，1969），两组试验其余参数化方案相同：积云对流参数化方案为 Kain-Fritsch 方案（Kain，2004），长波辐射为 RRTM 方案（Mlawer et al.，1997），短波辐射为 Dudhia 方案（Dudhia，1989），近地层为莫宁-奥布霍夫方案（Monin et al.，1954），陆面过程为 Unified Noah 方案（Tewari et al.，2004），边界层为 YSU 方案（Hong et al.，2006b）。

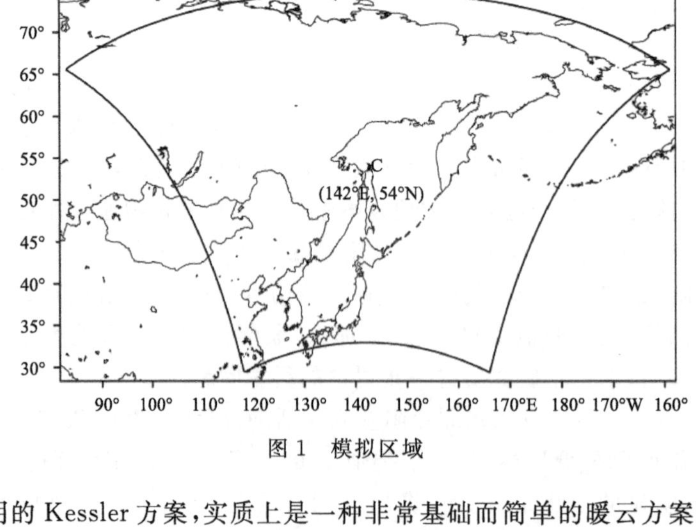

图 1　模拟区域

敏感性试验所采用的 Kessler 方案，实质上是一种非常基础而简单的暖云方案，主要用于云模式的理想试验研究，此方案所描述的物质较少，主要为水汽、云水和降雨 3 种物质，无任何冰相物质。对照试验所采用的 WSM6-Class 方案与 Kessler 方案相比最大的不同是其囊括了多种冰相物质，主要为雨水、水汽、雪水、霰水、冰水等，属于复杂冰相方案过程。不仅如此，WSM6-Class 方案还对计算的量级进行最优化处理，减小了对模式时间步长的敏感性。

3.2　模拟结果检验

3.2.1　500 hPa 和 200 hPa 平均位势高度场的对比分析

图 2 为此次阻塞高压过程(7 月 25 日—8 月 3 日) 500 hPa、200 hPa 平均位势高度场实况(FNL 资料)、对照试验模拟结果(CON)、敏感性试验模拟结果(SEN)，为了使结果更直观，将 500 hPa 和 200 hPa 层上闭合高压等值线用红色标记。对照试验模拟结果(图 2b)显示，500 hPa 平均位势高度场上，大气呈现出 Ω 环流形势，存在闭合的高压等值线，强度为 5840 gpm，中心位于(53°N，140°E)附近，在阻塞高压中心的上游有西北—东南方向的低压槽，低压中心的位势高度值约为 5680 gpm，在阻塞高压中心的下游有一偏西向的狭长低压带，中心值为 5780 gpm，与实况(图 2a)几乎相同，可见对照试验对此次阻塞高压过程的整体环流形势、高压中心值、高压中心位置、高压强度以及中心上下游的低压槽强度有很好的模拟效果。

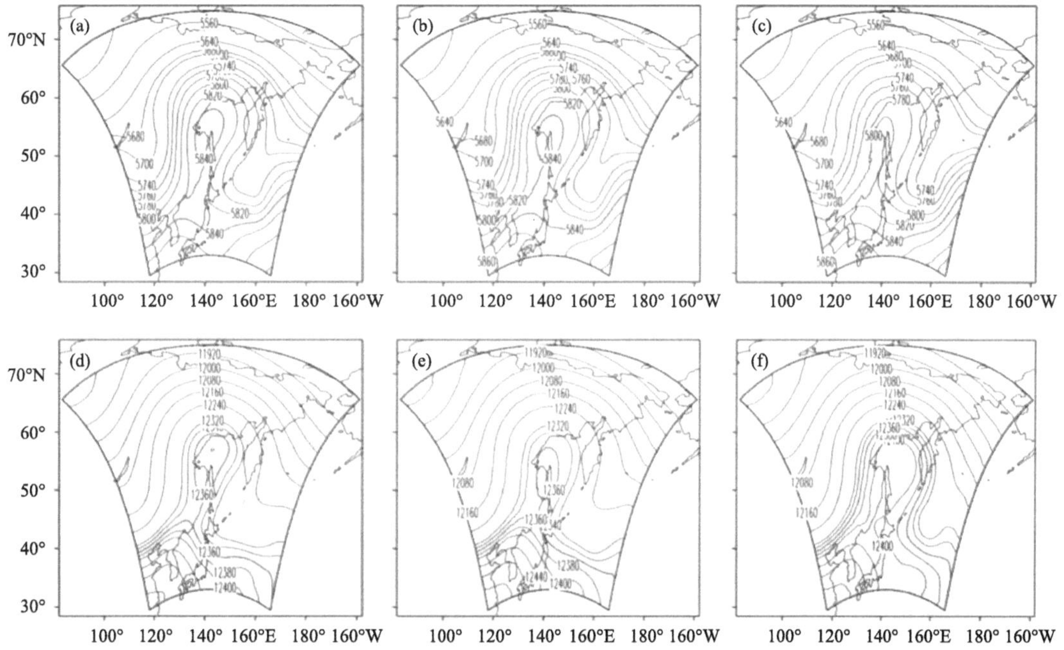

图 2　阻塞高压过程的平均位势高度场的实况(a、d)、对照试验模拟结果(b、e)和敏感性试验模拟结果(c、f)对比
(a—c，500 hPa；d—f，200 hPa；单位：gpm)

敏感性试验模拟结果(图 2c)显示，500 hPa 平均位势高度场虽然整体呈现出近似 Ω 环流形势，但阻塞高压中心值、高压位置、高压范围及上下游低压槽的强度较实况(图 2a)整体偏弱，无闭合的高压等值线，高压中心强度仅为 5800 gpm 且高压范围缩小。

图 2e—f 分别为此次阻塞高压过程 200 hPa 平均位势高度场实况、对照试验模拟结果、敏感性试验模拟结果，与 500 hPa 结果类似，对照试验在闭合高压等值线及上下游低压槽的强度方面整体的模拟结果更接近于实况，敏感性试验模拟无闭合高压中心，且较实况值高出 40 gpm，可见对照试验的模拟结果优于敏感性试验，另外对比 850 hPa 平均位势高度场(图略)可知，实况、对照试验模拟结果、敏感性试验模拟结果相似，这里不再赘述。

3.2.2　500 hPa 和 200 hPa 的剖面对比分析

图 3a—f 依次为实况、对照试验、敏感性试验的 500 hPa 位势高度场和 200 hPa 位势高度场沿 53°N 的经度-时间剖面图，图中阴影部分代表阻塞高压的外围廓线，一定程度上可近似表征阻塞高压的持续时间。对比图 3a—c 可知，500 hPa 实况及对照试验结果显示二者描述阻塞高压维持时间较为接近，在 8 月 2—3 日阻塞高压消亡，而敏感性试验模拟结果显示此次阻塞高压过程在 7 月 28—29 日就已经崩溃消亡，与阻塞高压实际维持时间出入很大。由此可见，在阻塞高压的维持时间方面，对照试验模拟结果更接近实况，敏感性试验模拟结果并不理想。由剖面图 3d—f 也可知，200 hPa 模拟结果与 500 hPa 相似，对照试验的模拟结果显示阻塞高压的持续时间与实况相近，而敏感性试验模拟结果在 8 月 3 日 18 时后阻塞高压还一直维持，与实况不符。

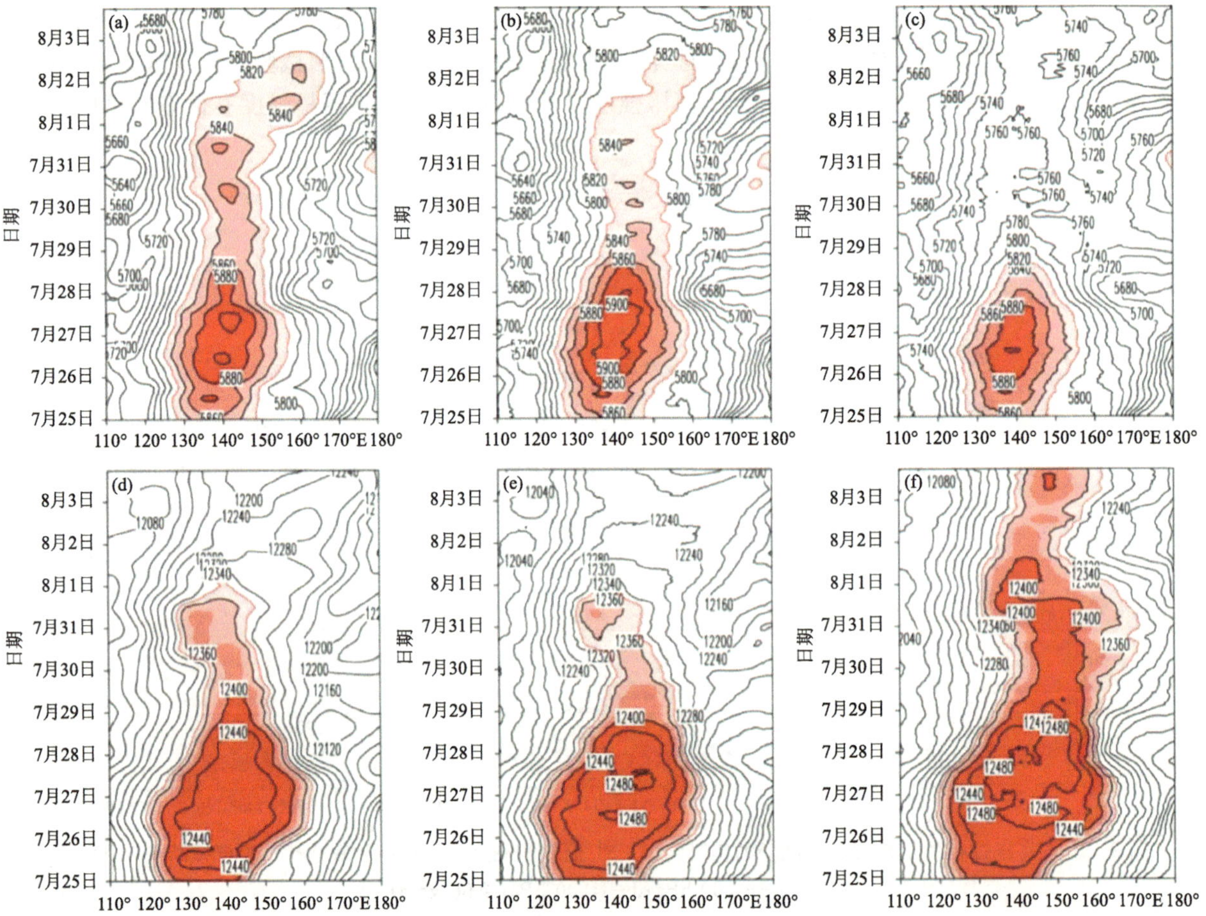

图 3　500 hPa(a—c)和 200 hPa(d—f)位势高度场(单位:gpm)沿着 53°N 的经度-时间剖面图
(a、d. 实况，b、e. 对照试验，c、f. 敏感性试验；阴影部分为阻塞高压活动过程中的外围廓线)

基于以上分析可知，对照试验在模拟阻塞高压的强度、维持时间、高压中心位置等方面与实况较为接近，能够较好地模拟出位势高度随时间的变化趋势，且能把位势高度场随时间演变特征较好地表现出来，较好地模拟了此次阻塞高压过程。敏感性试验中层(500 hPa)模拟强度偏弱，持续时间短，高层(200 hPa)模拟强度偏强，持续时间偏长，模拟效果不理想。此数值模拟试验验证了 WRF 对阻塞高压演变过程也有很好的模拟能力，这也为进一步分析提供了时空分辨率更高的数据。本研究将以 500 hPa 为代表层，分析其各项物理量的变化，简要分析两组试验模拟结果不同的原因。

4　不同微物理方案模拟结果差异分析

4.1　温度场差异分析

宏观上单位体积内空气柱温度升高或者降低会使气柱产生膨胀或者收缩，相应位势高度场也会伴随着出现升高或者降低的趋势。由于大气中加热场变化直接体现在温度的变化上，为了便于比较温度场的变化特征，将 FNL 资料中每层阻塞高压活动最外层闭合等压线内的区域平均温度定义为该层次上阻塞高压内的温度，对其进行时间平均(7 月 25 日至 8 月 3 日)后的温度定义为该层阻塞高压过程的平均温度，计算 FNL 资料及两组数值试验逐层阻塞高压活动区域的温度差值(温差)及温度距平，距平的正负分布仅表示低于或高于该层次上平均温度的程度，图 4 为两组数值试验与 FNL 在阻塞高压活动区域内的温差及温度距平的垂直分布随时间的变化。

从两组数值试验与 FNL 在阻塞高压活动区域内的温差随时间的变化(图 4a、b)可知，对照试验中，除 1000 hPa 可能受近地层因素影响较多外，其余层次的温差几乎均在－1～1 K，变化幅度较小，29 日前后在温差“0”线附近，呈现出整层温差分布逐渐相反的趋势，温差由“－”变为以“＋”为主(图 4a)。相较于对照试验，敏感性试验中温差梯度明显增强，温差分布多在－2～＋5 K，25—26 日 200 hPa 附近的温差甚至达到了 7 K，温差随时间变化幅度较大，29 日前后高低空的温差分布由“＋－”转变为“－＋”为主(图 4b)，可知在敏感性试验中，29 日前阻塞高压活动区域内高层相对大幅度变暖，中低层相对变冷，29 日后高层相对变冷，中低层相对变暖。相应的，在 29 日前高层位势高度场升高，中低层位势高度场下降，29 日后高层位势高度场下降，中低层位势高度场上升。另外，对比两组数值试验，对照试验的模拟结果比敏感性试验的模拟结果更接近于 FNL 资料，更符合实际温度分布。

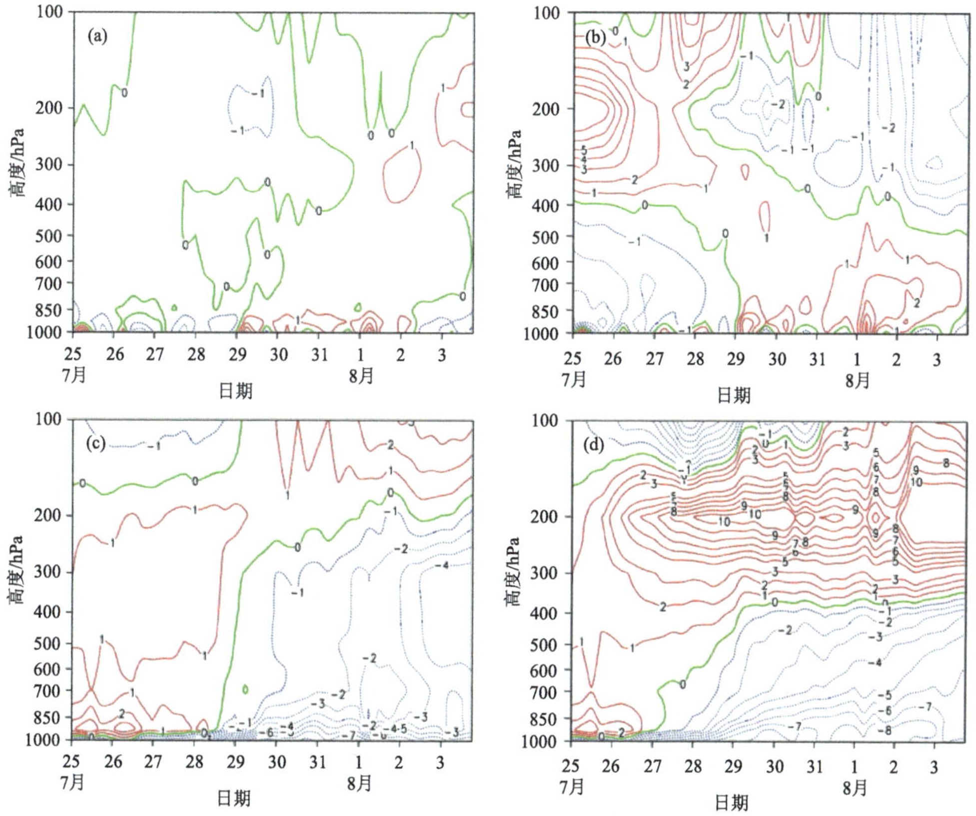

图 4　两组数值试验与 FNL 在阻塞高压活动区域内的温差及温度距平的垂直分布随时间的变化
(a. 对照试验与 FNL 的温差，b. 敏感性试验与 FNL 的温差，c. 对照试验温度距平，d. 敏感性试验温度距平；单位：K)

FNL 的阻塞高压活动区域内温度距平与对照试验(图 4c)整层温度距平的正负区域分布较为相似，29 日前，高层和中低层“温度距平”整体呈现出“－＋”的分布趋势，而 29 日后呈现出相反的分布趋势。敏感性试验(图 4d)的整层温度距平与 FNL 和对照试验存在着明显的区别，29 日前整层温度距平以“＋”为主，29 日后高空—低空的温度距平呈现出“＋－”的分布特征。相较于对照试验的模拟结果，敏感性试验在 400 hPa 以上有明显的温度正距平区，尤其是 200 hPa 附近温度距平的最大值超过了 10 K，而对照试验在此层次上的温度正距平最大值仅为 1 K 左右，说明在高层敏感性试验模拟温度高于对照试验。400 hPa 以下，两组试验都存在着温度负距平区，不同的是随着时间推移，敏感性试验温度负距平的“强度”高于对照试验，且这种趋势越到阻塞高压活动后期愈加明显。即敏感性试验对 400 hPa 以上的高层模拟比对照试验偏暖，400 hPa 以下比对照试验偏冷，冷暖幅度明显。如果把阻塞高压整体简单地比作一个气柱，在气柱中高层相对暖、中低层相对冷，会使得气柱高层气体膨胀，低层气体收缩，在一定程度上也会造成位势高度场下降，这也与前文中(图 2)在整个阻塞高压过程中敏感性试验模拟 500 hPa 附近位势高度场偏低，200 hPa 位势高度场模拟偏高相呼应。

由阻塞高压活动区域内温度差值和温度距平随时间的变化趋势可知，对照试验模拟结果更加接近实况，为更细致对比两组试验对阻塞高压过程的影响，用对照试验模拟结果输出的高分辨率精细资料代替实况资料，讨论分析导致两组试验模拟差异的原因。

4.2 非绝热加热差异分析

4.2.1 非绝热加热分布差异分析

大气中非绝热加热是维持大气环流运转和发展天气系统的主要热力强迫因子(高丽 等，2013)。阻塞高压活动区域内非绝热加热的变化对它的形成有重要作用，阻塞能否形成及阻塞过程的具体特点与这种热源强迫有很大关系(朱正心，1984；朱正心 等，1984)，前期一些数值试验也证明了如果不考虑这种强迫作用阻塞形势就会不存在(朱正心 等，1982；朱正心，1983)。

由前述可知，对照试验和敏感性试验在此次阻塞高压模拟过程中温度场差异较为明显，这说明二者的大气加热场不同。纬向不对称热力强迫用 Q^* 表示(朱正心，1984)。

$$Q^* = \frac{RQ}{c_p \sigma p_2^2 f_0} \tag{2}$$

式中，Q 为非绝热加热率；其他物理量为气象上常用的参数。由式(2)可知，纬向不对称热力强迫与非绝热加热关系密切，且同一地点，纬向不对称加热主要取决于非绝热加热。故计算阻塞高压实际活动区域内非绝热加热随时间的变化，探讨其对阻塞高压维持及崩溃的影响。

从阻塞高压活动区域内的非绝热加热随时间的变化(图 5)可知，对照试验(图 5a)非绝热加热日变化特征较为明显，尤其在 28 日前阻塞高压强度不断增强时期，整层大气的非绝热加热以正大值区为主，且中层500 hPa 最大非绝热加热中心值大于 6 K/d，说明此时大气整体上处于加热升温状态，有利于阻塞高压强度的增强。28—30 日整层大气有明显的非绝热加热负值区，从 200 hPa 附近一直延伸到地面，说明此时大气整体上处于冷却降温状态，阻塞高压强度稍有减弱，高压中心也略有移动。30 日后非绝热加热呈现出较为稳定的趋势，但量值较 28 日前弱。由阻塞高压实际活动情况可知，在 28 日前的非绝热加热正值区，阻塞高压在这段时间内闭合高压中心的强度不断增强，一度达到 5900 gpm，29—30 日非绝热加热的负值区间，阻塞高压中心强度减弱，但可能由于前期累积的热量较多，在此期间大气降温作用虽使得阻塞高压中心强度减弱，但不足以使此次阻塞高压过程崩溃。30 日后，非绝热加热又转为正值区，但强度较 28 日前弱，阻塞高压活动得以继续维持。此外，1—2 日、2—3 日同样存在着一个非绝热加热的负值狭长地带，而且 500 hPa 附近(邻近层次)非绝热加热的正值也在减弱，说明非绝热加热对大气升温在逐渐减弱，阻塞高压过程在此时处于临近崩溃阶段。

敏感性试验(图 5b)在阻塞高压活动区域内的非绝热加热随时间变化与对照试验分布较为类似，7 月

26—28 日 500 hPa 最大非绝热加热中心值小于 5 K/d，300～200 hPa 最大非绝热加热中心值小于 3 K/d。相较于同期的对照试验，大气升温幅度小，非绝热加热整体强度弱，此阶段大气累积的热量少，阻塞高压中心强度弱。28—30 日，阻塞高压内整层大气的非绝热加热为强的负值区，最大非绝热加热中心值小于 −6 K/d，大气冷却降温幅度大，不利于阻塞高压维持。此时模拟结果也显示，28—30 日，阻塞高压的中心强度也迅速减弱，阻塞高压过程逐渐崩溃消亡。

另外，由二者的非绝热加热的差值分布（图 5c）可进一步看出两组试验中非绝热加热的明显差异。1 日前，整层的非绝热加热的差值以正值为主，说明对照试验比敏感性试验的非绝热加热强，对大气的加热升温作用也明显，有利于阻塞高压的维持。

由以上分析可知，非绝热加热随时间的变化同阻塞高压过程中心强度和持续时间的变化有很好的时间对应关系。在非绝热加热的正值时段，大气升温，阻塞高压中心加强，反之，在非绝热加热的负值时段，大气降温，阻塞高压中心减弱甚至阻塞高压过程崩溃消亡。总体上，非绝热加热的变化趋势对此次阻塞高压过程的强度和持续时间有一定正反馈作用。

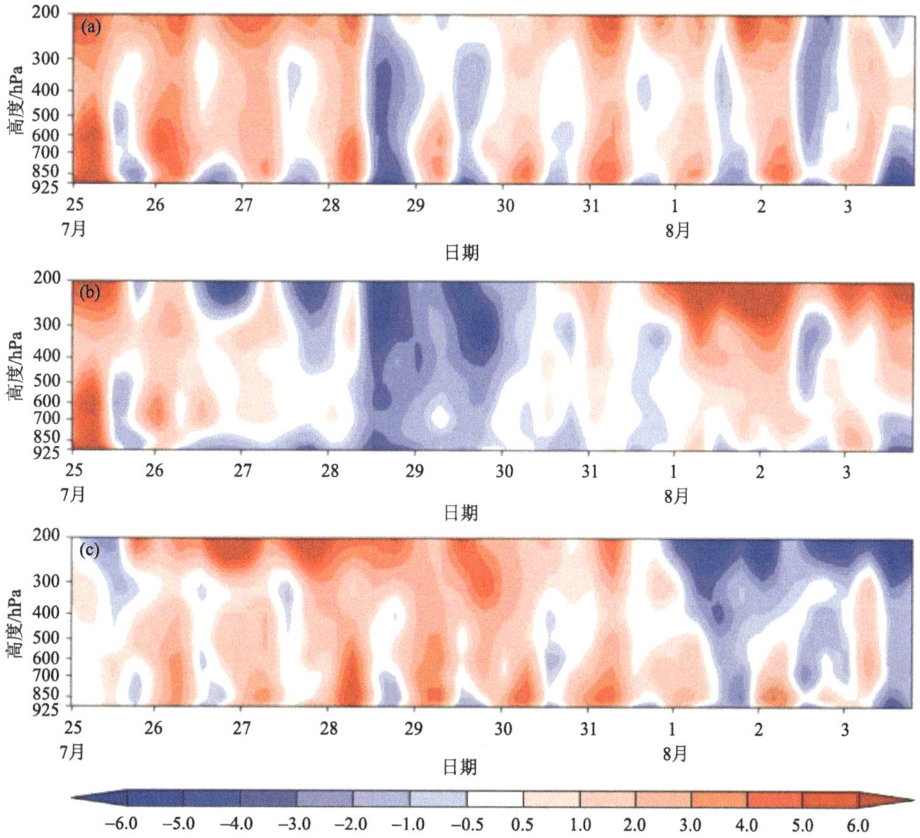

图 5　阻塞高压活动区域内非绝热加热（单位：K/d）随时间的变化

（a. 对照试验，b. 敏感性试验，c. 对照试验与敏感性试验的差值）

4.2.2　非绝热加热的各项差异分析

非绝热变化由局地变化项、水平平流项、垂直输送项组成，由前文分析可知，阻塞高压活动区域内的非绝热加热对阻塞高压的强度和维持时间有明显的影响，本节主要讨论两组试验的非绝热加热的差异主要由哪项引起。

两组试验在阻塞高压活动区域内非绝热加热各项的变化及其相应差值分布（图 6）表明，在阻塞高压过程中，对照试验的非绝热加热的局地变化项呈现出接近正、负相间的日变化趋势（图 $6a_1$），这是导致非绝热加热日变化特征明显的主要因素。敏感性试验的局地变化项（图 $6b_1$）存在较为明显的高层正值、中低层负值的分布特点，值得注意的是，28—30 日整层非绝热加热的局地变化项以负值为主，这与阻塞高压崩溃消亡的时间接近。由二者非绝热加热的局地变化项差值分布（图 6c_1）也可知，局地变化项在 400 hPa 以下以正、负相间为主，400 hPa 以上以负值为主，说明非绝热加热的局地变化项在 400 hPa 以下相对以加热大气为主，400 hPa 以上相对以冷却大气为主。

两组试验非绝热加热的水平平流项的主要加热和冷却区域集中在大气的高层（300 hPa 以上）以及低层（700 hPa 以下），对照试验中非绝热加热的水平平流项在两个主要集中区域均以正值为主（图 6a_2），说明水平平流项在高层和低层中均对大气起加热作用。敏感性试验的水平平流项在低层以正值为主，但强度明显弱于对照试验，对大气的加热作用不明显。30 日前高层的水平平流项以负值为主，对大气有冷却降温作用，30 日后转为正值（图 6b_2）。由二者的差值分布场（图 6c_2）可知，水平平流项在 300 hPa 以上及 700 hPa 以下对大气加热或冷却的作用明显强于中间层次，两侧中心加热最大值超过 2 K/d，是中间层次的 4 倍左右，1—2 日高层 300～200 hPa 出现了非绝热加热的负值，强度不断增强，并向低层传播。

从两组试验非绝热加热的垂直输送项的变化（图 6a_3、b_3）可见，对照试验中非绝热加热的垂直输送项在整个阻塞高压活动演变过程中除 29 日前后外几乎全部为正值（图 6a_3），敏感性试验在 1 日前非绝热加热的垂直输送项主要为高层负值、低层正值，负值起初是在高层 300 hPa 附近，随后扩展到整层大气中，延伸至地面，后又逐渐退到 300 hPa 附近（图 6b_3）。由两组试验的垂直输送项差值（图 6c_3）也可看出，1 日之前整层大气的垂直输送项以正值为主，大气相对被加热升温，相应阻塞高压的强度和持续时间也较强。垂

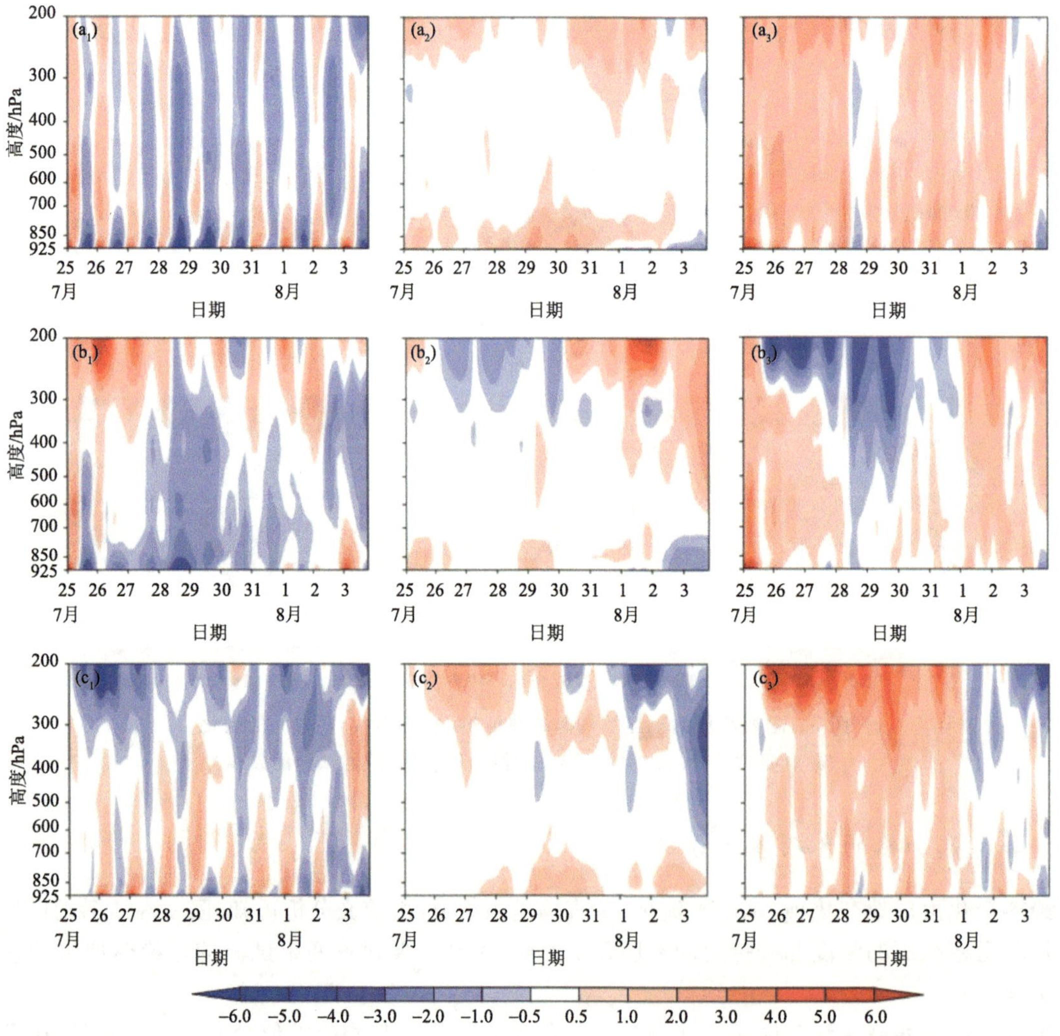

图 6 阻塞高压活动区域内两组试验非绝热加热的局地变化项（a_1、b_1）、水平平流项（a_2、b_2）、垂直输送项（a_3、b_3）的变化及其相应差值（c_1、c_3）（其中 a_1—a_3. 对照试验，b_1—b_3. 敏感性试验，c_1—c_3. 对照试验与敏感性试验的差值单位：K/d）

直输送项的差值分布场与前文中非绝热加热的整体差值分布场(图 5c)相似,说明在阻塞高压的演变过程中,垂直输送项对加热大气的作用显著。

综上所述,非绝热加热的变化对阻塞高压过程的强度和持续时间有很好的指示意义。非绝热加热整体的正、负相间的日变化特征主要由局地变化项引起,垂直输送项在整个阻塞高压过程中对大气的加热作用显著,水平平流项的作用主要集中在 300 hPa 以上及 700 hPa 以下。

4.3 云冰含量及微物理潜热释放的分析

微物理过程的潜热释放也会对非绝热加热产生影响,一般情况下,成云、降水等天气过程的发生、发展离不开动力过程和热力过程的影响,二者主要靠云内的微物理过程来实现(廖菲 等,2006)。微物理过程释放的潜热不仅对天气系统动力过程有影响,更能加速对流发展(Marwitz,1983)。以往研究结果表明,模式中引入了含有冰相过程微物理方案后,与未考虑冰相物理过程的微物理方案相比,其潜热变化会对云及某些动力过程产生较大的影响(Grabowski,1998)。冰相粒子的凝华、融化等过程必然伴随着热量的释放和吸收,从而引起大气温度的升降,以及热量的传递和释放。鉴于两种微物理方案最明显的区别就是冰相物理过程,本节主要分析云冰含量的背景场及阻塞高压活动过程内部及周围的潜热释放的差异可能对阻塞高压过程引起的影响。

从对照试验阻塞高压活动过程中 45°—60°N 区域内平均云冰含量随着高度的变化(图 7)可知,云冰主要分布在 700～300 hPa,在 400 hPa 层次上达到最大值,在阻塞高压较为强盛的 500～300 hPa,云冰含量较其他层次高。此外,根据图 8 可知,在 500 hPa 和 400 hPa 平面上,云冰含量大值区只存在于阻塞高压的外围,尤其是阻塞高压的上游基本流分支点处,这些冰相物质在相变过程中伴随的潜热释放和吸收在一定程度上会对周围的大气温度起到加热或冷却作用,微物理潜热释放的热量的传输也可能对阻塞高压外围的大气温度产生影响。本次阻塞高压过程主要活动在 45°—60°N 区域,结合阻塞高压在 500 hPa 层次上活动活跃,因此比较了此层次上两组数值试验在该纬度带的平均潜热释放的经度-时间变化(图 9),探讨两种方案中微物理过程释放的潜热及其差值分布情况。

整体上云冰的分布区域与微物理过程潜热释放的"大值区"相对应,两种微物理方案潜热释放大值区并不在阻塞高压的活动中心,而是在其外围(阻塞高压的上、下游),且对照试验的微物理过程潜热释放强度明显比敏感性试验强,从风场形势可知,微物理过程释放的热量被不断地输送到阻塞高压的活动中心

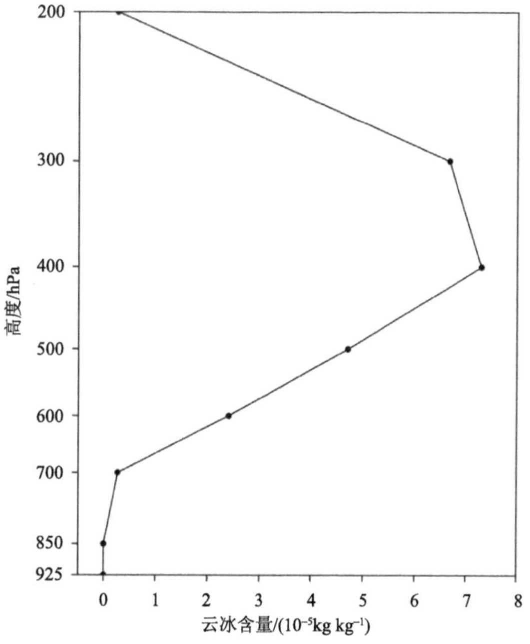

图 7 阻塞高压活动过程中 45°—60°N 区域内的平均云冰含量随高度的变化廓线

(图 9a、b)，说明微物理过程释放的能量在阻塞高压活动区域内是不断汇合的，这对位势高度维持具有一定的正反馈作用，这可能也是阻塞高压一种自维持机制。

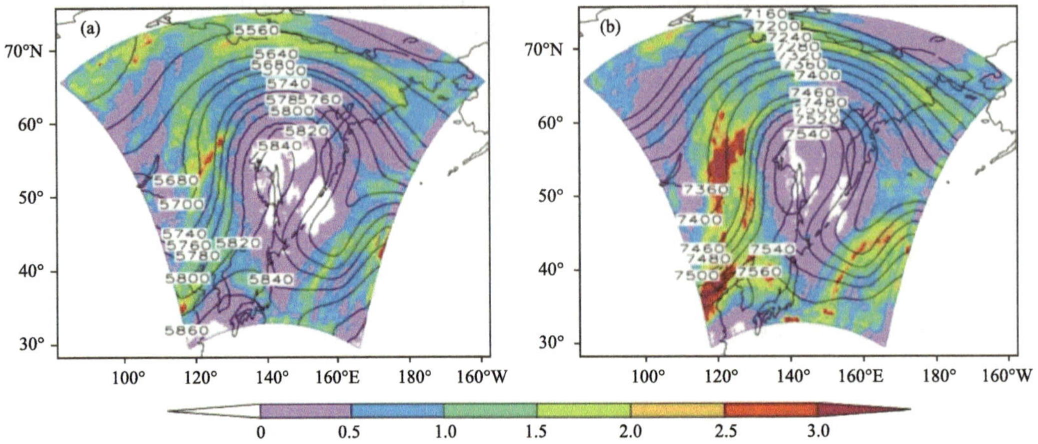

图 8　阻塞高压活动时间段内 500 hPa(a)、400 hPa (b)的平均云冰含量分布
(色阶为云冰含量，单位：10^{-5} kg kg^{-1}；等值线为位势高度，单位：gpm)

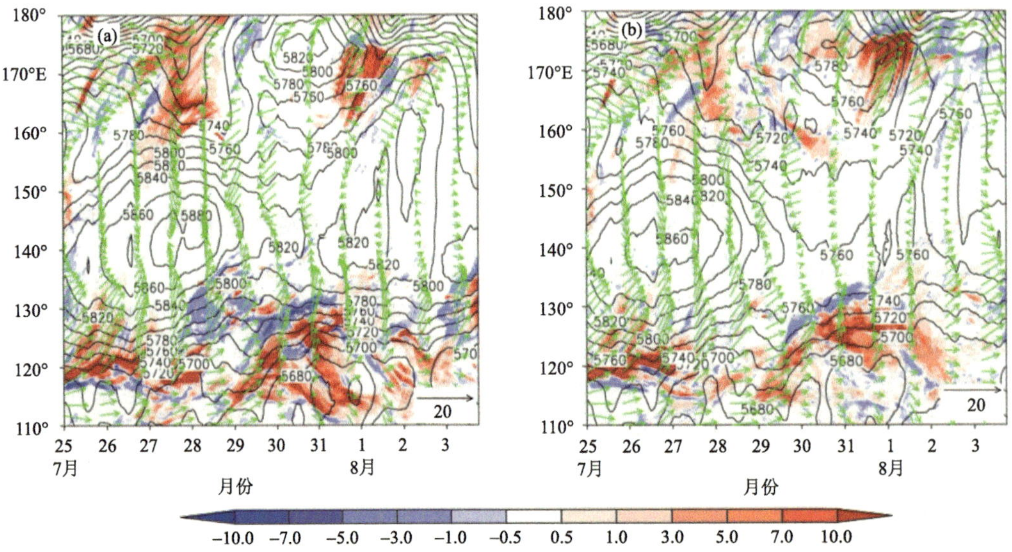

图 9　对照试验(a)与敏感性试验(b) 500 hPa 沿 45°—60°N 纬度带的平均微物理过程潜热释放随时间的分布
(等值线为位势高度场，单位：gpm；色阶为微物理过程潜热释放，单位：K/d；矢量为风场，单位：m/s)

此外，对照试验潜热释放的最大值中心，无论是正值部分(加热)还是负值部分(冷却)的面积及程度均较敏感性试验强，这其中的差异，可能主要是对照试验中考虑了多种冰相物质的原因。可见在阻塞高压这种大尺度环流系统的数值模拟中，大气中冰相物质的存在不容忽视。

为了更好地讨论微物理过程潜热释放对阻塞高压过程的影响，除纬向平均外，也应该讨论不同时间节点微物理过程潜热释放的水平分布，由前文分析可知在 28—29 日前后阻塞高压的非绝热加热存在一个明显的转折，故将 25—29 日作为第 1 时段，30 日至 8 月 3 日作为第 2 时段，分别计算两个时段内 500 hPa 上的平均微物理过程潜热释放量(图 10)。

由图 10 可知，无论是对照试验还是敏感性试验，微物理过程潜热释放的大值区主要分布在阻塞高压的外围，主要集中在阻塞高压的外围及上游分支点附近，这与图 8a 中的云冰含量的主要位置有很好的对应关系。虽然两组试验的微物理过程潜热释放的主要区域大致相同，但二者的微物理过程潜热释放量级和强度明显不同，在两个时段内，对照试验的平均微物理过程潜热释放明显比敏感性试验所累积的热量多，对大气的升温作用也强，尤其在阻塞高压的上游，释放的热量随着风场被源源不断输送到阻塞高压内部，一定程度上有利于阻塞高压的维持。

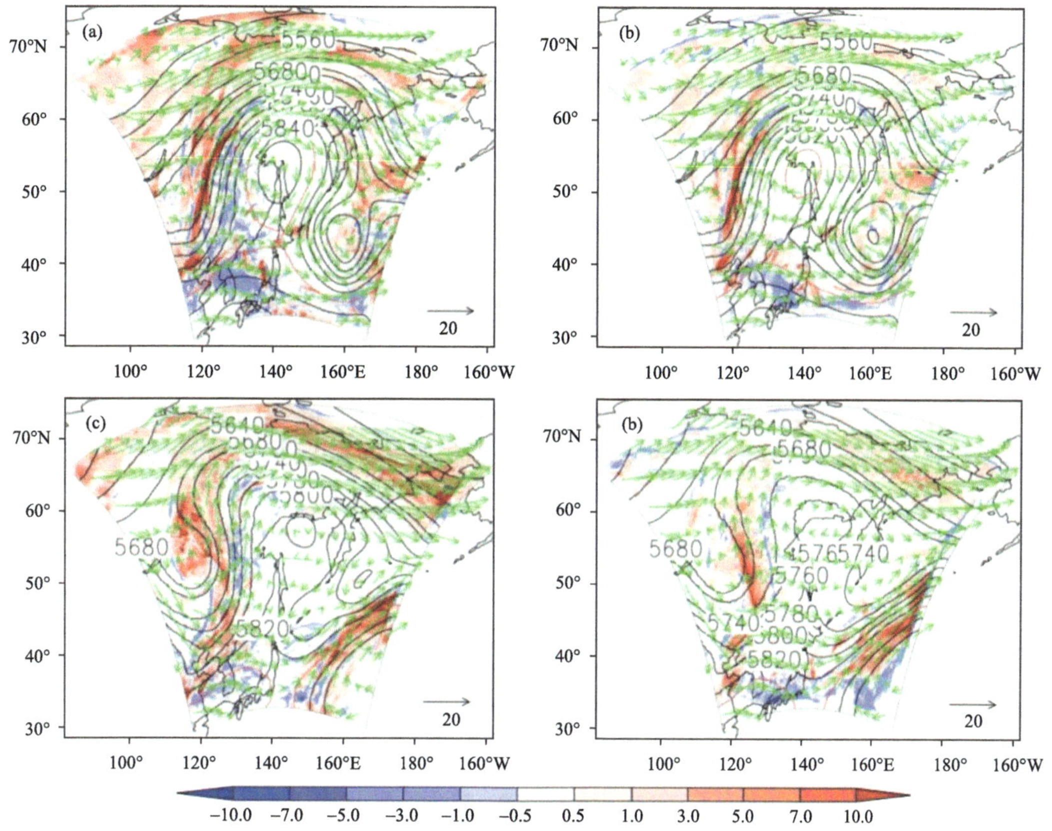

图 10　500 hPa 不同时段微物理潜热释放的平均分布

（a. 对照试验 7 月 25—29 日平均，b. 敏感性试验 7 月 25—29 日平均，c. 对照试验 7 月 30 日—8 月 3 日平均，d. 敏感性试验 7 月 30 日—8 月 3 日平均；等值线为位势高度场，单位：gpm；色阶为微物理过程潜势释放，单位：k/d；矢量为风场，单位：m/s）

5　结论与讨论

利用 WRF 模式，选用有无冰相物理过程的两种微物理方案，对 2011 年 7 月 25 日—8 月 3 日东北亚阻塞高压过程进行数值模拟，探讨了其模拟结果的差异，并简要地从非绝热加热、潜热释放角度，阐述了微物理过程对阻塞高压强度和维持时间变化的作用，得到以下结论。

（1）WRF 模式对阻塞高压过程的强度和维持时间有较好的模拟能力，无冰相物理过程的微物理方案模拟阻塞高压强度弱，持续时间短，有冰相物理过程的物理方案模拟结果更接近实况。有冰相物理过程的微物理方案对 500 hPa 和 200 hPa 整体阻塞高压的闭合环流形势、高压中心值、高压中心位置、高压强度以及高压中心上下游的低压槽强度的模拟均与实况接近，无冰相物理过程的微物理方案模拟结果在 500 hPa 和 200 hPa 位势平均图上虽然整体呈现出近似的环流形势，但在阻塞高压中心的强度和位置上与实况有较大差异，500 hPa 和 200 hPa 平均位势高度场上没有闭合的高值中心出现，且 500 hPa 阻塞高压强度弱，持续时间短，200 hPa 则相反。

(2)两种微物理方案对阻塞高压模拟结果的差异主要是由非绝热加热不同引起的，它对阻塞高压强度及维持时间具有很好的指示意义。数值试验显示，局地变化项、垂直输送项对东北亚阻塞高压形成、发展和维持作用很重要，水平平流项作用次之。

(3)微物理过程潜热释放对阻塞高压过程也有一定影响，数值模拟结果显示微物理过程潜热的释放在阻塞高压活动区域的外围不断向高压中心汇合，这可能也是阻塞高压局地自我维持的一个原因，但潜热释放的强度对照试验明显大于敏感性试验，这可能也是造成两种微物理方案模拟阻塞高压强度及维持时间不同的原因之一。

本研究选取一次东北亚阻塞高压演变过程并开展数值模拟试验，得出了大气中相变过程引起的非绝热加热作用对阻塞高压模拟有重要影响的相关结论，具有一定的代表性，揭示了东北亚阻塞高压形成与垂直局地加热有重要的机理联系。但本研究仅仅限于个例讨论研究，阐述不同微物理过程可能引起的对阻塞高压演变过程模拟结果的差异，主要涉及热力因素的分析，而实际过程中，除了热力因素外，阻塞高压内部的大气内动力过程、地形强迫、上下游能量传输、波流相互作用等因素是十分复杂的，有些问题还需进一步探索和研究。

参考文献

毕幕莹，丁一汇，1992.1980 年夏季华北干旱时期东亚阻塞形势的位涡分析[J].应用气象学报，3(2)：145-156.

陈希，沙文钰，李妍，2002.1998 年夏欧亚阻塞高压对西太平洋副热带地区大气环流影响的数值研究[J].热带气象学报，18(4)：368-373.

段廷扬，唐佑民，巩远发，等，1997.印度夏季风的年际变异与北半球大气环流的特征[J].气象学报，55(3)：371-378.

高丽，李建平，2013.非绝热加热对大气局地扰动位能的影响和机理[J].地球物理学报，56(10)：3255-3269.

李尚锋，廉毅，陈圣波，等，2012.东北初夏极端低温事件的空间分布特征及其成因机理分析[J].地理科学，32(6)：752-758.

廖菲，洪延超，郑国光，2006.影响云和降水的动力、热力与微物理因素的研究概述[J].气象，32(11)：3-11.

刘刚，沈柏竹，廉毅，等，2012.亚洲阻塞高压分类及其与东北冷涡活动和东北夏季低温的联系[J].地理科学，32(10)：1269-1274.

陆日宇，黄荣辉，1998.东亚—太平洋遥相关型波列对夏季东北亚阻塞高压年际变化的影响[J].大气科学，22(5)：727-734.

罗哲贤，1989.阻塞高压形成机制的数值研究[J].中国科学(B 辑)(6)：665-672.

潘婕，纪立人，布和朝鲁，2009.夏季欧亚中高纬持续流型的季节内气候特征[J].大气科学，33(2)：300-312.

孙力，郑秀雅，王琪，1994.东北冷涡的时空分布特征及其与东亚大型环流系统之间的关系[J].应用气象学报，5(3)：297-303.

汤懋苍，1957.亚洲东部的阻塞形势及其对天气气候的影响[J].气象学报，28(4)：282-293.

陶诗言，赵煜佳，陈晓敏，1958.东亚的梅雨期与亚洲上空大气环流季节变化的关系[J].气象学报，29(2)：119-134.

王毅，金荣花，代刊，等，2014.2008 年 1 月欧亚阻塞形势的 ECMWF 集合预报效果评估检验[J].大气科学学报，37(3)：257-267.

魏东，王亚非，董敏，2006.日本东部附近海域海温异常对鄂霍次克海高压的影响[J].气象学报，64(4)：518-526.

吴国雄，刘辉，陈飞，等，1994.时变涡动输送和阻高形成：1980 年夏中国的持续异常天气[J].气象学报，52(3)：308-320.

吴静，刁一娜，庄绪宗，2015.CMIP5 气候模式对北半球夏季阻塞的评估[J].中国海洋大学学报，45(2)：18-30.

徐爱华，潘华盛，1991.两大洋海表温度场形势变化与 6 月东亚阻塞高压建立的长期天气过程分析[J].海洋预报，8(4)：68-71.

徐祥德，解以扬，李培彦，1994.阻塞结构对近海台风异常路径影响的数值试验[J].应用气象学报，5(1)：104-108.

叶笃正，陶诗言，朱抱真，等，1962.北半球冬季阻塞形势的研究[M].北京：科学出版社：4-45.

张存杰，宋连春，李耀辉，2004.东亚地区夏季阻塞过程的研究进展[J].气象学报，62(1)：119-127.

张庆云，陶诗言，1998.亚洲中高纬度环流对东亚夏季降水的影响[J].气象学报，56(2)：199-211.

赵汉光，陈雪珍，1990.北半球阻塞高压的统计分析[J].气象，16(3)：3-7.

周宁芳，贾小龙，2018.NCEP CFSv2 对北半球夏季中高纬阻塞高压的预测检验[J].高原气象，37(2)：469-480.

周晓平，1957.亚洲中纬度区域阻塞形势的统计研究[J].气象学报，28(1)：75-85.

周学鸣，何金海，叶榕生，1995.乌拉尔阻塞高压影响亚洲夏季风环流和我国东部旱涝的数值试验[J].南京气象学院学报，18(1)：25-32.

朱正心，1983.地形与热力共同作用下超长波的非线性平衡态与阻塞形势[J].大气科学，7(1)：13-22.

朱正心，1984.阻塞动力机制中两个本质性的物理因子[J].大气科学，8(3)：233-241.

朱正心，朱抱真，1982. 纬向不对称热力强迫下超长波的非线性平衡态与阻塞形势[J]. 中国科学（B辑），12(4)：361-371.

朱正心，肖洁，1984. 阻塞形势动力机制的数值试验[J]. 气象学报，42(2)：129-138.

朱正心，崔秀兰，陈燕珠，1985. 北半球冬季阻塞过程的数值模拟及控制性试验[J]. 南京气象学院学报(1)：17-26.

CHARNEY J G, DEVORE J G, 1979. Multiple flow equilibria in the atmosphere and blocking[J]. J Atmos Sci, 36(7): 1205-1216.

DOLE R M, 1986. The life cycles of persistent anomalies and blocking over the North Pacific[J]. Adv Geophys, 29: 31-69.

DUDHIA J, 1989. Numerical study of convection observed during the winter monsoon experiment using a mesoscale two-dimensional model[J]. J Atmos Sci, 46(20): 3077-3107.

ELLIOTT R D, SMITH T B, 1949. A study of the effects of large blocking highs on the general circulation in the Northern-Hemisphere westerlies[J]. J Meteor, 6(2): 68-85.

GRABOWSKI W W, 1998. Toward cloud resolving modeling of large-scale tropical circulations: A simple cloud microphysics parameterization[J]. J Atmos Sci, 55(21): 3283-3298.

HARTMANN D L, GHAN S J, 1980. A statistical study of the dynamics of blocking[J]. Mon Wea Rev, 108(8): 1144-1159.

HONG S Y, LIM J O J, 2006a. The WRF Single Moment 6 Class Microphysics Scheme (WSM6)[J]. J Korean Meteor Soc, 42(2): 129-151.

HONG S Y, NOH Y, DUDHIA J, 2006b. A new vertical diffusion package with an explicit treatment of entrainment processes[J]. Mon Wea Rev, 134(9): 2318-2341.

JIA X L, YANG S, SONG W L, et al, 2014. Prediction of wintertime northern Hemisphere blocking by the NCEP climate forecast system[J]. J Meteor Res, 28(1): 76-90.

KAIN J S, 2004. The Kain-Fritsch convective parameterization: An update[J]. J Appl Meteor, 43(1): 170-181.

KESSLER E, 1969. On the Distribution and Continuity of Water Substance in Atmosphere Circulations[M]. Boston: American Meteorological Society: 84.

LU R Y, HUANG R H, 1996. Numerical simulation of the effect of the SST anomalies in the tropical western Pacific on the blocking highs over the Northeastern Asia[J]. Adv Atmos Sci, 13(4): 411-424.

MARWITZ J D, 1983. The kinematics of orographic airflow during sierra storms[J]. J Atmos Sci, 40(5): 1218-1227.

MATSUEDA M, MIZUTA R, KUSUNOKI S, 2009. Future change in wintertime atmospheric blocking simulated using a 20-km-mesh atmospheric global circulation model[J]. J Geophys Res, 114(D12): 114.

MLAWER E J, TAUBMAN S J, BROWN P D, et al, 1997. Radiative transfer for inhomogeneous atmospheres: RRTM, a validated correlated-k model for the longwave[J]. J Geophys Res, 102(D14): 16663-16682.

MONIN A S, OBUKHOV A M, 1954. Basic laws of turbulent mixing in the surface layer of the atmosphere[J]. Tr Akad Nauk SSSR Geophiz Inst, 24(151): 163-187.

OKAWA T, 1973. Growth mechanism of the Okhotsk high[J]. J Meteor Res, 25: 65-77.

PELLY J L, HOSKINS B J, 2003. How well does the ECMWF ensemble prediction system predict blocking? [J]. Quart J Roy Meteor Soc, 129(590): 1683-1702.

REX D F, 1950a. Blocking action in the middle troposphere and its effect upon regional climate. Ⅰ: An aerological study of blocking action[J]. Tellus, 2(3): 196-211.

REX D F, 1950b. Blocking action in the middle troposphere and its effect upon regional climate. Ⅱ: The climatology of blocking action[J]. Tellus, 2(4): 275-301.

SANDERS R A, 1953. Blocking highs over the eastern North Atlantic Ocean and Western Europe[J]. Mon Wea Rev, 81(3): 67-73.

SUMNER E J, 1954. A study of blocking in the Atlantic-European sector of the northern hemisphere[J]. Quart J Roy Meteor Soc, 80(345): 402-416.

TEWARI M, CHEN F, WANG W, et al, 2004. Implementation and verification of Unified Noah land surface model in the WRF Model[C]. Seattle: 20th Conference on Weather Analysis and Forecasting: 11-15.

TREIDL R A, BIRCH E C, SAJECKI P, 1981. Blocking action in the northern hemisphere: A climatological study[J]. Atmos Ocean, 19(1): 1-23.

WANG Y F, 1992. Effects of blocking anticyclones in Eurasia in the rainy season (Meiyu/Baiu Season)[J]. J Meteor Soc

Japan,70(5):929-951.

WATSON J C,COLUCCI S J,2002. Evaluation of ensemble predictions of blocking in the NCEP global spectral model[J]. Mon Wea Rev,130(12):3008-3021.

WHITE E B,CLARK N E,1975. On the development of blocking ridge activity over the central North Pacific[J]. J Atmos Sci,32(3):489-502.

YANAI M,ESBENSEN S,CHU J H,1973. Determination of bulk properties of tropical cloud clusters from large-scale heat and moisture budgets[J]. J Atmos Sci,30(4):611-627.

Climate Characteristics of Abnormal Double-Blocking Activities over the Ural Mountains and Sea of Okhotsk*

LIU Gang[1] WANG Tao[2] YANG Xueyan[3] WANG Yishu[2,4] YANG Xu[1] CUI Yan[2]

(1 Jilin Institute of Meteorological Sciences, Laboratory of Middle-High Latitude Circulation System and East Asian Monsoon Research, Changchun 130062; 2 Shenyang Regional Climate Center, Shenyang 110166; 3 Jilin Meteorological Observatory, Changchun 130062; 4 Shenyang Agricultural University, Shenyang 110866)

Abstract: By using NCEP-NCAR daily reanalysis data for June—August of 1948—2009, the synoptic situation of the double-blocking high pressure process over the Ural Mountains—Sea of Okhotsk region was identified objectively, and the climatic characteristics and dynamic mechanism during the double blockings were also investigated. The results suggest that the Urals-Okhotsk double-blocking high experienced obviously an abrupt change around 1977. After the change, the occurrence frequency and number of sustained days of the blocking high reduced significantly. Moreover, the zonal spacing between the Urals blocking and the Okhotsk blocking also decreased, while their meridional locations did not change much. The double blockings became strengthened in general, and the blocking high over the Sea of Okhotsk extended northward evidently. However, during the process without double-blocking activities, the intensity and location of the double-blocking high showed no noticeable variation before and after the abrupt change. In addition, dynamic diagnosis shows that after the abrupt change, the energy of the double-blocking high mainly propagated vertically from low to high levels, and transferred horizontally from low to high latitudes; whereas, previously, before the abrupt change, this occurred from high to low latitudes. Meanwhile, the development of a low pressure system was fueled between the double blockings, which meant that the double-blocking activity was beneficial to the maintenance and development of the Northeast China cold vortex. In short, diagnosis of the wave energy has provided us with a better explanation of the life cycle characteristics of this double-blocking phenomenon before and after 1977.

Keywords: objective identification; blocking high; dynamic diagnosis; characteristics analysis

1 Introduction

Blockings are a characteristic type of meridional circulation, whose establishment, collapse, and retreat are often accompanied by abrupt changes in large or even hemispheric-scale circulation systems. At the same time, they can cause weather and climate anomalies in surrounding and nearby areas, and

* 本文原载于 *Journal of Meteorology*, 2017, 31(4): 694-707

sometimes even severely disastrous weather (Zhu et al. ,2007). Blockings possess noticeable features in time, space, and geographical distribution, and are closely correlated with the occurrence of the Meiyu season in China, summer drought in North China, and floods, cold-air outbreaks, and other disasters (Zhao et al. ,1990; Li et al. ,2012). Scientists at home and abroad have long been aware of the importance of blockings in Asia, and have thus carried out a series of studies on these systems to further our knowledge on them. Early research in this regard was mainly in the form of statistical analyses, while more recent work has focused on the underlying mechanisms involved and their numerical simulation (Elliot et al. ,1949; Rex, 1950; Sumner, 1954; Tang, 1957; Ye et al. ,1962; Shi et al. ,1984; Shi, 1989, 2002; Xu et al. ,1994; Liu et al. ,1995; Gao et al. ,1998; Li et al. ,2004; Duan, 2012; Liu et al. ,2012).

Statistical analysis has revealed that, geographically, blocking high activities that significantly affect summer weather in China primarily concentrate in the areas of the Sea of Okhotsk, the Ural Mountains, and Lake Baikal (Tang, 1957; Bi et al. ,1992). In terms of occurrence frequency, the annual mean blocking frequency over the Ural Mountains area is the highest, while in summer, the Okhotsk blocking happens most frequently (Li et al. ,2012). Regarding the temporal distribution of blockings in Eurasia, the blocking activity in summer lasts longest on average; and meanwhile, spring and summer blockings are most frequently seen, and the number of autumn blockings is the smallest (Huang et al. ,2002). In June, blocking activities occur mostly in the Ural Mountains and the Sea of Okhotsk region, mainly in the form of double blockings (Li et al. ,2004). Alongside the above, many scientists have also become interested in researching the formation mechanisms of blocking highs. As early as in 1949, Ye Duzheng (1962), from the viewpoint of energy dispersion, investigated the forming and fading mechanism of blocking highs for the first time. Following his pioneering work, other scholars launched a series of studies, establishing many kinds of theories, such as multi-equilibrium theory, soliton and dipole theory, synoptic-scale disturbance excitation, and so on. Accordingly, progress regarding the formation and maintenance of blocking situations quickly reached a very high level, thus laying solid foundations for later studies (Rossby, 1950; Charney et al. ,1979; Zhu et al. ,1982; Miao, 1984; Shutts, 1986; Holopainen et al. ,1987; Bi et al. ,1992; Wu et al. ,1994).

As research in this field moved forward, scientists found that, in the case of a single blocking high, the Ural Mountains-Sea of Okhotsk blocking high performs a very important function in the evolution of atmospheric circulations in the eastern part of China, as well as the entire northeast of Asia. As one of the major meteorological disasters in China, floods are mainly affected by the blocking-high activities over the Ural Mountains and Sea of Okhotsk (Li et al. ,2012). Besides, the occurrence of low temperature in summer and cold vortex activities in Northeast China have been found to be closely related to the Okhotsk blocking high (Wang, 2002; Wang et al. ,2005; Liang et al. ,2009; Liu et al. ,2012). Similarly, when double blockings appear over the Asian region, their impact on the weather and climate in China is particularly significant. Shi (2007) pointed out that, in years when the number of days of double blockings in June and July is obviously greater in the Eurasian continent, the quantity of precipitation over the Yangtze River valley is also much more than normal. Yao et al. (2005), through analyzing rainstorms in the Huaihe River basin during June—July 2003, found that the establishment of high-latitude double-blocking type highs, which are strong and stable, is beneficial to maintaining a relatively stable state of the subtropical high, creating persistent and abnormal rainfall over the Huaihe River basin, and thus strengthening the westward extension of the subtropical high. Wang et al. (2010) analyzed the anomalous features of the western Pacific subtropical high during periods of torrential rain in the Huaihe River basin along with abnormally high temperature across the Jiangnan region (the area south of the Yangtze

River) and South China during June—July 2007, and found that the establishment of mid-high latitude double-blocking type highs helps the strengthening and development of the western Pacific subtropical high. Yu et al. (2006) revealed that, when there are blocking highs over Europe and the Sea of Okhotsk, that is, the double-blocking type high in the Eurasian mid-high latitudes, the Huaihe River basin will receive flood-inducing rainfall and the precipitation amount over the whole of China will be greater than usual. However, if only the Okhotsk blocking high, that is, an "East Blocking" type high, exists, the Yangtze River valley and Huaihe River basin will suffer from flood-inducing rainfall while North and South China will experience less rainfall than usual. Gong et al. (2004) studied the circulation characteristics in mid-high latitudes in the summer of 2003 and the pentad variation of a blocking-high index over the three regions of the Ural Mountains, the Sea of Okhotsk, and Lake Baikal, and found that, from late June to early July, mid-high latitudes in East Asia are under the control of a "double-blocking" situation, which causes sustained (for more than a month) and concentrated heavy rainfall over the Huaihe River basin. Therefore, we can see that the configuration of flow regimes, in various local domains in Asia, not only have an important influence on Northeast Asia (Shen et al., 2011), but even on the anomalies of summer weather and climate over the whole of China (Ding, 2005).

To date, most investigations on the double-blocking high system have been separate in their approach, with few taking a more holistic perspective on the basis of case studies. Based on previous research results, we began by analyzing the blocking-high activities over the Ural Mountains (named "Urals" for short) and the Sea of Okhotsk (named "Okhotsk" for short) and, under the co-existence of the two blocking highs, investigated the evolution and climatological changes of the Urals-Okhotsk double-blocking type activities. In this work, we regarded the double-blocking disturbance as a quasi-stationary process. Through case analyses, we explored the atmospheric energy dispersion and propagation characteristics of quasi-stationary planetary waves in the double-blocking situation, so as to provide a reference for further revealing the effect of Urals-Okhotsk double-blocking activities in the atmospheric circulation of the Eurasian mid-high latitudes.

2 Data and methods

The data used in this study were the NCEP/NCAR 2.5°×2.5° daily geopotential height, wind, and temperature data for the Northern Hemisphere during 1948—2009 (Kalnay et al., 1996). Based on the definition of blocking-high synoptic meteorology and its existence condition given by Rex (1950) and Zhu et al. (2007), the criteria for mid-high-latitude blocking high processes in the Northern Hemisphere are that: a high pressure center exists in the 500 hPa geopotential height field; the daily moving range of the pressure center is no more than 12.5 degrees of longitude; and the high pressure center maintains for at least 3 days. According to these criteria, the Urals-Okhotsk double-blocking high process over 45°—75°N, 40°—160°E during June—August of 1948—2009 was retrieved and analyzed.

Relying on the classification of the Eurasian blocking high established by Liu et al. (2012), we objectively identified the Urals-Okhotsk double-blocking high process, obtaining its formation, development, and dissipation processes, and determining its onset, end, and maintenance times, the cumulative number of days, the frequency, and the distribution of the double-blocking activities through statistical analysis. Characteristics of the meteorological variables of the double-blocking configuration were analyzed synthetically. In addition, two typical double-blocking cases were selected (one that maintained for 8 days from 29 June to 6 July 1964, and the other that lasted for 11 days from 14 to 24 July 2008), and the wave activity flux diagnosis method (Plumb, 1985) was used to analyze the dynamic characteristics of the establishment and maintenance of the double-blocking high.

3 Distributional features of the Urals-Okhotsk double blockings

The number of days and frequency of the Urals-Okhotsk regional blocking high, from June to August 1948—2009, were statistically analyzed and the Mann-Kendall (MK) method was used to test the interannual variation of the number of days and occurrence frequency of the double blockings, separately (Fig. 1). As can be seen from Fig. 1a, the average number of days of double blockings is characterized by a remarkable abrupt change around 1977. Before 1977, the average number of days of double-blocking activities was 16.7 days, whereas after 1977, the number was only 5.8. To test the stability of the abrupt change, we used the STARS (Sequential t-test Analysis of Regime Shift) method (Rodionov, 2004, 2006). According to the different cut lengths, the regime shift index (RSI) values under different parameters were calculated, and the finding was that the abrupt change signal of the double-blocking process in 1977 is credible. Fig. 1b shows the results of the MK variation test for the interannual change in the frequency of the double-blocking configuration, revealing that the occurrence frequency of the double-blocking activity also experienced an abrupt change around 1977.

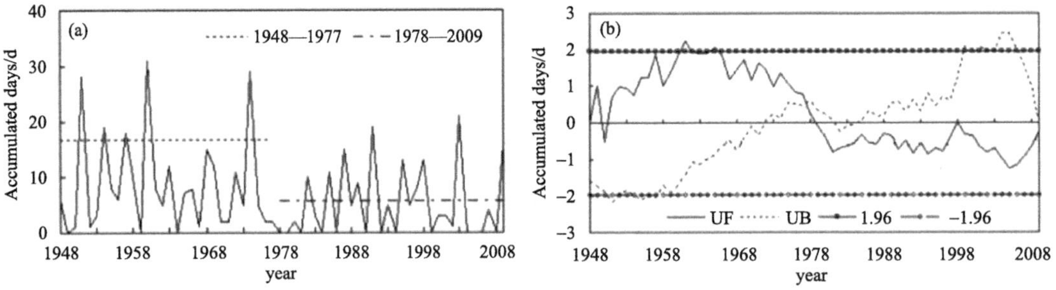

Fig. 1 (a) Interannual variations of accumulated days and (b) the frequency revealed by the MK method for the Urals-Okhotsk blockings from June to August 1948—2009. In (b), UF (UB) is forward (backward) sequential statistic, and the horizontal lines of ±1.96 indicate values passing the 95% confidence level

Based on the timing of the change in double-blocking activity, the difference between the blocking high intensity (maximum value in the 500 hPa height field) and the mean latitude and longitude positions before and after the abrupt change was analyzed (Fig. 2). Based on the variation of the blocking high intensity (Figs. 2a and 2d), the intensity of the Urals (Okhotsk) blocking high strengthened significantly from an average of 5796.6 gpm (5749.2 gpm) before the abrupt change, to 5817.6 gpm (5774.9 gpm) after the change. The changes in the longitude and latitude locations, where the blocking-high activities occurred (Figs. 2b, 2e, 2c and 2f), illustrate unobvious variation. The Urals blocking shifted eastward slightly after the abrupt change, while the Okhotsk blocking moved westward. The longitude position of the Urals (Okhotsk) blocking activity was on average 56.3°E (117.9°E) before the change, but moved eastward to 57.2°E (westward to 111.7°E) after the change. Thus, through analyzing the evolutionary characteristics of the Urals-Okhotsk double-blocking activity, we have discovered the timing of an abrupt change (i. e., around 1977) and the differences of the double-blocking high before and after that change. Next, taking the year when the abrupt change occurred as the boundary, we used composite analysis to investigate the climatological characteristics and differences of the double-blocking highs before and after the abrupt change.

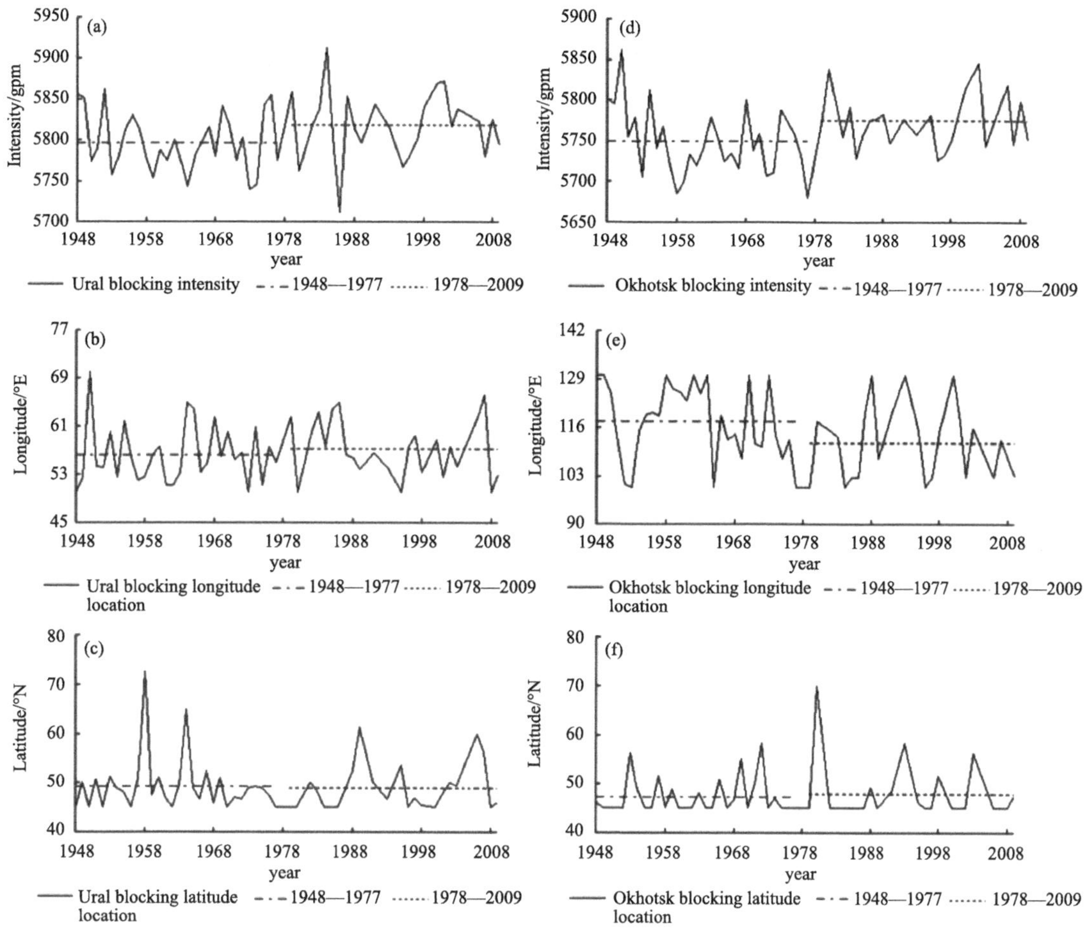

Fig. 2　Variations of (a) intensity (gpm), (b) longitude location (°E), and (c) latitude location (°N) of the blocking highs located over the Ural Mountains from June to August 1948—2009. (d—f) As in (a—c), but for the Sea of Okhotsk

3.1　Distribution features of height and wind fields

Fig. 3 shows the composite distribution of the 500 hPa height and wind fields of the Urals-Okhotsk double-blocking activities before and after the abrupt change. Before the abrupt change (Fig. 3a), the area from the Caspian Sea to the Ural Mountains and Sea of Okhotsk had two strong high-pressure ridges, with the 5700 gpm contour extending to the north near 65°N and an anticyclonic center near the Sea of Okhotsk. Thus, at 500 hPa, the blocking-high situation over the Ural Mountains was more remarkable than the Okhotsk high. After the abrupt change (Fig. 3b), the peripheral 5700 gpm contour of the Urals-Okhotsk blocking moved near to 70°N, showing that the Urals-Okhotsk high's ridge extended northward to higher latitudes. The Urals blocking high became more intensive, with the 5760 gpm contour of the center intensity extending to 50°N and the area further north, and the center longitude moving near 60°E. Meanwhile, the 5700 gpm contour of the Okhotsk blocking extended obviously to the northeast, and the anticyclone center became stronger. The composite height difference distribution (Fig. 3c) shows the area north of the Caspian Sea and Lake Balkhash, the northern section of the Sea of Okhotsk and the eastern Siberia region as the maximum negative-value areas of difference, at −40 and −60 gpm, respectively, both passing the 95% confidence level. Therefore, it can be seen that after the abrupt change, the double-blocking activities reduced significantly in terms of the number of days, but the blocking high in the double-blocking activity area was considerably stronger.

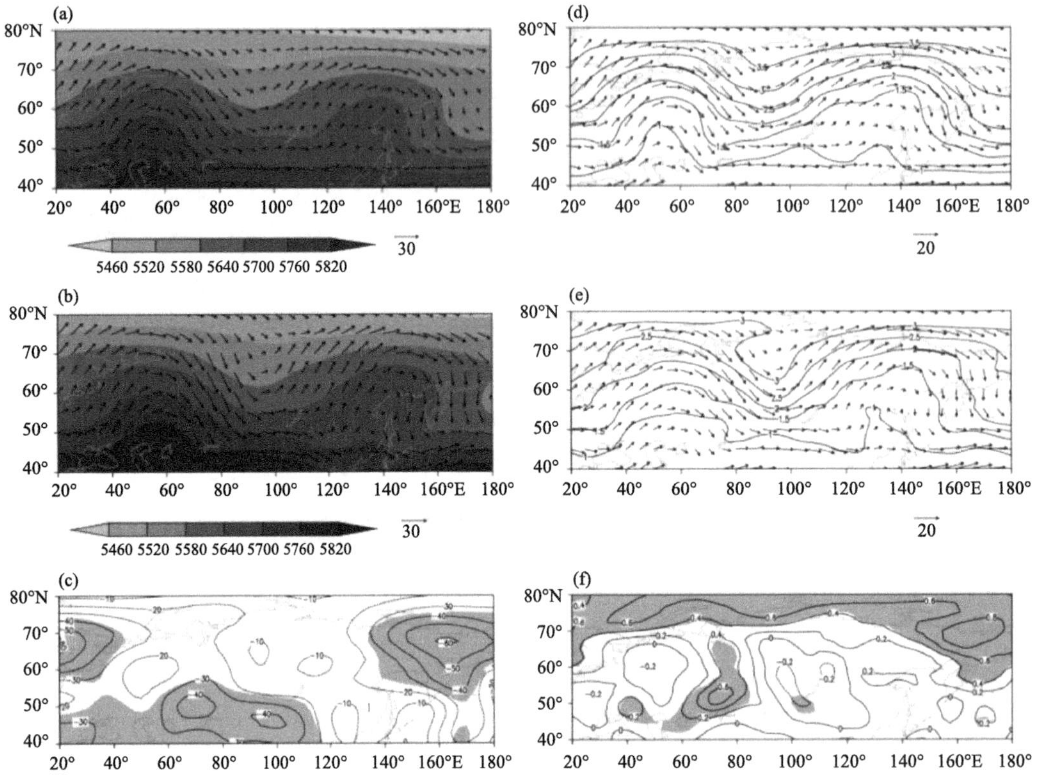

Fig. 3　Average height (gpm) and wind fields (m s) for (a) 1948—1977 and (b) 1978—2009, and (c) the composite difference of height (gpm) during the Urals-Okhotsk double-blocking activity from June to August 1948—2009. (d—f) As in (a—c), but for the 325 K isentropic potential vorticity (PVU) and wind field (m s^{-1}). The composite difference refers to the value before the abrupt change minus that after the change. Shading indicates areas that passed the 95% confidence level

3.2　Distribution features of the 325 K isentropic potential vorticity

To further investigate the atmospheric circulation features of the double-blocking activities before and after the abrupt change, we analyzed the variational features of the isentropic potential vorticity (IPV) during the Urals-Okhotsk double-blocking events. In the Northern Hemisphere, the value of the isentropic surface is usually 315 K in winter and 325 K in summer. On this basis, we present the 325 K isentropic surface vorticity and wind field features on the double-blocking days from June to August before and after the abrupt change (Fig. 3). Before the change (Fig. 3d), the 1-PVU contour over the Ural Mountains lay near 55°N and the 1.5-PVU contour near 60°N; whereas, in the Sea of Okhotsk area, the 1.5-PVU contour was located near 65°N. Furthermore, all contours were saddle-backing toward high latitudes, showing characteristics of minimum potential vorticity, and the wind field was in the clockwise direction, strengthening the development of the anticyclonic circulation in this region.

After the abrupt change (Fig. 3e), the 1-PVU isoline near the Ural Mountains retreated southward to the vicinity of 53°N, and the 1.5-PVU isoline near 60°N. In comparison, the 1.5-PVU contour over the Sea of Okhotsk area shifted little. The distribution of the difference values (Fig. 3f) shows large areas of the Caspian Sea-Balkhash Lake and its northern region as having a positive center of difference, having passed the 95% confidence level. This indicates that the potential vorticity in this region diminished significantly after the abrupt change while the anticyclonic circulation strengthened, contributing to the maintenance and intensification of the Ural Mountains blocking high. In the region of the Sea of Okhotsk and to the north, the 325 K IPV difference also shows a positive distribution.

3.3 Dynamic diagnosis

Fig. 4 displays the 500 hPa wave activity flux and its divergence distribution before and after the abrupt change of the double blockings over the Ural Mountains and the Sea of Okhotsk region. In general, wave energy propagated from upstream to downstream, and the characteristics of energy transfer from lower layers to middle layers were consistent, which was favorable for the maintenance of the blocking situation. Specifically, the wave energy coming from the upstream and lower level (Figs. 4a and 4b) converged in the western section (40°—55°N, 120°—140°E) and diverged in the eastern section (55°—70°N, 140°—155°E) of the two blocking highs before the abrupt change. Thus, the wave energy had consistent convergence features from low to high levels, which was conducive to the maintenance of deep systems. However, after the abrupt change, wave energy convergence area in the upstream of the Urals blocking high was reduced, causing the convergence center to shift westward, and making the divergence area in the downstream enlarged and the divergence intensity weakened, and then the divergence expanded northward (Fig. 4c). As seen from Fig. 4d, the upstream convergence of the Urals blocking high strengthened obviously, with its area expanding northward and the downstream divergence being more significant; the convergence of wave energy in the western part of the Okhotsk blocking high was significantly strengthened, and meanwhile the divergence in its eastern part also strengthened (Figs. 4c and 4d). Thus, the configuration of wave energy propagation after the abrupt change helped enhance the intensity of the double-blocking high and its extension northward. The major difference in energy propagation before and after the abrupt change lies in the fact that the path of energy dispersion from the Ural Mountains to the Sea of Okhotsk turned from the mid-low latitudes to high latitudes, which is of great significance in our quest to understand the dynamical mechanisms involved in the double-blocking high

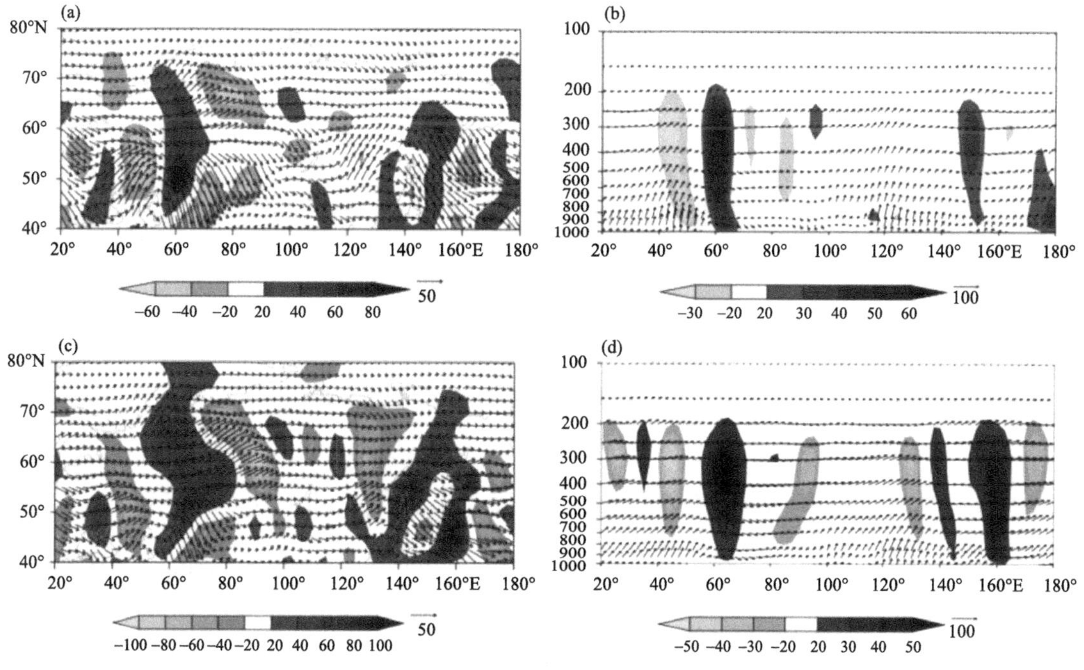

Fig. 4 Composite analysis of wave activity flux (vectors; m s^{-1}) and its divergence (shading; m s^{-1}) during the Urals-Okhotsk blocking activity at 500 hPa from June to August 1948—2009: (a, b) average during 1948—1977 and (c, d) average during 1978—2009. (b, d) Meridional average cross-section along 45°—75°N

activities before and after the abrupt change.

To highlight the situational characteristics of the Urals-Okhotsk double blockings, we further analyzed the composite distributional characteristics of the 500 hPa height and wind fields, and the 325-K IPV and wind field in the case of no double-blocking activities, based on atmospheric circulations with/without double blockings, before and after the abrupt change (Fig. 5). It is apparent that obviously large- or low-value centers do not exist in the composite height field (IPV) on the days without double blockings, regardless of the occurrence of abrupt changes, and all the isolines are distributed evenly in the latitudinal direction, except that the Sea of Okhotsk height field extends slightly to the north after the abrupt change. Meanwhile, the wind field essentially matches with the height field, whose zonal degrees are bigger but meridional degrees are smaller. Fig. 6 shows the composited distributions of the 500 hPa wave activity flux and its divergence in the process without double blockings before and after the abrupt change. From Fig. 6, we can see that, before and after the abrupt change, wave energy spread from high to low latitudes, and also from low to high levels. In addition, the convergence and divergence intensity of the wave energy were weaker in comparison.

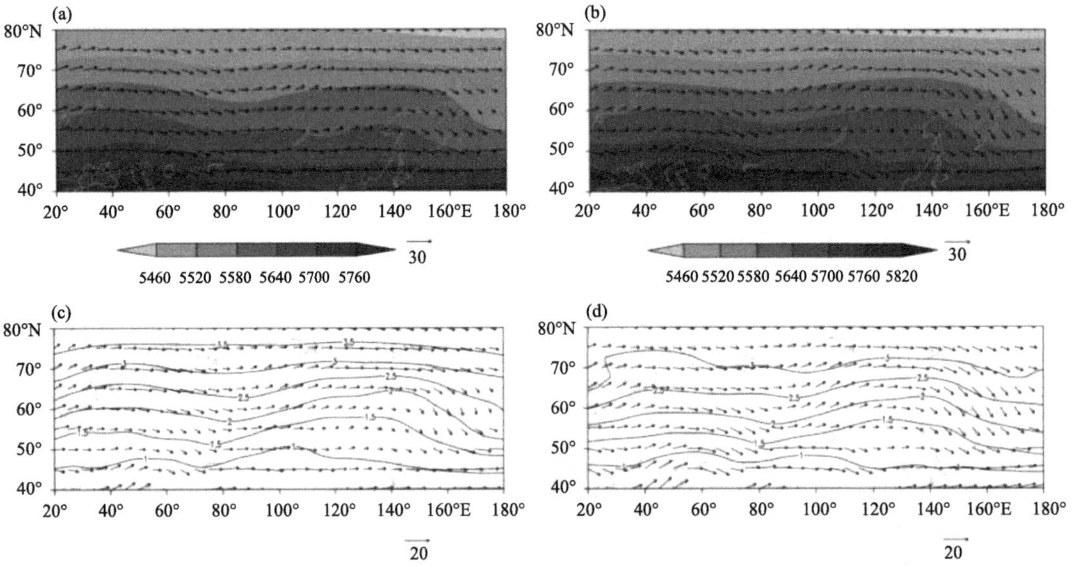

Fig. 5　Average height (gpm) and wind field ($m\ s^{-1}$) during (a) 1948—1977 and (b) 1978—2009 for the process without double-blocking activities from June to August 1948—2009. (c, d) As in (a, b), but for the 325-K isentropic potential vorticity (IPV) and wind field ($m\ s^{-1}$).

4　Case analysis of the Urals-Okhotsk double blockings

4.1　Analysis of the height and wind fields before the abrupt change

Fig. 7 reveals the characteristics of meteorological variables at 500 hPa during the Urals-Okhotsk double blockings from 29 June to 6 July 1964. As can be seen from Fig. 7, the double blockings were established on 29 June (Fig. 7a). Affected by the high-latitude low-value system in the upstream of the Ural Mountains, a weak anticyclonic center appeared in the north of the Ural Mountains (60°—70°N, 50°—70°E). The 5680 gpm contour extended northward to 70°N and a strong high ridge formed in the area southwest of the contour and over western Black Sea. The 5760 gpm contour extended close to 55°N.

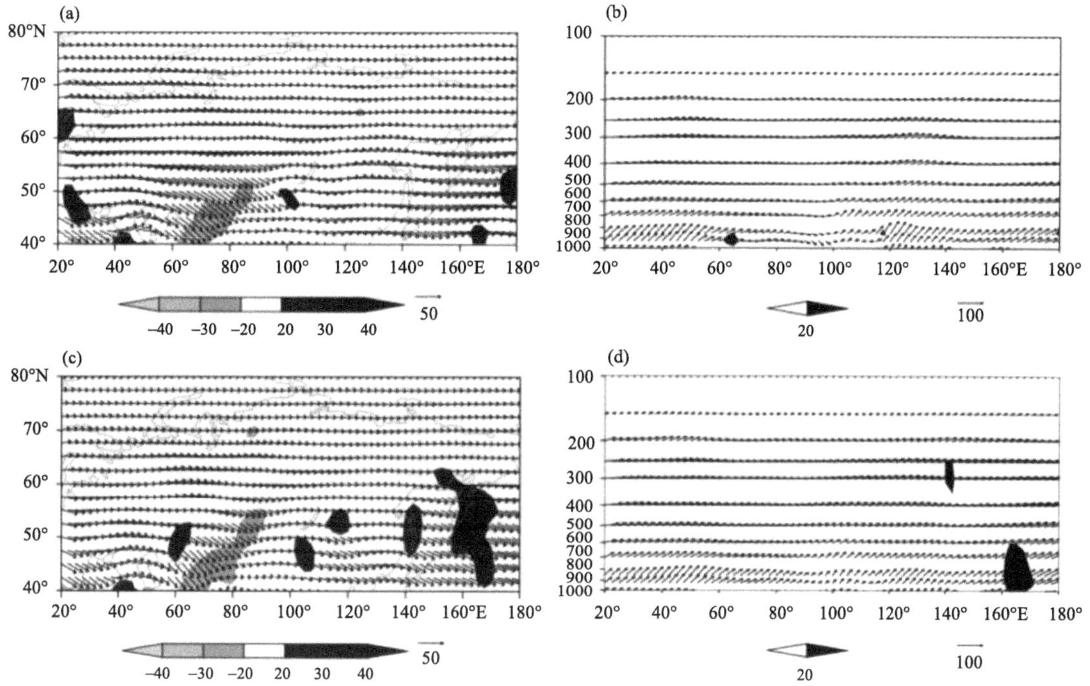

Fig. 6 Composite analysis of wave activity flux (vectors; m s^{-1}) and its divergence (shading; m s^{-1}) at 500 hPa during the process without double-blocking activities from June to August 1948—2009: (a,b) average during 1948—1977 and (c,d) average during 1978—2009. (b,d) Meridional average cross section along 45°—75°N

The area north of the Sea of Okhotsk featured a closed anticyclonic circulation, with the Chersky Mountains at the center, and in its southwest part, the Northeast Asian coast had a strong high ridge, with its 5760 gpm ridge line going northward to 55°N. To the west of the Da Hinggan Mountains, China, there was a strong cyclonic center (Northeast China cold vortex). On 2 July (Fig. 7b), due to the eastward movement of the low-value system in the upstream of the Ural Mountains, the Ural anticyclone moved further eastward. Affected by the terrain, the Urals blocking high attained a quasi-stationary state, causing the area and central intensity of the anticyclone to increase continuously. As a result, the central intensity exceeded 5760 gpm and a closed high-pressure center was born, making the Urals blocking high maintain and strengthen further. The sustained Urals blocking generated strong northerly winds on its leeward slope, providing energy for the development and maintenance of the downstream low-value system, which was stable and less dynamic. On the one hand, the southwest airflow in front of the trough powered the maintenance and development of the Okhotsk blocking high and, on the other hand, the existence of the Okhotsk blocking high made the low-value system maintain for a long time. On 6 July (Fig. 7c), the upstream low-value system gradually weakened and the Urals blocking high dispersed, causing the northerly winds on the leeward slope to disappear and the downstream low-value system to weaken and collapse. Besides, the southwest air flow in front of the trough faded, gradually disappearing, and along with it the Okhotsk blocking high weakened. Finally, the Urals-Okhotsk double blocking situation weakened and collapsed.

4.2 Diagnostic analysis of wave activity flux in a case before the abrupt change

Through the preliminary analysis of the characteristics of meteorological variables, we obtained the evolutionary features of the atmospheric circulation during the process of double-blocking activity. Next, we further diagnosed and analyzed the energy propagation features of the double-blocking case via the wave activity flux method. Fig. 8 shows the 500 hPa stream function field, wave activity flux, and divergence during the process of establishment, development, and weakening of the double-blocking case. On 29 June (Figs. 8a and 8b), the Urals Sea of Okhotsk region was covered by a large-value zone of stream function, representing anticyclonic activity. The wave energy from the upstream low-value system of the Ural Mountains spread to the anticyclonic center of the mountains, causing convergence, whilst downstream of the Ural Mountains was a weak divergence zone.

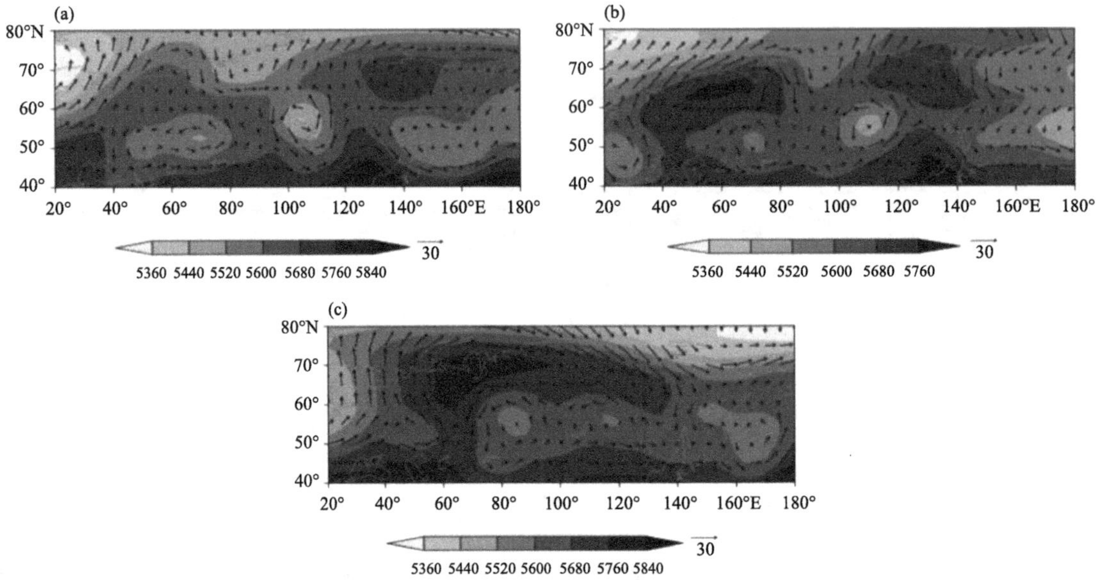

Fig. 7　Characteristics of meteorological variables at 500 hPa during the Urals-Okhotsk blocking activity from 29 June to 6 July 1964: (a) 29 June, (b) 2 July, and (c) 6 July. Shaded area denotes height field in gpm, and vectors denote wind field in m s^{-1}

Such propagation characteristics of the wave energy were conducive to the maintenance and development of the Urals blocking. The upstream wave energy of the Ural Mountains was divided into two branches, one north and another south, by the Urals blocking, dispersing to the downstream. The energy was transmitted to low latitudes more obviously, spreading eastward along 50°—60°N, and finally dispersing downstream to the Sea of Okhotsk area via the Northeast China cold vortex. The wave propagation process provided kinetic energy not only for the development and maintenance of the Northeast China cold vortex but also, indirectly, for the downstream Okhotsk blocking activities. Over the Sea of Okhotsk region, however, it was wave energy convergence that kept the system developing further and maintaining. On 2 July (Figs. 8c and 8d), the double blocking stepped into the developing period, and the large-value stream function center in the Ural Mountains and Okhotsk Sea increased further. As the upstream low-value system of the Ural Mountains deepened further and moved eastward, the wave propagation originating from the upstream became much stronger and broader, forming a ribbon of energy transmission along a route from the Caspian Sea to Volga River and Beloye More. This converged in the Urals blocking area and continued to propagate downstream along the blocking edge, so that the Urals blocking high developed further. The Northeast China cold vortex remained stable, and a southwest—northeast energy transmission belt was established in the fourth quadrant of the Northeast China cold vortex activities, enhancing the Okhotsk blocking situation. By 6 July (Figs. 8e and 8f), the low-value system in the Ural Mountains upstream was crippled, causing the energy source to die down. Although the Urals blocking high remained, the central energy became dispersed, and thus the wave energy propagating downstream was significantly weakened. As a result of the cutoff of the energy source, the Northeast China cold vortex was gradually undermined, and ultimately disappeared. The intensity of the Okhotsk blocking high then developed in a similar way and finally, the situation of the double-blocking high was seriously impacted until its collapse. Therefore, during the whole life cycle of double-blocking activity, the low-value system energy source in the upstream of the Urals blocking high, and its downstream dispersion, played crucial roles in the establishment and maintenance of the double-blocking high, while the existence of the Okhotsk high was affected by the low-level wave energy (Figs. 8b, 8d and 8f).

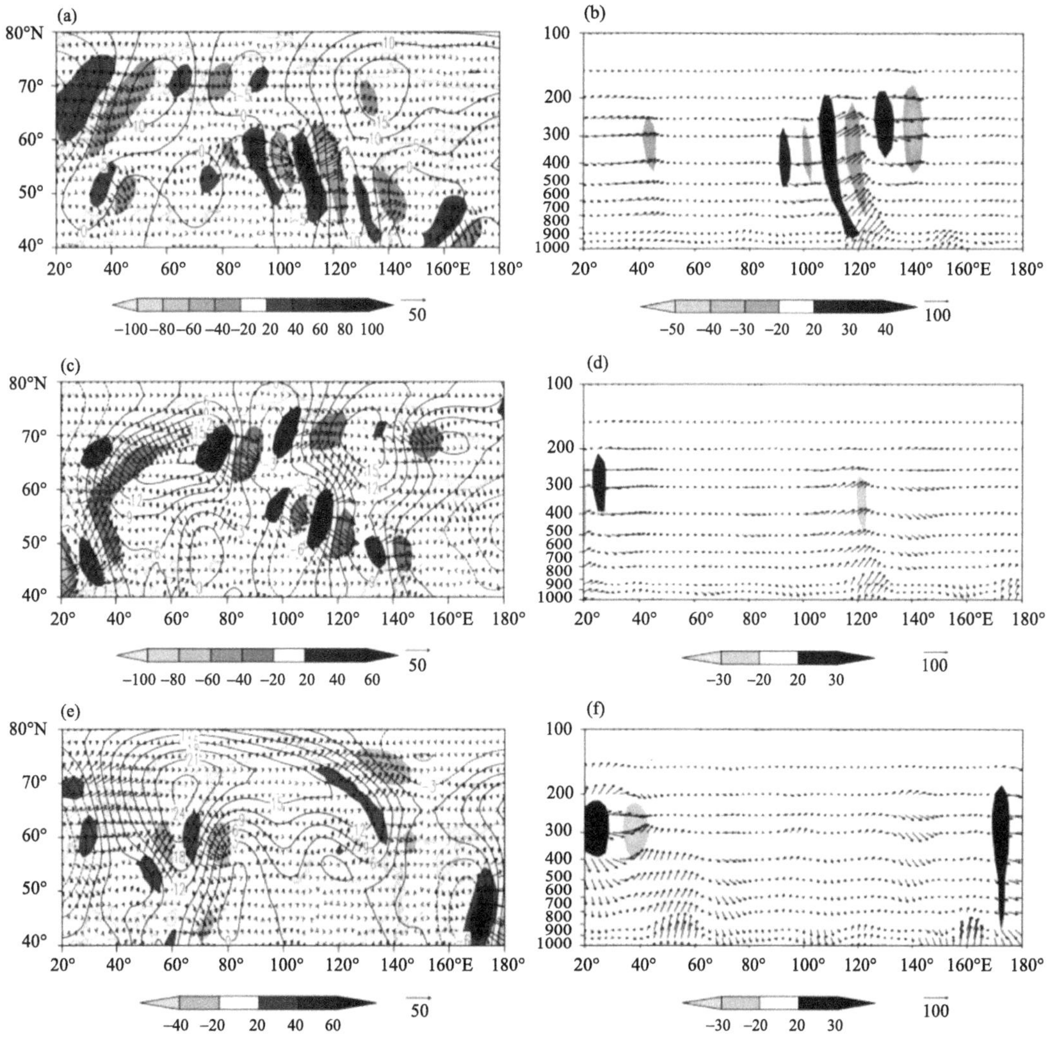

Fig. 8　Characteristics of the stream function field (contours; m s^{-1}), wave activity flux (vectors; m s^{-1}), and its divergence (shading; m s^{-1}) at 500 hPa during the Urals-Okhotsk blocking activity from 29 June to 6 July 1964: (a, b) 29 June, (c, d) 2 July, and (e, f) 6 July. (b, d, f) Meridional average cross section along 45°—75°N.

4.3 Analysis of the height and wind fields in a case after the abrupt change

A double-blocking case after the abrupt change (14—24 July 2008) was selected for comparative analysis with the case before the abrupt change. Fig. 9 shows the evolution of meteorological variables during the generation, development, and weakening of the double-blocking case after the change. When the double-blocking high was established on 14 July (Fig. 9a), there was a high pressure ridge over the northwest of the Caspian Sea, and the 5840 gpm contour extended northward, close to 60°N. In the vicinity of East Siberia Sea of Okhotsk area, a closed high-pressure center was clearly present, whose 5840 gpm contour extended northward to 75°N, and thus a cyclonic center formed between the two highs. During the development stage of the double blocking on 19 July (Fig. 9b), the upstream low-value system shifted eastward and southward, pushing the Caspian Sea high pressure to move eastward and northward. Then, obstructed by terrain, the Urals blocking high was generated. The high center remained stable and less dynamic, but its central intensity increased gradually, causing the Urals blocking high to develop further, and a prominent closed high-pressure center to appear. At the same time, the downstream cold vortex system stepped eastward and southward, and together with the Okhotsk high, formed an eastwest positive-negative distributional pattern. Such a situation was conducive to the stable maintenance of the cold vortex system, and the southerly airflow in the front of the cold vortex was helpful for the development and maintenance of the Okhotsk blocking high. The modulation effect of the two made the Okhotsk blocking move southward and the 5840 gpm contour to shift to 65°N, forming a high-pressure center. On 24 July (Fig. 9c), as the low-value system in the upstream of the Ural Mountains weakened, the Urals blocking high gradually weakened, retreating southward, and the 5840 gpm contour moved southward to around 60°N. Affected by this situation, the downstream vortex shifted northward, causing the dynamic conditions for the Okhotsk high to gradually weaken. The 5840 gpm contour moved southward to 55°N, the closed center of high pressure disappeared, and the double-blocking situation faded, before ultimately collapsing.

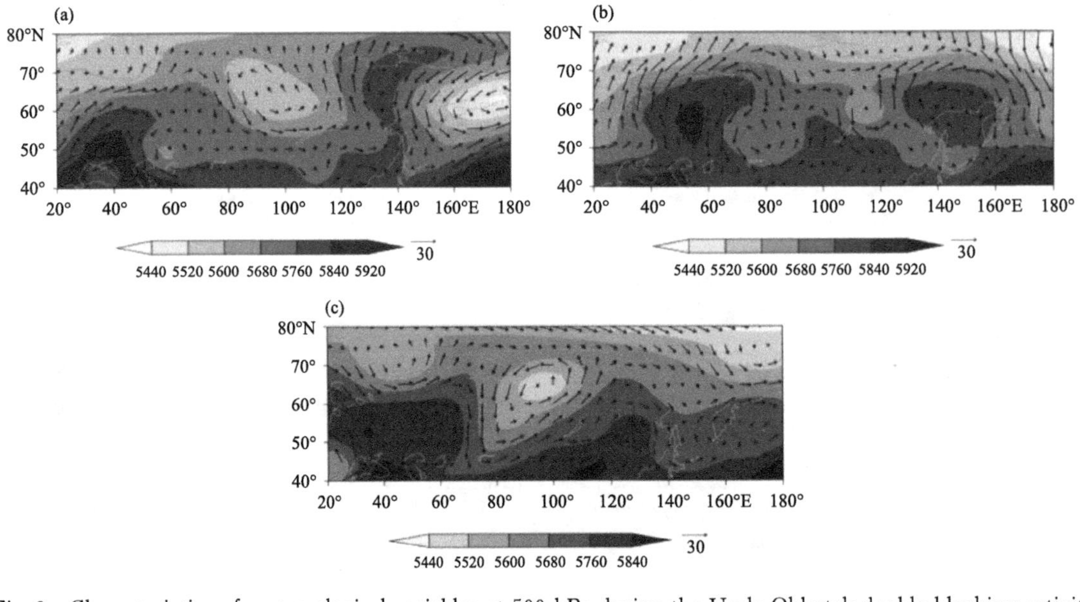

Fig. 9　Characteristics of meteorological variables at 500 hPa during the Urals-Okhotsk double-blocking activity of 14—24 July 2008: (a) 14 July, (b) 19 July, and (c) 24 July (shading: height field in gpm; vectors: wind field in m s^{-1}).

4.4　Diagnostic analysis of wave activity flux in the case after the abrupt change

Fig. 10 shows the 500 hPa stream function field, wave activity flux, and its divergence characteristics for the course of establishment, development, and weakening of the double-blocking case during 14—24 July 2008. The double-blocking high was established on 14 July (Figs. 10a and 10b), when high-value zones of stream function existed over the region north to the Black Sea in the Ural Mountains upstream, and in the eastern Siberia region north to the Okhotsk Sea. The high-value zones were characterized by anticyclonic circulation, which derived from the wave energy propagation of the low-value system in the Ural Mountains upstream, to the western section of the Ural Mountains anticyclone, producing convergence, and thus contributing to the formation and development of the downstream anticyclone. Wave energy divergence occurred in the southeast side of the stream function high-value zone, driving the downstream energy in the blocking system to transmit to the southeast. In fact, similar characteristics were found in the central Siberia depression and eastern Siberia high, but the divergence (convergence) varied in intensity. Over the Ural Mountains and the Okhotsk Sea area, the wave energy divergence in the downstream was equivalent to the convergence intensity in the upstream, which was favorable for the maintenance of the system. Viewed from the wave energy propagation direction: 1) a Rossby wave train propagated from the Urals upstream to the area with Urals blocking-high activities, and further dispersed to the northeast of the Caspian Sea; and 2) the Rossby wave generated in the southeast side of the Ob River downstream was transmitted to the west side of the Siberia vortex, and then to the Okhotsk blocking area through the dispersion of this system. The corresponding longitude-height profile (Fig. 10b) shows that all of the three systems (two ridges and one trough) were very profound, the wave energy convergence (divergence) from low to high levels possessed consistent propagation characteristics, and the center was located near 300—500 hPa. Meanwhile, the wave energy from low levels spread to the blocking-high area of the Ural Mountains and the Okhotsk Sea. On 19 July (Figs. 10c and 10d), impacted by the southeast movement of the Ural Mountains upstream low-value system, the wave energy spread to the Urals blocking area from the upstream of the Urals blocking high and generated a strong convergence, strengthening the Urals blocking high and lifting it northward. In its southeast side, divergence occurred, propagating eastward into the vortex on the northwest side of Lake Baikal, which kept the vortex developing stably and moving to the southeast. In the Urals blocking-high area, the intensity and range of wave energy convergence exceeded the divergence strength on the eastern side of the Urals blocking, which was good for the Urals blocking high to maintain and develop. On the west side of the Okhotsk blocking, there was still wave energy convergence, but the divergence on the east side was relatively weak. This helped to further the development of the Okhotsk blocking high. The corresponding longitude-height profile (Fig. 10d) was the same as the configuration of the system on 14 July (Fig. 10b), which means that the low-level wave energy spread to the double-blocking area of the Ural Mountains and the Sea of Okhotsk, and the wave energy divergence (convergence) center lay in the vicinity of 250～400 hPa. On 24 July (Figs. 10e and 10f), the low-value system in the upstream of the Ural Mountains weakened gradually, making the energy source for the Urals blocking high die away. The wave energy convergence in the west side of the Urals blocking weakened, while the divergence in the east side strengthened, which was unfavorable for the persistence of the Urals blocking high. The wave energy in the downstream low-value system mainly featured intense convergence, but the energy dispersion to the downstream of the Okhotsk Sea region weakened gradually, which was not beneficial to the maintenance and development of the Okhotsk blocking. Thus, the Okhotsk blocking weakened and the double-blocking situation faded until its collapse.

The corresponding longitude-height profile (Fig. 10b) shows that all of the three systems (two ridges and one trough) were very profound, the wave energy convergence (divergence) from low to high levels possessed consistent propagation characteristics, and the center was located near 300—500 hPa. Meanwhile, the wave energy from low levels spread to the blocking-high area of the Ural Mountains and the Sea of Okhotsk. On 19 July (Figs. 10c and 10d), impacted by the southeast movement of the Ural Mountains upstream low-value system, the wave energy spread to the Urals blocking area from the

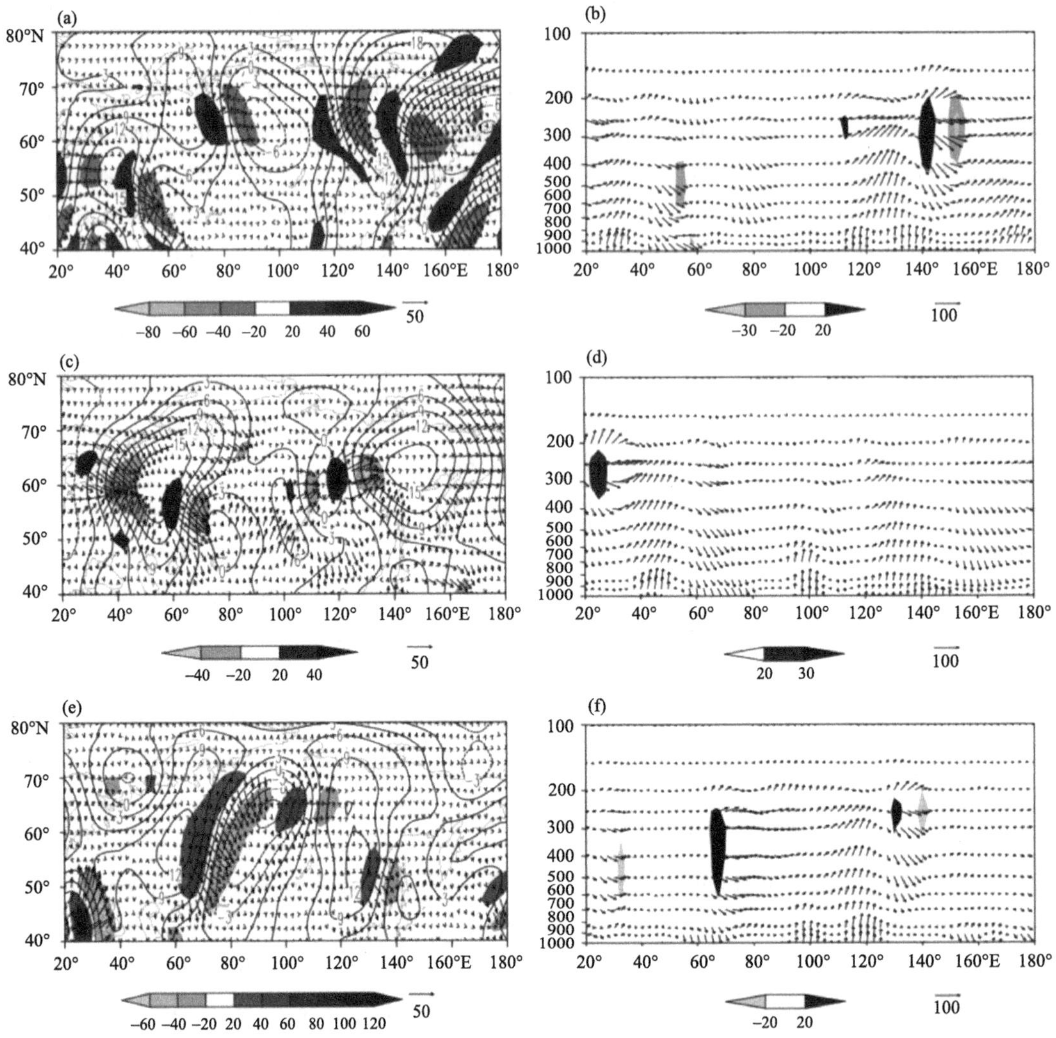

Fig. 10 Characteristics of the stream function field (contours; m s^{-1}), wave activity flux (vectors; m s^{-1}), and its divergence (shading; m s^{-1}) at 500 hPa during the Urals-Okhotsk double-blocking activity of 14—24 July 2008: (a, b) 14 July, (c, d) 19 July, and (e, f) 24 July. (b, d, f) Meridional average cross section along 45°—75°N.

up-stream of the Urals blocking high and generated a strong convergence, strengthening the Urals blocking high and lifting it northward. In its southeast side, divergence occurred, propagating eastward into the vortex on the northwest side of Lake Baikal, which kept the vortex developing stably and moving to the southeast. In the Urals blocking-high area, the intensity and range of wave energy convergence exceeded the divergence strength on the eastern side of the Urals blocking, which was good for the Urals blocking high to maintain and develop. On the west side of the Okhotsk blocking, there was still wave energy convergence, but the divergence on the east side was relatively weak. This helped to further the development of the Okhotsk blocking high. The corresponding longitude-height profile (Fig. 10d) was the same as the configuration of the system on 14 July (Fig. 10b), which means that the low-level wave energy spread to the double-blocking area of the Ural Mountains and the Sea of Okhotsk, and the wave energy divergence (convergence) center lay in the vicinity of 250—400 hPa. On 24 July (Figs. 10e and 10f), the low-value system in the upstream of the Ural Mountains weakened gradually, making the energy source for the Urals blocking high die away. The wave energy convergence in the west side of the Urals blocking weakened, while the divergence in the east side strengthened, which was unfavorable for the persistence of the Urals blocking high. The wave energy in the downstream low-value system mainly featured intense convergence, but the energy dispersion to the downstream of the Sea of Okhotsk region weakened gradually, which was not beneficial to the maintenance and development of the Okhotsk blocking. Thus, the Okhotsk blocking weakened and the double-blocking situation faded until its collapse.

5 Conclusions and discussion

By identifying the Eurasian blocking high-pressure system objectively and automatically, we statistically analyzed the climatological characteristics of the Ural Mountains—Okhotsk Sea double-blocking processes over the summer months of 1948—2009, before and after an abrupt change in 1977, and then dynamically diagnosed the possible causes for such blocking flow activities. Our conclusions are as follows.

(1) The Urals-Okhotsk double-blocking activity experienced a significant abrupt change around 1977. After the abrupt change, the number of days of double-blocking activity was reduced greatly, but its intensity became much stronger; especially, the northern boundary of the Okhotsk blocking high extended northward noticeably. The zonal spacing between the Urals blocking high and the Okhotsk high became smaller, but the meridional positions of their activity ranges did not change much. The IPV analysis results before and after the abrupt change verified the above conclusion.

(2) The low-level wave energy from the upstream region of the Ural Mountains fueled the double-blocking activities, and then dispersed downstream via the Urals blocking, benefiting the formation and development of the downstream vortex. Meanwhile, the long-term maintenance of the Okhotsk blocking was also an important factor for the maintenance of the vortex. The energy propagation characteristics from the Ural Mountains to the Sea of Okhotsk altered from a mid-low-latitude spread before the abrupt change to a high-latitude spread after the change. On days without double-blocking highs, regardless of the abrupt change, all wave energies were conveyed from high to low latitudes and from low to high levels, and the intensity of convergence and divergence was also weak.

(3) Case analysis suggested that the Urals-Okhotsk double-blocking activity modulated the development and maintenance of the Northeast China cold vortex. The existence of a double-blocking high was conducive to the development and maintenance of the Northeast China cold vortex and in turn, the stable and sustained Northeast China cold vortex helped promote the maintenance of the Urals-Okhotsk blocking high pressure.

In short, an examination of the wave energy propagation can help to better explain the energy propagation characteristics of the establishment, development, and dissipation of Urals-Okhotsk blocking high situations. However, some problems remain, which need to be discussed further. For example, given the abrupt change of double-blocking activity around 1977, what might be its relationship to the known climate change event of 1976—1977 (in association with the abrupt and obvious deepening of the Aleutian low)? Is it a temporal coincidence or an inherent mechanism? Does the double-blocking flow pattern reflect a climatic feature or a kind of low frequency variability? Actually, it is difficult to reveal the differences in double-blocking activity before and after the abrupt change through analysis of a single case only. In addition, how do the Urals blocking and Okhotsk blocking interact with one another? All these problems remain to be discussed in future research.

References

BI M Y, DING Y H, 1992. A study of budget of potential vorticity of blocking high during the drought period in summer of 1980[J]. J Appl Meteor Sci, 3(2): 145-156. (in Chinese)

CHARNEY J G, DEVORE J G, 1979. Multiple flow equilibria in the atmosphere and blocking[J]. J Atmos Sci, 36(7): 1205-1216, doi: 10.1175/1520-0469(1979)036<1205:MFEITA>2.0.CO;2.

DING Y H, 2005. Advanced Synoptic Meteorology 2nd ed[M]. Beijing: China Meteorological Press: 585. (in Chinese)

DUAN C F, 2012. Climate characteristics, effects and its influencing factors of cold vortex over Northeast China[D]. Nanjing: Nanjing University of information Science&Technology: 188-194. (in Chinese)

ELLIOT T R D, SMITH T B, 1949. A study of the effects of large blocking highs on the general circulation in the Northern Hemisphere westerlies[J]. J Meteor, 6(2): 68-85, doi: 10.1175/1520-0469(1949)006<0068:ASOTEO>2.0.CO;2.

GAO S T, ZHU W S, DONG M, 1998. On the wave flow interaction in the low frequency atmospheric variation: Blocking pattern[J]. Acta Meteor Sinica, 56(6): 665-680. (in Chinese)

GONG Z S, WANG Y G, XU L, 2004. Middle-high latitude circulation and rainfall of Huaihe River basin in the summer of 2003[J]. Meteor Monthly, 30(2): 30-33. (in Chinese)

HOLOPAINEN E, FORTELIUS C, 1987. High-frequency transient eddies and blocking[J]. J Atmos Sci, 44(12): 1632-1645, doi: 10.1175/1520-0469(1987)044<1632:HFTEAB>2.0.CO;2.

HUANG F, JIANG Z N, 2002. Study on the statistical characteristics of atmospheric blocking in the Eurasia and its relationship with the summer rainfall over the east of China[J]. J of Ocean Univ of Qingdao, 32(2): 186-192. (in Chinese)

KALNAY E, KANAMITSU M, KISTLER R, et al, 1996. The NCEP/NCAR 40-year reanalysis project[J]. Bull Amer Meteor Soc, 77(3): 437-471, doi: 10.1175/1520-0477(1996)077<0437:TNYRP>2.0.CO;2.

LI F, DING Y H, 2004. Statistical characteristic of atmospheric blocking in the Eurasian high-mid latitudes based on recent 30-year summers[J]. Acta Meteor Sinica, 62(3): 347-354. (in Chinese)

LI Y, WANG S G, JIN R H, et al, 2012. Abnormal characteristics of blocking high during durative low temperature, snowfall and freezing weather in southern China[J]. Plateau Meteor, 31(1): 94-101. (in Chinese)

LIANG H, WANG Y, GUO Z Q, 2009. The teleconnection relationship between the northeast cold vortex and the subtropical high, the Okhotsk high in summer[J]. Sci Meteor Sinica, 29(6): 793-796. (in Chinese)

LIU G, SHEN B Z, LIAN Y, et al, 2012. The sorts of 500 hPa blocking high in Asia and its relations to cold vortex and aestival low temperature in Northeast China[J]. Sci Geographica Sinica, 32(10): 1269-1274. (in Chinese)

LIU H, WU G X, ZENG Q C, 1995. On maintenance of blocking anticyclones of Northern Hemisphere. Part I: Quasi-strophic and ertel potential vorticity analysis[J]. Acta Meteor Sinica, 53(2): 177-185. (in Chinese)

MIAO J H, 1984. Contributions of the nonlinear interactions among waves to the formation, persistence, and decay of the blocking[J]. Acta Meteor Sinica, 42(1): 35-45. (in Chinese)

PLUMB R A, 1985. On the three-dimensional propagation of stationary waves[J]. J Atmos Sci, 42(3): 217-229, doi: 10.1175/1520-0469(1985)042<0217:OTTDPO>2.0.CO;2.

REX D F, 1950. Blocking action in the middle troposphere and its effect upon regional climate. I: An aerological study of blocking action[J]. Tellus, 2(3): 196-211.

RODIONOV S N, 2004. A sequential algorithm for testing climate regime shifts[J]. Geophys Res Lett, 31(9): 1-4.

RODIONOV S N, 2006. The problem of red noise in climate regime shift detection[J]. Geophys Res Lett, 31: 1-12.

ROSSBY C G, 1951. On the dynamics of certain types of blocking waves[J]. J Chinese Geophys Soc, 2: 1-13.

SHEN B Z, LIN Z D, LU R Y, et al, 2011. Circulation anomalies associated with interannual variation of early-and late-summer precipitation in Northeast China[J]. Sci China Earth Sci, 54(7): 1095-1104.

SHI N, 1989. A multi-statistical analysis of the Southern Oscillation (SO) and its relation to the mean monthly atmospheric circulation at 500 hPa in the Northern Hemisphere[J]. Acta Meteor Sinica, 47(4): 457-466. (in Chinese)

SHI N, 2002. Multivariate Analysis Method in Meteorological Research and Forecasting 2nd ed[M]. Beijing: China Meteorological Press: 192. (in Chinese)

SHI N, SUN L P, SHEN J B, 1984. Canonical correlation method and its application in weather analysis and forecasting[J]. J

Nanjing Inst Meteor,7(2):251-256. (in Chinese)

SHI X J,2007. Blocking in Eurasia and its relation to the precipitation in the Yangtze River valley during boreal summer[D]. Nanjing:Nanjing University of Information Science&Technology. (in Chinese)

SHUTTS G T,1986. A case study of eddy forcing during an Atlantic blocking episode[J]. Adv Geophys,29(part II):135-162,doi:10. 1016/S0065-2687(08)60037-0.

SUMNER E J,1954. A study of blocking in the Atlantic-European of the Northern Hemisphere[J]. Quart J Roy Meteor Soc, 80(345):402-416,doi:10. 1002/(ISSN)1477-870X.

TANG M T,1957. On the blocking situation of the eastern Asia and its climatic effects[J]. Acta Meteor Sinica,28(4):282-293. (in Chinese)

WANG J,2002. Global circulation anomalous spatial character of cold/warm years in summer in northeastern China and long-period variation[J]. Sci Meteor Sinica,22(4):394-401. (in Chinese)

WANG L J,HE J H,SHI D,et al,2010. Analysis of impacts of northeast cold vortex processes on Meiyu rainfall period over Yangtze-Huaihe River basin[J]. Trans Atmos Sci,33(1):89-97. (in Chinese)

WANG X Q,MA S Q,XI Z X,et al,2005. Method study on long-term forecast of summer freezing damage in Northeast China[J]. J Catastrophol,20(1):36-39. (in Chinese)

WU G X,LIU H,CHEN F,et al. ,1994. Transient eddy transfer and formation of blocking high—On the persistently abnormal weather in the summer of 1980[J]. Acta Meteor Sinica,52(3):308-320. (in Chinese)

XU X D,ZHAO T L,1994. Numerical experiment on the formation of the dipole system and blocking structure[J]. Acta Meteor Sinica,52(1):107-111. (in Chinese)

YAO X P,YU Y B,LIU H Z,2005. Characteristics of the subtropical anticyclone during the abnormal rainfall period over the Huaihe River region in 2003[J]. J Trop Meteor,21(4):393-401. (in Chinese)

YE D Z,TAO S Y,ZHU B Z,et al,1962. The Study on Blocking Situation in the Northern Hemisphere Winter[M]. Beijing: Science Press:1-10. (in Chinese)

YU S Q,LIN X C,2006. Characteristics of two general circulation patterns during floods over the Changjiang—Huaihe River valley[J]. Acta Meteor Sinica,64(5):605-613. (in Chinese)

ZHAO H G,CHEN X Z,1990. The statistical analysis of atmospheric blocking in the Northern Hemisphere[J]. Meteor Monthly,16(3):3-7. (in Chinese)

ZHU Q G,LIN J R,SHOU S W,et al,2007. The Principles and Methods of Synoptic Meteorology 4th ed[M]. Beijing:China Meteorological Press:186-192. (in Chinese)

ZHU Z X,ZHU B Z,1982. Nonlinear equilibrium state and blocking situation of superlong waves under zonally asymmetric thermodynamic forcing (A)[J]. Sci China (Ser. B)25(4):361-371. (in Chinese)

冬季中国东北极端低温事件环流背景特征分析*

李尚锋[1,2,3]　姜大膀[1,3]　廉毅[2]　尹路婷[4]

(1 中国科学院大气物理研究所,北京 100029;2 吉林省气象科学研究所长白山气象与气候变化吉林省重点实验室/中高纬度环流系统与东亚季风研究开放实验室,长春 130062;3 中国科学院大学,北京 100049;4 吉林省气象服务中心,长春 130062)

摘　要:利用1961—2014年CN05.2逐日温度数据,对冬季东北极端低温事件进行了定义,并按其发生时冷空气对中国东部(105°E以东)的影响范围,将其分为第一类和第二类极端低温事件,其中前者局限在东北,而后者则扩展至中国东部大部分地区。分析表明,在年代际时间尺度上,第一类极端低温事件的强度减弱,而第二类则增加;对持续天数而言,第二类极端低温事件在减少,而第一类在20世纪90年代以前也持续减少,但是90年代之后急剧增加;2月总的极端低温事件发生天数最多,其在90年代以前要远大于12月和1月,且在90年代以前总体在减少,以后则增加;850 hPa风场分析显示,第二类极端低温事件中来自贝加尔湖的西北路径冷空气比第一类的要强,而来自鄂霍次克海的东北路径冷空气则相反;在300 hPa的E-P通量散度场上,这两类极端低温事件中东北都处于波动能量辐散中心,第二类极端低温事件发生时罗斯贝波波动能量传播比第一类时的要弱,第一类发生时则纬向传播比较明显,而且波动中心值都比较大;在位势高度距平场的谐波分析中,长波槽同位相扰动叠加在超长波槽上更易导致极端低温事件的发生。

关键词:极端低温事件;中国东北;冬季;罗斯贝波;谐波分析

1　引言

冬季,中国北方受冷空气影响频繁,常常出现区域性大风、低温雨雪天气,有时还伴有雨凇或霜冻过程,可形成灾害性天气(张宗婕 等,2012),给国民经济和人民生产生活造成损失。东亚冬季风强烈地向南暴发会产生寒潮,通常情况下侵袭中国的寒潮绝大多数溯源于北极地区(陶诗言,1959)。总体上,影响中国的寒潮冷空气主要有三条路径:一是新地岛以西的北冰洋洋面,冷空气经巴伦支海、苏联的欧洲地区进入中国;二是新地岛以东的北冰洋洋面,冷空气经过喀拉海、太梅尔半岛、苏联地区进入中国;三是冰岛以南的大西洋洋面,冷空气经苏联的欧洲南部地区或地中海、黑海、里海进入中国(Ding,1990;张培忠 等,1999)。通常情况下,冷空气沿这三条路径进入西伯利亚中部(43°—65°N,70°—90°E)的寒潮关键区(陶诗言,1959),滞留、聚集和加强,在适当的天气形势下南下爆发侵入中国。

北半球冬季极端天气事件以强冷空气活动和寒潮暴发最为常见(易明建 等,2013),其中,中高纬度环流异常可使侵入东亚的冷空气活动异常,是导致中国冬季气温异常的直接原因(郭其蕴,1994;武炳义 等,1999;陈海山 等,2001)。近10 a来,北半球范围内冬季区域性极端低温及其伴随的冰冻、暴风雪事件频繁发生,且影响范围大,持续时间长,如:2005年冬季,山东威海遭受了百年一遇的罕见持续性暴风雪(隋素丽 等,2007);2006年冬季,美国和加拿大频繁遭受暴风雪和低温严寒(李威 等,2007);2008年年初,中国南方出现大范围低温雨雪冰冻天气(布和朝鲁 等,2008;丁一汇 等,2008;Bueh et al.,2011a);2009年冬

* 本文原载于《大气科学》,2018,42(5):963-976

季，美国和欧洲多国经历多次大雪和寒流，波兰气温降到－20 ℃，而巴伐利亚的气温降到了欧洲有记录以来最低的－33.6 ℃（李威 等，2010）；2010 年冬季，美国南部和东北部遭受暴风雪袭击，日本、韩国、中国乃至印度和孟加拉国等亚洲国家也受到了低温寒流的席卷（司东 等，2012）；2012、2013 年冬季，中国东北冬季平均气温突破过去 34 个冬季历史极值，有 25 个台站出现极端低温（苗青 等，2016）。

有关极端低温事件的定义、时空特征、环流背景和成因机理方面的研究，中国学者近年来做了大量工作，取得了一系列成果。对于极端低温事件的定义，从采用早期气象业务系统中参照寒潮定义方法以某一具体温度值为阈值的判断标准，到采用温度阈值百分位方法（Jones et al.，1999），使得判据更加适用与合理。基于上述方法，Zhang 等（2011）对单站持续性极端低温事件进行了重新定义：日最低气温低于常年值（常年值＝当日及前后 5 d，共 11 d 的气候平均值），且日最低气温按大小排序（由小到大排序）小于第 10 个百分位值，连续日数超过 5 d。由于不同地区冷空气活动的频率和影响范围存在较大不同，在对区域持续性极端低温事件的定义上相关学者也做了适当调整（Peng et al.，2011；王晓娟 等，2013；李尚锋 等，2014a）。同时，还有学者根据区域性极端低温事件的定义，建立了识别极端低温事件的指标体系，并应用于业务预报中，该体系包括以下几个部分：极端低温阈值的确定、空间区域的识别、空间区域的连续性过程提取（龚志强 等，2012）。除了对极端低温事件的时空特征进行分析外，许多学者对引起这类事件的环流因子也进行了探讨。Li 等（2016）揭示，春季来自贝加尔湖的西北路径冷空气对东北强极端低温事件起主要作用，而东北弱极端低温事件的主要气流则来自东北路径的鄂霍次克海地区冷空气。Bueh 等（2011b）对引起中国冬季区域性极端低温事件的环流做了系统剖析，指出中国冬季区域性低温事件的一个重要环流特征是在对流层中层存在一对欧亚大陆尺度的、沿东北至西南方向倾斜的槽脊。另外，高纬度的平流层与冬季极端低温也有关联，平流层极涡异常偏强或者偏弱能相应地引起冬季地面温度变化（易明建 等，2009；施宁 等，2015）。

众所周知，东北位于中国东部的最北面，易受极地冷空气侵袭，加之那里既是国家主要的重工业基地，亦是战略粮食储备地，对该地区极端低温事件进行研究尤为必要。过去对东北低温的分析主要是集中在季节和月时间尺度上开展（东北低温科研协作组，1979；廉毅 等，1998；沈柏竹 等，2011；李尚锋 等，2014b），对天气尺度的极端低温事件研究较少，且早期对极端低温事件的研究主要侧重于单站低温的强度和持续时间等方面。本文针对东北进行冷空气影响区域和持续时间相结合的研究，相比单站分析而言，区域性研究可以探讨影响东北极端低温过程的冷空气路径，分析何种路径冷空气影响频次高，研究何种路径冷空气引起的降温更为剧烈以及各类路径冷空气对东北以外地区（尤其是中国东南部地区）是否有影响。另外，东北的冬季极端低温同全国其他区域极端低温有无联系，是东北的独有区域性事件，还是全国极端低温的前哨？另外，通过对东北区域性极端低温事件环流异常的研究，有助于揭示其成因机理，对于东北极端低温的预测有一定的实际意义。

2　资料与方法

2.1　数据资料

所采用的数据包括：(1)吴佳等（2013）根据 2416 个中国地面气象台站的逐日观测记录，通过插值程序建立起来的一套 1961—2014 年中国区域水平分辨率为 0.5°×0.5°的格点数据集 CN05.2，这里使用的是日平均气温资料；(2) NCEP/NCAR（Kalnay et al.，1996）再分析数据，包括位势高度场和水平风场，选取长度为 1961—2014 年，水平分辨率为 2.5°×2.5°，垂直方向为 1000～70 hPa 共 13 层。东北定义为（38°—55°N，116°—137°E），冬季为当年 12 月、翌年 1 月和 2 月。

2.2　冬季东北极端低温事件的定义

冬季东北一次极端低温事件需要同时满足下述两个条件：一是任一格点上日平均气温距平 $\overline{T}_d \leqslant \beta$、持续日数≥3 d,其中 β 为该格点冬季日平均气温距平的第 10 个百分位的值(按由小到大的排列顺序);二是东北达到条件一的格点总数大于该区域格点总数的 50%,并且定义第一天达到上述标准的日期为极端低温事件的开始日期,峰值日期定义为该次降温过程中东北达到阈值要求的格点 $\overline{T}_d$ 的区域平均值最小的日期,峰值强度定义为该次极端低温过程中峰值日达到阈值要求的格点 $\overline{T}_d$ 的区域平均值。

将达到上述条件的极端低温事件统称为总的极端低温事件(简称:TECEs),为了研究东北极端低温事件与中国东部其他地区极端低温的关系,以中国 105°E 以东的地区作为研究对象,若在一次东北极端低温事件发生过程中,该区域内达到标准(≤该格点 $\overline{T}_d$ 的第 10 个百分位值)的格点数超过该区域格点总数的 50%,并且发生天数≥1 d,则将该类极端低温事件定义为第二类极端低温事件(简称:第二类),否则称为第一类极端低温事件(简称:第一类)。简而言之,第一类表示的是仅在东北发生的极端低温事件,第二类表示的是东北发生极端低温事件的同时,中国东部其他地区也发生了大范围的极端低温事件。关于极端低温事件的开始日期、持续时间、峰值日期、峰值强度和类型等情况,详见表 1。

同 Peng 等(2011)定义的冬季极端低温事件日历相比,表 1 中第二类极端低温事件发生了 20 次,这 20 次事件在 Peng 等(2011)的日历中能找到 14 次;同 Bueh 等(2011b)定义的冬季持续性低温事件日历相比,表 1 中 20 次第二类极端低温事件在其日历中发生了 17 次。由于本文使用的温度资料(水平分辨率为 0.5°×0.5°的 CN05.2 数据)与 Peng 等(2011)和 Bueh 等(2011b)使用的(中国 756 个气象测站温度资料)不同,且判断标准亦有差别;加之本文定义的第二类极端低温事件前提是东北局地发生极端低温后,冷空气南移发生的,而 Peng 等(2011)和 Bueh 等(2011b)定义的低温事件则没有该项限制,故而低温日历有部分差异。

表 1 1961—2014 年 46 次冬季极端低温事件的开始日期、持续时间、峰值日期、峰值强度和类型

序号	开始日期	持续时间/d	峰值日期	峰值强度/℃	类型
1	1964-02-11	3	1964-02-11	−8.8	第二类
2	1964-02-19	5	1964-02-22	−8.1	第二类
3	1966-01-18	3	1966-01-18	−8.8	第一类
4	1966-12-21	6	1966-12-25	−8.9	第二类
5	1967-02-09	5	1967-02-11	−8.5	第二类
6	1967-12-06	3	1967-12-07	−9.5	第一类
7	1967-12-26	4	1967-12-28	−9.2	第二类
8	1968-01-29	3	1968-01-30	−8.2	第二类
9	1968-02-19	3	1968-02-19	−7.4	第二类
10	1969-01-27	5	1969-01-29	−8.5	第二类
11	1969-02-01	5	1969-02-05	−8.3	第一类
12	1969-02-13	3	1969-02-15	−10.0	第一类
13	1969-02-18	11	1969-02-21	−12.9	第二类
14	1970-01-01	4	1970-01-03	−10.5	第一类
15	1970-02-21	3	1970-02-22	−8.4	第一类
16	1971-02-26	3	1971-02-27	−11.4	第一类
17	1974-01-27	3	1974-01-28	−7.1	第一类
18	1974-02-24	3	1974-02-25	−8.1	第二类
19	1976-12-25	4	1976-12-26	−10.7	第二类

续表

序号	开始日期	持续时间/d	峰值日期	峰值强度/℃	类型
20	1976-12-30	5	1977-01-02	−9.7	第二类
21	1977-01-26	6	1977-01-30	−10.0	第二类
22	1977-02-12	5	1977-02-14	−10.0	第一类
23	1978-02-10	7	1978-02-13	−9.0	第二类
24	1979-01-30	3	1979-01-31	−10.1	第二类
25	1980-02-05	3	1980-02-06	−8.5	第二类
26	1980-12-27	3	1980-12-27	−6.6	第一类
27	1981-02-24	4	1981-02-25	−10.7	第二类
28	1983-02-10	5	1983-02-12	−9.8	第一类
29	1983-02-18	3	1983-02-19	−6.9	第一类
30	1984-02-02	5	1984-02-06	−8.3	第二类
31	1985-02-16	5	1985-02-17	−6.9	第二类
32	1985-12-06	6	1985-12-08	−10.9	第二类
33	1987-01-09	5	1987-01-12	−8.0	第一类
34	1990-01-21	7	1990-01-25	−10.0	第一类
35	1991-02-21	3	1991-02-22	−9.0	第一类
36	1998-12-01	3	1998-12-02	−8.7	第一类
37	2000-12-23	4	2000-12-25	−8.2	第一类
38	2001-01-10	7	2001-01-11	−9.5	第一类
39	2001-02-02	5	2001-02-06	−8.9	第一类
40	2001-02-10	4	2001-02-11	−8.1	第一类
41	2002-12-08	3	2002-12-09	−7.7	第一类
42	2004-12-21	3	2004-12-21	−7.1	第一类
43	2010-12-13	3	2010-12-14	−6.5	第一类
44	2012-12-06	3	2012-12-07	−7.6	第一类
45	2012-12-23	3	2012-12-23	−9.4	第一类
46	2013-01-07	4	2013-01-08	−7.8	第一类

为了探讨两类极端低温事件发生频次的异同，对其空间分布特征做了分析。在 1961—2014 年冬季整个时段内(图略)，第一类极端低温事件发生频次最高的地区主要位于黑龙江东南部、吉林和辽宁。此外，其冷空气影响次频繁地区为河北省和山东半岛一带。第二类极端低温事件发生期间，其冷空气作用频次较高的地区为 105°E 以东的中国地区，其发生频次最高的地区为黑龙江东南部、吉林、辽宁、河北、山东、安徽和江苏等地。

本文采用准地转二维 E-P 通量(Edmon et al.,1980)，来诊断罗斯贝波的活动，计算公式如下：

$$\begin{cases} F_\phi = - a\cos\phi\,\overline{u'v'} \\ F_p = fa\cos\phi\,\dfrac{\overline{u'\theta'}}{\theta_p} \end{cases} \tag{1}$$

式中，F_ϕ、F_p 分别为由于波动效应，单位质量空气南北方向的涡动角动量输送和涡动热量输送；u、v 分别为纬向风和经向风；$u'=u-\bar{u}$，$v'=v-\bar{v}$；a 为地球半径；θ 为位温；f 为科氏参数；ϕ 为纬度。

为了研究极端低温事件发生期间超长波和长波活动特点，采用了谐波分析，用波数 1～3 波之和代表超长波活动（王绍武，1964；吴晓红 等，1996），波数 4～6 波之和代表长波活动（廉毅 等，2010）。

3　东北极端低温事件的年代际变化

为了探讨冬季东北极端低温事件是否有年代际变化，对总的、第一类和第二类极端低温事件的峰值强度和发生天数的年代际特征分别做了分析。由图 1 可见，在整个时段内，总的和第一类极端低温事件的峰值强度均随时间减弱，第二类极端低温事件从 1960 年代到 1970 年代为增加、1970 年代到 1980 年代减弱、1980 年代后则不再发生；第二类极端低温事件峰值强度要比第一类的强，这说明能够影响到中国东部更南方的这种大范围冷空气活动，其在东北造成的低温过程要比仅发生在东北局地的低温过程更加强烈。

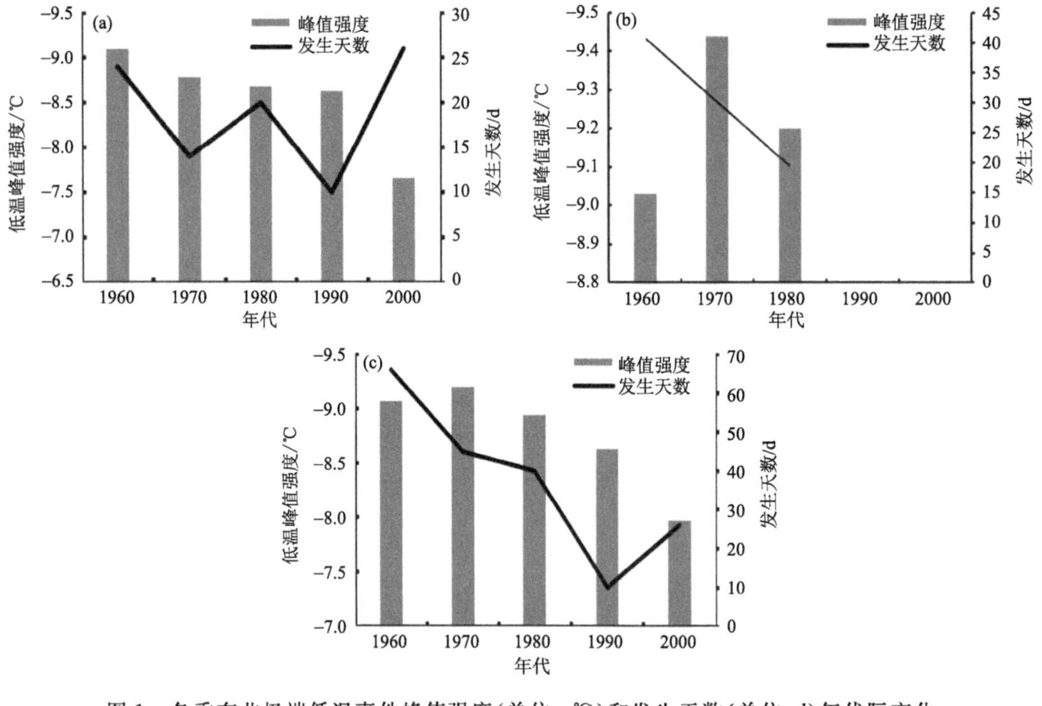

图 1　冬季东北极端低温事件峰值强度（单位：℃）和发生天数（单位：d）年代际变化
(a)第一类；(b)第二类；(c)总的极端低温事件

就冬季东北极端低温事件发生天数的年代际变化（图 2）而言，总的极端低温事件在 20 世纪 90 年代以前减少明显，但 90 年代以后发生天数开始增加。第一类极端低温事件的发生天数在 60—80 年代期间少变，90 年代以后则明显增加。第二类极端低温事件的发生天数在 60—90 年代急剧减少，90 年代后则不再发生。上述特征表明，90 年代以后，随着全球变暖的影响，冬季中国东部这种大范围的极端低温事件出现甚少（Peng et al.，2011），但是区域尺度的极端低温事件的发生频次则在增加。

根据东北极端低温事件冬季各月份发生天数的年代际变化（图 3）可知，2 月总的极端低温事件发生天数最多，它在 20 世纪 90 年代以前发生天数要远大于 12 月和 1 月，且在 90 年代以前总体减少，90 年代以后开始增加，这与 1 月的情形比较一致。在整个时间段上，12 月总的极端低温事件发生天数减少比较明显，发生天数最低的时段是 80 年代，之后开始增加，说明 12 月总的极端低温事件发生天数一直比较稳定，这与 12 月东北冬季温度变率相对于 1 月和 2 月较为稳定有关（图略）；而 2 月是冬春季节间的过渡性月份，期间低中高纬度系统间相互作用频繁，导致温度变化比较剧烈，使得该月是基于温度距平场上定义的极端低温事件发生频率最高的月份。

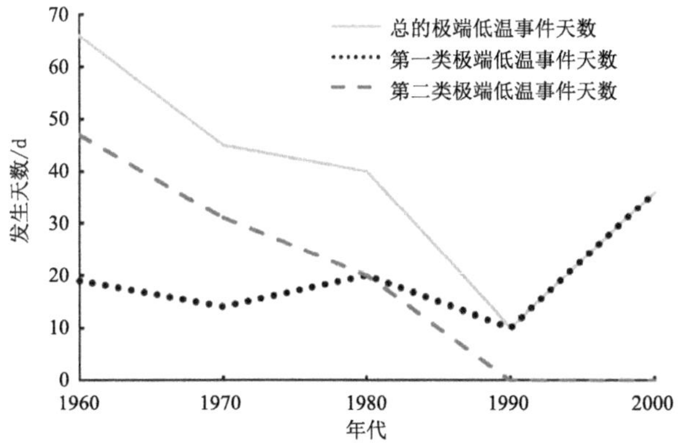

图 2　冬季总的第一类和第二类极端低温事件发生天数的年代际变化

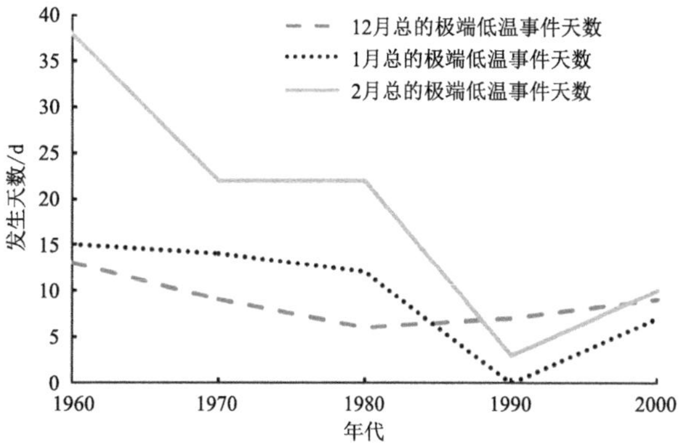

图 3　冬季各月总的极端低温事件发生天数的年代际变化

冬季各月第一类和第二类极端低温事件发生天数的年代际变化(图 4)显示，在 20 世纪 60 年代和 70 年代，冬季所有月份第二类极端低温事件发生天数要远高于第一类。另外，2 月第二类的发生天数要大于 12 月和 1 月；80 年代期间 12 月只发生了第二类极端低温事件，1 月只发生了第一类极端低温事件，2 月第一类和第二类极端低温事件均有发生，但 2 月第二类极端低温事件的发生天数要多于第一类；而 20 世纪 90 年代和 21 世纪 00 年代期间第二类极端低温事件基本没有出现。同时，90 年代至 00 年代期间，在所有月份中第一类极端低温事件的发生天数增加，表明全球变暖背景下，发生像 60 年代和 70 年代期间那种大范围的极端低温事件的概率在降低，但是区域性极端低温事件的发生天数增加得却很明显。这与先前研究中认为东北冬季气温在 80 年代存在增暖突变相一致(刘实 等，2009)。

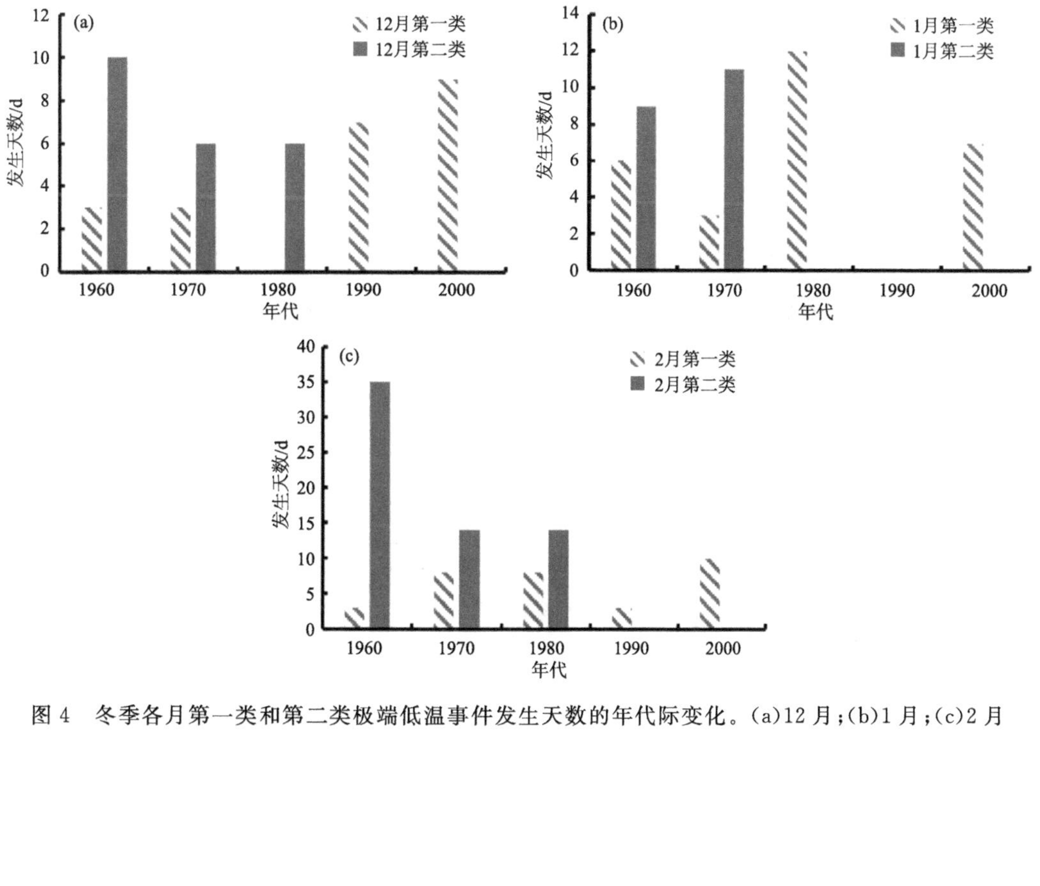

图 4　冬季各月第一类和第二类极端低温事件发生天数的年代际变化。(a)12 月；(b)1 月；(c)2 月

4　关键环流系统特征

为了研究第一类和第二类极端低温事件发生过程中冷空气活动的异同，利用合成分析方法探讨了它们发生、峰值和整个过程中的对流层低层 850 hPa 风场的变化特征。图 5 表明，引起这两类极端低温事件的主要气流来源相同，均是由来自贝加尔湖方向的西北气流（简称西北路径冷空气）和来自鄂霍次克海方向的东北气流（简称东北路径冷空气）在 55°N 附近区域汇合入侵东北。但是，影响第一类极端低温事件的西北气流来自中心位于叶尼塞河中游地区的阻塞高压（图 5a、c、e），而第二类极端低温事件的西北气流则来自位于乌拉尔山至贝加尔湖间的反气旋；鄂霍次克海上空维持的异常气旋环流，引导东北风向南侵入

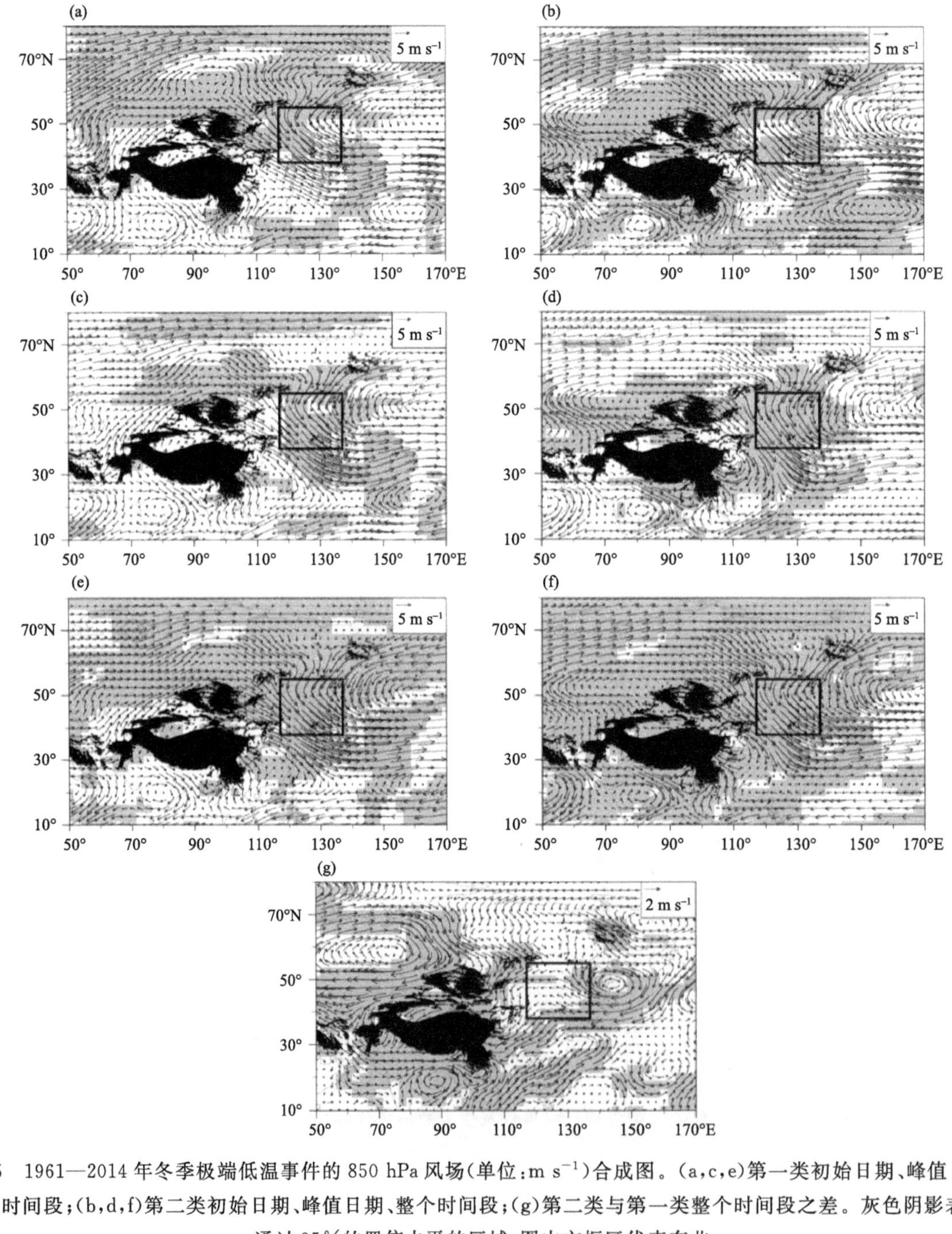

图 5　1961—2014 年冬季极端低温事件的 850 hPa 风场（单位：m s^{-1}）合成图。(a,c,e)第一类初始日期、峰值日期、整个时间段；(b,d,f)第二类初始日期、峰值日期、整个时间段；(g)第二类与第一类整个时间段之差。灰色阴影表示为通过 95% 的置信水平的区域，图中方框区代表东北

中低纬度地区，使得中国东部大部分出现低温(图 5b、d、f)，是这两类事件中东北气流的主要来源。

另外，从这两类事件的风场差值场(图 5g)可以看到，鄂霍次克海上空出现一个异常的反气旋环流，说明第二类极端低温事件中鄂霍次克海低压系统比第一类要弱，同时也说明第二类极端低温事件中东北路径冷空气比第一类要弱；乌拉尔山至叶尼塞河中游地区亦出现了一个异常的反气旋环流，说明第二类极端低温事件中西北路径冷空气比第一类要强。

由对流层中层 500 hPa 上极端低温事件位势高度距平合成图(图 6)可见，在亚洲中高纬度地区北正南负波列的分布为第一类和第二类极端低温事件中最明显的特点，即：乌拉尔山地区为很强的阻塞高压，而亚洲地区维持低压，这种典型的北高南低型表明大气的经向度大，有利于冷空气大规模南下，进而导致受其影响的地区出现极端低温，该结构特征与 2008 年中国南方大范围低温冰冻灾害的 500 hPa 位势高度距平场特点相似，但 2008 年阻塞高压更强且持续时间长(Bueh et al.，2011a)。第一类和第二类极端低温事件的不同点也比较明显：第一类极端低温事件的波列主要为贝加尔湖至鄂霍次克海的较强的正距平区，其中心位于贝加尔湖附近，负距平中心始终位于中国东北上空(图 6a—c)，而第二类极端低温事件的波列

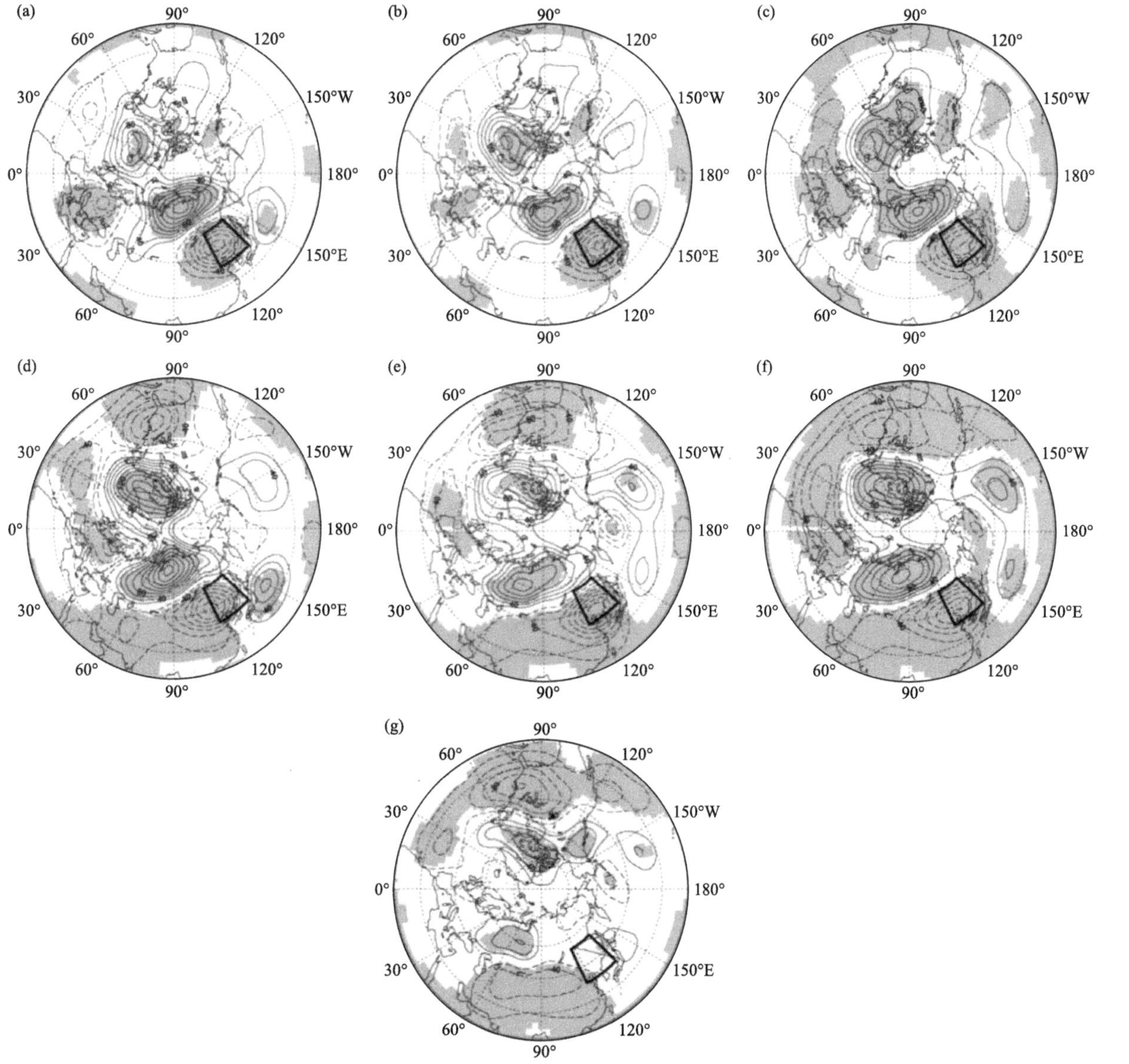

图 6　1961—2014 年冬季极端低温事件的 500 hPa 位势高度距平(等值线间隔：20，单位：gpm)合成图。(a，b，c)第一类初始日期、峰值日期、整个时间段；(d，e，f)第二类初始日期、峰值日期、整个时间段；(g)第二类与第一类整个时间段之差。灰色阴影表示为通过 95%的置信水平的区域，图中方框区代表东北

则主要为乌拉尔山至贝加尔湖西北侧的较强的正距平区，中心位于乌拉尔山，东北虽为负距平区控制，但是负距平中心处在东北的西南侧(图 6d—f))；第二类与第一类极端低温事件的差值图(图 6g)更能突出显示它们的差异，第二类极端低温事件中的乌拉尔山阻塞高压较第一类要更稳定和强大，同时，东北上空覆盖的负距平中心也比第一类要强，这种配置使得大气南北经向梯度更大，冷空气较易入侵到中国更南部的地区，而第一类极端低温事件中的鄂霍次克海上空的槽则比第二类要更深，强的槽前西南气流使得高纬度入侵的冷空气仅仅能影响到东北和朝鲜半岛，这与图 5 所示的鄂霍次克海上空对流层低层的大气环流型相匹配。

为进一步研究第一类和第二类极端低温事件发生过程中能量的传播特征，采用合成分析探讨了这两类事件所对应的对流层中高层(300 hPa)罗斯贝波波动能量频散变化，试图找出各自的主要物理过程。冬季第一类和第二类极端低温事件发生、峰值以及整个过程的 300 hPa 上 E-P 通量、散度及流函数场的合成图(图 7)表明，第一类和第二类极端低温事件发生时东北上空均被负的流函数场所覆盖，其位于塞班岛上空的南部中心为一个正的流函数场，从北到南这种北负南正的波列在第一类和第二类事件中都很明显，说明在这两类极端低温事件发生时，东北上空都是冷空气的堆积区。同时，在这两类极端低温事件中，东北上空都处于弱的波作用通量辐散中心，该中心处于上下游的辐合、辐散波列传播路径之上；不同之处在于，在极端低温事件发生、峰值以及整个过程中，第一类极端低温事件时在东亚 30°N 以北地区上空波作用通量辐散、辐合中心都呈东西向交替分布，且明显比第二类强，表明第一类极端低温事件 E-P 通量波列分布利于该区域的纬向环流加速，这种传播形态使得能量更容易向下游频散，不利于大范围和长时间的维持(图 7a、c、e)；第二类极端低温事件中向下游西北太平洋上空传播的波动能量比第一类事件要弱得多(图 7b、d、f)，该特点在两者差值图上表现得很明显(图 7g)。

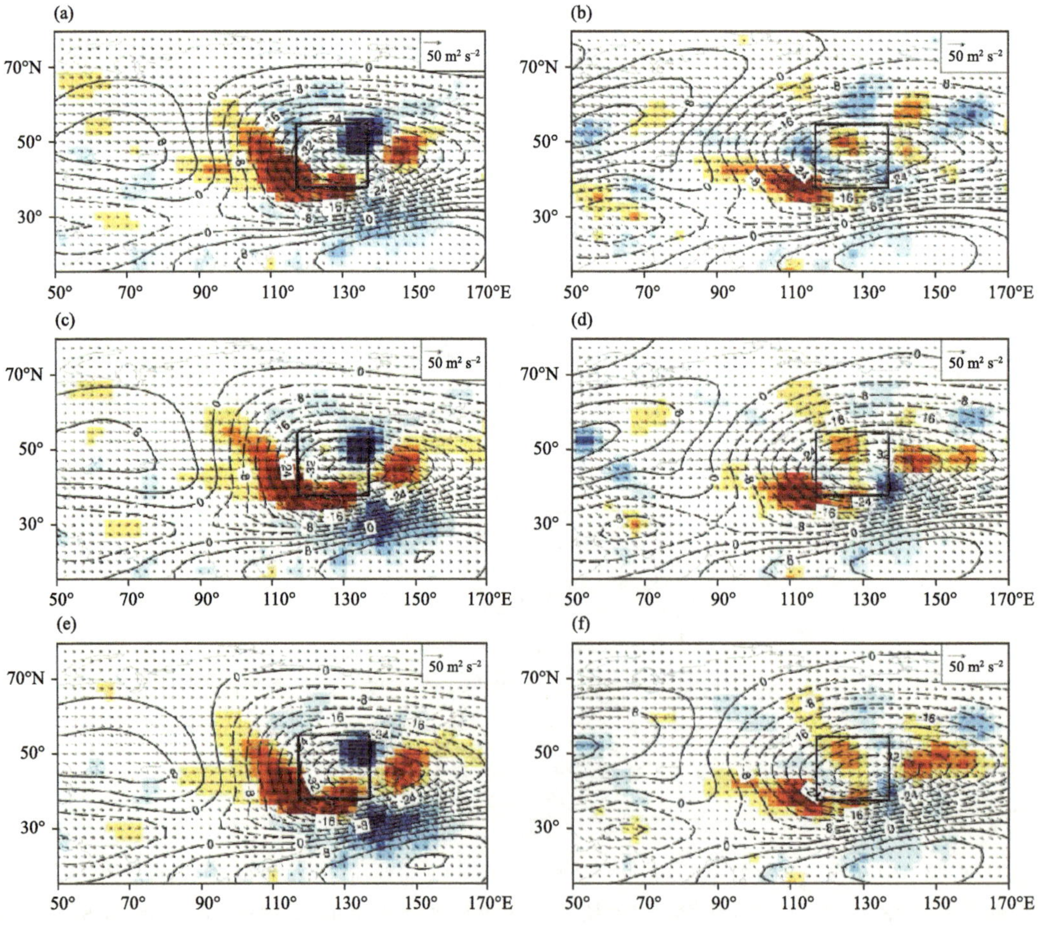

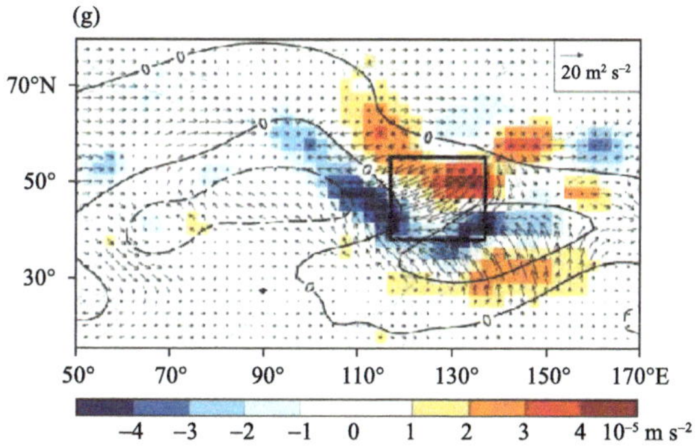

图 7　1961—2014 年冬季极端低温事件的 300 hPa 上 E-P 通量(箭头,单位:$m^2 s^{-2}$)、散度(阴影,单位:10^{-5} m s^{-2})及流函数场(等值线,单位:10^6 $m^2 s^{-1}$)合成图。(a,c,e)第一类初始日期、峰值日期、整个时间段;(b,d,f)第二类初始日期、峰值日期、整个时间段;(g)第二类与第一类整个时间段之差。图中方框区代表东北

王绍武(1964)指出,谐波分析可以定量客观地计算出各种波的位置和振幅,对研究不同尺度波在大气环流变化中的作用是一种较好的方法。本文利用这一方法尝试找出这两类事件中超长波和长波相互作用关系的异同,以便于更清晰地剖析第一类和第二类极端低温事件的发生过程。从冬季 500 hPa 位势高度距平谐波合成场(图 8)上可见:第一类和第二类极端低温事件中超长波有相同的结构特点(图 8a、b),北半球超长波为 2 波分布,为稳定型,超长波脊一个位于北太平洋的中东部至北美洲的太平洋西海岸上空,另一个自北大西洋横跨中亚地区上空;超长波槽较弱的一个位于北美洲大陆的西北侧上空,另一个横跨东亚至阿留申群岛上空,表明东亚大槽稳定,十分有利于冷空气活动在东亚地区。第一类和第二类极端低温事件中长波结构异同点均较为明显:从东亚经北太平洋至北美洲大陆上空的负—正—负—正长波槽脊分布较为相似;但是,从乌拉尔山至东亚上空的长波脊槽分布在两类事件中不尽相同,最显著的区别在于,第一类极端低温事件的长波脊位于巴尔喀什湖至贝加尔湖附近,第二类则出现在乌拉尔山附近,其次,第一类的长波槽主要位于东北亚上空,而第二类的长波槽则明显偏西,控制着从贝加尔湖至中国华北、华东,乃至华中的宽广东亚地区上空(图 8d、e)。

第一类和第二类极端低温事件中超长波和长波的 500 hPa 位势高度差值图显示,东北上游 60°—90°E 的中高纬度存在明显的北高南低异常(图 8c),这是影响第二类极端低温事件的典型中高层环流型,而图 10f 中并没有这样的特征,表明第二类极端低温事件更多地受到超长波的影响,第一类中长波作用较明显。并且,超长波更具有驻波性质,传播慢,频散慢,有利于冷空气堆积与活动,这与图 7 的分析结果相一致。而第一类极端低温事件时,东北上空为异常槽的超长波和长波的叠加,使得该槽较第二类的要深,也进一步揭示第一类比第二类极端低温事件发生范围小的重要谐波特征,这与图 6 所示的结果吻合。综上可知,位势高度负距平场中异常槽的长波和超长波的扰动叠加,是形成极端低温事件的基础,这种同位相扰动叠加能够诱发极端低温事件发生,该结论与李尚锋等(2012)研究夏季极端低温事件时的超长波、长波的扰动叠加特点相同。

5　结论和讨论

本文利用 CN05.2 日平均温度数据,给出了冬季东北极端低温事件定义,并将其按影响范围分为第一类极端低温事件和第二类极端低温事件两种,通过对两类极端低温事件峰值强度和发生天数的年代际特点、对应的环流背景以及能量频散等的研究,得到以下主要结论。

(1)1961—2014 年,在年代际尺度上,第一类极端低温事件峰值强度减弱,而第二类的增大;总的极端低温事件发生天数在 20 世纪 90 年代以前总体减少,但此后增多,第二类极端低温事件发生天数显著减少并在 90 年代以后基本消亡,第一类极端低温事件发生天数前期比较稳定但在 90 年代以后急剧增加。

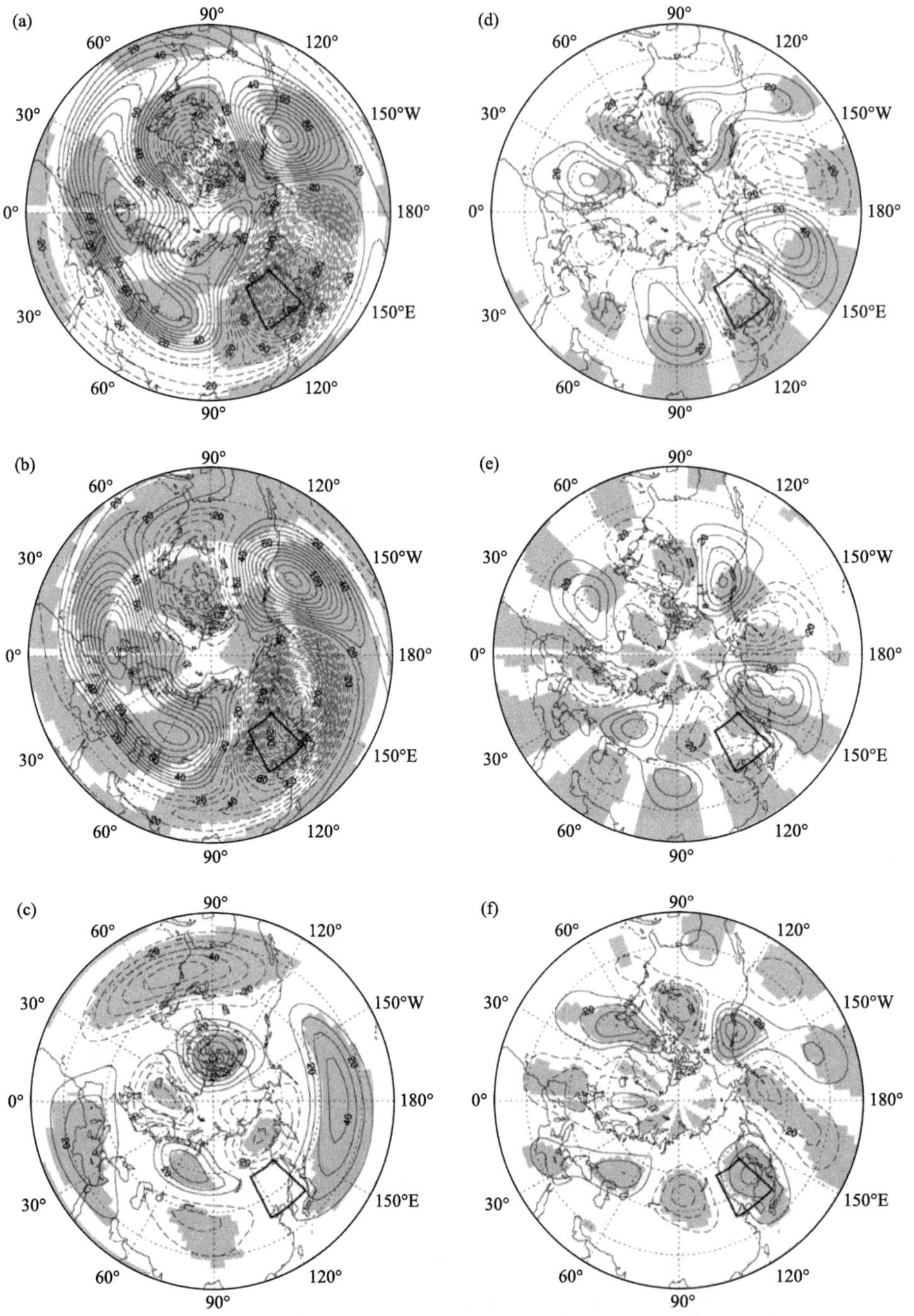

图 8　1961—2014 年冬季极端低温事件发生时的 500 hPa 位势高度距平谐波合成场(等值线间隔：10，单位：gpm)。(a)第一类和(b)第二类超长波；(d)第一类和(e)第二类长波；第二类与第一类(c)超长波和(f)长波之差。灰色阴影表示为通过 95%的置信水平的区域，图中方框区代表东北

(2)就冬季各月而言，2 月总的极端低温事件发生天数最多，其在 90 年代以前要远大于 12 月和 1 月，且在 90 年代以前总体减少而后为增加，这与 1 月的情形比较一致；在整个时间段内，12 月最低频次出现在 80 年代，较 60 年代稍微减少，而后开始增加；在 60 年代和 70 年代，冬季所有月份第二类极端低温事件发生频次要远大于第一类事件，2 月第二类极端低温事件发生天数要大于 12 月和 1 月。

(3)第一类和第二类极端低温事件发生时，流函数场显示东北都是冷空气堆积区，在 E-P 通量散度场上东北都处于波动能量辐散大值区中心；相比较而言，第二类事件中波动能量传播比第一类要弱得多，第一类事件则纬向传播比较明显，而且波动中心值都比较大，这种方式的传播使得能量更容易频散，不利于

大范围和长时间维持。

(4)第二类极端低温事件中西北路径冷空气比第一类的要强,东北路径冷空气则相反,在第一类极端低温事件中的强度要大于第二类;在500 hPa上位势高度距平场上,长波槽扰动叠加在超长波槽上,这种同位相扰动叠加更易导致极端低温事件的发生。

需要说明的是,本文从冬季东北地区极端低温事件的角度出发,按其影响范围分为第一和第二类极端低温事件两类,但是,对于中国其他地区发生大范围的极端低温事件而东北没有出现的情况在本文并没有讨论,这两种事件间的关系还应细化并分别进行讨论,这些问题有待于进一步研究。

参考文献

布和朝鲁,纪立人,施宁,2008.2008年初我国南方雨雪低温天气的中期过程分析I:亚非副热带急流低频波[J].气候与环境研究,13(4):419-433.

陈海山,孙照渤,2001.一个反映中国冬季气温异常的指标——东亚区域西风指数[J].南京气象学院学报,24(4):458-466.

丁一汇,王遵娅,宋亚芳,等,2008.中国南方2008年1月罕见低温雨雪冰冻灾害发生的原因及其与气候变暖的关系[J].气象学报,66(5):808-825.

东北低温科研协作组,1979.东北地区冷夏、热夏长期预报的初步研究[J].气象学报,37(3):44-58.

龚志强,王晓娟,崔冬林,等,2012.区域性极端低温事件的识别及其变化特征[J].应用气象学报,23(2):195-204.

郭其蕴,1994.东亚冬季风的变化与中国气温异常的关系[J].应用气象学报,5(2):218-225.

李尚锋,廉毅,陈圣波,等,2012.东北初夏极端低温事件的空间分布特征及其成因机理分析[J].地理科学,32(6):752-758.

李尚锋,孙钦宏,姚耀显,等,2014a.东北夏季极端低温天气事件的定义及其冷空气路径分析[J].地理科学,34(2):249-256.

李尚锋,应爽,姚耀显,等,2014b.东北夏季月低温事件的定义及大气环流年代际特征分析[J].气象与环境学报,30(3):38-45.

李威,朱艳峰,2007.2006年全球重大天气气候事件概述[J].气象,33(4):108-111.

李威,蔡锦辉,郭艳君,等,2010.2009年全球重大天气气候事件概述[J].气象,36(4):106-110.

廉毅,安刚,1998.东亚季风El Niño与中国松辽平原夏季低温关系初探[J].气象学报,56(6):724-735.

廉毅,布和朝鲁,谢作威,等,2010.初夏东北冷涡活动异常与北半球环流低频变化[J].大气科学,34(2):429-439.

刘实,闫敏华,隋波,2009.东北三省冬季气温变化的有关研究进展[J].气候变化研究进展,5(6):357-361.

苗青,巩远发,邓锐捷,等,2016.北半球中高纬度低频振荡对2012/2013年冬季中国东北极端低温事件的影响[J].大气科学,40(4):817-830.

沈柏竹,刘实,廉毅,等,2011.2009年中国东北夏季低温及其与前期海气系统变化的联系[J].气象学报,69(2):320-333.

施宁,布和朝鲁,2015.中国大范围持续性极端低温事件的一类平流层前兆信号[J].大气科学,39(1):210-220.

司东,李修仓,任福民,等,2012.2011年全球重大天气气候事件及其成因[J].气象,38(4):480-489.

隋素丽,贺玉平,张明芳,等,2007.威海市"2005年12月"连续暴雪特性初步分析[J].海洋预报,24(1):32-37.

陶诗言,1959.十年来我国对东亚寒潮的研究[J].气象学报,30(3):226-230.

王绍武,1964.北半球500百帕月平均环流特征及演变规律的研究:超长波[J].气象学报,34(3):316-328.

王晓娟,龚志强,沈柏竹,等,2013.近50年中国区域性极端低温事件频发期的气候特征对比分析研究[J].气象学报,71(6):1061-1073.

武炳义,黄荣辉,1999.冬季北大西洋涛动极端异常变化与东亚冬季风[J].大气科学,23(6):641-651.

吴佳,高学杰,2013.一套格点化的中国区域逐日观测资料及与其他资料的对比[J].地球物理学报,56(4):1102-1111.

吴晓红,王绍武,1996.对流层大气环流的甚低频振荡[J].气象学报,54(4):427-436.

易明建,陈月娟,周任君,等,2009.2008年中国南方雪灾与平流层极涡异常的等熵位涡分析[J].高原气象,28(4):880-888.

易明建,陈月娟,周任君,等,2013.亚洲东部冬季地面温度变化与平流层弱极涡的关系[J].大气科学,37(3):668-678.

张培忠,陈光明,1999.影响中国寒潮冷高压的统计研究[J].气象学报,57(4):493-501.

张宗婕,钱维宏,2012.中国冬半年区域持续性低温事件的前期信号[J].大气科学,36(6):1269-1279.

BUEH C,SHI N,XIE Z W,2011a. Large-scale circulation anomalies associated with persistent low temperature over southern China in January 2008 [J]. Atmos Sci Lett, 12 (3): 273-280, doi: 10.1002/asl.333.

BUEH C, FU X Y, XIE Z W, 2011b. Large-scale circulation features typical of wintertime extensive and persistent low temperature events in China [J]. Atmos Oceanic Sci Lett, 4 (4): 235-241, doi: 10.1080/16742834.2011.11446935.

DING Y H, 1990. Build-up, air mass transformation and propagation of Siberian high and its relations to cold surge in East Asia [J]. Meteor Atmos Phys, 44 (1-4): 281-292, doi: 10.1007/BF01026822.

EDMON H J, HOSKINS B J, MCINTYRE M E, 1980. Eliassen-Palm cross sections for the troposphere [J]. J Atmos Sci, 37 (2): 2600-2616, doi: 10.1175/1520-0469(1980)037<2600:EPCSFT>2.0.CO;2.

JONES P D, HORTON E B, FOLLAND C K, et al, 1999. The use of indices to identify changes in climatic extremes [J]. Clim Change, 42 (1): 131-149, doi: 10.1023/A:1005468316392.

KALNAY E, KANAMITSU M, KISTLER R, et al, 1996. The NCEP/NCAR 40-year reanalysis project [J]. Bull Amer Meteor Soc, 77 (3): 437-472, doi: 10.1175/1520-0477(1996)077<0437:TNYRP>2.0.CO;2.

LI S F, JIANG D B, LIAN Y, et al, 2016. Interdecadal variations of cold air activities in Northeast China during springtime [J]. J Meteor Res, 30 (5): 645-661, doi: 10.1007/s13351-016-5912-6.

PENG J B, BUEH C, 2011. The definition and classification of extensive and persistent extreme cold events in China[J]. Atmos Oceanic Sci Lett, 4 (5): 281-286, doi: 10.1080/16742834.2011.11446943.

ZHANG Z J, QIAN W H, 2011. Identifying regional prolonged low temperature events in China [J]. Adv Atmos Sci, 28 (2): 338-351, doi: 10.1007%2Fs00376-010-0048-6.

夏季亚洲阻塞高压识别及其对中国东北气候异常的可能影响:不同再分析资料对比*

刘刚[1,2,3]　徐士琦[1,4]　廉毅[1,2]

(1 中高纬度环流系统与东亚季风研究开放实验室,长春 130062;2 吉林省气象科学研究所,长春 130062;
3 吉林省气象信息网络中心,长春 130062;4 吉林省气候中心,长春 130062)

摘　要:基于阻塞高压(阻高)客观识别方法,利用 1979—2016 年夏季(6—8 月)NCEPⅠ、NCEPⅡ逐日再分析资料和 ERA-interim 逐 6 h 再分析资料对识别结果进行对比分析;并以 D 类(130°—160°E)阻高为例,讨论其对 6 月中国东北地区气候的可能影响。结果表明:NCEPⅠ和 NCEPⅡ再分析资料对阻高活动天数、发生频次及年代际变化的识别结果差异较小,而 ERA-interim 与前两种资料的结果差别较大。3 种再分析资料下,夏季各类阻高活动天数均与 500 hPa 高度场存在相应的显著相关区,且形态相近。但前两种资料对于各类阻高的表征结果较为一致,而 ERA-interim 再分析资料对各类阻高面积和范围的表征偏小。6 月 D 类阻高活动天数与东北地区气温和降水关系密切,D 类阻高活跃年,大气环流以经向型为主,东北地区低层低温、暖平流,高层高温、冷平流的结构指示大气层结不稳定,且东北上空为异常低压环流控制,上升气流较强,有利于 6 月东北地区出现低温多雨天气。鄂霍次克海地区是 6 月罗斯贝波的重要来源地之一,而 6 月 D 类阻高的形成可能与海-陆温差有关。

关键词:阻塞高压;客观识别;资料对比;中国东北

1　引言

阻塞高压(阻高)是发生在中高纬度地区西风带中的大尺度环流系统,多以暖性正压结构为主,常在某些固定区域内出现,可维持较长时间。作为中高纬度地区典型的持续性环流异常,阻高一旦出现,往往通过上下游效应带来严重的天气、气候灾害。汤懋苍(1957)对天气、气候与亚洲东部阻塞形势的关系进行了细致研究,陶诗言等(1958)通过个例分析发现中国梅雨在很大程度上受到乌拉尔山和鄂霍次克海阻高的影响;叶笃正等(1962)指出,乌拉尔山阻高的建立和崩溃对东亚寒潮有重要影响。随着观测手段和气象资料、研究方法等的不断丰富和完善,对亚洲阻高影响天气、气候的研究越来越深入。众多研究表明,欧亚中高纬度不同地区出现阻塞形势可以造成中国南方(胡钰玲 等,2017;赖珍权 等,2017;杜小玲 等,2014;顾伟宗 等,2016)、华北(张培忠 等,1996;孙安健 等,2000;李春 等,2003;刘樱 等,2016)和东北(张庆云 等,2001;刘刚 等,2012;苏丽欣 等,2015;苗青 等,2016)等不同区域低温冰冻冷害及暴雨洪涝灾害。例如,1954 年(陈汉耀,1957)、1991 年(王亚非 等,1998)、1998 年(李维京,1999)、1999 年(陈菊英 等,2001)、2010 年(王小玲 等,2013)及 2017 年(郑志海 等,2018)中国夏季气候异常均与中高纬度阻高的活动有关。陶诗言(1980)对中国暴雨类型进行总结,发现 10 类暴雨类型中有 6 类受阻高的影响。赵振国(1999)从影响中国夏季旱涝的因子中发现,中国夏季雨带的五大主要影响因素分别是赤道东太平洋海温(东)、青藏高

* 本文原载于《气象学报》,2019,77(2):303-314

原积雪(西)、位势高度场异常(西)、亚洲季风(南)、阻高(北)和西太平洋副热带高压(副高)(中)。

在对阻高进行的研究中,首要问题就是如何对阻高过程进行识别,这在当今阻高气候变化特征研究和现代天气预报客观监测业务中显得尤为重要。Elliott 等(1949)以气压距平的大小和持续性为客观标准,确定阻塞过程日期和时间;Rex(1950)定义阻高时指出,西风带上必须观测到有从上游纬向型向下游经向型的明显转换,500 hPa 上有维持 10 d 以上的西风急流且存在分支,急流间的跨度必须大于 45 个经距;此后,许多学者通过人工查阅天气图的方法给出阻塞形势定义(Brezowsky et al.,1951;汤懋苍,1957;周晓平,1957;Sumner,1959)。但因方式、方法不同,统计结果也不尽相同,同时耗费大量人力和时间,不便于进行长期统计。2000 年以来,越来越多的学者关注阻高的客观识别研究,并取得了一定成果。李峰等(2004)识别结果表明,夏季亚欧中、高纬度阻高出现在乌拉尔山和贝加尔湖东部地区的频次最高;刘刚等(2012)指出,夏季阻高在贝加尔湖活动最为频繁,鄂霍次克海次之,乌拉尔山最弱;张婷婷等(2009)采用计算机识别方法检索欧亚大陆中高纬度阻高,得出计算机程序的识别方法可以替代人工识别的结论。由此可见,即使同为客观识别,由于资料或方法的不同,识别结果也存在差异。

气象工作者在进行大气科学研究时,选取资料至关重要。利用计算机对阻高进行客观识别时可以应用的再分析资料种类繁多,且再分析资料的形成过程十分复杂,资料来源、模式参数化方案和资料同化系统等因素都会影响再分析资料的结果(Bromwich et al.,2004),客观评估其对大气真实状态描述的差异及准确程度成为应用前必须回答的关键问题。从上述研究看,目前还没有一项工作能较完整地对比不同再分析资料对阻高的客观识别结果是否一致。为此,本研究利用多种再分析资料,对比分析不同资料对夏季亚洲阻高的识别结果,揭示不同资料中夏季亚洲阻高的异同,并在此基础上讨论其与中国东北气候的联系。

2 资料和方法

2.1 资料

选取 NCEP/NCAR(NCEPⅠ)逐月高度场、风场、温度场等再分析资料,NCEP/NCAR(NCEPⅠ)、NCEP/DOE(NCEPⅡ)逐日高度场再分析资料,以及 ERA-interim 逐 6 h 高度场再分析资料(文中将该套资料处理为逐日资料),以上资料水平网格分辨率为 2.5°×2.5°,垂直方向 17 层。此外,还利用国家气候中心提供的东北地区 109 站逐日气温和降水量观测资料。所有资料时段均取 1979—2016 年夏季(6—8 月)。

2.2 方法

针对亚洲区域(50°—160°E)特点,依据 Rex(1950)提出的阻高天气学定义和朱乾根等(2001)总结出的阻高存在条件,对中高纬度阻高过程给出如下标准(刘刚 等,2012):(1)40°—75°N 内,500 hPa 位势高度场上有高压中心。(2)高压中心日移动不超过 12.5 个经距。(3)高压中心维持不少于 3 d。根据亚洲阻高出现在乌拉尔山、西西伯利亚、贝加尔湖和鄂霍次克海的地理分布情况,分为 A 类(50°—70°E)、B 类(70°—90°E)、C 类(90°—130°E)和 D 类(130°—160°E)。

为研究阻高活动的动力学机制,计算了波作用通量,采用 Wang 等(1994)根据 Plumb(1985)改进的球坐标系下三维波活动通量诊断方程:

$$\boldsymbol{F}=\begin{pmatrix}F_\lambda\\F_\varphi\\F_Z\end{pmatrix}=\frac{p}{p_S}\cos\varphi\times\begin{Bmatrix}v'^2-\dfrac{1}{2\Omega a\sin2\varphi}\dfrac{\partial(v'\Phi')}{\partial\lambda}\\-u'v'+\dfrac{1}{2\Omega a\sin2\varphi}\dfrac{\partial(u'\Phi')}{\partial\lambda}\\\dfrac{2\Omega\sin\varphi}{S}\left[v'T'-\dfrac{1}{2\Omega a\sin2\varphi}\dfrac{\partial(T'\Phi')}{\partial\lambda}\right]\end{Bmatrix}\tag{1}$$

式中，φ、λ、Φ、a、Ω 分别为纬度、经度、位势、地球半径和地球自转速率；T、p、p_S 分别为气温、气压、1000 hPa；上标"—"和"′"分别为纬圈平均和纬向偏差。$S=\frac{\partial \overline{T}}{\partial z}+\frac{\kappa \overline{T}}{H}$ 为静力稳定度，$z=-H\ln\frac{p}{p_0}$ 为 $\lg p$ 垂直坐标，$\kappa \approx \frac{2}{7}$ 是气体常数与定压比热之比，$H=8$ km 为标高。风场满足准地转平衡，$u=-\frac{1}{fa}\frac{\partial \Phi}{\partial \varphi}$ 为纬向风，$v=\frac{1}{fa\cos\varphi}\frac{\partial \Phi}{\partial \lambda}$ 为经向风，f 为科里奥利常数。

3　夏季亚洲阻高的识别

3.1　阻高活动天数和频次

根据上述阻高的定义标准，利用 3 种再分析资料统计 1979—2016 年夏季亚洲地区阻高出现次数。NCEPⅠ和 NCEPⅡ再分析资料统计结果分别为 1022 次和 1024 次；ERA-interim 再分析资料结果明显较少，仅 885 次。图 1 给出 3 种资料统计的 1979—2016 年夏季亚洲阻高活动过程次数随维持天数的变化曲线。可见亚洲阻高以生命期 3～7 d 居多，维持时间越长，发生次数越少。其中，NCEPⅠ和 NCEPⅡ统计出阻高维持生命期在 3～7 d 的过程分别为 914 次和 893 次，分别占总次数的 89.43%和 87.21%；ERA-interim 统计出现 786 次，占总次数的 89.12%。

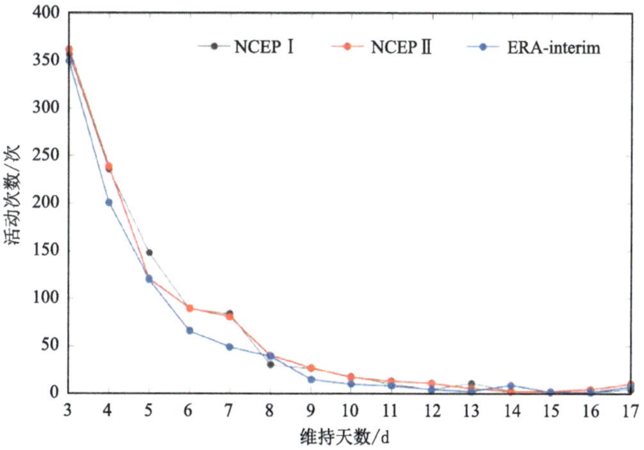

图 1　阻高活动过程次数随维持天数的变化曲线

图 2 为 1979—2016 年夏季亚洲阻高活动过程累计天数在各经度位置的分布。3 种资料均显示，夏季亚洲阻高活动具有明显的地域差异，高压中心主要出现在 50°—60°E、100°—127.5°E 和 132.5°—150°E，分别对应着乌拉尔山、贝加尔湖和鄂霍次克海区域，即阻高在这些区域活动更为频繁。其中，NCEPⅠ结果显示，阻高最多出现在 135°E 的鄂霍次克海区域，共 175 d；而 NCEPⅡ和 ERA-interim 分别在 115°E 和 120°E 的贝加尔湖区域出现累计天数峰值，分别为 201 d 和 209 d。在 65°—95°E，阻高活动较少。其中，NCEPⅠ结果表明，在 87.5°E 附近的西西伯利亚，阻高累计活动仅有 40 d，平均每年出现 1 d；NCEPⅡ和 ERA-interim 统计的阻高中心在 67.5°N 和 65°E 的乌拉尔山地区活动最少，分别为 52 d 和 38 d。

3.2　阻高与亚洲环流的关系

对 1979—2016 年逐年夏季阻高活动与 500 hPa 平均位势高度场(图 3)进行相关分析发现，前两种资料对于各类阻高的表征结果较为一致，而 ERA-interim 再分析资料对各类阻高面积和范围的表征较小。总体而言，3 种资料下二者之间存在显著相关且形态相近，初步证明客观识别结果的科学性和准确性。从图 $3a_1$—c_1 可见，当乌拉尔山为高值中心时，显著的负相关区分别位于中高纬度的欧洲大陆和贝加尔湖、

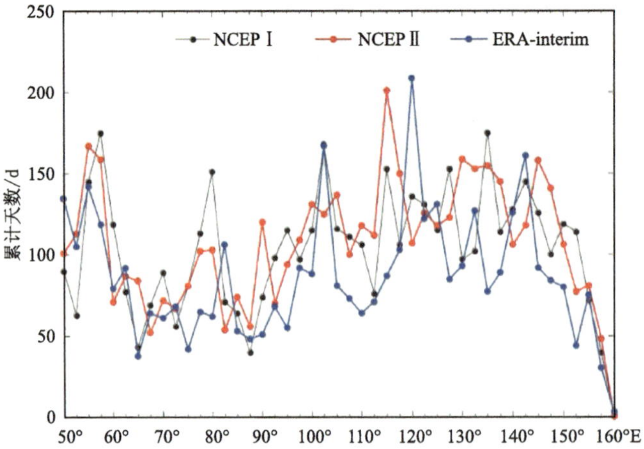

图 2　阻高活动过程累计天数随经度的变化曲线

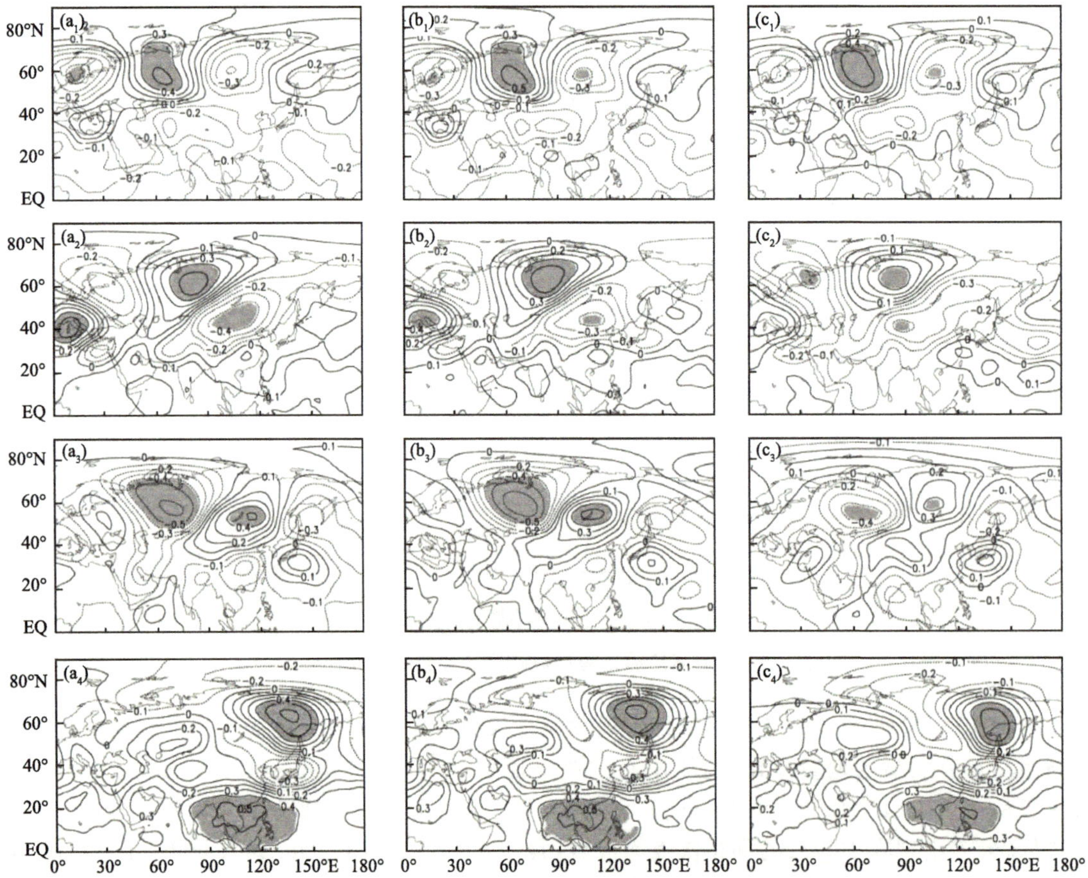

图 3　各类阻高逐年累计天数与 500 hPa 平均位势高度场的相关系数分布

(a_1—a_4. NCEP Ⅰ, b_1—b_4. NCEP Ⅱ, c_1—c_4. ERA-interim；a_1—c_1、a_2—c_2、a_3—c_3、a_4—c_4 分别表示 A、B、C、D 类；阴影区为达到 99% 的置信水平)

中纬度的青藏高原及其西部，而鄂霍次克海为正相关区，这在亚洲中高纬度形成了东、西向的相关波列，同时在西亚地区也产生了一个南、北向的相关波列。从图 3a_2—c_2 可见，B 类阻高出现时，相关系数较大的区域形成了沿着西欧经过东欧和西西伯利亚到达贝加尔湖的一个大圆，类似 Hoskins 等(1981)提出的准静止罗斯贝波列的传播路径。图 3a_3—c_3 中的 C 类阻高出现时对应的中高纬度相关系数显著区域与 B 类阻高近乎相反；同时，中国东北地区常在 C 类阻高控制范围内，而中、低纬度上中国南方为低值区，这表明中国东北和南方地区的天气气候与 C 类阻高密切相关。图 $3a_4$—c_4 表明，当鄂霍次克海为高压脊控制时，日本附近的中纬度北太平洋地区为显著负相关，而显著的正相关区出现在低纬度西太平洋，这表明 D 类阻高对东亚南部南、北向相关波列的形成和副高有重要影响。

3.3　阻高活动的年代际变化

1979—2016 年 3 种再分析资料统计出阻高年平均活动累计天数，均表现为 C 类最多、D 类次之、B 类最少的特征，前两种资料统计的阻高天数较第 3 种资料明显偏多。其中，NCEPⅠ再分析资料结果显示，A、B、C 和 D 类阻高平均每年出现天数分别为 20、14、40、27 d；NCEPⅡ再分析资料结果分别为 21、15、40、29 d；ERA-interim 再分析资料结果分别为 17、14、32、24 d。对比 3 种资料下各类阻高逐年活动累计天数滤波分析后的变化曲线（图 4），可见 A 和 B 类阻高在 20 世纪 80 年代最为活跃，90 年代最弱，21 世纪最初 10 年增强；C 和 D 类阻高在 20 世纪 80 年代活动天数最少，90 年代最多，21 世纪初又有所减少。这与刘刚等（2012）研究结论类似。各类阻高的年代际变化显著，且在一定程度上存在此消彼长的制约关系。

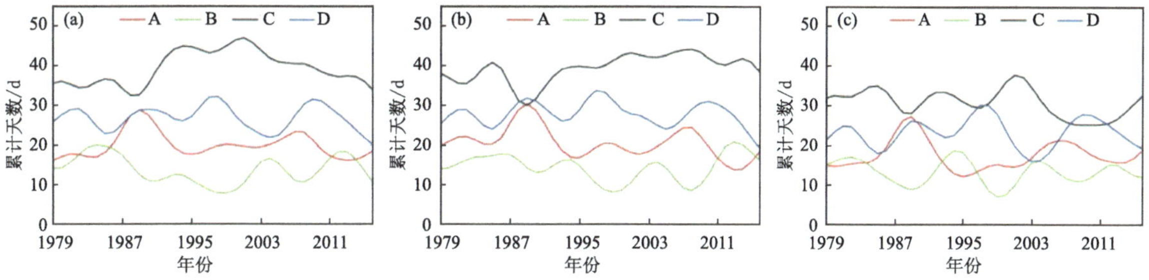

图 4　各类阻高活动累计天数年际变化的滤波分析
（a. NCEPⅠ；b. NCEPⅡ；c. ERA-interim）

4　阻高与 6 月中国东北气候异常的联系

Nakamura 等（2004）研究表明，鄂霍次克海阻高的形成机制在 5、6、7 月均存在很大差异；沈柏竹等（2011）指出，中国东北地区初夏（5、6 月）和盛夏（7、8 月）降水的影响系统不同。可以预期，即使在夏季，各类阻高在不同月份的动力学效应也有所不同。因此，下文将主要分析初夏（6 月）阻高与同期中国东北气候的联系。

4.1　阻高与 6 月中国东北气温、降水量的相关

对 6 月中国东北地区的气温和降水量进行经验正交函数（EOF）分解，前两模态所占解释方差分别为 84.74%、9.13%和 69.18%、9.28%，均通过 north 显著性检验，能够很好地代表 6 月东北地区气温和降水的主要分布。其中，第 1 模态表现为全区一致高温多雨，第 2 模态表现为北部低温多雨、南部高温少雨的特征。统计 6 月各类阻高活动总天数与东北地区气温和降水 EOF 分解后前两模态时间序列相关系数的关系（表 1），可以看出，3 种再分析资料均显示 D 类阻高活动天数与东北地区气温和降水的相关最为显著，D 类阻高多（寡）年对应 6 月东北低（高）温、多（少）雨。张庆云等（1998）指出，鄂霍次克海高压异常对东亚中、高纬度位势高度及副高的变化有重要作用。李艳等（2010）对影响中国天气的关键区阻高进行统计时发现，鄂霍次克海阻高是夏季雨涝灾害的一个重要环流背景。刘刚等（2012）研究表明，鄂霍次克海阻高异常有利于偏东北路径的冷空气入侵东北亚及中国东北地区。这些研究成果揭示出鄂霍次克海阻高对东亚气候异常的重要作用。根据表 1 显示结果，在 D 类阻高与东北气温和降水的关系中，NCEPⅠ再分析资料的相关系数最大。因而，文中以下内容将以 NCEPⅠ再分析资料为例，讨论 D 类阻高对东北气候的可能影响。

表 1　6 月各类阻高逐年累计天数与东北气温、降水量 EOF 分解后的前两模态时间序列的相关系数

阻高类型	A1	B1	C1	D1	A2	B2	C2	D2	A3	B3	C3	D3
气温 TEOF1	−0.13	−0.02	0.24	−0.49**	−0.25	−0.03	0.38*	−0.39*	−0.20	−0.03	0.39**	−0.35*
气温 TEOF2	−0.18	0	−0.27	0.23	−0.05	−0.10	−0.19	0.18	−0.08	0.20	−0.06	0.30
降水 TEOF1	0.02	0.22	0.02	0.24	−0.03	0.34*	−0.08	0.18	0.05	0.23	−0.04	0.17
降水 TEOF2	−0.06	−0.05	−0.24	0.60**	0.16	−0.15	−0.20	0.48**	0	0.07	−0.22	0.55**

注：* 和 ** 分别为达到 95%和 99%的置信水平；A1—D1 为 NCEP Ⅰ 再分析资料结果，A2—D2 为 NCEP Ⅱ 再分析资料结果，A3—D3 为 ERA-interim 再分析资料结果。

4.2　D 类阻高多寡年的环流异常

定义 6 月亚洲阻高在鄂霍次克海周围发生天数 I_D 多(少)于该年各类阻高平均天数 I 的为 D 类阻高多(寡)年(表 2)，下面将对这些年份的相关要素进行合成分析。从 6 月 NCEP Ⅰ 再分析资料下 D 类(简称 D_1)阻高多寡年 200、500、1000 hPa 层上位势高度和温度的合成差值场可以得出，6 月的合成场中，鄂霍次克海地区低层有暖高压，对流层中、上层的反气旋中心位于低层反气旋中心的西北部，异常高压场表现为斜压结构(图略)。Nakamura 等(2004)指出，这可能由于 6 月鄂霍次克海和西部陆面较大的海-陆温差所致。阻高偏多年，高层 200 hPa 东北地区上空气温明显偏高，鄂霍次克海周围气温低；而低层 1000 hPa 上东北地区为明显低温，鄂霍次克海周围气温高。东北地区低层温度高、高层温度低的配置指示大气层结不稳定，有利于降水的产生。

表 2　6 月 D_1 阻高多寡年选取

指数	年份
$I_{D_1}>I_1$	1979、1980、1981、1982、1983、1986、1988、1989、1990、1992、1993、1996、1997、1998、1999、2002、2006、2007、2009、2012、2013、2014、2016
$I_{D_1}<I_1$	1984、1985、1987、1991、1994、1995、2000、2001、2003、2004、2005、2008、2010、2011、2015

注：I_1 为阻高平均天数。

对流层中，大气温度的垂直递减率(以下简称气温直减率)是决定大气静力稳定度的主要因素。上冷下暖的温度平流垂直差异常导致气温直减率增大，使得大气层结趋于不稳定(朱禾 等，2016)。为此，对 6 月 D_1 阻高多寡年 500 和 850 hPa 矢量风及温度平流的差值场进行合成(图略)。可以得出，500 hPa，鄂霍次克海上空为反气旋环流，西部强烈的暖平流使得等压面升高，有利于 D 类阻高的加强和维持。强大的 D 类阻高同时有利于其西侧槽的发展和加深，加强了对流层中层东北地区冷空气南下和偏北气流的输送，东北地区上空以冷平流、气旋性环流为主。低层 850 hPa，鄂霍次克海附近强烈的冷平流使得地面加压、位势高度升高，有利于低层反气旋环流的加强和维持。东北地区以从内陆而来的偏北气流为主，由于夏季陆面温度较高，自北向南的暖平流加强，有利于地面减压，使得东北地区位势降低。东北地区高空为冷平流而低空为暖平流，大气层结趋于不稳定，有利于对流性天气的产生。

除水平环流场外，垂直环流也可以反映大气运动在高低层间的联系。图 5 给出 40°—70°N 纬向平均的位势高度和垂直速度的合成差值垂直剖面。欧洲大陆至乌拉尔山及鄂霍次克海上空的异常高压系统延伸至对流层整层，这与大气高低层辐散场及系统配置相吻合；1000～100 hPa，乌拉尔山及鄂霍次克海周围均存在深厚的下沉气流，尤以鄂霍次克海周围的下沉气流最为强烈。而中国东北地区则以异常低压和上升气流为主，低压中心位于 250 hPa 附近，上升运动从 1000 hPa 一直延伸至 300 hPa 以上。

通过上述对多种要素场的分析看到，D 类阻高活跃年，大气环流以经向型为主，加强了中高纬度冷空气南下和偏北气流的输送。东北地区低层低温、暖平流，高层高温、冷平流，同时伴随强烈的垂直上升运动。这种配置结构十分有利于对流性天气出现，一旦水汽条件充沛，极易产生降水。

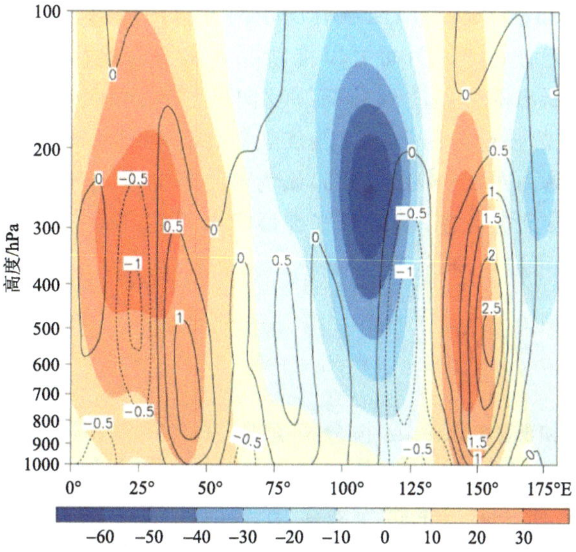

图 5　6 月 D_1 阻高多寡年 40°—70°N 平均的位势高度(单位:gpm)和垂直速度（等值线,单位:10^{-2} Pa s^{-1},其中正值表示下沉、负值表示上升运动)的合成差值垂直剖面

4.3　D 类阻高波作用通量

为进一步揭示 6 月 D 类阻高活动的动力学机制,根据 Wang 等(1994)的方法计算了波作用通量的水平和垂直分量。D_1 阻高多寡年 500 hPa 波作用通量和位势高度差值场表明,中、高纬度,欧洲大陆至乌拉尔山及鄂霍次克海地区为正异常高度场控制,中国东北地区及西北太平洋上空为负异常高度场,纬向存在高度异常中心"＋－＋－"的分布型。东北地区上空有罗斯贝波向西北辐合的趋势,说明波扰能量在该处被阻截,加强了东北地区的局地环流异常,有利于低压系统维持。在鄂霍次克海地区有一个准静止罗斯贝波向东南方向传播,波活动通量方向与大圆路径类似。

Plumb(1985)指出,尽管罗斯贝波通常呈准水平传播,但有时垂直方向分量亦不可忽略,因为其强迫源可能来自低层大气向上的波作用通量。为此,计算了沿 140°E 的纬度-高度剖面(图 6a)和沿 55°N 的经度-高度剖面(图 6b)合成差值场。可见 50°—60°N、125°—160°E,1000～400 hPa 出现较强的上升通量。大多数垂直通量向上传播,但在 160°E 以东 700～200 hPa 有较大的向东传播的波作用通量。

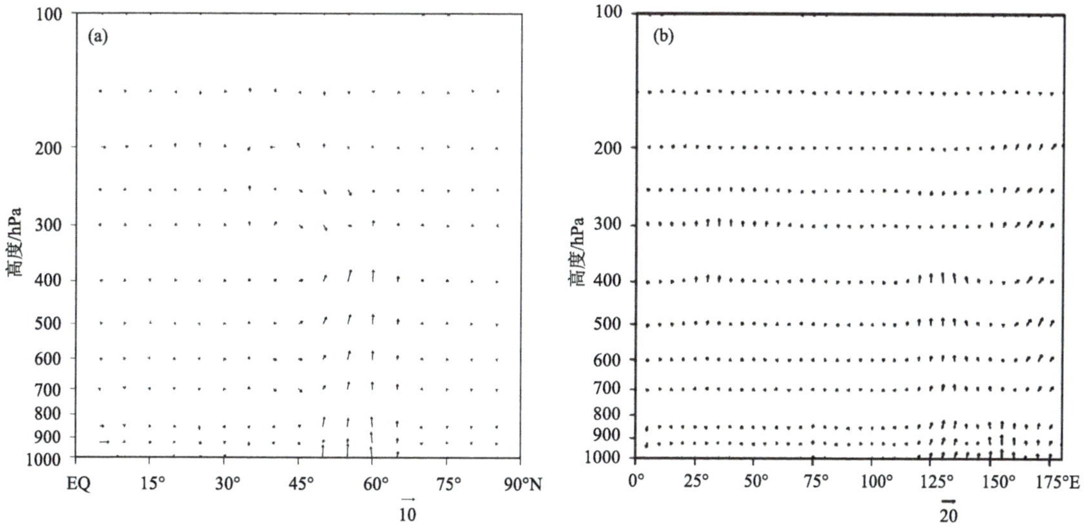

图 6　6 月 D_1 阻高多寡年波作用通量(矢线,单位:m^2 s^{-2})在(a)140°E 的纬度-高度剖面和(b)55°N 的经度-高度剖面合成差值场

由图 6 可知，6 月，500 hPa 鄂霍次克海区域有向东南方向传播的水平波动通量，1000～400 hPa 强烈的上升通量也出现在鄂霍次克海周围。因而可以推断鄂霍次克海地区是 6 月罗斯贝波的重要来源地之一。这与 Wang 等(1994)诊断出的初夏波列从高纬度向低纬度地区传播的结果类似。Plumb (1985)研究发现，准静止波传播最主要的强迫机制为非绝热加热。Okawa(1973,1976)指出，初夏东西伯利亚大陆与白令海峡之间存在强大的热力对比；Kato(1985)认为 6 月华北以北地区气温上升迅速。Wang(1992)对梅雨季节期间欧亚大陆反气旋研究发现，初夏反气旋最常出现在(50°—70°N,131°—150°E)地区。这些事实表明，6 月 D 类阻高的形成可能与海-陆温差有关。

5　结论与讨论

基于 NCEPⅠ、NCEPⅡ和 ERA-interim 再分析资料，对 1979—2016 年夏季亚洲阻高特征进行了初步对比分析，并以 D 类阻高为例，讨论其对 6 月中国东北地区气候的可能影响。结果表明：

(1)夏季亚洲阻高以生命期 3～7 d 的过程居多，维持时间越长，发生次数越少。NCEPⅠ和 NCEPⅡ再分析资料对阻高活动天数、发生频次及年代际变化的识别结果差异较小，而 ERA-interim 与前两种资料的结果差别较大。各类阻高年平均活动累计天数均表现为“C 类最多、D 类次之、B 类最少”的特征，且具有明显的年代际变化。

(2)3 种再分析资料下，夏季各类阻高活动天数均与 500 hPa 高度场存在相应的显著相关区，且形态相近。但前两种资料对于各类阻高的表征结果较为一致，而 ERA-interim 再分析资料对各类阻高面积和范围的表征偏小。

(3)6 月 D 类阻高活动天数与东北地区气温和降水关系密切，即 D 类阻高多(寡)年对应 6 月东北低(高)温、多(少)雨。从环流形势上看，D 类阻高活跃年，大气环流以经向型为主，加强了中、高纬度冷空气南下和偏北气流的输送。东北地区低层低温、暖平流，高层高温、冷平流的结构指示大气层结不稳定，且东北上空为异常低压环流控制，上升气流较强，这种大气环流的配置有利于对流性天气出现。

(4)6 月 D 类阻高偏多年，500 hPa 上鄂霍次克海地区有东南方向传播的水平波动通量，1000～400 hPa 强烈的上升通量也出现在鄂霍次克海周围。因而，可以推断鄂霍次克海地区是 6 月罗斯贝波的重要来源地之一，而 6 月 D 类阻高的形成可能与海-陆温差有关。

值得注意的是，这里分析的亚洲阻高年代际变化与整个北半球阻高活动的年代际变化是否具有类似的特征？二者存在怎样的联系？此外，本研究仅对 D 类阻高总天数的多寡年进行分析，并未对孤立子、偶极子等情况分开讨论。阻高有时会同时出现在不同地区，且常伴随东北冷涡活动(刘宗秀 等，2002；廉毅 等，2010；刘刚 等，2012，2015)。两大系统之间的关系及其不同的配置方式对东北地区乃至中国、甚至亚洲气候会产生怎样的影响？这些问题均有待深入研究。

参考文献

陈汉耀，1957. 1954 年长江淮河流域洪水时期的环流特征[J]. 气象学报，28(1):1-12.

陈菊英，王玉红，王文，2001. 1998 及 1999 年乌山阻高突变对长江中下游大暴雨过程的影响[J]. 高原气象，20(4):388-394.

杜小玲，高守亭，彭芳，2014. 2011 年初贵州持续低温雨雪冰冻天气成因研究[J]. 大气科学，38(1):61-72.

顾伟宗，陈丽娟，左金清，等，2016. 多因子协同作用对 1992 年和 1998 年黄淮地区夏季降水异常的影响[J]. 大气科学，40(4):743-755.

胡钰玲，赵中军，康彩燕，等，2017. 中国南方 2008 年与 2016 年两次低温雨雪冰冻天气过程对比分析[J]. 冰川冻土，39(6):1180-1191.

赖珍权，翟丽萍，张禹德，等，2017. 2016 年初广西罕见的一次低温雨雪冰冻灾害天气分析[J]. 自然灾害学报，26(1):156-164.

李春，孙照渤，2003. 中纬度阻塞高压指数与华北夏季降水的联系[J]. 南京气象学院学报，26(4):458-464.

李峰，丁一汇，2004. 近 30 年夏季亚欧大陆中高纬度阻塞高压的统计特征[J]. 气象学报，62(3):347-354.

李维京,1999.1998年大气环流异常及其对中国气候异常的影响[J].气象,25(4):20-25.

李艳,金荣花,王式功,2010.1950—2008年影响中国天气的关键区阻塞高压统计特征[J].兰州大学学报(自然科学版),46(3):47-55.

廉毅,布和朝鲁,谢作威,等,2010.初夏东北冷涡活动异常与北半球环流低频变化[J].大气科学,34(2):429-439.

刘刚,沈柏竹,廉毅,等,2012.亚洲阻塞高压分类及其与东北冷涡活动和东北夏季低温的联系[J].地理科学,32(10):1269-1274.

刘刚,廉毅,颜鹏程,等,2015.5—8月东北冷涡客观识别分类及北半球大气环流特征分析[J].地理科学,35(8):1042-1050.

刘樱,郭品文,冯涛,2016.华北地区冬季持续性异常低温事件与大气低频振荡活动的关系[J].大气科学学报,39(3):370-380.

刘宗秀,廉毅,高枞亭,等,2002.东北冷涡持续活动时期的北半球500 hPa环流特征分析[J].大气科学,26(3):361-372.

苗青,巩远发,邓锐捷,等,2016.北半球中高纬度低频振荡对2012/2013年冬季中国东北极端低温事件的影响[J].大气科学,40(4):817-830.

沈柏竹,林中达,陆日宇,等,2011.影响东北初夏和盛夏降水年际变化的环流特征分析[J].中国科学:地球科学,41(3):402-412.

苏丽欣,廉毅,李尚锋,等,2015.东北区夏季低温事件概率空间分布与亚洲阻塞流型域及其冷空气活动路径[J].地理科学,35(2):251-258.

孙安健,高波,2000.华北平原地区夏季严重旱涝特征诊断分析[J].大气科学,24(3):393-402.

汤懋苍,1957.亚洲东部的阻塞形势及其对天气气候的影响[J].气象学报,28(4):282-293.

陶诗言,1980.中国之暴雨[M].北京:科学出版社.

陶诗言,赵煜佳,陈晓敏,1958.东亚的梅雨期与亚洲上空大气环流季节变化的关系[J].气象学报,29(2):119-134.

王小玲,丁一汇,2013.2010年夏季欧亚异常阻高演变过程及对天气气候的影响[J].气象,39(9):1089-1095.

王亚非,宋永加,1998.1991年夏东亚阻塞高压形成的诊断分析[J].气象学报,56(2):212-224.

叶笃正,陶诗言,朱抱真,等,1962.北半球冬季阻塞形势的研究[M].北京:科学出版社.

张培忠,杨素兰,1996.阻塞高压活动的气候变化及其对中国某些地区旱涝的影响[J].气象学报,54(5):633-640.

张庆云,陶诗言,1998.亚洲中高纬度环流对东亚夏季降水的影响[J].气象学报,56(2):199-211.

张庆云,陶诗言,张顺利,2001.1998年嫩江、松花江流域持续性暴雨的环流条件[J].大气科学,25(4):567-576.

张婷婷,智协飞,白永清,等,2009.欧亚阻塞高压的计算机识别[J].科技信息(15):407,409-410.

赵振国,1999.中国夏季旱涝及环境场[M].北京:气象出版社.

郑志海,王永光,2018.2017年夏季北半球大气环流特征及对我国天气气候的影响[J].气象,44(1):199-205.

周晓平,1957.亚洲中纬度区域阻塞形势的统计研究[J].气象学报,28(1):75-85.

朱乾根,林锦瑞,寿绍文,等,2001.天气学原理和方法[M].北京:气象出版社.

朱禾,王秀明,2016.影响气温直减率变化的物理过程讨论[J].气象科技进展,6(5):51-54.

BREZOWSKY H,FLOHN H,HESS P,1951. Some remarks on the climatology of blocking action[J]. Tellus,3(3):191-194.

BROMWICH D H,FOGT R L,2004. Strong trends in the skill of the ERA-40 and NCEP-NCAR reanalyses in the high and midlatitudes of the Southern Hemisphere,1958—2001[J]. J Clim,17(23):4603-4619.

ELLIOTT R D,SMITH T B,1949. A study of the effects of large blocking highs on the general circulation in the Northern-Hemisphere westerlies[J]. J Meteor,6(2):68-85.

HOSKINS B J,KAROLY D J,1981. The steady linear response of a spherical atmosphere to thermal and orographic forcing [J]. J Atmos Sci,38(6):1179-1196.

KATO K,1985. On the abrupt change in the structure of the Baiu front over the China continent in late May of 1979[J]. J Meteor Soc Japan,63(1):20-36.

NAKAMURA H,FUKAMACHI H,2004. Evolution and dynamics of summertime blocking over the Far East and the associated surface Okhotsk high[J]. Quart J Roy Meteor Soc,130(599):1213-1233.

OKAWA T,1973. Growth mechanism of the Okhotsk high[J]. Res Lett,25:65-77. (in Japanese)

OKAWA T,1976. Behaviour of Okhotsk High and 500 hPa level flow pattern in the Baiu season[J]. J Meteor Res,28:253-261. (in Japanese)

PLUMB R A,1985. On the three-dimensional propagation of stationary waves[J]. J Atmos Sci,42(3):217-229.

REX D F,1950. Blocking action in the middle troposphere and its effect upon regional climate: Ⅱ. The climatology of blocking action[J]. Tellus,2(4):275-301.

SUMNER E J,1959. Blocking anticyclones in the Atlantic-European sector of the Northern Hemisphere[J]. Met Mag,88: 300-311.

WANG Y F,1992. Effects of blocking anticyclones in Eurasia in the rainy season(Meiyu/Baiu season)[J]. J Meteor Soc Japan,70(5):929-951.

WANG Y F,YASUNARI T,1994. A diagnostic analysis of the wave train propagating from high-latitudes to low-latitudes in early summer[J]. J Meteor Soc Japan,72(2):269-279.

第二部分

东亚—西北太平洋夏季副热带季风

东亚夏季风在中国东北区建立的标准、日期及其主要特征分析*

廉毅　沈柏竹　高枞亭　安刚　唐晓玲

（吉林省气象科学研究所，长春 130062）

摘　要：利用中国东北区 80 个测站，1980 年以来近 20 a 逐日降水和同期 NCEP/NCAR 逐日再分析资料，提出了东亚夏季风在中国东北区建立和撤退的标准，即取 850 hPa 候平均资料，作沿 122.5°E 的时间-纬度剖面图，将 θ_{se} 的 336 K 和南风≥4 m/s 等值线同时越过 40°N 的候定义为东亚夏季风在中国东北区建立的日期（候），持续及累计的总候数为影响时间；该年 θ_{se} 的 336 K 等值线永久撤离到 40°N 以南的候为东亚夏季风从中国东北区撤退的日期；如果某年 θ_{se} 的 336 K 和南风 4 m/s 等值线没有同时北移越过 40°N 的候出现，确定为东亚夏季风在中国东北区没有建立的年份。文中相应给出 1980—2000 年建立、持续及累积和撤退的时间年历表，经计算东亚夏季风在中国东北区建立的平均日期为第 41 候（7 月第 5 候），累计平均影响候数为 3.6 候（18 d 左右），开始撤退的平均日期为第 45 候（8 月第 3 候）；东亚夏季风在中国东北区建立前后的候降水量增加和撤退前后的候降水量递减十分显著，建立时水汽场由辐散转变为辐合。定义了中国东北区强夏季风的年份（建立的时间早，且累计候≥4 候的年份）为 1981、1988、1990、1994 年，与国内学者定义的强夏季风年基本一致；在中国东北区没有建立东亚夏季风的年份为 1980、1983、1987、1991、1993、1997 年，与国内学者定义的弱夏季风年亦基本一致。通过分析东亚夏季风在中国东北区建立和不建立的典型年份流场和加热场，发现两者差异显著：建立的年份西北太平洋副热带高压偏西和偏北（大约 7 个经、纬距），东亚中高纬度地区没有阻塞形势；不建立年则相反。夏季强季风年（1994 年）的 200 hPa 加热场正距平中心偏向南印度洋赤道地区，西太平洋暖池相对是冷源，中国东部沿海地区辐散风 V_x 距平梯度较大，而没有建立夏季风的 1980 年北半球海陆热力差异相对减小，低层更加不明显。

关键词：东亚夏季风；东北区；建立标准

1　引言

自从 1934 年竺可桢研究中国大陆夏季风以来，许多学者对中国夏季风的性质、季风进退、季风雨带活动及其与大气环流的联系等进行了大量的研究。1980 年以来有关亚洲季风的研究取得了重要进展。陈隆勋（1984）、金祖辉等（1982）和 Tao 等（1987）发现亚洲地区存在着两支季风：印度季风和东亚季风，既有联系，又有独立性。施尚文等（1982）指出，在中国北方出现夏季风期间，200 hPa 上的副热带西风急流已北退至 40°N 以北，高层东风急流在 25°N 以南建立。因此，北方地区的夏季风可称为副热带季风。陶诗言（1996）指出，印度季风和东亚季风的主要差异在于印度季风属于热带性质的季风，而东亚季风是副热带性质的季风，受中高纬度的影响比较大，东亚季风比印度季风要复杂得多。喻世华等（1986）指出东亚季风和印度季风基本上以 100°E 为界。Zhu 等（1986）进一步指出东亚季风可划分为南海季风（热带季风）和副

* 本文原载于《气象学报》，2003，61(5)：548-558

热带季风，即南海—西太平洋热带季风区和大陆—日本副热带季风区，热带季风区和副热带季风区可以南亚高压为界。Tao 等(1987)更明确提出，东亚存在一个东亚季风环流系统，并确定了该季风系统的成员，东亚季风系统与印度季风系统既相互独立又相互作用。这个结果突破了长期以来东亚季风研究中的一个误解，即中国大陆的降水和旱涝或东亚季风是受印度季风控制或只是印度季风的一种延伸，实际上，东亚季风系统在整个亚洲季风区暴发最早，它有自己独立的冷热源驱动，并产生独特的降水型，如东亚著名的梅雨(丁一汇 等,1986)。

郭其蕴(1983)认为，东亚大陆地区的季风主要由纬向海陆热力差异所形成，计算 10°—50°N 各纬度上 110°E 与 160°E 海平面月平均气压差值，将差值≤－5 hPa 的各种纬度值累加起来取绝对值然后标准化，并首次提出了一种夏季风指数(SMI)。为了表征副热带夏季风的进退活动，不少学者都着眼于寻找副热带夏季风活动的指标，汤明敏等(1983)和周玉孚(1983)使用 1000 hPa 上 344～348 K 等 θ_{se}线，或 850 hPa 上 336～340 K 等 θ_{se}线作为副热带夏季风的进退指标。施尚文等(1982)用 850 hPa 上 332 K 等 θ_{se}线作为北方地区夏季风影响期的临界值等。就温湿指标而言，鉴于副热带夏季风是高温高湿的气流，朱乾根等(1989)提出用等 θ_{se}线不如用等压面上的等露点线，如利用 850 hPa 上 14 ℃等露点线表征副热带季风的前沿更能反映副热带夏季风活动的实际情况。赵汉光等(1994)引用郭其蕴(1983)的方法，建立了冬季风强度指数。施能等(1996)为了消去不同格点均方差不均匀造成的影响，定义东亚季风指数为在 20°—50°N 范围内每 5 个纬度分别在 110°E 和 160°E 格点上标准化的海平面气压差之和，并将所得的求和值再进行一次标准化处理。张庆云等(1998)将东亚热带季风槽区(10°—20°N,100°—150°E)与东亚副热带地区(25°—35°N,100°—150°E)850 hPa 风场 6—8 月平均的 U 分量距平差定义为东亚夏季风指数(EASMI)。

孙秀荣等(2001)从季风形成的本质出发，尝试提出了一个指标，即东亚季风区(27°—35°N,105°E 以东的大陆)内的地温和副热带西北太平洋(15°—30°N,120°—150°E)的海表温度之差，用华南地区(27°N 以南,105°E 以东的大陆)的地温和南海(5°—18°N,105°—120°E)的海表温度之差表示南北向海陆热力差，来表示东亚热带、副热带东西向海陆热力差异和南北向海陆热力差异的强弱变化，定义了一个东亚海陆热力差指数，该指数较好地反映了东部季风区夏季降水和异常状况。强指数年，雨带偏北，江淮流域和长江中下游明显干旱，华南、华北降水偏多，弱指数年反之。

陈隆勋(1984)指出，1979 年 6 月影响中国大陆的季风主要有两支：一支是来自印度和孟加拉湾的西南季风，另一支是副热带高压西侧转向的西南季风，两支气流各附有一个 θ_{se}大值带。廉毅等(1997)曾指出吉林省主要受副热带高压西侧转向的西南季风影响，还在文献(廉毅 等,1998)中发现 7—8 月 850 hPa 候平均图上 336 K(作为西南副热带季风的前沿)线，沿 120°E θ_{se}时间-纬度剖面图上越过 40°N 的时间达 2 个候以上，最北界≥45°N,8 月份 316 K 线没越过 45°N 以南，当年夏季气温较常年偏高，反之，夏季气温偏低。本文在此基础上进一步明确定义了东亚夏季风在中国东北区建立的标准、日期，还分析了加热场、流场和水汽输送方面的基本特征。

2　资料和方法

文中所用的地面资料包括吉林、辽宁、黑龙江和内蒙古东部四省(区)80 个测站 1980 年以来近 20 a 逐日降水资料；同期的 NCEP/NCAR 全球再分析逐日再分析资料。

水汽通量散度和辐散风 V_x(加热场)采用丁一汇(1989)的方法进行计算。

3　东亚夏季风在东北区建立的标准和日期

3.1　定义

利用 NCEP 资料和部分典型年份的探空资料，分别在 850 hPa θ_{se}和南风分量候平均图上沿 120°E，

122.5°E 和 125°E 作时间-纬度剖面图，发现区别不大，最后选定 850 hPa 候平均图沿 122.5°E 的时间-纬度剖面上，将 θ_{se}的 336 K 和 $Vs \geqslant 4$ m/s(南风分量)等值线同时越过 40°N 的候，定义为东亚夏季风在东北区建立的日期，持续或累计的候数为影响时间；当该年 θ_{se}的 336 K 等值线永久撤离到 40°N 以南的候为东亚夏季风从中国东北区撤退的日期；如果某年 θ_{se}的 336 K 和 $Vs \geqslant 4$ m/s 等值线(南风分量)没有同时北移越过 40°N 的候出现，确定为东亚夏季风在中国东北地区没有建立的年份(表 1)。

从表 1 可以看出，东亚夏季风在中国东北区建立(沿 122.5°E 越过 40°N)的平均日期为第 41 候(不包括没有建立的年份)，即 7 月第 5 候，与 Tao 等(1987)的东亚夏季风推进的平均日期 7 月 20 日是比较接近的，累计的平均候数为 3.6 候，开始撤退的平均日期为第 45 候，即 8 月第 3 候。建立候的均方差为 1.948 候，持续的均方差为 1.708 候，撤退候的均方差为 1.275 候。

表 1　东亚夏季风在东北区建立、影响和撤退的标准

年份	东亚夏季风建立		东亚夏季风影响		θ_{se}(336 K)达最北界日期/候	东亚夏季风撤退
	θ_{se}(336 K)越过 40°N 日期/候	南风 /(m s^{-1})	θ_{se}(336 K)持续时间/候	θ_{se}(336 K)累计时间/候		θ_{se}(336 K)退回 40°N 以南日期/候
1980	—	—	0	0	—	—
1981	37	≥4	3	4	47	43
1982	43	≥4	3	3	49	46
1983	—	—	0	0	—	—
1984	43	≥43	3	3	44	46
1985	41	≥4	3	3	43	44
1986	42	≥4	1	1	41	43
1987	—	—	0	0	—	—
1988	40	≥4	4	6	50	47
1989	40	≥4	2	2	43	44
1990	38	≥4	6	6	42	44
1991	—	—	0	0	—	—
1992	41	≥4	2	2	43	43
1993	—	—	0	0	—	—
1994	38	≥4	8	8	50	46
1995	42	≥4	3	4	46	45
1996	42	≥4	2	2	43	44
1997	—	—	0	0	—	—
1998	40	≥4	3	4	50	45
1999	44	≥4	2	2	51	46
2000	40	≥4	2	3	50	46
平均	41	≥4	3.1	3.6	46	45
均方差	1.948			1.708		1.275

注："—"表示东亚夏季风没有越过 40°N，东亚夏季风在中国东北区没有建立，下同。

3.2　1980—2000 年 850 hPa 候平均 θ_{se}和 V 风速分量沿 122.5°E 的平均时间-纬度剖面特征

从图 1 可见，850 hPa 候平均 θ_{se}的 336 K 线于 41～42 候越过了 40°N(122.5°E)，西南副热带季风的前沿平均状态于 7 月下旬可达到中国东北地区的南部，V 风速分量≥1 m/s 的外围线于 40 候开始越过 40°N，43 候(8 月第 1 候)达到最北界为 46°N，从 45 候即 8 月中旬中期开始，南风分量完全撤退到 40°N 以南。表明 1980—2000 年中国夏季西南副热带季风，影响中国东北地区最长的时间为 1～2 个候，纬度在 40°—42°N。

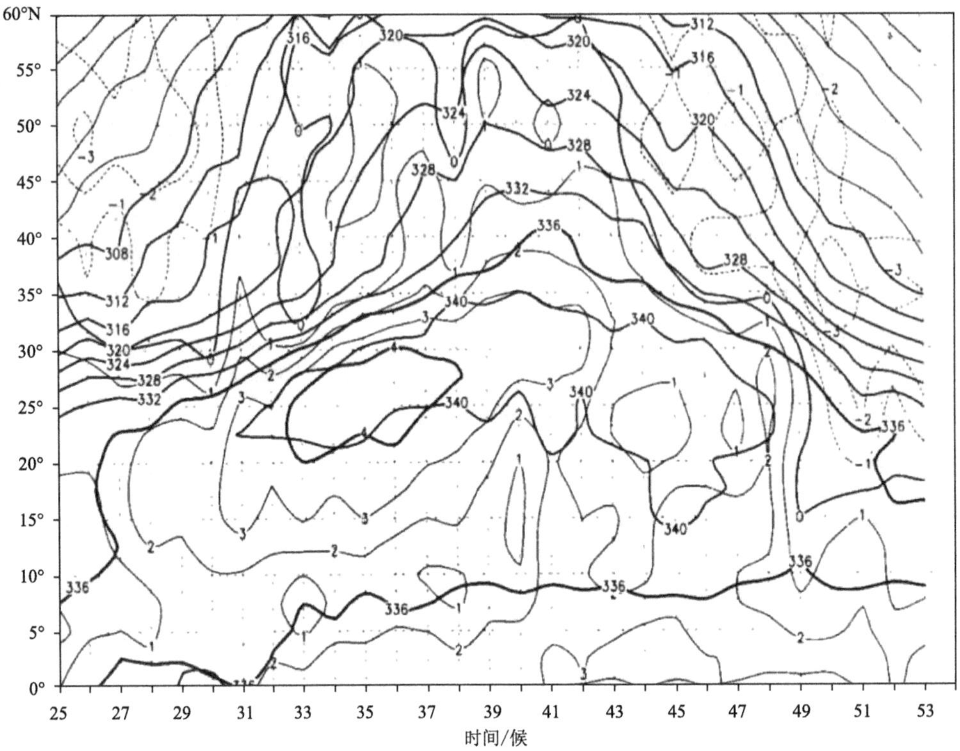

图 1　1980—2000 年 850 hPa 的 θ_{se} 和 V 分量沿 122.5°E 逐候时间-纬度剖面
（细实线为南风风速等值线，虚线为北风风速等值线，间隔 1 m/s；粗实线为 θ_{se} 等值线，间隔 4 K）

3.3　东亚夏季风在东北区建立和撤退的特征

3.3.1　东亚夏季风建立和撤退的东北区候降水量特征

图 2 为 1980—1995 年夏季风建立 15 个候的前后 3 个候的东北区 80 个测站的平均降水量，从图 2 可见，建立候的降水量接近 50 mm，为前 1 候降水量 24 mm 的 2 倍多，建立第 2 候的降水量也稳定在 40 mm (8 mm/d)。图 3 为 1980—1995 年夏季风撤退 15 候的前后 3 个候东北区 80 个测站平均降水量，趋势与建立时正相反，结束前 1 个候降水量接近 40 mm，撤退时则迅速降至 25 mm 以下。

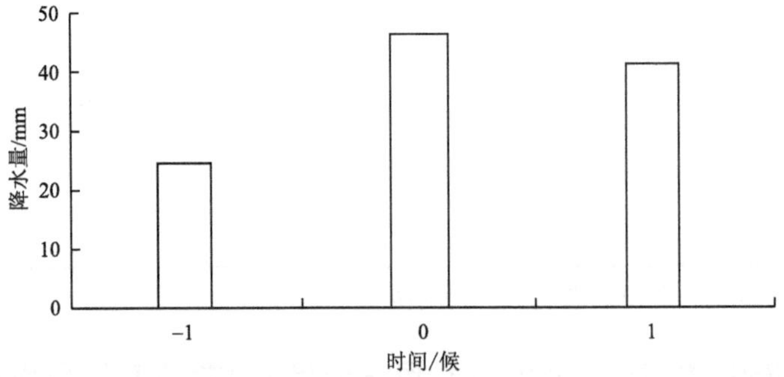

图 2　1980—1995 年东亚夏季风建立年建立候及其前后东北区候平均降水量
（横坐标值“－1”表示东亚夏季风在东北区建立的前 1 候，“0”表示东亚夏季风在东北区建立的候，“1”表示东亚夏季风在东北区建立的第 2 候）

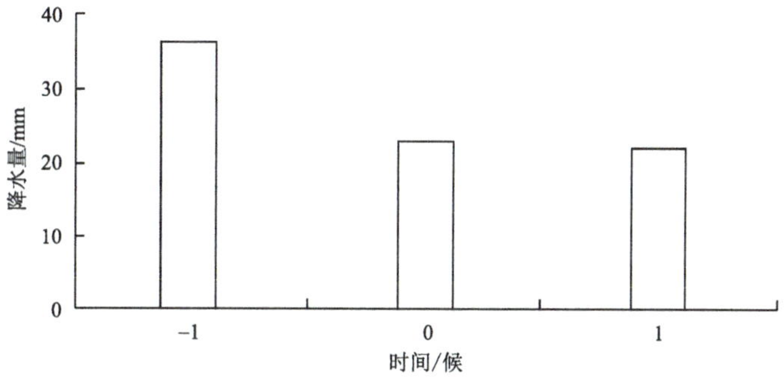

图 3 1980—1995 年东亚夏季风建立年撤退候及其前后东北区候平均降水量
(横坐标说明同图 2)

3.3.2 东亚夏季风建立早、持续时间长和影响最强的 1994 和 1988 年

尽管东亚夏季风在东北区建立后持续的平均时间是短暂的,但具体到某一年却不尽相同,各具特点。

3.3.2.1 θ_{se} 场特征

从表 1 可见,东亚夏季风在东北区建立最早、最强的典型年份为 1994 年。从图 4 可见,1994 年 850 hPa θ_{se} 候平均值为 336 K 等值线和 $V \geqslant 4$ m/s 等风速线,在 7 月第 2 候(38 候)就越过了 40°N,整个 7 月至 8 月中呈南北舌状突进到东北区,并且停留达 8 个候,最北界已达 49°N。1994 年是自 1980 年以来东亚夏季风在东北区维持时间最长和影响最强的年份。

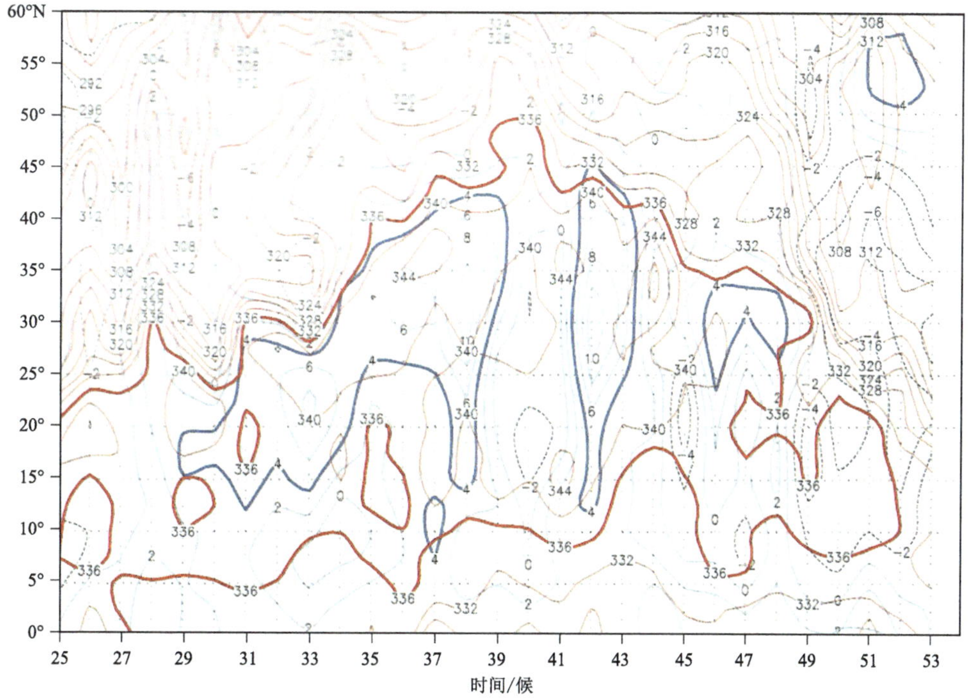

图 4 1994 年 850 hPa 的 θ_{se}、V 分量沿 122.5°E 逐候时间-纬度剖面图
(蓝实线为南风风速等值线,虚线为北风风速等值线,间隔 2 m/s;粗蓝实线为风速 4 m/s 等值线。红色实线为 θ_{se} 等值线,间隔 4 K,粗红实线为 336 K 等 θ_{se} 线)

3.3.2.2 东亚夏季风在东北区建立时流场、水汽输送特征分析

从表 1 可见,1988 年东亚夏季副热带季风影响东北区的时间和强度仅次于 1994 年,图 5a 和图 5b 分别为 1988 年东亚夏季风在东北区建立的前 1 候和建立候 850 hPa 平均流场图,从图 5a 和 5b 对比可见,

850 hPa 候平均图上，1988 年东亚夏季风在东北区建立的前 1 候（第 39 候，图 5a）低层副热带高压西北侧转向的西南风尚停留在 40°N 附近（122.5°—140.0°E），而建立时的第 40 候（图 5b），则因副热带高压的西进和北跳，贝加尔湖东侧低涡减弱，使副热带高压西北侧转向的西南风与西风带西南风合并加强为低空南风急流迅速挺进到 60°N 附近（120°—140°E）。

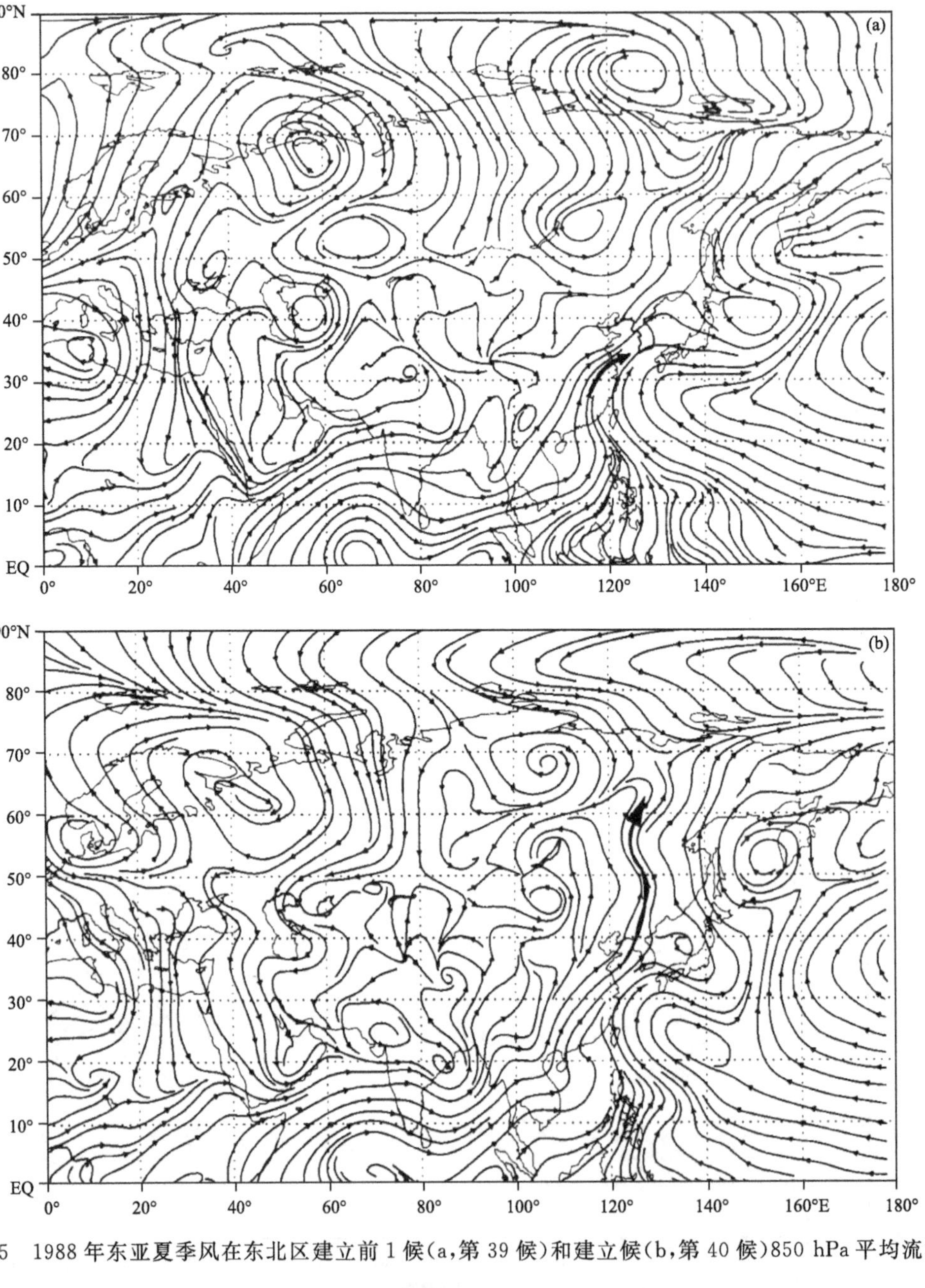

图 5　1988 年东亚夏季风在东北区建立前 1 候（a，第 39 候）和建立候（b，第 40 候）850 hPa 平均流场

从图 6a 和图 6b 对比可见，850 hPa 候平均图上，1988 年东亚夏季风在东北区建立的前 1 候（第 39 候，图 6a）在 120°E 附近，低层的副热带高压西南侧的水汽辐合带呈南北走向，强中心位于中国华东地区，而东亚夏季风在东北区建立时的第 40 候（图 6b），在（30°N，120°E）附近形成了一支伸向东北方向（30°—48°N，120°—137°E）的水汽辐合带，该水汽辐合带向南一直与中国南海及西太平洋暖池的水汽辐合带贯通，南北方向达 40 多个纬距，与 θ_{se} 的 336 K 线伸向中国东北区的舌状高值带相匹配，清楚地表明东亚夏季风在东北区域建立的高温高湿气团属性。

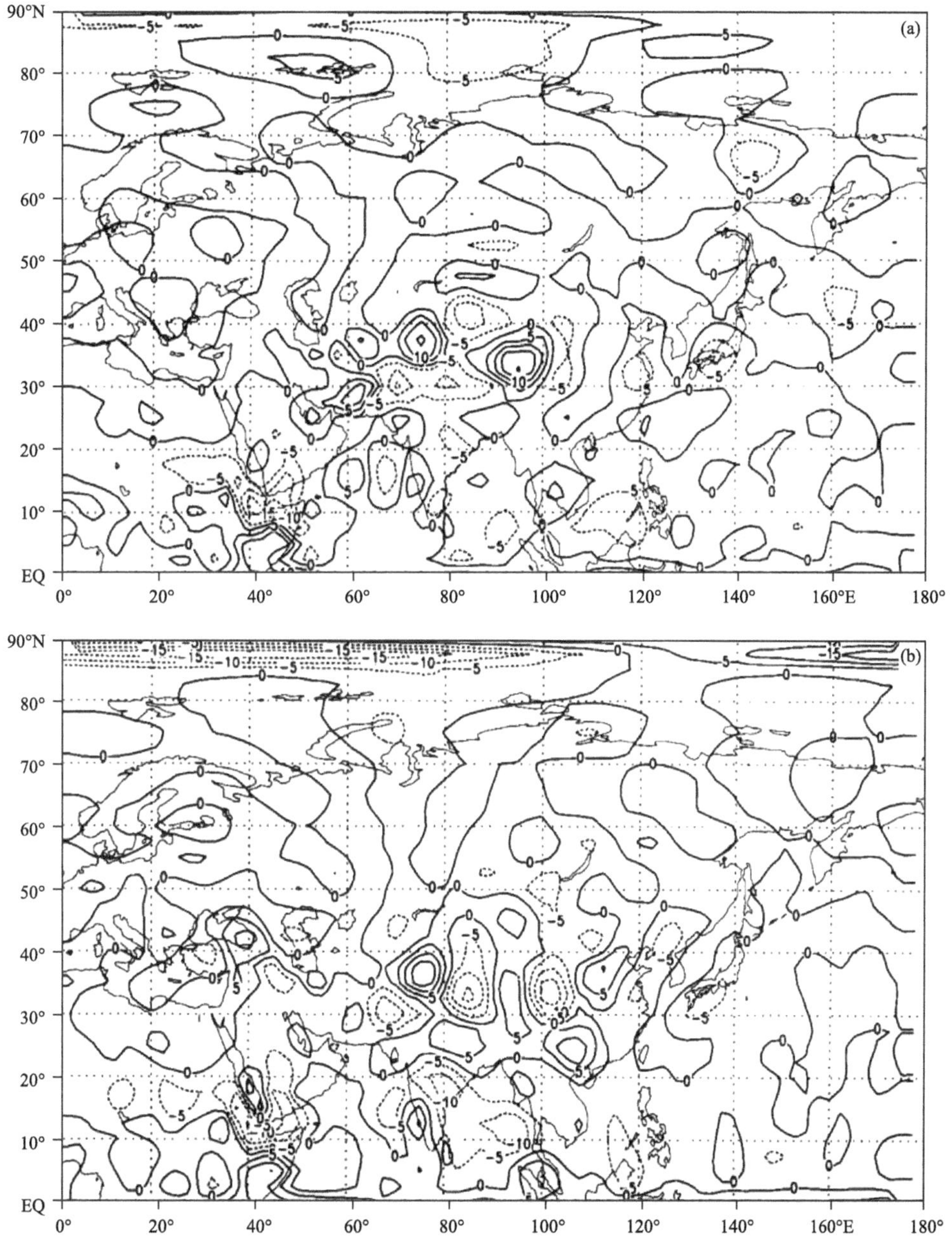

图 6　1988 年东亚夏季风在东北区建立前 1 候(a,第 39 候)和建立时(b,第 40 候)
850 hPa 水汽散度场,单位:10^{-8}g (cm hPa s)$^{-1}$

3.3.3　东亚夏季风建立最晚、持续时间较短的 1999 年

3.3.3.1　θ_{se}场特征

图 7 为 1999 年 850 hPa 候平均 θ_{se}沿 122.5°E 的时间-纬度剖面,从图 7 可见,850 hPa 候平均 θ_{se}的 336 K 等值线虽然在第 40 候就越过了 40°N,但 850 hPa 候平均沿 122.5°E 的风场尚处于北风控制下,这表明影响东北区的气团属性为大陆性暖气团控制,不具有副热带季风性质。θ_{se}的 336 K 等值线于 8 月第 2 候(44 候)再次突进 40°N 以北的东北区,位于 46°N 附近,南风风速达 4～6 m/s,持续了 2 个候,即第 46 候撤回 40°N 以南,是自 1980 年以来东亚夏季风在东北区建立最晚和持续时间较短,也是影响较弱的年份(表 1)。

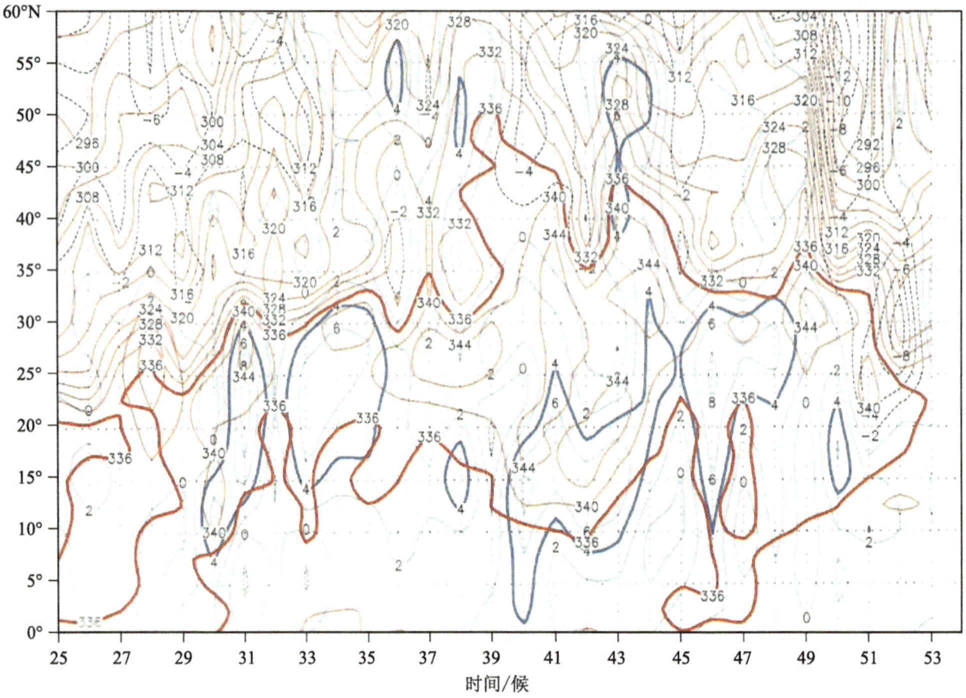

图 7　1999 年 850 hPa 候平均 θ_{se} 沿 122.5°E 时间-纬度剖面

（蓝实线为南风风速等值线，虚线为北风风速等值线，间隔 2 m/s；粗蓝实线为风速，单位 4 m/s 等值线。红色实线为 θ_{se} 等值线，间隔 4 K，粗红实线为 336 K 等 θ_{se} 线）

3.3.3.2　建立时流场特征

从图 8a、图 8b 与图 5a、图 5b 比较可见，东亚夏季风于 8 月在东北区建立的 1999 年与 7 月在东北区建立的 1994 年相比，副高脊线偏北，达到 35°N 以北（120°—140°E），1999 年东亚夏季风东北区建立前和建立后随着副热带高压前沿经日本海明显向西伸展，原在日本海上空的西南气流西移到中国东北地区。

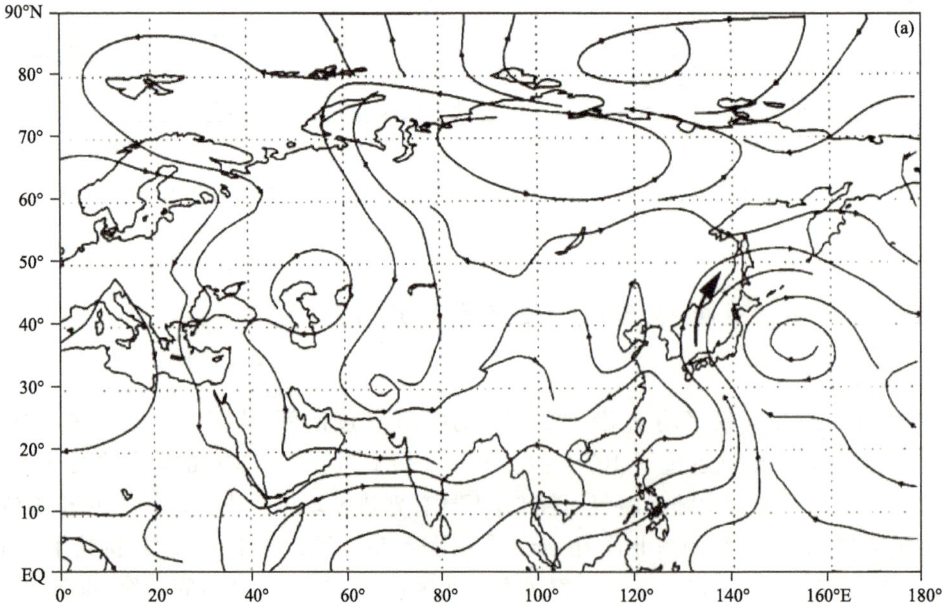

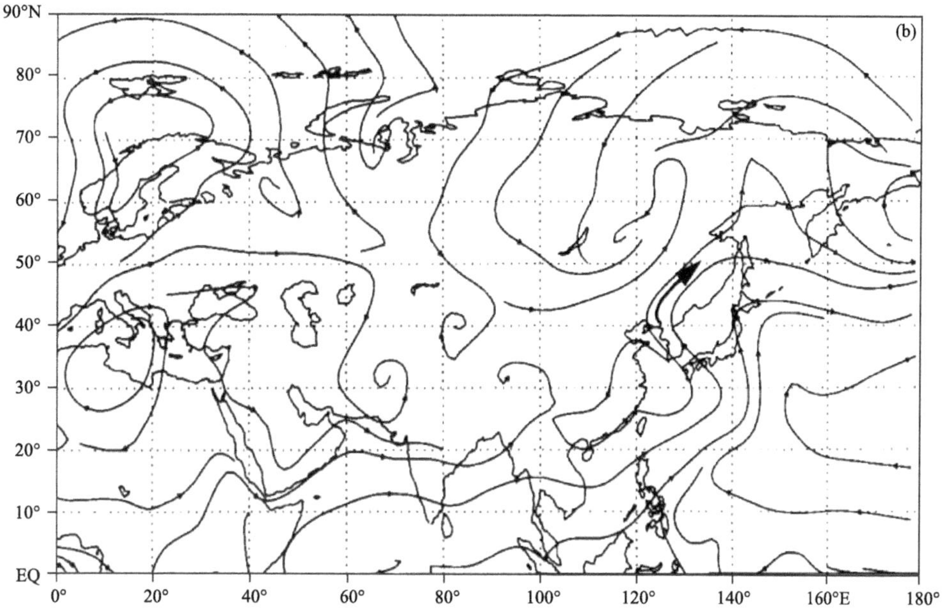

图 8　1999 年东亚夏季风在东北区建立前 1 候(a,第 43 候)和建立候(b,第 44 候)850 hPa 平均流场

3.4　东亚夏季风在东北区建立日期和持续时间与夏季风强弱的相关分析

从表 2 可见,文中定义的东亚夏季风在东北区没有建立的 1980、1983、1987、1991、1993、1997 年 6 a 与国内有关学者所定义的弱夏季风年基本一致,只有 1997 年有差别,也就是说对中国副热带地区所定义的夏季风弱的年份,自然也很难影响到中国中高纬度的东北地区;本文定义的强夏季风年(即建立的时间早、累计候≥4 候的年份)1981、1988、1990、1994 年与国内有关学者所定义的强夏季风年基本一致,只是对 1988 年的强弱定义有所不同。

表 2　东亚夏季风在东北区建立日期和持续时间与东亚夏季风强弱的相关(廉毅 等,1997;孙力 等,2003)

年份	建立日期/候	累计/候	最北界/°N	撤退日期/候	施能(1996)	孙秀荣等(2001)	张庆云等(1998)	孙力等(2003)
1980 ○	—	0	—	—	弱	弱	弱	弱
1981 ▲	37	4	47	43	强	强	强	强
1982	43	3	49	46	正常	弱	强	正常
1983 ○	—	0	—	—	弱	弱	弱	弱
1984	43	3	44	46	较强	较弱	强	强
1985	41	3	43	44	强	较强	强	强
1986	42	1	44	43	弱	较弱	强	正常
1987 ○	—	0	—	—	弱	弱	较弱	正常
1988 ▲	40	6	58	47	强	正常	弱	弱
1989	40	2	49	44	强	弱	弱	弱
1990 ▲	38	6	46	47	强	较强	较强	正常
1991 ○	—	0	—	—	弱	较弱	正常	弱
1992	41	2	43	43	较强	较弱	较弱	弱
1993 ○	—	0	—	—	强	弱	弱	弱
1994 ▲	38	8	50	46	强	强	强	强
1995	42	4	43	45	强	较弱	弱	强
1996	42	2	43	44		较弱	弱	弱
1997 ○	—	0	—	—		正常	强	弱
1998	40	4	50	45		弱	弱	弱
1999	44	2	51	46		弱		弱
2000	40	3	50	46				较弱

注:○表示文中定义的东亚夏季风在东北区没有建立的年份;▲表示文中定义的东亚夏季风在东北区建立而且是强季风年。表中空白内容代表未在参考文献中查到相关结果。

4　东亚夏季风在东北区建立与否的典型流场和加热场差异分析

4.1　流场差异

从图 9a 和图 9b 对比可以发现，有几点显著差别：1994 年西北太平洋反气旋(120°—160°E)中心位于日本岛南端 30°N 以北的区域，而 1980 年反气旋中心位于 25°N 附近，比 1994 年偏南 7 个纬距而且偏东 7～8 个经距；东亚夏季风在东北区建立的 1994 年东亚(110°—140°E)中高纬度地区没有阻塞形势，而东亚夏季风在东北区没有建立的 1980 年则刚好相反；1994 年在青藏高原地区有一个次反气旋中心对印度的季风气旋东北侧的西南季风起到了明显的牵制作用，致使中国西南地区盛行东北风，而 1980 年则相反，印度和孟加拉湾的西南季风则十分活跃。

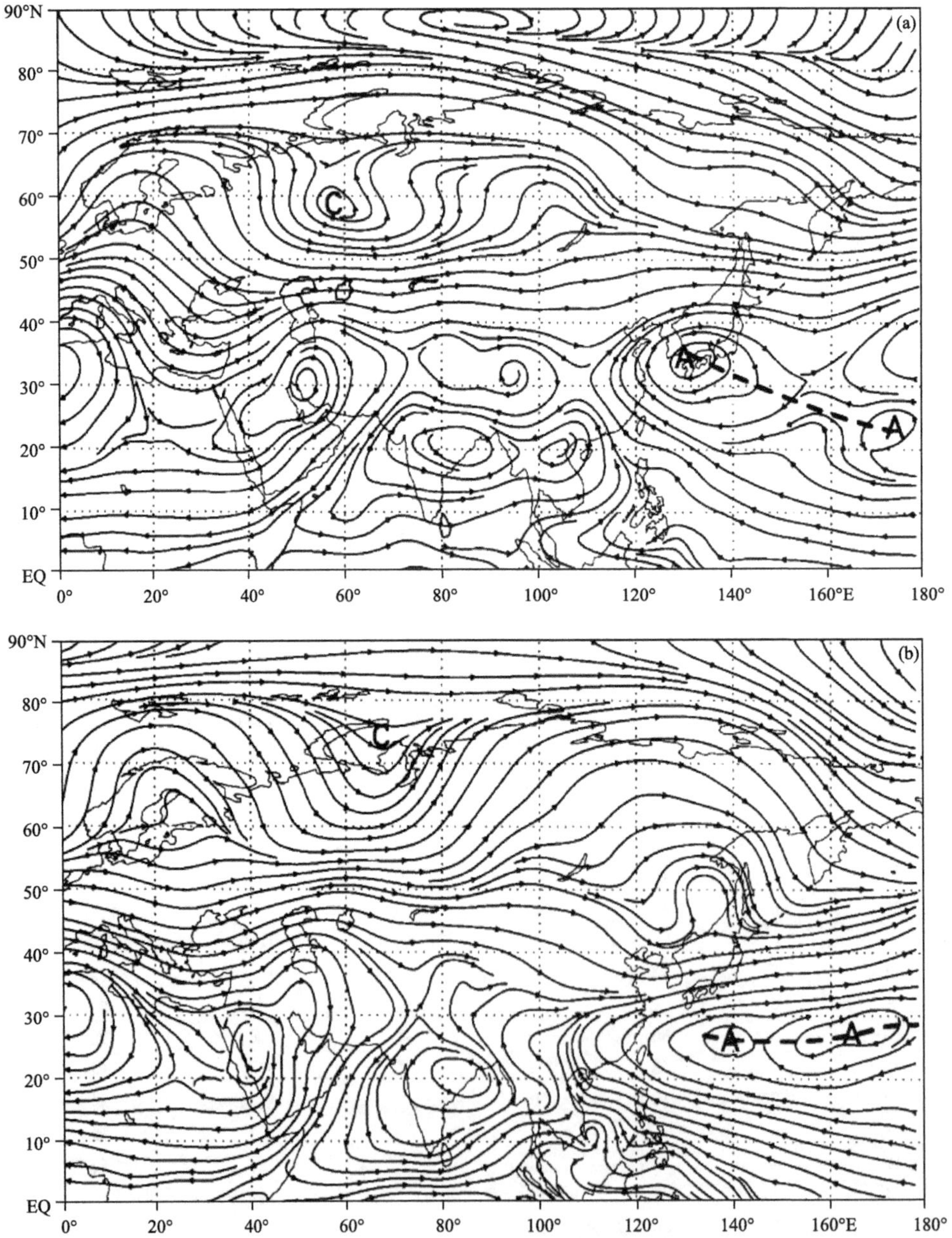

图 9　东亚夏季风在中国东北区建立最早、最强的 1994 年(a)和没有建立的 1980 年(b)7—8 月 500 hPa 平均流场
(图中“C”表示为气旋活动中心，“A”为反气旋活动中心，粗黑色虚线为反气旋活动中心连线)

4.2　加热场的差异

众所周知，速度势的梯度就是辐散风场，对流层上层的辐散中心是与上升运动和大气热源相对应的。从图10a可见，1994年200 hPa北太平洋上空绝大部分为V_x距平的负值区，欧洲和亚洲的西部为V_x距平的正值区，与印度洋的正值区连成一片，尤其在南印度洋靠近赤道地区为V_x强大距平正值区，西太平

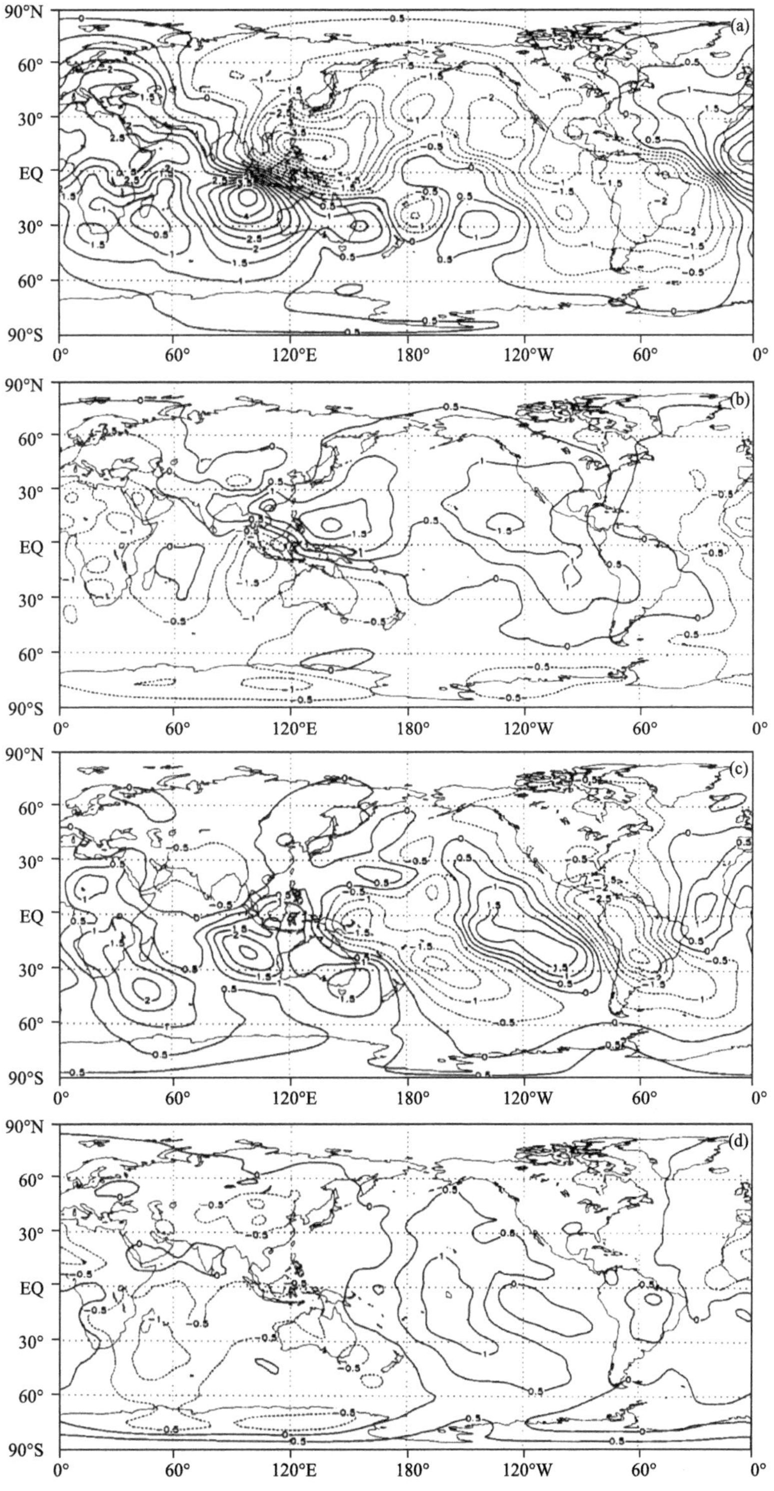

图10　V_x月平均距平分布

（a和b分别为1994年7月200 hPa和850 hPa；c和d分别为1980年7月200 hPa和850 hPa；单位为$10^6\,m^2\,s^{-1}$）

洋暖池附近(120°—140°E)为 V_x 距平负值中心(相对冷源),从印度尼西亚的西北部经中国南海到菲律宾之间为 V_x 距平等值线密集带,形成自西南向东北方向强的辐散风场。图 10b 所显示的低层则与 200 hPa 相反分布,而图 10c 则表明东亚夏季风没有建立的 1980 年欧亚大陆和北太平洋大部分区域上空 V_x 距平不甚明显,图 10d 所显示的低层更加不明显。

5 结论和讨论

(1)本文定义的东亚夏季风在中国东北地区建立的标准,即 850 hPa 候平均 θ_{se} 和南风风速沿 122.5°E 时间-纬度剖面图上,θ_{se} 的 336 K 等值线和 Vs(南风风速≥4 m/s)等值线均越过 40°N 的候为建立时间,退回 40°N 的最后 1 候为撤退候,所给出的 1980—2000 年的建立、持续和撤退的时间年历表表明,建立的平均日期为第 41 候,与陶诗言(1996)和陈隆勋(1984)所给出的 7 月 20 日比较接近,建立、持续和撤退期均方差较华北、长江和华南要小,在 1.9~1.2 候,因为持续的时间短,平均为 3.6 候,开始撤退的平均日期为第 45 候(8 月第 3 候)。东亚夏季风在中国东北区建立前后候降水量增加和撤退前后候降水量递减是十分显著的,建立时低层水汽辐散场变成辐合场。

(2)本文定义的东亚夏季风在中国东北区没有建立的 1980、1983、1987、1991、1993、1997 年 6 a 与国内有关学者所定义的弱夏季风年基本一致,只有 1997 年有差别,这表明所定义的中国副热带地区夏季风弱的年份,自然也很难影响到中国中高纬度的东北地区。反之,定义的中国东北区强夏季风年(即建立的时间早、累计候≥4 的年份)的 1981、1988、1990、1994 年也与国内有关学者所定义的强夏季风基本一致,只是对 1988 年的强弱定义有所不同。

(3)东亚夏季风在中国东北区建立和不建立年的流场、加热场都存在显著差异:西北太平洋的反气旋中心无论在低层还是在中层,位置偏北和偏西重要的差别,夏季风强劲的 1994 年 500 hPa 比夏季风没有建立的 1980 年副热带反气旋中心偏北 7 个纬距,偏西 7~8 个经距;1994 年东亚没有中高纬度阻塞形势而 1980 年则很明显;1994 年在中国青藏高原地区有一个次反气旋中心对印度的季风气旋东北侧的西南季风起到明显的牵制作用,致使中国西南地区盛行东北风,而 1980 年则相反,印度和孟加拉湾的西南季风十分活跃。加热场 1994 年与 1980 年差异亦很明显,1994 年 200 hPa 北太平洋上空绝大部分为 V_x 距平的负值区,欧洲和亚洲的西部为 V_x 距平的正值区,与印度洋的正值区连成一片,尤其在南印度洋靠近赤道地区为 V_x 强大距平正值区,西太平洋暖池附近(120°—140°E)为 V_x 距平负值中心(相对冷源),从印度尼西亚的西北部经中国南海到菲律宾之间为 V_x 距平等值线密集带,形成自西南向东北方向强的辐散风场。低层则与 200 hPa 相反分布,而没有建立的 1980 年欧亚大陆和北太平洋大部分区域上空 V_x 距平不甚明显,表明海陆热力差异相对减小。

(4)从表 1 可以看出,El Niño 年对中国东北地区夏季风建立的影响似乎没有什么规律性,20 世纪 80 年代以来的 6 a El Niño 中,有 3 a 东亚夏季风在东北区没有建立,如 1991、1993、1997 年;有 1 个强劲年如 1994 年;还有 1982 年和 1986 年为夏季风影响一般年份。

参考文献

陈隆勋,1984. 东亚季风系统的结构及其中期变动[J]. 海洋学报(6):744-785.

丁一汇,1989. 天气动力学中的诊断分析[M]. 北京:科学出版社:280.

丁一汇,马鹤年,1986. 东亚季风的研究现状[M]//何金海. 亚洲季风研究的新进展. 北京:气象出版社:2-3.

郭其蕴,1983. 东亚夏季风强度指数及其变化的分析[J]. 地理学报,38(3):207-217.

何金海,丁一汇,高辉,等,2001. 南海夏季风建立日期的确定与季风指数[M]. 北京:气象出版社:123.

金祖辉、陈隆勋,1982. 夏季东亚季风环流系统的中期变化及印度季风系统的相互关系[C]//全国热带夏季风学术会议文集编写组. 全国热带夏季风学术会议文集. 昆明:云南人民出版社:204-215.

廉毅,安刚,1998. 东亚季风、El Niño 与中国松辽平原夏季低温关系初探[J]. 气象学报,56(6):720-735.

廉毅,安刚,王琪,等,1997. 吉林省 40 年来气温和降水的变化[J]. 应用气象学报,8(2):204.

施能,1996. 近40年东亚冬季风强度的多时间尺度变化特征及其与气候的关系[J]. 应用气象学报,7(2):175-182.

施能,朱乾根,关彬贵,1996. 近四十年东亚夏季风及中国夏季大尺度天气气候异常[J]. 大气科学,20(5):575-583.

施尚文,巢俊民,1982. 中国北方地区的夏季风[C]//全国热带夏季风学术会议文集编写组. 全国热带夏季风学术会议文集. 昆明:云南人民出版社:72-85.

孙力,安刚,唐晓玲,2003. 东北亚地区夏季850 hPa南风异常与东北旱涝的关系[J]. 大气科学,27(3):425-434.

孙秀荣,陈隆勋,何金海,2001. 东亚海陆热力差指数与东亚夏季风强度关系探讨[M]//何金海. 南海夏季风建立日期的确定与季风指数. 北京:气象出版社,96-108.

汤明敏,黄士松,1983. 1979年中国东部夏季风的进展[C]//全国热带夏季风学术会议文集编写组. 全国热带夏季风学术会议文集. 昆明:云南人民出版社:15-30.

陶诗言,1996. 季风研究中有待解决的问题[M]//国家自然科学基金委. 现代大气科学前沿与展望. 北京:气象出版社:35-36.

喻世华,颜景容,1986. 1979年东亚夏季风环流建立过程的分析[J]. 热带气象,2(1):55-61.

张庆云,陶诗言,1998. 夏季东亚热带和副热带季风与中国东部汛期降水[J]. 应用气象学报,9(Z):16-23.

赵汉光,张先恭,1994. 东亚季风和中国夏季雨带的关系[J]. 气象,22(4):8-12.

周玉孚,1983. 中国东部地区季风进退与雨季起讫的关系[C]//全国热带夏季风学术会议文集编写组. 全国热带夏季风学术会议文集. 昆明:云南人民出版社:311-317.

朱乾根,杨松,1989. 东亚副热带季风的北进及其低频振荡[J]. 南京气象学院学报,12(3):249-257.

TAO S Y,CHEN L X,1987. A review of research on the East Asian Summer Monsoon in China[M]//Chang C P,Krishnamurti T N. In Monsoon Meteorology. Oxford:Oxford University Press:60-92.

ZHU Q G,HE J X,WANG P X,1986. A study of circulation differences between East-Asian and Indian Summer Monsoons with their interaction[J]. Adv Atmos Sci,3:446-477.

关于确定东亚夏季风强度指数的探讨*

廉毅 沈柏竹 高枞亭

（吉林省气象科学研究所，长春 130062）

摘　要：利用作者曾定义的东亚夏季风在中国东北地区（122.5°E，40°N）的建立标准，根据相同的方法，分别计算了沿 112.5°、117.5°和 122.5°E 上 20°N 及以北每隔 5 个纬度东亚夏季风建立、持续和撤退时间（候），将某年持续和多年平均持续候数相比的标准化值，定义为一种沿某一经圈上某一纬度的东亚夏季风强度指数 $I_{SM\Phi}$，还分析了该指数与中国夏季降水量场和 500 hPa 高度场的相关。结果表明：（1）沿 117.5°E 经度上，东亚夏季风在 20°、25°、30°、35°、40°N 建立的平均日期分别为 27.26、28.54、34.43、37.12、37.65 候，撤退的平均日期分别为 54.44、53.69、51.85、48、46.76 候。其中（117.5°E，20°N）代表南海的中北部，确定的该区夏季风建立、撤退的平均日期分别为 27.26 和 54.44 候，与国内学者公认的 5 月第 4 候（28 候）和 10 月第 1 候（55 候）相当吻合。（2）沿 112.5°、117.5°、122.5°E 的同一纬度上，东亚夏季风建立的平均日期并不相同，西边先于东边建立，每隔 5 个经度相差 1～2 候，而撤退的平均日期（30°N 及以北）分布则相反，东边先撤退。（3）沿 117.5°E、30°N 和 35°N 的 $I_{SM\Phi}$ 和沿 122.5°E、40°N 的 $I_{SM\Phi}$ 均与中国华北和东北地区大部 7—8 月降水量呈正相关，与长江上游和下游地区呈负相关，显著性水平 $\alpha \geqslant 0.05$，表明 $I_{SM\Phi}$ 与中国大陆夏季主要雨带有很好的相关性。（4）$I_{SM\Phi}$ 所反映的中国大陆副热带地区东亚夏季风强弱既与西太平洋副热带高压又与中高纬度的鄂霍次克海高压相关密切，而中国东部副热带以北的中纬度地区东亚夏季风强弱与中高纬度地区的大气环流形势相关更为密切。（5）探讨确定随经度和纬度而变化的东亚夏季风指数 $I_{SM\Phi}$，将有可能避免幅员辽阔的中国大陆由一个夏季风强度指数来衡量而掩盖了东亚夏季风在向北推进和向南撤退时的多样性，将有利于进一步研究东亚夏季风进退、强度季内和年际变化规律。

关键词：东亚夏季风；指数；建立和撤退；中国夏季降水

1 引言

自 1934 年竺可桢（1934）研究中国大陆夏季风以来，许多学者已对东亚夏季风对中国不同纬度影响的性质、季风进退、季风雨带活动及其与大气环流的联系等进行了大量的研究。关于季风指数的研究涉及如何定量描述夏季风的强度和年际变率，也是季风研究中的一个基本问题，郭其蕴（1983）认为东亚大陆地区的季风主要由纬向海陆热力差异所形成，利用 10°—50°N 各纬度上海平面气压做出110°E 气压减 160°E 气压差（月平均）值，对差值≤−5 hPa 的各种纬度值累加起来取绝对值然后标准化，首次提出一种夏季风指数 SMI。为了表征副热带夏季风进退活动，不少学者都着眼于寻找副热带活动的指标，汤明敏等（1983）使用 1000 hPa 上 344～348 K 等 θ_{se} 线，或 850 hPa 上 336～340 K 等 θ_{se} 线为副热带夏季风进退指标。施尚文等（1982）用 850 hPa 的 332 K 等 θ_{se} 线作为北方地区夏季风影响期的临界值等。就温湿指标而言，鉴于副热带夏季风是高温高湿的气流，朱乾根等（1989）提出与其用等 θ_{se} 线不如用等露点线，如利用

* 本文原载于《气象学报》，2004，62（6）：782-789

850 hPa 14 ℃等露点。赵汉光等(1994)引用郭其蕴(1983)的方法,建立了冬季风强度指数,施能等(1996)取 10°—50°N 范围内 5 个纬度的 110°、160°E 的标准化的海平面气压差之和的序列的标准化值为东亚季风指数。陈隆勋等(1999)、孙秀荣等(2002)从季风形成的本质出发,定义了一个东亚海陆热力差指数。祝从文等(2000)将东西向海平面气压差与低纬度高、低层纬向风切变相结合,定义了东亚夏季风指数,较好地反映了东亚冬、夏季风变化。张庆云等(2003)将东亚热带季风槽(10°—20°N,100°—150°E)与东亚副热带地区(25°—35°N,100°—150°E)6—8 月平均的 850 hPa 风场的纬向风距平差,定义为东亚夏季风指数(EASMI)。总之,关于季风指数的确定大体有 3 种方法,即以季风区降水量的变化为标准、以高低空风场为标准和使用较多的风场结合温湿指标的方法(高辉 等,2001)。

廉毅等(2003)在文献中曾提出东亚夏季风在中国东北地区建立的标准,即取 850 hPa 候平均资料,作沿 122.5°E 的时间-纬度剖面,将 θ_{se}的 336 K 等值线和南风 4 m/s 等值线同时越过 40°N 的候定义为东亚夏季风在中国东北地区建立的日期(候),相应给出 1980—2000 年的建立、持续和撤退出东北地区的时间年历表,发现该标准与国内专家学者所分析的东亚夏季风强弱标准有较好的对应性(Tao et al.,1987)。本文尝试将此标准应用在东亚夏季风影响的其他地区上,分别计算沿 112.5°、117.5°、122.5°E 上,20°N 及以北每隔 5 个纬度东亚夏季风建立、持续和撤退时间(候),该结果与国内专家学者所分析的东亚夏季风进退日期亦有较好的对应性。本研究探讨确定随经度和纬度而变化的东亚夏季风指数,试图避免由一个夏季风指数来衡量所有纬度的夏季风强度而掩盖了东亚夏季风在向北推进和向南撤退时的多样性,以便有利于进一步研究东亚夏季风强弱的季内和年际变化规律。

2 资料和方法

本文所用全国 151 站(因某些站点资料年限较短,故用到 160 个测站中的 151 个测站)1961—2000 年的逐月降水量取自国家气候中心。

从 1948—2002 年 NCEP/NCAR 全球再分析资料中获取 850 hPa 候平均资料,作沿 112.5°、117.5°、122.5°E 的时空-纬度剖面,当 θ_{se}的 336 K 等值线和南风≥4 m/s 等值线同时越过某一纬度时,定义为该纬度东亚夏季风建立的日期(候),至此之后,θ_{se}的 336 K 持续候数与多年平均持续候比值的标准化定义为当年的相应纬度的季风强度指数,计算指数集与全国 151 个测站的 6 月、7 月、8 月、7—8 月、6—8 月降水量和 500 hPa 高度场的线性相关。

3 夏季风指数 $I_{SM\Phi}$的定义及与中国夏季降水的相关关系

本文定义沿某一经圈上某一个纬度夏季风强度指数 $I_{SM\Phi}$值为:

$$I_{SM\Phi_j} = \frac{S_j - \overline{S}}{\sqrt{\sum_{j=1948}^{2002} \frac{(S_j - \overline{S})^2}{n}}}$$

式中,$S_j = \frac{h_j}{\overline{h}}$;$\overline{S} = \frac{\sum_{j=1948}^{2002} S_j}{n}$;$h_j$ 为第 j 年东亚夏季风建立之后 θ_{se}的 336 K 越过某一纬度持续候数;$\overline{h}$ 为 1948—2002 年 θ_{se}的 336 K 越过某一纬度平均持续候数;S_j 为第 j 年某一纬度 θ_{se}的 336 K 的持续候数与平均持续候数的比值;$n=55$,即 1948—2002 年样本数;$I_{SM\Phi}$为沿某一经圈上某一纬度夏季风强度指数。

取 850 hPa 候平均资料,作时间-纬圈剖面图,将某年 θ_{se}的 336 K 和南风风速≥4 m/s 等值线同时越过某一个纬度后,θ_{se}的 336 K 的持续候数与平均持续候数相比的标准化值定义为 $I_{SM\Phi}$。本文仅举出沿 117.5°E、30°N 和 35°N 及沿 122.5°E、40°N 夏季风指数与全国 151 个测站 7—8 月降水相关系数分布。

沿 117.5°E、30°N 的东亚夏季风强度指数 $I_{SM\Phi}$与中国 151 个测站 7—8 月降水总量相关系数分布上有

2 片高相关区(相关系数≥30 或≤−30 的灰色区域相应 $\alpha \geq 0.05$),一是最大片正相关区位于华北东部和北部、东北地区中南部,另一片负相关区位于长江流域下游和浙江北部。沿 117.5°E、35°N 的东亚夏季风强度指数 $I_{SM\Phi}$与中国 151 个测站 7—8 月降水总量相关系数分布有 3 片高相关区(显著性水平 $\alpha \geq 0.05$)。第 1 大片正相关区依然是华北和东北地区中南部、中部,相关区域明显向西伸展和向东北地区中部延伸。第 2 片负相关区依旧是长江流域下游和浙江北部,并向西延伸。第 3 片负相关区位于新疆的南部。沿 122.5°E、40°N 的东亚夏季风强度指数 $I_{SM\Phi}$与中国 151 个测站 7—8 月降水总量相关系数分布有 4 片高相关区,华北和东北地区分为 2 片正相关区,第 3、第 4 片负相关区,还仍然分别位于长江流域下游和新疆的南部地区。

从 850 hPa 候平均图上 θ_{se}的 336 K 等值线和南风风速≥4 m/s 等值线确定东亚夏季风指数 $I_{SM\Phi}$的时间序列(图 1)可见:沿 117.5°E、25°N 东亚夏季风指数 $I_{SM\Phi}$ 1948—2002 年演变(图 1a)趋势为自 20 世纪 90 年代以来进入一个较弱时期,指数为正的只有 4 a,而负指数达 8 a;沿 117.5°E、30°N 的 $I_{SM\Phi}$自 70 年代中期开始减弱(图 1b);沿 117.5°E、35°N 和沿 122.5°E、40°N 夏季风指数(分别为图 1c、图 1d),自 60 年代中期开始就进入一个较弱时期,尤其是沿 122.5°E、40°N 的夏季风指数减弱的更为明显,负夏季风指数年份占绝对优势。另外,沿 117.5°E,25°、30°和 35°N,以及沿 122.5°E、40°N 的 $I_{SM\Phi}$年际变化也不尽相同,呈现多样性。

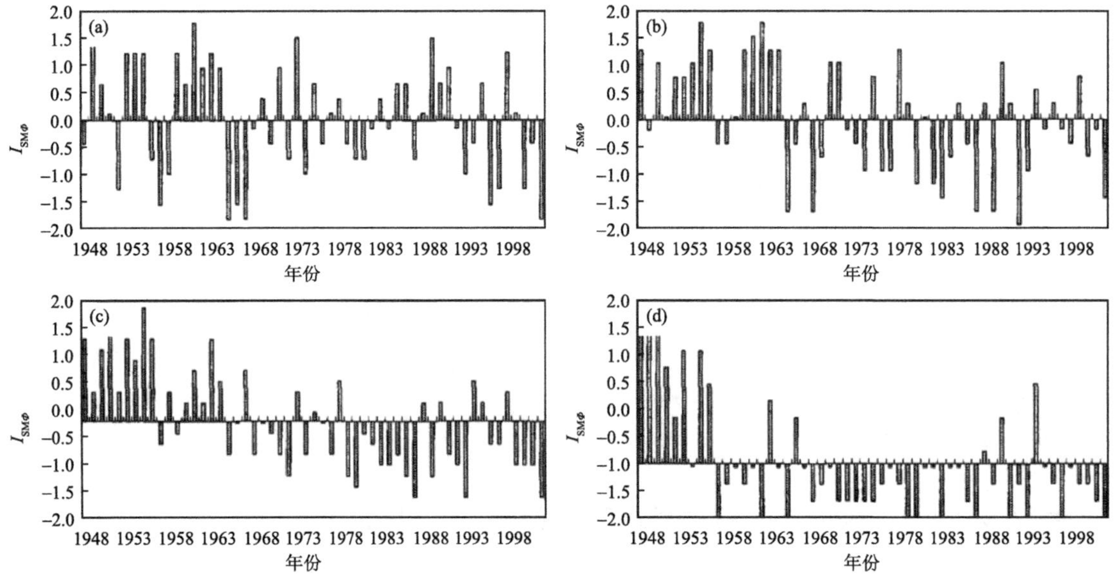

图 1 东亚夏季风指数 $I_{SM\Phi}$时间序列

(a,b,c 分别为沿 117.5°E,25°、30°、35°N;d 为沿 122.5°E,40°N)

4 夏季风指数 $I_{SM\Phi}$与 500 hPa 高度场的同期相关

沿 117.5°E、25°N 典型 $I_{SM\Phi}$强指数减去弱指数年 8 月 500 hPa 高度场距平差值分布(图 2)表明,脊线位于 30°N 的副高控制区(120°—150°E)为较强的正距平区,在鄂霍次克海地区为较强的负距平区。

图 3 为沿 117.5°E、35°N 和沿 122.5°E、40°N 东亚夏季风指数 $I_{SM\Phi}$与 7—8 月 500 hPa 高度场的相关系数分布,表明沿 117.5°E、35°N,沿 122.5°E、40°N 东亚夏季风指数 $I_{SM\Phi}$均与中高纬度地区的新地岛至鄂霍次克海上空 500 hPa 高度场有一个宽达 30～40 纬度的高显著度正相关带状区(显著性水平 $\alpha \geq 0.05$),其南侧为宽广的弱的负相关区。

图 2　117.5°E,25°N 强、弱典型年 $I_{SM\Phi}$ 8 月 500 hPa 高度场距平差值

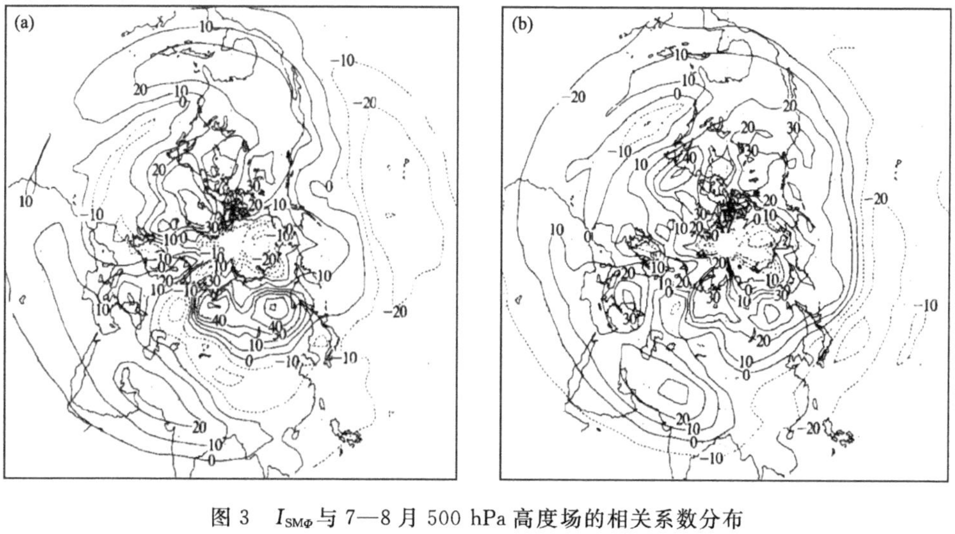

图 3　$I_{SM\Phi}$ 与 7—8 月 500 hPa 高度场的相关系数分布
(a. 沿 117.5°E,35°N;b. 沿 122.5°E,40°N;相关系数扩大了 100 倍,间隔为 10)

5　东亚夏季风建立和撤退的平均日期

本文计算了沿 112.5°、117.5°和 122.5°E 的 20°、25°、30°、35°、40°N 东亚夏季风建立、持续、撤退的年历表,仅举出 30°N 东亚夏季风建立、持续和撤退的年历表。

从表 1 可见,沿 117.5°E 东亚夏季风在 30°N 建立的平均日期为第 34.55 候(相当 6 月 22—23 日),持续和累计时间分别为 13.82 和 15.98 候,持续 2～2.5 个月,撤退的平均日期为第 51.85 候(9 月 19 日左右)。从表 2 可见,沿 117.5°E 东亚夏季风在 35°N 建立的平均日期为第 37.12 候(7 月 6 日),持续和累计时间分别为 7.6 和 9.23 候,大约持续 1 个半月,撤退的平均日期为第 48 候(8 月 31 日)。沿 122.5°E 东亚夏季风在 40°N 建立的平均日期为第 39.36 候(7 月 17 日前后),持续和累计时间分别为 4 和 4.75 候,在中国东北地区持续 20～23 d,撤退的平均日期为第 45.20 候(8 月 15 日)。沿 112.5°、117.5°和 122.5°E 的同一纬度上,东亚夏季风建立的平均日期并不相同,西边先于东边建立,每隔 5 个经度相差 1～2 候,而撤退的平均日期(30°N 及以北)分布则相反,东边先撤退。

表 1　由 5—10 月 850 hPa 候平均图沿 117.5°E θ_{se}336 K 等值线和南风风速≥4 m/s 确定过 30°N 东亚夏季风建立、持续、撤退和夏季风指数年历表

年份	θ_{se}过 30°N /候	南风 /(m/s)	θ_{se}持续 /候	θ_{se}累计 /候	θ_{se}至最北界 /候	θ_{se}退回 30°N 以南日期/候	季风相对强弱值	持续候或持续平均候标准化/($I_{SM\Phi}$)	累计候或累计平均候标准化/($I_{SM\Phi}$)
1948	26	≥4	19	24	56	54	强	1.27	1.95
1949	28	≥4	13	22	56	53	正常	−0.20	1.46
1950	29	≥4	18	21	56	52	强	1.03	1.22
1951	34	≥4	14	17	54	52	正常	0.04	0.25
1952	30	≥4	17	23	59	53	强	0.78	1.71
1953	31	≥4	17	19	58	51	强	0.78	0.73
1954	26	≥4	18	22	60	50	强	1.03	1.46
1955	29	≥4	21	23	58	53	强	1.77	1.71
1956	26	≥4	19	23	58	53	强	1.27	1.71
1957	37	≥4	12	12	47	49	弱	−0.45	−0.97
1958	35	≥4	12	15	52	53	弱	−0.45	−0.24
1959	36	≥4	14	14	58	50	正常	0.04	−0.48
1960	32	≥4	19	20	56	>55	强	1.27	0.98
1961	32	≥4	20	21	58	>55	强	1.52	1.22
1962	30	≥4	21	23	46	54	强	1.77	1.71
1963	26	≥4	19	23	58	53	强	1.27	1.71
1964	34	≥4	19	19	58	53	强	1.27	0.73
1965	40	≥4	7	9	44	51	弱	−1.68	−1.70
1966	35	≥4	12	15	48	51	正常	−0.45	−0.24
1967	36	≥4	15	15	56	51	正常	0.29	−0.24
1968	36	≥4	7	13	58	53	弱	−1.68	−0.72
1969	39	≥4	11	13	46	>55	弱	−0.69	−0.72
1970	36	≥4	18	18	57	54	强	1.03	0.49
1971	32	≥4	18	18	44	50	强	1.03	0.49
1972	37	≥4	13	13	44	50	正常	−0.20	−0.72
1973	37	≥4	12	12	45	49	正常	−0.45	−0.97
1974	37	≥4	10	11	51	52	弱	−0.94	−1.21
1975	39	≥4	17	17	47	>55	强	0.78	0.25
1976	40	≥4	10	10	44	50	弱	−0.94	−1.45
1977	36	≥4	10	13	52	52	弱	−0.94	−0.72
1978	32	≥4	19	19	44	51	强	1.27	0.73
1979	35	≥4	15	17	44	53	正常	0.29	0.25
1980	35	≥4	9	14	46	50	弱	−1.18	−0.48
1981	36	≥4	14	14	45	50	正常	0.04	−0.48

续表

年份	θ_{se}过 30°N /候	南风 /(m/s)	θ_{se}持续 /候	θ_{se}累计 /候	θ_{se}至最北界 /候	θ_{se}退回 30°N 以南日期/候	季风相对强弱值	持续候或持续平均候标准化/($I_{SM\Phi}$)	累计候或累计平均候标准化/($I_{SM\Phi}$)
1982	41	≥4	9	9	48	50	弱	−1.18	−1.70
1983	40	≥4	8	12	44	54	弱	−1.43	−0.97
1984	38	≥4	11	12	56	51	弱	−0.69	−0.97
1985	38	≥4	15	15	43	53	正常	0.29	−0.24
1986	35	≥4	12	14	57	52	正常	−0.45	−0.48
1987	37	≥4	7	12	42	50	弱	−1.68	−0.97
1988	36	≥4	15	15	51	51	正常	0.29	−0.24
1989	31	≥4	7	17	51	53	弱	−1.68	0.25
1990	31	≥4	18	18	47	52	强	1.03	0.49
1991	29	≥4	15	21	55	51	正常	0.29	1.22
1992	40	≥4	6	11	44	52	弱	−1.92	−1.21
1993	38	≥4	10	12	56	53	弱	−0.94	−0.97
1994	35	≥4	16	16	48	51	强	0.54	0.00
1995	38	≥4	13	13	43	51	正常	−0.20	−0.72
1996	34	≥4	15	18	44	53	正常	0.29	0.49
1997	37	≥4	13	13	56	50	正常	−0.20	−0.72
1998	35	≥4	12	14	57	52	正常	−0.45	−0.48
1999	33	≥4	17	18	55	52	强	0.78	0.49
2000	35	≥4	11	14	46	50	弱	−0.69	−0.48
2001	36	≥4	13	13	44	49	正常	−0.20	−0.72
2002	41	≥4	8	10	54	52	弱	−1.43	−1.45
平均	34.55		13.82	15.98	50.98	51.85			

表 2　沿 112.5°E、117.5°E 和 122.5°E 上 20°N 及以北东亚夏季风建立和撤退的平均日期

纬度/N	建立的平均日期						撤退的平均日期					
	112.5°E		117.5°E		122.5°E		112.5°E		117.5°E		122.5°E	
	候	时间	候	时间	候	时间	候	时间	候	时间	候	时间
20°	25.96	5月10日	27.26	5月19日	29.56	5月28日	54.61	10月3日	54.44	10月2日	53.74	9月29日
25°	26.76	5月14日	28.54	5月23日	30.76	6月4日	53.85	9月29日	53.69	9月28日	52.93	9月25日
30°	32.44	6月12日	34.43	6月22日	34.81	6月24日	51.91	9月20日	51.85	9月19日	50.81	9月14日
35°	35.65	6月28日	37.12	7月6日	38.25	7月11日	49.16	9月6日	48.00	8月21日	47.25	8月26日
40°	35.61	6月28日	37.65	7月8日	39.36	7月17日	48.64	9月3日	46.79	8月24日	45.20	8月15日

6　结论和讨论

(1)作者曾定义的东亚夏季风在中国东北地区建立的标准,即取 850 hPa 候平均资料,作沿 122.5°E 时间-纬度剖面图,将 θ_{se}的 336 K 等值线和南风 4 m/s 等值线同时越过 40°N 的候定义为东亚夏季风在中国东北地区建立的日期(候),持续及累计的总候数为影响时间,该年 θ_{se}的 336 K 等值线永久撤离到 40°N 以南的候为东亚夏季风从中国东北地区撤退的日期。根据相同方法,分别计算了沿 112.5°、117.5°和

122.5°E 经度上，20°N 及以北每隔 5 个纬度东亚夏季风建立、持续及累计和撤退时间(候)，将某年持续和多年平均持续候数相比的标准化值，定义为一种沿某一经圈上某一纬度的东亚夏季风强度指数 $I_{SM\Phi}$。沿 117.5°E 经度上，东亚夏季风在 20°、25°、30°、35°、40°N 建立的平均日期分别为 27.76、28.54、34.43、37.12、37.65 候，撤退的平均日期分别为 54.44、53.69、51.85、48、46.76 候。其中(117.5°E，20°N)可以代表南海的中北部地区，本文所确定的该区夏季风建立和撤退的平均日期分别为 27.76 和 54.44 候，与国内学者公认的 5 月第 4 候和 10 月第 1 候相当吻合。

(2)沿 112.5°、117.5°和 122.5°E 的同一纬度上，东亚夏季风建立的平均日期并不相同，西边先于东边建立，每隔 5 个经度相差 1～2 候，这有可能与副热带高压的西伸脊点平均经度位置 5—8 月间逐渐东撤有关；撤退的平均日期(30°N 及以北)分布则相反，东边先撤退。

(3)沿 117.5°E，30°N 东亚夏季风强度指数 $I_{SM\Phi}$与中国 151 个测站 7—8 月降水总量相关系数分布有 2 片高相关区(显著性水平 $\alpha\geqslant0.05$)，一是最大片正相关区位于华北东部和北部、东北地区中南部，另一片负相关区位于长江流域下游和浙江北部；沿 117.5°E，35°N 东亚夏季风强度指数 $I_{SM\Phi}$与中国 151 个测站 7—8 月降水总量相关系数分布有 3 片高相关区(显著性水平 $\alpha\geqslant0.05$)，第 1 大片正相关区依然是华北和东北地区中南部、中部，相关区域明显向西伸展和向东北地区中部延伸，第 2 片负相关区依旧是长江流域下游和浙江北部，并向西延伸，第 3 片负相关区位于新疆的南部；沿 122.5°E，40°N 东亚夏季风强度指数 $I_{SM\Phi}$与中国 151 个测站 7—8 月降水总量相关系数分布有 4 片高相关区(显著性水平 $\alpha\geqslant0.05$)，华北和东北地区分为 2 片正相关区，第 3、第 4 片负相关区，还仍然分别位于长江流域下游和新疆的南部地区。表明该指数与中国大陆夏季主要雨带有很好的相关性。$I_{SM\Phi}$在中国不同纬度上的时间序列表明，东亚夏季风年际和年代际变化是不尽相同的，25°N 的 $I_{SM\Phi}$是从 20 世纪 90 年代才开始变弱的，30°N 自 70 年代中期开始减弱，而 35°和 40°N 自 60 年代中期就开始进入了一个长时间的较弱期。另外，沿 117.5°E，30°、35°N，以及沿 122.5°E，40°N 的 $I_{SM\Phi}$年际变化亦呈现多样性。

(4)$I_{SM\Phi}$与同期东亚大气环流形势有着较高的相关性。沿 117.5°E，25°N 典型强与弱的 $I_{SM\Phi}$的 8 月 500 hPa 高度距平差值场表明，在副热带高压区(120°—150°E，30°N 附近)呈现较强的正距平，在鄂霍次克海地区呈现较强的负距平；沿 117.5°E，35°N 和沿 122.5°E，40°N 的 $I_{SM\Phi}$均与新地岛至鄂霍次克海上空 7—8 月 500 hPa 高度场有一个宽达 10～20 纬度的正相关带状区(显著性水平 $\alpha\geqslant0.05$)，其南侧为宽广的弱的负相关区。表明 $I_{SM\Phi}$所反映的中国大陆副热带地区东亚夏季风强弱既与西太平洋副热带高压又与中高纬度的鄂霍次克海高压相关密切。而副热带地区以北的中纬度地区东亚夏季风强弱主要与中高纬度大气环流形势场相关密切，其物理解释是，当新地岛至鄂霍次克海一带为高压带时，极地冷空气被阻断于极圈附近，有利于副热带暖湿空气的北上。

(5)$I_{SM\Phi}$夏季风指数具有明确的物理观测基础，不直接涉及季风发生的机制和根源，既考虑了 θ_{se}在湿绝热过程中具有的准守恒性，能代表来自副热带地区气团的属性，又考虑了季风的南风候平均风速要达到一定的强度(风速≥4 m/s)，以 θ_{se}符合标准的持续候数与多年平均持续候数比值的标准化，作为描述季风强弱的最直观的标准，具有较好的实测代表性，所以，本文定义的影响中国东部和东北地区的夏季风指数，有可能避免由一个夏季风强度指数来衡量幅员辽阔的中国大陆所有纬度的季风强度，因而掩盖了夏季风在向北推进、建立和向南撤退的多样性，亦更有利于进一步研究东亚夏季风进退、强弱的季内变化和年际变化规律。

参考文献

陈隆勋，刘洪庆，王文，等，1999. 南海及其邻近地区夏季风爆发的特征及其机制的初步研究[J]. 气象学报，57(1)：16-29.

高辉，何金海，徐海明，2001. 关于确定南海夏季风建立日期的讨论[M]//何金海，丁一汇，高辉，等. 南海夏季风建立日期的确定与季风指数. 北京：气象出版社，1-41.

郭其蕴，1983. 东亚夏季风强度指数及其变化的分析[J]. 地理学报，38(3)：207-217.

廉毅，沈柏竹，高枞亭，等，2003. 东亚夏季风在中国东北区建立的标准、日期及其主要特征分析[J]. 气象学报，6(5)：

548-558.

施能,1996.近40年东亚冬季风强度的多时间尺度变化特征及其与气候的关系[J].应用气象学报,7(2):175-182.

施能,朱乾根,关彬贵,1996.近四十年东亚夏季风及中国夏季大尺度天气气候异常[J].大气科学,20(5):573-583.

施尚文,巢俊民,1982.中国北方地区的夏季风[C].//全国热带夏季风学术会议.全国热带夏季风学术会议文集编写组.昆明:云南人民出版社:72-85.

孙秀荣,陈隆勋,何金海,2002.东亚海陆热力差指数及其与环流和降水的年际变化关系[J].气象学报,60(2):164-172.

汤明敏,黄士松,1983.1979年中国东部夏季风的进退[C].//全国热带夏季风学术会议文集编写组.全国热带夏季风学术会议文集.昆明:云南人民出版社:311-317.

张庆云,陶诗言,陈烈庭,2003.东亚夏季风指数的年际变化与东亚大气环流[J].气象学报,61(5):559-568.

赵汉光,张先恭,1994.东亚夏季风和中国夏季雨带的关系[J].气象,22(4):8-12.

朱乾根,杨松,1989.东亚副热带季风的北进及其低频振荡[J].大气科学学报,12(3):249-258.

竺可桢,1934.东南季风与中国之雨量[J].地理学报,1(1):1-27.

祝从文,何金海,吴国雄,2000.东亚季风指数及其与大尺度热力环流年际变化关系[J].气象学报,58(4):391-402.

TAO S Y,CHEN L X,1987. A review of Research on the East Asian Summer Monsoon in China[M]//Chang C P,Krishnamurti T N. Oxford:Oxford University Press Monsoon Meteorology:60-92.

东亚夏季副热带季风主要活动路径的特征分析*

刘刚[1,2] 沈柏竹[1,3] 廉毅[1,3] 曹玲[4] 刘平[2] 包书新[5]

(1 中高纬度环流系统与东亚季风研究开放实验室,长春 130062;2 吉林省气象科技服务中心,长春 130062;
3 吉林省气象科学研究所,长春 130062;4 吉林省气象影视制作中心,长春 130062;
5 中国人民解放军 91422 部队气象台,烟台 265200)

摘 要:利用 1948—2002 年 NCEP/NCAR 全球再分析资料,根据东亚夏季副热带季风建立和活动范围的气候学时空分布特征,对其进行了分类研究;揭示了不同类型的东亚夏季副热带季风与代表大气环流形势的 850 hPa 风场之间的关系。得出结论:(1)根据东亚夏季副热带季风在我国建立并向北跃进特征的不同,将东亚夏季副热带季风分为三种类型:偏西类、中间类、偏东类,给出了 1948—2002 年东亚夏季副热带季风分类年历表。(2)850 hPa 南风风场的特征与东亚夏季副热带季风向北跃进的特征具有比较好的对应关系。

关键词:东亚夏季副热带季风;主要活动路径;南风风场

1 前言

我国位于东亚季风区,是世界上著名的季风活动区之一。东亚夏季副热带季风的异常变化对我国天气和气候以及生态环境影响非常大东部地区 5—9 月的汛期降水分布、雨带移动以及旱涝灾害的发生在很大程度上受到东亚夏季副热带季风的控制,与东亚夏季副热带季风的进退过程以及位置变化有着密切关系。东亚夏季副热带季风季节内变化主要指东亚夏季副热带季风的暴发、活跃、中断以及撤退的变化过程。陈隆勋等(1982)利用 1980 年夏季的资料,计算了 850 hPa 等压面上 40 和 14 d 周期振荡的纬向风和经向风交叉谱的位相分布,发现这两种振荡是从南半球向赤道地区变化。He 等(1990)、Wang 等(1992)也详细讨论了亚洲季风区低频振荡的经向传播特征,在过去的研究中有种种迹象表明东亚沿海存在着一个南北向的波列,Huang(1992)和 Nitta(1987)称之为东亚波列或 P-J 波列。李崇银(1993)的研究也表明在东亚夏季副热带季风区存在明显经向传播的低频振荡。Murakam 等(1984)也揭示了在南亚和西太平洋地区的 30～60 d 振荡还存在显著的经向传播。在以往国内外学者对东亚夏季副热带季风季节内振荡的研究中,都只是对东亚夏季副热带季风在经向上变化的特征进行分析。这从某一方面能够反映出东亚夏季副热带季风的振荡特点,但当不同年东亚夏季副热带季风在向北跃进过程中到达同一纬度时,就无法反映出各年东亚夏季副热带季风的不同特点。本研究就是从这一点出发,研究东亚夏季副热带季风在向北跃进的过程中在不同经度位置的特征。

2 特征分类

根据廉毅等(2003,2004)所定义的东亚夏季副热带季风 $I_{SM\phi}$ 指数计算方法,当假相当位温 θ_{se} 的 336 K 等值线和 $Vs \geqslant 4$ m/s(南风分量)等值线同时越过某一纬度时,定义为该纬度(一般指$\geqslant$20°N)的地区东亚

* 本文原载于《吉林大学学报(地球科学版)》,2008,38(Z1):196-199

夏季副热带季风建立的日期(候)。本文利用1948—2002年NCEP/NCAR全球再分析资料绘出了逐年东亚夏季副热带季风建立日期经-纬图。1962、1986、1994年的东亚夏季副热带季风建立日期经-纬图中数值为候数(图略),图中的实线代表假相当位温 θ_{se} 的336 K等值线和南风≥4 m/s等值线同时达到的位置。从中可以看出,在1962年的第26候东亚夏季副热带季风沿着112.5°E方向在25°N附近建立,并沿着这一经度向北阶段性跳跃,在第30候(5月26—31日)向北到达40°N,此时向北跃进的舌状仍处在112.5°E附近。同时可以看出1986年、1994年东亚夏季副热带季风前进路线分别沿着107.5°、122°E的方向自20°N开始向北跃进的。

根据东亚夏季副热带季风在我国建立和活动范围的时空分布基本模态,东亚夏季副热带季风向北推进的平均经向位置在112.5°E附近,将1948—2002年共计55个总样本年的东亚夏季副热带季风活动路径划分为偏西类、中间类、偏东类共三类(表1)。即季风活动的等时线向北突出的舌状位于110°E以西的为偏西类,表明在同一纬度110°E以西地区的东亚夏季副热带季风先建立;季风活动的等时线向北突出的舌状位于110°—115°E的为中间类,表明在同一纬度110°—115°E地区先建立;季风活动的等时线向北突出的舌状位于115°E以东的为偏东类,表明在同一纬度115°E以东地区的东亚夏季副热带季风先建立。对55个总样本年的特征分类结果见表1,从表1可以得出,55个总样本年中偏西类12 a,占总样本数的22%;中间类35 a,占总样本数的64%;偏东类8 a,占总样本数的15%。

表1　1948—2002年东亚夏季副热带季风东西向类型分类

偏西类(110°E以西)年份	中间类(110°—115°E)年份		偏东类(115°E以东)年份
1951	1949	1971	1948
1959	1950	1973	1972
1966	1952	1974	1978
1968	1953	1975	1990
1976	1954	1977	1991
1981	1955	1979	1994
1984	1956	1980	1996
1986	1957	1982	2001
1992	1958	1983	
1997	1960	1985	
1999	1961	1987	
2000	1962	1988	
	1963	1989	
	1964	1993	
	1965	1995	
	1967	1998	
	1969	2002	
	1970		

注:表中空白为没有相应类型典型年。

3　信号场分析

通过对NCEP/NCAR全球逐日850 hPa风场格点资料的处理,计算得出850 hPa的经向风分量的逐候资料,得到偏西类风场与55个总样本年风场的候距平场。偏西类特征年与55个总样本年在850 hPa经向风的候距平经-纬图,南风分量中规定南风为正,北风为负。在第26、27、28候的图上(图略),可以看出在(100°—110°E,30°—40°N)范围内的距平为零,相应偏东位置为负距平,并且到28候负距平范围扩大并且距平为负值的绝对值也有所增加;在到达29候时(100°—110°E,30°—40°N)范围内的距平由零值转为正值,同时开始向北移动;说明在这个时候该位置的南风风量有所增加并开始向北传播;同时偏东位置的负距平值绝对值也明显增大，范围也同时扩大，说明在该范围南风较弱，并且呈继续走弱趋势。因此可以得出东亚夏季副热带季风偏西类特征年的平均建立日期（第 28 候）前后，南风风场同样具备在 100°—110°E、30°—40°N 位置的南风分量较 110°E 以东位置的南风风量的数值大，并且在第 29 候出现了突变。在该位置风速的迅速增加，与偏西类特征年东亚夏季副热带季风平均建立日期发生在第 28 候并自此向北跳跃的特点，无论是在时间上还是在空间位置上有着比较好的对应关系。对于中间类与偏东类年东亚夏季副热带季风，同样可以得到其振荡特征与相应的低层风场的形态特征的较好对应关系。

4 结论

通过对东亚夏季副热带季风的振荡特征分析，利用东亚夏季副热带季风建立和活动范围的气候学时空分布特征，对东亚夏季副热带季风进行了相关分析得出以下结论。

(1)在国内第一次提出了沿东亚夏季副热带季风的经向跃进路径进行纬向上的分类，给出了分类年历表。

(2)东亚夏季副热带季风建立及向北推进的振荡特征与 850 hPa 南风风场的形态特征具有很好的一致性。东亚夏季副热带季风向北振荡移动的轨迹特征反映着 850 hPa 南风风场的变化特征；反之，850 hPa 南风风场的变化特点又可以用来说明东亚夏季副热带季风的移动特点。

参考文献

陈隆勋，罗绍华，沈如桂，1982. 夏季亚洲季风环流的结构及其与大气环流季节变化的关系 [M]. 北京：科学出版社，92-99.

李崇银，1993. 大气低频振荡 [M]. 北京：气象出版社：310.

廉毅，2003. 东亚夏季风在中国东北区建立的标准、日期及主要特征分析 [J]. 气象学报，61(5)：548-558.

廉毅，沈柏竹，高枞亭，2004. 关于确定东亚夏季风强度指数的探讨 [J]. 气象学报，62(6)：782-789.

HE J H, SONG Y, 1990. Meridional propagation of East-Asian low frequency oscillation and mid-latitude low-frequency waves[J]. Acta Meteor Sinica(special issue), 5(2): 51-59.

HUANG R H, 1992. The East Asia Pacific pattern teleconnection of summer circulation and climate anomaly in East Asia [J]. Acta Meteor Sinica, 6(1): 25-36.

MURAKAM T, NAKAZAWA T, HE J H, 1984. On the 40—50 day oscillations during the 1979 Northern Hemisphere summer, phase propagation[J]. J Meteor Soc Japan, 62(9): 440-468.

NITTA T, 1987. Convective activities in the tropical western Pacific and their impact on the northern hemisphere summer circulation[J]. J Meteor Soc, 65(12): 373-390.

WANG B, DING Y H, 1992. An overview of the Madden-Julian oscillation and its relation to monsoon and mid-latitude circulation[J]. Adv Atmos Sci, 9(5): 93-111.

北太平洋涛动与东亚夏季风的相关研究*

廉毅

（中高纬度环流系统与东亚季风研究开放实验室，长春 130062；吉林省气象科学研究所，长春 130062）

摘　要：利用 1948—2002 年 NCEP/NCAR 全球再分析资料、海平面气压场资料、月平均海温资料和国家气候中心 1961—2000 年 151 站逐月降水量资料、西北太平洋夏季副热带高压指数资料，采用作者曾定义的东亚夏季风强度指数（$I_{SM\Phi}$）和北太平洋涛动（NPO）指数（NPO_L和 NPO_W(wallaee)），用相关和奇异值分解（SVD）方法，分析了 NPO 区指数、海平面气压场、500 hPa 高度场和北太平洋海温（SST）场与 $I_{SM\Phi}$、西北太平洋夏季副热带高压指数和中国大陆夏季降水量场的相关关系。结果表明：①沿 117.5°E 的 30°N 和 35°N 的 $I_{SM\Phi}$与当年冬季的 NPO_W呈负相关，而前一年冬季的 NPO_W与当年沿 117.5°E 的 30°N 的 $I_{SM\Phi}$也呈负相关，表明当东亚副热带和中纬度地区夏季风强劲时，冬季 NPO 将呈现北高南低负位相阶段为主的形势；而如果前一年冬季 NPO 呈现北低南高的正位相阶段形势时，当年东亚副热带地区（30°N）的夏季风偏弱。②前一年冬季 NPO_L和 NPO_W与中国大陆夏季降水量场相关分布很相似，与中国的西北地区、长江中下游以南地区呈正相关，与东北地区大部、江淮地区和华南地区东部及西南地区呈负相关。③前一年冬季 NPO 区 500 hPa 高度场与中国大陆夏季降水量场 SVD1 呈由北向南的“＋－”相关波列，也就是当前一年冬季 NPO 区呈负位相涛动阶段时，中国东北地区大部、华南东部沿海、长江以北和淮河流域夏季降水量偏多，长江以南、东北地区西部、华北大部和西北地区夏季降水量偏少。④前一年冬季海平面气压场与中国大陆夏季降水量场的 SVD1 同前一年冬季 NPO 区 500 hPa 与中国大陆夏季降水量场的 SVD1 很相似。⑤前一年冬季 NPO 区正、负位相与中国大陆夏季降水量场的 SVD1 同前一年冬季 ENSO 位相与中国大陆夏季降水量场的 SVD1 具有耦合性特征。⑥前一年冬季 NPO_L5 个正、负位相典型年所对应的当年夏季 850 hPa 流场和 200 hPa V_χ、差值合成场显示，前一年冬季海平面气压场正、负位相涛动阶段的差异，将强迫出滞后半年的东亚大气环流形势和物理场的某些遥响应特征是：西太平洋副高显著偏南，200 hPa V_χ的正距平中心亦偏南，反映影响东亚夏季风的加热场偏南。

关键词：NPO 指数；东亚夏季风指数；遥响应

1　引言

自 20 世纪 50 年代以来，国内外学者比较关注中高纬度的切断和阻塞异常环流的研究，80 年代初，国外学者揭示了北大西洋涛动（NAO）和北太平洋涛动（NPO）有关高纬度与中纬度气压场间涛动的统计概念模型（Wallace et al.，1981）。由此以来，极地与中纬度环流主要的振荡特征及异常对气候变化的影响研究开始活跃（Rogers，1984；李崇银 等，1999）。近 30 a 来，亚洲季风研究有许多成果，认识到厄尔尼诺-南方涛动（ENSO）事件及暖池地区的对流活动与东亚夏季风的活动及其降水异常有重要联系是其中的重要成果之一（丁一汇 等，1996，2004），这表明东亚夏季风与北太平洋中低纬度海-气环流系统之间的相互作

* 本文原载于《地理科学》，2007，27(Z1)：19-27

用，这种相互作用对东亚夏季风及中国气候变化影响亦越来越引起中国学者的关注（丁一汇 等，2000；王会军，2000），有的学者从 NPO 活动规律入手，研究其与中国区域气候变化的相关特征（刘宗秀 等，2002，2003），Mantua 等（1997）研究了北太平洋年代际振荡（PDO）与热带太平洋的相关性，也引起了国内学者（杨修群 等，2004）的关注，并指出中纬度海-气相互作用的关键环节及物理过程并没有被认识清楚，以往研究对于中纬度影响热带的环节只考虑了海洋而几乎忽略了大气。

本文利用作者曾定义的东亚夏季风强度指数（$I_{SM\Phi}$）（刘宗秀 等，2002）和 NPO 指数（NPO_L 和 NPO_W（Wallace））（Wallace et al.，1981；刘宗秀 等，2003），分析了 NPO 区指数、海平面气压场、500 hPa 高度场和北太平洋海温（SST）场与 $I_{SM\Phi}$、中国大陆夏季降水量场的相关关系，并对气候变率和相关机制进行了有益的探讨。

2　资料和方法

利用 1948—2002 年 NCEP/NCAR 全球再分析资料、海平面气压场资料和 NOAA 月平均海温资料；国家气候中心的全国 151 站 1961—2000 年的逐月降水量资料。

$I_{SM\Phi}$的计算方法参见文献（廉毅 等，2003，2004），NPO_W 的计算方法参见文献（Wallalce et al.，1981），计算了 $I_{SM\Phi}$和 NPO_W 的线性相关系数；将北太平洋涛动区的范围（25°—65°N，130°—170°E）扩大为（25°—70°N，140°E—150°W）；定义 NPO_L 为 25°—40°N 与 50°—65°N 区域的平均气压差标准差，计算方法参见相关文献（刘宗秀 等，2003）。

辐散风（加热场）和垂直速度等物理诊断量采用文献（丁一汇，1989）的方法进行计算。

3　东亚夏季风与 NPO 的相关性

3.1　$I_{SM\Phi}$与 NPO_W的相关特征

利用 NCEP 逐日探空资料分别计算了沿 117.5°E 的 25°、30°、35°N 和沿 122.5°E 的 40°N 共 4 个纬度的 $I_{SM\Phi}$（1949—1999 年），利用 NCEP 海平面气压场资料计算了 NPO_W四季和年的涛动指数（1948—1998 年），还计算了 $I_{SM\Phi}$与 NPO_W的超前和滞后的相关系数。

从表 1 可以看出，沿 117.5°E 的 30°N 和 35°N 的 $I_{SM\Phi}$与当年冬季 NPO_W的相关系数≤−0.28（表 1 中数据为相关系数×100，以下类同），置信水平＞95％，沿 117.5°E 的 35°N 的 $I_{SM\Phi}$还与同季的夏季 NPO_W呈正相关，与秋季 NPO_W呈负相关，相关系数分别＞0.28 和＜−0.28，置信水平＞95％。

表 1　$I_{SM\Phi}$与当年 NPO_W的相关系数（×100）

纬度/°N	春	夏	秋	冬	年
25	−2.51	11.13	−9.90	−23.11	−8.31
30	−9.06	25.56	−22.93	−43.04	−14.45
35	8.51	33.94	−33.91	−35.75	−0.13
40	−6.55	14.70	−21.58	−25.36	−5.81

从表 2 可以看出，前一年冬季的 NPO_W与当年沿 117.5°E 的 30°N 的 $I_{SM\Phi}$呈负相关，相关系数＜−0.28，显著性水平 α＞0.05。

表 2　$I_{SM\Phi}$指数与前一年 NPO_W的相关系数(×100)

纬度/°N	春	夏	秋	冬	年
25	−9.00	−11.30	−22.68	−7.53	−18.60
30	14.34	19.67	−13.91	−29.14	5.93
35	0.88	14.72	−27.36	−20.90	−10.52
40	0.54	10.09	−5.47	−13.65	3.70

3.2　前一年冬季 NPO_W与西北太平洋夏季副热带高压指数的相关特征

利用 NCEP 资料分别计算了前一年冬季 NPO_W与当年夏季西北太平洋副热带高压指数的相关系数。

从表 3 可以看出，前一年冬季 NPO_W与当年 6 月、7 月和 8 月以及夏季(6—8 月)的西北太平洋副热带高压指数相关十分显著。其与西北太平洋副热带高压夏季面积和强度指数呈正相关，相关显著性水平 $\alpha>0.01$；与西北太平洋副热带高压夏季西脊点呈负相关，相关显著性水平 $\alpha\approx0.05$，其中，前一年冬季 NPO_W与 6 月的西北太平洋副热带高压指数相关的显著性水平 $\alpha>0.05$。

表 3　前一年冬季 NPO_W与当年夏季西北太平洋副热带高压指数相关系数(×100)

月份	面积	强度	脊线	北界	西脊点
6	40.05	36.95	−10.68	−6.18	−34.80
7	46.81	45.37	9.33	8.04	−10.01
8	43.42	42.19	0.14	2.77	−20.70
夏季(6—8 月)	47.58	46.43	0.18	1.53	−27.51

3.3　NPO_W和 NPO_L指数与中国大陆夏季降水的相关特征

从 1961—2000 年前一年冬季 NPO_L和 NPO_W与中国大陆 151 个气象站夏季(6—8 月)降水量的相关系数分布中可以看出(图略)，两种指数与中国大陆 151 个气象站夏季降水量的相关系数分布很相似。在西北和华北地区、长江中下游以南地区为正相关区，在东北地区大部、江淮地区、华南地区东部和西南地区为负相关区。

从前一年 NPO_L与中国大陆 151 个气象站 8 月降水量相关系数的分布可见，华北地区东北部、东北地区为负相关区；华北地区东北部、东北地区的中西部和中东部地区的显著性水平 $\alpha>0.05$；中国长江流域和新疆的广大地区为正相关区，长江上游地区的显著性水平 $\alpha>0.05$。

3.4　NPO 区 500 hPa 高度场与中国大陆夏季降水量场的隔季相关特征

将 NPO 区 500 hPa 高度场(140°E—150°W)与中国大陆夏季降水量场做隔季(半年尺度)的奇异值分解(SVD)，结果表明(表 4)，奇异值分解收敛速度很快，前 4 对耦合模态的累积方差贡献可以达到 90%，第 1 耦合模态占有较大的协方差，达到 57%，第 2 对耦合模态次之，为 20%，场与场之间相关显著性水平 $\alpha>0.01$($\alpha_{0.01}=0.39$)。这表明 NPO 区前一年冬季 500 hPa 高度场与中国大陆的夏季降水量场异常之间具有较密切的时空相关性。

表 4　NPO 区前一年冬季 500 hPa 高度场与中国大陆夏季降水量场前 4 对奇异向量的方差贡献和相应的模态相关系数

项目	第 1 对	第 2 对	第 3 对	第 4 对
方差贡献	0.57	0.20	0.08	0.05
相关系数	0.84	0.86	0.75	0.74

NPO 区前一年冬季 500 hPa 高度场与中国大陆夏季降水量场的第 1 对奇异向量(SVD1)同性相关系

数的时空分布显示，当 500 hPa 高度场呈由北向南“＋－”波列时，即 NPO 为负位相涛动阶段，与其相对应的中国大陆夏季降水量场分布有以下特点：东北地区大部、华南东部沿海、广西大部、海南和西南地区，特别是长江以北和江淮流域呈正值分布，表明夏季这些地区降水量偏多；长江以南、东北地区西部、华北地区大部和西北地区呈负值分布，表明这些地区夏季降水量偏少。反之，如 NPO 区前一年冬季 500 hPa 高度场呈由北向南“－＋”波列时，也就是 NPO 为正位相涛动阶段时，中国大陆的夏季降水量场分布与上述分布相反。

图 1a 和图 1b 分别为 NPO 区前一年冬季 500 hPa 高度场与中国大陆夏季降水量场的 SVD1 时间系数。由图 1 可见，两个场的系数趋势分布基本一致：前一年冬季 500 hPa 高度场时间系数在 20 世纪 60 年代至 70 年代前期以正异常为主，80 年代至 90 年代末期以负异常为主，中国大陆夏季降水量场时间系数的年代际变化类同，只是个别年份不一致。

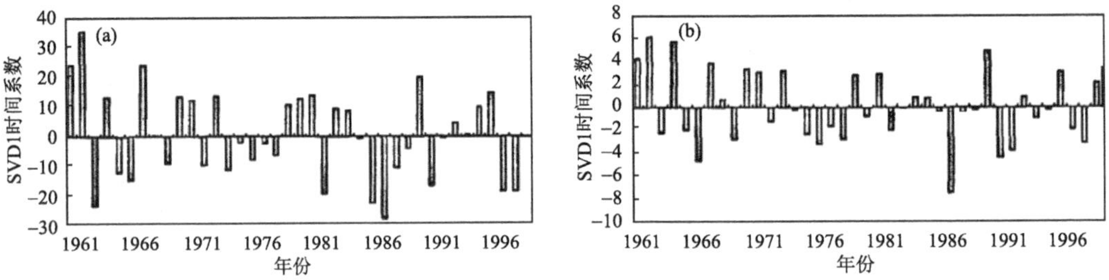

图 1　NPO 区前一年冬季 500 hPa 高度场(a)与中国大陆夏季降水量场(b)的 SVD1 时间系数

3.5　NPO 区前一年冬季海平面气压场和前一年冬季北太平洋 SST 场与中国大陆夏季降水量场隔季相关的耦合特征

分别将 NPO 区前一年冬季海平面气压场和前一年冬季北太平洋 SST 场与中国大陆夏季降水量场做隔季（半年尺度）的奇异值分解（SVD），结果（表 5、表 6）显示，前一年冬季海平面气压场与中国大陆夏季降水量场的奇异值分解速度比前一年冬季北太平洋 SST 场与中国大陆夏季降水量场的奇异值收敛速度快，表 5 中前 4 对耦合模态的累积方差贡献率可达 90%，而在表 6 中则为 69%，表 5 比表 6 中的第一对耦合模态占总方差的比例大，分别为 50% 和 36%，场与场的相关显著性水平 $\alpha>0.01(\alpha_{0.01}=0.39)$。

表 5　前一年冬季海平面气压场和中国大陆夏季降水量场前 4 对奇异向量的方差贡献和相应的模态相关系数

项目	第 1 对	第 2 对	第 3 对	第 4 对
贡献率	0.50	0.23	0.09	0.08
相关系数	0.84	0.85	0.74	0.78

表 6　前一年冬季北太平洋 SST 场和中国大陆夏季降水量场前 4 对奇异向量的方差贡献和相应的模态相关系数

项目	第 1 对	第 2 对	第 3 对	第 4 对
贡献率	0.36	0.14	0.11	0.08
相关系数	0.83	0.83	0.77	0.86

NPO 区前一年冬季海平面气压场和北太平洋 SST 场与中国大陆夏季降水量场的第 1 对奇异向量的时空分布图（图略）中，当 NPO 区前一年冬季海平面气压场在北半球呈由北向南的“－＋”波列时，即 NPO 为正位相涛动阶段时，与其对应的中国大陆的夏季降水量场的分布为，东北地区大部、长江以北与江淮流域、华南大部与西南地区呈负值分布，表明这些地区降水量偏少；当北太平洋 SST 场中东赤道太平洋为负值呈 La Niña 型分布时，与其对应的中国大陆的夏季降水量场的分布与第 1 对奇异向量中国大陆夏季降水量场同性相关系数分布呈现惊人的相似，只是正、负值相反，东北地区的东部和华南地区的东部沿海呈

正值分布，比第 1 对奇异向量中国大陆夏季降水量场同性相关系数分布的负值区面积要小一些，也就是表明当前一年冬季 NPO 海平面气压场呈由北向南的“＋－”波列和北太平洋 SST 场呈 La Niña 分布时，与中国大陆夏季降水量场耦合模态是非常相似的，反之，前一年冬季 NPO 区海平面气压场呈由北向南的“－＋”波列和北太平洋 SST 场呈 El Niño 型分布时，与中国大陆夏季降水量场耦合模态亦是非常相似的。

4　东亚夏季风和 NPO 相关物理量诊断分析

4.1　流场特征

图 2 为前一年冬季 NPO_L 5 个正、负位相典型年所对应的当年夏季 850 hPa 流场差值合成场。由图 2 可见，西太平洋副热带高压脊线偏南（位于 20°N 以南），影响中国 40°N 以北地区的南风主要是西风带系统所为，表明前一年冬季 NPO_L 为正位相时，预示夏季中国大陆中东部西太平洋副热带高压西侧的偏南风既偏弱又偏南。

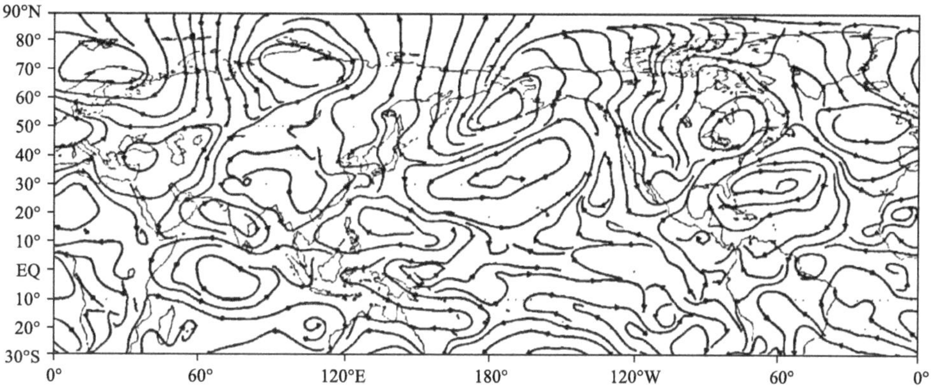

图 2　前一年冬季 NPO_L 5 个正、负位相典型年所对应的当年夏季 850 hPa 流场差值合成场

4.2　加热场特征

图 3 为前一年冬季 NPO_L 指数 5 个正、负位相典型年所对应的当年夏季 200 hPa 速度势（V_χ）距平合成差值场。从图 3 可知，东亚地区 200 hPa V_χ 为负距平区所控制，而相对是热源的 V_χ 正距平中心位于赤道及南半球，与东亚地区较近的有两个，一个位于 30°S、80°E 印度洋的中部地区，一个位于西太平洋暖池南部 125°E 赤道附近。

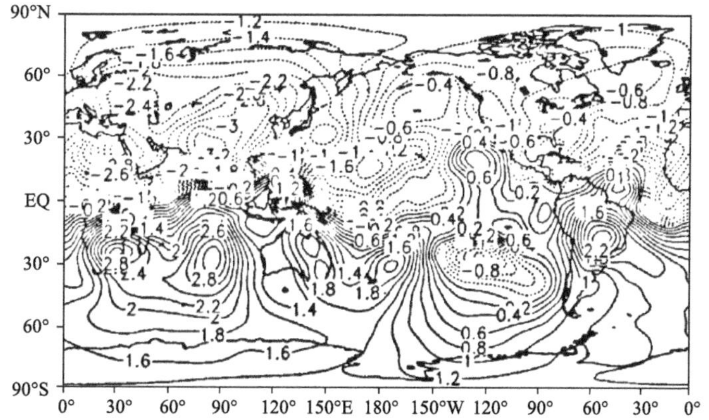

图 3　前一年冬季 NPO_L 指数 5 个正、负位相典型年所对应的当年夏季 200 hPa V_χ 距平合成差值场（$\times10^6 m^2 s^{-1}$）

5 结论

(1)沿 117.5°E 的 30°N 和 35°N 的 $I_{SM\Phi}$与当年冬季的 NPO_W呈负相关(显著性水平 $\alpha>0.05$),前一年冬季的 NPO_W与当年沿 117.5°E 的 30°N 的 $I_{SM\Phi}$也呈负相关(显著性水平 $\alpha>0.05$),其物理意义是当东亚副热带和中纬度地区夏季风强劲时,冬季 NPO 将呈现北高南低负位相阶段为主的形势;而如果前一年冬季 NPO 呈现北低南高的正位相阶段形势时,当年东亚副热带地区(30°N)夏季风比较弱。前一年冬季 NPO_W与东亚夏季风的重要环流系统之一的西北太平洋副热带高压指数相关亦很显著。前一年冬季 NPO_W与 6—8 月各月及夏季的西北太平洋副热带高压面积和强度指数呈正相关,但与西脊点呈负相关(其中 6 月和夏季的相关显著性水平 $\alpha\geqslant0.05$),其物理意义表明,当前一年冬季 NPO_W呈正位相阶段形势时,夏季副热带高压遥响应特征虽然表现为面积偏大和指数偏强,但又表现为副热带高压西脊点偏东,不利于东亚夏季风沿中国大陆东部向北挺进。

(2)NPO 指数、海平面气压和 500 hPa 高度场与中国大陆夏季降水场相关显著。前一年冬季 NPO_L和 NPO_W与中国大陆 151 个气象站夏季降水量的相关系数分布十分相似,均与中国的西北和华北地区、长江中下游以南地区呈正相关,与东北地区大部、江淮地区、华南地区东部及西南地区呈负相关。

(3)前一年冬季 NPO 区 500 hPa 高度场(140°E—150°W)与中国大陆夏季降水量场隔季(半年尺度)的 SVD 结果表明,奇异值分解收敛速度很快,前 4 对耦合模态的累积方差贡献可达 90%,第 1 对耦合模态占有比较大的协方差,达到 57%,场与场之间相关置信水平>99%($\alpha_{0.01}=0.39$);当前一年冬季 500 hPa 高度场呈由北向南"+-"波列时,即 NPO 为负位相阶段,中国东北地区大部、华南东部沿海、广西大部、海南和西南地区,特别是长江以北和江淮流域夏季的降水量偏多,长江以南、东北地区西部、华北地区大部和西北地区夏季的降水量偏少。反之,如 NPO 区前一年冬季 500 hPa 高度场呈由北向南"-+"波列时,即 NPO 为正位相涛动阶段时,中国大陆的夏季降水量分布与上述分布相反。前一年冬季 NPO 区海平面气压场与中国大陆夏季降水量场隔季(半年尺度)的 SVD 结果显示,第 1 对耦合模态同 500 hPa 高度场与中国大陆夏季降水量场的第 1 对耦合模态十分相似,只是正、负值分布正好相反,但得出的结论是相同的,即当前一年冬季 NPO 区海平面气压场呈由北向南"-+"波列时即 NPO 区为正位相涛动阶段时,中国大陆夏季降水量场分布为中国东北区大部、华南东部沿海、广西大部、海南和西南地区,特别是长江以北和江淮流域夏季降水量偏少,而长江以南、东北地区西部、华北大部和西北地区夏季降水量偏多,反之,如 NPO 区前一年冬季海平面气压场分布呈由北向南"+-"波列时,即 NPO 为负位相涛动阶段时,中国大陆夏季降水量分布与 NPO 区正位相涛动阶段所对应的分布相反。

(4)前一年冬季北太平洋 SST 场与中国大陆夏季降水量场 SVD1 同前一年冬季 NPO 区 500 hPa 高度场和海平面气压场与中国大陆夏季降水量场 SVD1 具有耦合性特征。前一年冬季北太平洋 SST 场与中国大陆夏季降水量场 SVD1,呈 La Nina(El Nino)分布表明,与之对应的中国大陆夏季降水量场分布与 NPO 区500 hPa 高度场和海平面气压场的负位相(正位相)相对应的中国大陆夏季降水量场十分相似。从而揭示了 NPO 对东亚夏季风的影响不是孤立的,并具有与 ENSO 共同影响东亚夏季风的耦合特征。

(5)前一年冬季 NPO_L5 个正、负位相典型年所对应的当年夏季 850 hPa 流场和 200 hPa 速度势距平差值合成场分别表明,前一年冬季 NPO 区海平面气压场正、负涛动位相阶段的差异,将强迫出滞后半年的东亚大气环流形势和物理场某些遥响应的特征是:当年夏季 850 hPa 西太平洋副热带高压偏南;200 hPa 速度势正距平中心亦偏南,一个位于印度洋中部(30°S),一个位于西太平洋暖池南部 125°E 赤道附近,反映影响东亚夏季风的加热场偏南。

参考文献

丁一汇,1989. 天气动力学中的诊断分析[M]. 北京:科学出版社:280.

丁一汇,李崇银,何金海,等,2004. 南海季风试验与东亚夏季风[J]. 气象学报,62(5):561-586.

丁一汇,罗勇,2000.亚洲季风的中长期预报[C]//国家自然科学基金委员会地球科学部.21 世纪初大气科学回顾与展望.北京:气象出版社:98-100.

丁一汇,马鹤年,1996.东亚季风的研究现状[C]//何金海,丁一汇,陈隆勋.亚洲季风研究的新进展.北京:气象出版社:1-14.

李崇银,李桂元,1999.北大西洋涛动和北太平洋涛动的演变与 20 世纪 60 年代的气候突变[J].科学通报,44 (16):1765-1769.

廉毅,沈柏竹,高枞亭,2004.关于确定东亚夏季风强度指数的探讨[J].气象学报,62(6):782-789.

廉毅,沈柏竹,高枞亭,等,2003.东亚夏季风在中国东北区建立的标准、日期及其主要特征分析[J].气象学报,61 (5):548-558.

刘宗秀,廉毅,高枞亭,等,2002.东北冷涡持续活动时期的北半球 500 hPa 环流特征分析[J].大气科学,26(3):361-372.

刘宗秀,廉毅,沈伯竹,等,2003.北太平洋涛动区 500 hPa 高度场季节变化特征及其对中国东北区降水的影响[J].应用气象学报,14(5):553-561.

王会军,2000.东亚季风气候预测的困难性及应加强研究的几个科学问题[C]//国家自然科学基金委员会地球科学部.21 世纪初大气科学回顾与展望.北京:气象出版社:52-54.

杨修群,朱益民,谢倩,等,2004.太平洋年代际振荡的研究进展[J].大气科学,28(6):979-992.

MANTUA N J, HARE S R, ZHANG Y, et al, 1997. A Pacific interdecadal climate oscillation with impacts on salmon production[J]. Bull Ame Meteor Soc, 78:1069-1079.

ROGERS J C, 1984. The association between the North Atlantic oscillation and the Southern oscillation in the Northern Hemisphere [J]. Mon Wea Rev, 112:1999-2015.

WALLACE J M, GUTZLER D S, 1981. Teleconnections in the geopotential height field during the Northern Hemisphere Winter [J]. Mon Wea Rev, 109:784-811.

确定东亚—西北太平洋地区夏季副热带季风建立和活动范围的一种方法*

廉毅[1,2] 沈柏竹[1,2] 高枞亭[1,2] 李尚锋[1,2]

(1 中高纬度环流系统与东亚季风研究开放实验室,长春,130062;
2 吉林省气象科学研究所,长春 130062)

摘　要:利用 1948—2002 年 NCEP/NCAR 全球再分析资料,1958—2002 年 ECWMF 再分析资料,国家气候中心的 618 个测站 1961—2005 年的逐日降水量资料,采用作者确定的东亚夏季风(EASM)建立标准,探讨东亚夏季风在南海暴发以后继续向北和向东推进活动范围、建立、持续和撤退的时空分布特征,结果表明:(1) 给出东亚—西北太平洋地区的东亚夏季副热带季风(EASSM)的定义,计算其建立时间(候)和空间分布,发现 EASSM 除继续向北推进的方向外,在 20°—25°N 还有明显向东的分支,以及向西稍偏北的分支,107.5°E 以西的季风建立等时线梯度十分密集,而以东的等时线梯度很小,季风建立的等时线总体以 110°—115°E 为轴心、明显向中国华北和东北地区凸起的近于折线的分布,表明 EASSM 先在中国大陆建立,而同纬度的中国近海和西北太平洋地区的夏季副热带季风后建立。(2) 撤退的等时线则近于与建立的等时线相反分布,沿 120°E 的方向凸向南,表明中国东北地区和近海地区先撤退。(3) 给出东亚—西北太平洋地区 EASSM 建立的概率等值线的空间分布,尝试定义了小于 35%概率等值线的地区为其影响的边缘地带,北边界始于黑龙江省北部 48°N 附近,向东南经由黑龙江省南部和俄罗斯滨海省,穿过日本海止于日本北海道南部,其西边界由东北向西南沿中蒙边境止于青藏高原。20 世纪 90 年代北部边缘地带与多年平均比偏南 3～5 个纬度,而西部边缘变动较小,只向东偏了 1 个经度。

关键词:东亚;西北太平洋地区;夏季副热带季风;建立和撤退;活动范围

1 引言

Ramage(1971)给出了亚洲-澳大利亚季风(A-AM)区覆盖的范围(30°S—40°N,40°—170°E)。陈隆勋(1984)、金祖辉等(1982)和 Tao 等(1987)提出了东亚地区存在一个独立于印度季风环流系统(Krishnamurti et al.,1976)的东亚季风环流系统(EAMCS),EAMCS 包含了南半球澳洲季风、南海—西太平洋热带季风、中国大陆副热带季风和高纬度环流的相互作用。Zhang 等(1996)认为东亚夏季风(EASM)是指东亚地区(20°—40°N,100°—140°E)的副热带季风。Murakami 等(1994)提出了西北太平洋季风(WNPSM),主要指超过菲律宾海和它附近的海上季风。Lau 等(1997)给出了印度夏季风和东亚夏季风以及 WNPSM 建立的平均日期,其北界可达中国大陆 40°N 和日本的 40°N 附近。之后,Lau 等(2000)在研究亚洲季风区域组成成分的动力和边界强迫特点一文中,提出亚洲夏季风可分成南亚季风(SAM)和东南亚季风(EAM)两个子系统。2004 年 11 月世界气象组织在中国杭州召开的 IWMⅢ季风工作会议上,首次将东亚季风建议为东亚—西北太平洋季风的称谓(East Asian-Western North Pacific monsoon),它是

* 本文原载于《气象学报》,2007,65(4):503-510

亚洲-澳大利亚季风(A-AM)的一部分,定义 105°E 以东和赤道至 45°N 之间的区域作为A-AM 副系统。尽管定义的名称有所不同,但共同关心的是不论哪个称谓的季风,它的影响范围的确定是很重要的科学问题,一方面从其活动范围可分清性质,另一方面对季风的年际和年代际变化的研究也是很有意义的,进而为预测季风活动的强弱奠定基础。近几年,东亚夏季风活动的西界和北界的研究不仅引起大气科学而且还引起相邻学科学者的关注,因为东亚季风边缘研究有利于加深认识季风以及中国生态系统在过渡带的特性(徐袁 等,2003;汤绪 等,2005;杨建平 等,2003),季风环流的强弱变化控制着气候界线空间摆动的范围和方向(杨建平 等,2005),以及涉及中国北方农牧交错带的地理界定及其生态问题(赵哈林 等,2002),更重要的是夏季风的活动边界也反映了夏季风的年际和年代际变化,它与东亚主要雨带的季节进程及旱涝密切相关(丁一汇 等,2004)。

季风定义决定着季风影响的大致区域,而有关季风边界的确定则是一件复杂且细致的研究工作,目前对其边界的确定大体可以分 3 种:(1) 动力边界,以风的变化为主,不仅是指同层风向的显著变化,还包含高低层风的切变(陈桦 等,2006)。(2) 热力边界,以气团的性质,如用 θ_{se} 的某一阈值作为气团属性外包线的抵达范围来表明季风可抵达区域。(3) 大都以降水量或水汽输送和收支作为指标对季风雨季的开始给予定义(高辉 等,2001;Sun et al.,2007)。

Zhu 等(1986)明确指出,南海季风为热带季风性质,中国大陆夏季风为副热带季风(SM),王慧等(2005)详细分析了西北太平洋季风区(5°—25°N,120°—160°E)的气候学特征。本文依据 850 hPa 候平均南风风速和假相当位温(θ_{se})的阈值定义东亚—西北太平洋地区(EA-WNP)的东亚夏季副热带季风(EASSM)建立和撤退时间(候)的标准(廉毅 等,2003,2004),计算了 102.5°—147.5°E 每隔 5 个经度,20°—50°N 每隔 5 个纬度的东亚和西北太平洋的 EASSM 建立、持续和撤退时间(候),为 EA-WNP 的 EASSM 活动范围确定了观测计算的一个标准,本文还给出了其建立的概率分布,以一定的阈值确定其活动的西和北边界带。

2 资料和方法

本文利用 1948—2002 年 NCEP/NCAR 全球再分析资料、1958—2002 年欧洲中期天气预报中心(ECMWF)的再分析资料计算 850 hPa 候平均资料,作沿 102.5°、105°、107.5°、112.5°、117.5°、122.5°、127.5°、132.5°、137.5°、142.5°、147.5°E 的时间-纬度剖面图,当假相当位温 θ_{se} 的 336 K 等值线和南风 ≥4 m/s 等值线同时越过某一纬度时,定义为该纬度的地区东亚夏季风建立的日期(候)(廉毅 等,2003,2004),本文将东亚—西北太平洋地区 20°N 及以北地区 θ_{se} 的 336 K 持续跨越该纬度的候数与多年平均持续候数比值的标准化值,定义为当年的沿某一经圈上相应纬度的夏季副热带季风强度指数:

$$I_{\text{SM纬}j} = \frac{(S_j - \overline{S})}{\sqrt{\sum_{j=1948}^{2002} \frac{(S_j - \overline{S})^2}{n}}}$$

$$S_j = \frac{h_j}{\overline{h}}$$

式中,$n=55$(1948—2002 年样本数);h_j 为某一年东亚夏季副热带季风建立之后 θ_{se} 的 336 K 等值线越过某一纬度持续候数;$\overline{h}$ 为 1948—2002 年 θ_{se} 的 336 K 等值线越过某一纬度平均持续候数;S_j 为某一年某一纬度 θ_{se} 的 336 K 的持续候数(h_j)与平均持续候数($\overline{h}$)的比值,其平均值为:

$$\overline{S} = \frac{\sum_{j=1948}^{2002} S_j}{n}$$

所用 618 个测站 1961—2005 年逐日降水量资料取自国家气候中心。

3 东亚夏季副热带季风建立、持续和撤退的时空分布特征

3.1 建立的时空分布特征

从利用NCEP资料计算的东亚大陆和西北太平洋地区1948—2002年东亚夏季副热带季风(EASSM)(20°N及以北)建立的平均日期(候)(图略)可见,东亚夏季副热带季风(EASSM)于26候(5月上旬)自20°N,110°E建立以来,继续推进建立的等时间(候)线的分布和主要前进方向,其特征:(1)东亚大陆至中国近海和西北太平洋EASSM等时线分布,总体呈"凸"状,36~40候等时线在中国大陆35°N附近明显向北突起,107.5°E以东的等时线自中国大陆至海洋区呈西北—东南走向,在海洋区域近于纬圈平行。表明107.5°E以东的东亚大陆EASSM建立时间显然早于同纬度的中国近海和西北太平洋海域,东亚大陆的EASSM建立的等时线梯度密集,西北太平洋的等时线梯度小而且比较均匀(20°—25°N除外)。(2)在(20°—25°N,102.5°—147.5°E)区域,以107.5°E为轴心等时线向东、西两个方向呈经向分布,107.5°E以西的等时线梯度很大,其前进方向以西北方向为主,34候(6月中旬)西端点位于(25°N,102.5°E),越过该点后等时线前进方向转向北,止于(35°N,102.5°E)的40候(7月中旬末)等时线的西端点。107.5°E以东的等时线梯度很小,等时线的主要前进方向向东,止于34候等时线(25°N,142.5°E)。(3)(25°N,102.5°—147.5°E)及以北地区,EASSM建立的等时线有两个主要前进方向:一个是东亚大陆EASSM主要向北的推进方向,32候等时线向北突出的部分先到达30°N以北之后,向北突出的等时线逐渐转向北北东方向,34候(6月中旬)到达长江流域,38候(7月上旬末)等时线凸向华北地区以北,到达(42°N,112.5°E),比同纬度的陆地要早1~2候。40候(7月中旬末)等时线与38候分布类同,已推进到中国东北地区,以及中国西北地区的东部。40候等时线转向东扩展到整个中国东北地区,西边的40候等时线穿越中国西北地区东部和蒙古东部地区,呈东北—西南向的斜线止于中国的(37°N,102.5°E)。东边的40候等时线近于南北向贯穿中国东北地区中部,经朝鲜半岛中部入日本海。另一个等时线前进方向是沿25°N向东,该等时线止于(26°N,147.5°E),表明EASSM可能达到西北太平洋副热带地区的东界。

从1961—2005年中国618个测站候降水量差值分布(图略)可见,第26候(即东亚夏季副热带季风在中国大陆建立的第1候)的降水量比第25候增加8~10 mm的地区为广西的20°—25°N,110°E以西的地区。第30候的降水量比第29候显著增加15~20 mm的为长江中游地区(28°—32°N,110°E),而115°—120°E的15~20 mm正差值区则位于长江中下游以南的浙江南部、江西省和福建省的北部地区。第35候与34候的降水量正差值15~20 mm区域位于长江中下游地区,其外包线5 mm凸向西北方向的河南省和山西省南部地区;第37候与第36候降水量的正差值5~10 mm区域中心位于(35°N,115°E)。1961—2005年中国618个测站的候降水量增加区域与1948—2002年东亚大陆和西北太平洋地区东亚夏季副热带季风建立的平均日期第26候、30候、35候、37候东亚夏季副热带季风先建立区域是相当吻合的。

3.2 平均持续时间的空间分布特征

利用NCEP资料计算的东亚夏季副热带季风在东亚大陆、中国近海和西北太平洋地区建立并且平均持续的时间(候)的空间分布(图略)可见,其持续的等时线与建立的等时线分布大体相似。从大陆至海洋的等时线呈西北—东南走向,表明东亚大陆持续的时间比同纬度海上的持续时间长。中国大陆华北及以南EASSM平均持续时间均在6个候(1个月)以上,35°N以北等时线以112.5°E为轴心向东、西两侧呈递减趋势。中国东北地区的EASSM持续等时线明显凹向华北地区,中国东北地区的持续时间与朝鲜半岛东北部、日本海和日本的本州岛中北部相近,大都为2~4候。中国的长江流域及以南地区EASSM持续时间大都在10候以上,25°N以南的大陆和海域平均持续的等时线近于呈经向分布。118°E以东的海域等时线梯度很大,等时线从18候迅速衰减到132°E的6候。中国大陆的广西和广东省持续20候以上,为东亚地区夏季副热带季风影响时间最长的地区。

3.3 平均撤退的时空分布特征

从利用 NCEP 资料计算的东亚大陆、中国近海和西北太平洋地区 1948—2002 年 EASSM 撤退的平均日期(候)的空间分布(图略)可见，EASSM 撤退的等时线和建立的等时线分布大体相反，25°N 以北，等时线从中国东北地区凹向中国黄海和东海，为撤退等时线主要撤退方向，表明中国东北地区和中国近海地域先于同纬度中国内蒙古东部、华北、华中和日本岛地区撤退，沿 35°N 自西向东横跨 54～49 候等时线，撤退时间可相差 6 候，中国东北地区和中太平洋地区 EASSM 撤退最早，为 44～46 候(8 月上旬至中旬)，中国华北地区平均日期为 8 月末(48 候)，长江流域撤退日期为 50～52 候(9 月上旬至中旬)，日本的大部地区和朝鲜半岛撤退日期为 46～48 候(8 月中旬至 8 月末)，中国的广西和广东地区为 54 候，撤退的时间最晚。

4 东亚夏季副热带季风建立的不同地带的划分

从利用 NCEP 和 ECMWF 资料计算的东亚大陆和西北太平洋地区 1948—2002 年 EASSM 建立的范围图中可见(图略)，西边界为 102.5°E，与青藏高原的东边界基本一致，南边界从 20°N 开始，北边界为 50°N，超过 50°N 计算已没有太大的意义，东边界到 147.5°E。东亚大陆—西北太平洋地区夏季副热带季风建立的概率为 25%的等值线，西界点为(37°N，102.5°E)，东界点为(43°N，147.5°E)，近于呈向北凸的折线。概率为 35%的等值线近于与 25%等值线平行，稍向南移大约一个纬距，称小于 35%概率等值线的地区为 EA-WNP 的 EASSM 影响的边缘地带，其西边界由黑龙江省北部 48°N 向西南方向沿中蒙边境止于青藏高原的东北边界[地形高度 3000 m，(36°N，103°E)附近]，其北边界始于黑龙江省 48°N 附近向东南经由黑龙江省南部和俄罗斯滨海省，穿过日本海止于日本北海道南部(42°N，145°E 附近)。概率为 65%的等值线，西界点为(34°N，104°E)，北界点为(43°N，117°E)，东界点为(35°N，147.5°E)，近于成向北凸的折线。将概率≥35%和≤65%的地区称之为 EA-WNP 的 EASSM 中度影响的地区，或称拉锯式影响地区。>65%概率等值线的地区称之为东亚夏季副热带季风(EASSM)主要影响地区，即 EA-WNP 的 EASSM 主要活动区。从利用 ECMWF 资料计算的东亚大陆和西北太平洋地区 EASSM 建立的范围可见，与利用 NCEP 资料计算的东亚大陆和西北太平洋地区 EASSM 建立的范围在 120°E 以东结果大体相近，但是 EASSM 在 120°E 以西中国大陆的主要活动区比利用 NCEP 资料计算的东亚大陆和西北太平洋地区 EASSM 建立的范围要偏东一些，特别是青藏高原对 EASSM 的边界效应比利用 NCEP 资料计算的东亚大陆和西北太平洋地区 EASSM 建立的范围反映的更为偏东。为了便于同期比较，还用 NCEP 1958—2001 年资料作了“东亚大陆和西北太平洋 1958—2001 年 EASSM 建立范围”(与利用 NCEP 资料计算的东亚大陆和西北太平洋地区 EASSM 建立的范围基本类同)。

5 EASSM 西部和北部边缘地带的平均位置年代变化

由 EA-WNP 20 世纪 50—90 年代及其各个年代 EASSM 的西部和北部边缘地带平均变化可见，90 年代西部边缘地带比多年平均偏东 1 个多经度，而北边缘位置却偏南 3～5 个纬度，表明 90 年代 EA-WNP 的 EASSM 北部边缘地带比西部边缘地带变化大。

6 结论和讨论

(1)本文按作者曾定义的东亚夏季风建立的标准，以某一经度和纬度的 850 hPa 候平均图上 θ_{se} 达到 336 K 和南风≥4 m/s 为阈值，利用 NCEP 资料给出东亚—西北太平洋地区的东亚夏季副热带季风(EASSM)的定义，计算了 1948—2002 年 EA-WNP 的 EASSM 建立的平均时间(候)等时线空间分布，表明 EASSM 自南海北部(20°N)建立以后，在 35°N 以南主要沿 110°E 稍偏西的方向向北推进，广西要稍早于广东，长江中游地区要早于长江中下游地区，在 35°N 以北，华北西南部稍早于华北东部，最后止于中国东北地区(于第 42 候)。经分析中国 618 个测站 1961—2005 年 EASSM 建立前后的候降水量正差值区与上述结论相匹配；除了继续向北推进的方向外，还有向西稍偏北的分支，107.5°E 以西的季风建立的等时线十分密集，而以东的等时线梯度很小，这与 107.5°E 以西青藏高原的地形高度高和阻挡有关。在 20°—25°N 还有明显向东的分支，这与西太平洋副热带高压从 5 月下旬逐渐向东撤和北进有关。总的分布表明，EASSM 先在中国大陆建立，而同纬度的中国近海和西北太平洋地区的夏季副热带季风后建立。与梁潇云等(2005)关于青藏高原对亚洲夏季风爆发位置及强度影响结果相近。

(2)撤退的日期等时线则近于与建立的日期等时线的相反分布，沿 120°E 的方向凸向南。表明中国东北地区和近海地区先撤退。

(3) 以 EA-WNP 的 EASSM 建立的概率等值线的空间分布，设阈值确定 EA-WNP 的 EASSM 活动范围。尝试定义＜35％概率等值线的地区为其影响的边缘地带，其西边界由黑龙江省北部 48°N 附近向西南方向沿中蒙边境止于青藏高原的东北边界；北边界始于黑龙江省北部 48°N 附近向东南经由黑龙江省南部和俄罗斯滨海省穿过日本海，止于日本北海道南部(42°N，145°E)，建立概率≥35％和≤65％地区为 EA-WNP 的 EASSM 中度影响地区，或称拉锯式影响地区，主要位于东北地区中部，建立时间大都在第 40 候左右(7 月中旬)，影响的候大都为 4～5 候，撤退时间大都在第 45 候(8 月中旬)以前；建立概率＞65％的地区为 EA-WNP 的 EASSM 主要活动地区，建立的候大都在第 38 候(7 月上旬)，影响时间≥6 候，撤退时间大都在第 46 候(8 月中旬末)以后。EA-WNP 的 EASSM 西部和北部边缘地区年代平均位置表明，北部边缘位置 20 世纪 90 年代比多年平均偏南 3～5 个纬度，而西部边缘地区只比多年平均偏东 1 个纬度，与廉毅等(2005)所分析的中国东北地区中南部 1961—2000 年年降水量线性负倾向，而西北地区中部为正倾向结果基本一致。

(4)用 ECMWF 的再分析资料计算的结果表明，与 NCEP 的结果在 120°E 以东分布大体相近，而以西差异较大，利用 ECMWF 计算的 EASSM 在中国大陆的主要活动区要比 NCEP 计算的范围偏小和偏东，可能是 ECMWF 的客观分析对青藏高原的边界效应反映较为强烈，认为 NCEP 计算结果比 ECMWF 更接近东亚夏季风活动实际。

(5)本文提出了以观测计算每个经度和纬度夏季副热带季风建立时间(候)，给出建立概率分布的阈值，来确定 EA-WNP 的 EASSM 活动的西和北边界带的一种方法，有可能为进一步研究夏季风的季内振荡提供较客观的参照标准。

参考文献

陈桦，丁一汇，何金海，2006. 亚洲夏季风指数的重新评估与季风的长期变化[J]. 气象学报，64(6)：770-789.

陈隆勋，1984. 东亚季风系统的结构及其中期变化[J]. 海洋学报(6)：744-785.

丁一汇，李崇银，何金海，等，2004. 南海季风试验与东亚夏季风[J]. 气象学报，62(5)：561-586.

高辉，等，2001. 关于确定南海夏季风建立日期的讨论[M]//何金海. 南海夏季风建立日期的确定与季风指数. 北京：气象出版社：1-41.

金祖辉，陈隆勋，1982. 夏季东亚季风环流系统的中期变化及印度季风系统的相互关系[C]//全国热带夏季风学术会议文集编写组. 全国热带夏季风学术会议文集. 昆明：云南人民出版社：204-215.

廉毅，沈柏竹，高枞亭，2003. 东亚夏季风在中国东北区建立的标准、日期及主要特征分析[J]. 气象学报，61(5)：548-558.

廉毅，沈柏竹，高枞亭，等，2004. 关于确定东亚夏季风强度指数的探讨[J]. 气象学报，62(6)：782-789.

廉毅，沈柏竹，高枞亭，等，2005. 中国气候过渡带干旱化发展趋势与东亚夏季风、极涡活动相关研究[J]. 气象学报，63(5)：740-749.

梁潇云，刘屹岷，吴国雄，2005. 青藏高原对亚洲夏季风爆发位置及强度的影响[J]. 气象学报，63(5)：799-805.

汤绪，钱维宏，梁萍，2005. 东亚夏季风边缘带的气候特征[J]. 高原气象，25(3)：375-381.

王慧，丁一汇，何金海，2005. 西北太平洋夏季风的气候学研究[J]. 气象学报，63(4)：418-430.

徐袁，钱维宏，2003. 东亚季风边缘活动带研究综述[J]. 地理学报，58(Z1)：138-146.

杨建平，丁永建，陈仁升，等，2003. 近50年中国干湿气候界线波动及其成因初探[J]. 气象学报，61(3)：364-373.

杨建平，丁永建，陈仁升，等，2005. 亚洲季风与中国干湿、农牧气候界线之关系[J]. 地理科学，25(4)：441-447.

赵哈林，赵学勇，张铜会，等，2002. 北方农牧交错带的地理界定及其生态问题[J]. 地球科学进展，17(5)：739-747.

KRISHNAMURTI T N, BHALME H N, 1976. Oscillation of a monsoon system. Part I: Observational aspects[J]. J Atmos Sci, 33(10): 1937-1954.

LAU K M, YANG S, 1997. Climatology and interannual variability of the southeast Asian summer monsoon[J]. Adv Atmos Sci, 14(2): 141-162.

LAU K M, KIM K M, YANG S, 2000. Dynamical and boundary forcing characteristics of regional components of the Asian summer monsoon[J]. J Climate, 13(14): 2461-2482.

MURAKAMI M, MATSUMOTO J, 1994. Summer monsoon over the Asian continent and Western North Pacific[J]. J Meteor Soc Japan, 72(5): 719-745.

RAMAGE C S, 1971. Monsoon Meteorology[M]. New York: Academic Press: 296.

SUN L, SHEN B Z, GAO Z T, et al, 2007. The impacts of moisture transport of East Asian monsoon on summer precipitation in Northeast China[J]. Adv Atmos Sci, 24(4): 606-618.

TAO S Y, CHEN L X, 1987. A Review of Research on the East Asian Summer Monsoon in China[M]// CHANG C P, KRISHNAMURTI T N. Monsoon Meteorology. Oxford: Oxford University Press: 60-92.

WANG B, LI T, DING Y H, et al, 2004. East Asian-Western North Pacific monsoon a distinctive component of the Asian-Australian monsoon system[C]// The 3rd International Workshop on Monsoon (IWM-Ⅲ). The Reviews on the Current State of Research/Forecasting of Monsoon. Hangzhou: 27-68.

ZHANG R H, AKIMASA S, MASAHIDO K, 1996. Impact of El Niño on the East Asian monsoon: A diagnostic study of the "86/87" and "91/92" events[J]. J Meteor Soc Japan, 74(1): 49-62.

ZHU Q G, HE J H, WANG P X, 1986. A study of circulation differences between East Asian and Indian summer monsoons with their interaction[J]. Adv Atmos Sci, 3(4): 466-477.

东亚夏季副热带季风活动范围研究进展*

李尚锋[1,2] 姚志平[3]

(1 中高纬度环流系统与东亚季风研究开放实验室，长春 130062；2 吉林省气象科学研究所，长春 130062；
3 吉林省气象信息技术保障中心，长春 130062)

摘　要：综述了东亚夏季副热带季风活动范围研究成果，分析了东亚夏季副热带季风活动范围变化的可能原因，对今后如何进一步开展这方面研究提出了一些建议，这对于东亚夏季副热带季风的进退、季风边缘带的气候变化、中国北方干旱化和生态环境的演变等科学问题的研究都有借鉴作用。季风活动的范围是一个动态区域，东亚夏季副热带季风活动的边界可用其活动所能到达地区的概率来划分。东亚夏季副热带季风活动的范围和边界的变化不仅与热带季风向北暴发的强弱有关，还与极地和中高纬度的环流系统，以及青藏高原对季风的动力、热力作用都有密切的关系，今后应加强这方面的研究工作，尤其要加强定量描述东亚夏季副热带季风变化的研究工作，这将有助于进一步科学界定东亚夏季副热带季风活动范围。
关键词：东亚夏季副热带季风；季风指数；季风边界；半干旱气候

1 引言

季风通常是指近地面冬、夏盛行风向变化近于相反(风向偏转大于 120°)且气候特征迥异的现象。经典季风学说认为由海陆热力差异以及其他加热不同所造成的冬、夏对流层低层盛行风向的显著转换是季风形成的根本原因。Ramage(1971)给季风做了如下定义：①1 月和 7 月盛行风向的变化有 120°；②1 月与 7 月盛行风向的频率超过 40%；③至少在 1 月和 7 月中有一个月的平均合成风超过 3 m/s；④在 5°经纬度矩形内，1 月和 7 月这两个月份中每个月气旋与反气旋的交替出现至少每两年一次。Webster(1987)提出一个更普遍的季风定义：冬、夏季风的季节性反转和干湿期的季节性交替出现，一般来说主要是根据地面盛行风向的冬、夏季转换确定季风区。曾庆存等(1998)和王安宇等(1990)指出，凡是冬、夏盛行风(风向和风速)差异显著(即季节变化大)的地区均应视为季风区。经典的季风区具有特定的含义(张家诚，1983)，可称之为狭义的季风区，而根据季节变化强烈与否定义的季风区则可称之为广义的季风区(曾庆存 等，1998)。

最近几十年来，随着季风研究的深入，人们对亚洲季风的认识取得了重大进展。有研究指出，东亚地区存在一个独立于印度季风环流系统(IMCS)的东亚季风环流系统(EAMCS)(金祖辉 等，1983；陈隆勋，1984；Tao et al.，1987；Krishnamurti et al.，1976)，EAMCS 包含了南半球澳洲季风、南海—西太平洋热带季风、中国大陆副热带季风和高纬度环流的相互作用。还有研究认为，东亚季风又可分为南海—西太平洋热带夏季风和中国大陆—日本的副热带夏季风(朱乾根 等，1985)；南海季风为热带季风性质，中国大陆夏季风为副热带季风(SM)(Zhu et al.，1986)。Murakami 等(1994)提出西北太平洋季风(WNPSM)主要指超过菲律宾海和它附近的海上季风。有学者认为东亚夏季风(EASM)是指东亚地区(20°—40°N，100°—140°E)之间的副热带季风(Zhang et al.，1996)。Lau 等(1997)给出了印度夏季风(ISM)和 EASM 以及

* 本文原载于《地理科学》，2007，27(Z1)：111-117

WNPSM 建立的平均日期，其北界可达中国大陆 40°N 和日本的 40°N 附近；提出亚洲夏季风可分成南亚季风(SAM)和东南亚季风(EAM)两个子系统(Lau et al.,2000)。由此可见，亚洲不同区域季风性质的称谓与其活动范围休戚相关。

近几年来，东亚夏季副热带季风(EASSM)活动范围的西界和北界的研究不仅引起大气科学而且还引起相邻交叉学科学者的关注，因为东亚季风边缘研究有利于加深认识季风以及中国生态系统在过渡带的特性(徐袁 等,2003;汤绪 等,2006;杨建平 等,2003)，季风环流的强弱变化控制着气候界线空间摆动的范围和方向(杨建平 等,2005)，以及涉及中国北方农牧交错带的地理界定及其生态问题(赵哈林 等,2002)。本文综述了 EASSM 影响范围的研究进展，对今后如何进一步开展这方面研究提出一些建议，这对于季风进退、季风边缘带的气候变化、中国北方干旱化和生态环境演变等多学科交叉科学问题的研究都有一定的促进作用。

2　东亚副热带季风的性质和称谓的由来

南亚(印度)夏季风与 EASM 是两个相对独立的亚洲季风子系统。前者主要盛行于印度洋、阿拉伯海、孟加拉湾和印度次大陆，后者主要盛行于东亚、东南亚东部及西太平洋。由于印度季风属于热带性质的季风，特别是印度夏季风，它几乎是一个单纯的热带环流系统；而 EASM 除了包含热带季风性质外，还包含副热带季风性质(Tao et al.,1987)。

朱乾根等(2000)用大气环流的正斜压分解的方法，并从对流层低层流场的正斜压流型角度出发对全球季风进行了分析后指出，亚洲热带地区(包括南亚地区、南海—西太平洋地区)的季风主要是由斜压分量流场的季节变化所产生，属于斜压流型季风区；东亚副热带季风区实际风向的季节反转是正斜压流场共同作用的结果，属于正斜压混合型季风区。周兵等(2003)在分析大尺度季风环流与对流活动存在内在联系的基础上，针对东亚副热带季风关键区——长江中下游地区降水主要集中在夏季的 6—7 月，考虑高低空越赤道气流在东亚季风环流系统中的重要性，以及季风区水汽经向输送等事实，提出能否采用经向风垂直切变来构造东亚区域季风指数，用于反映该地区季风环流特征。

陈际龙等(2006)的研究表明，由于受西太平洋副热带高压和中纬度西风急流的共同影响，东亚季风区纬向气流的高低层配置相对复杂；用对流层低层和对流层高层风场的垂直切变的不同将东亚季风区以 25°N 为界分为两个区域，认为 25°N 以南地区对应南海—西太平洋热带夏季风；25°N 以北地区对应中国大陆—日本副热带夏季风。

Tao 等(1987)指出，亚洲季风首先在南海地区爆发，一般平均在 5 月中旬，比印度季风要早一个月。在中国，往往把发生在南海地区的夏季风称为南海季风，也属于热带季风性质。南海季风的爆发标志着东亚季风的来临和中国华南前汛期即将开始，因而它具有重要的预报意义(Ding,1992;Ding et al.,1996)。亚洲季风在南海爆发之后，将经过两次阶段性北进和两次停滞，最后在 7 月中旬夏季风可以到达华北和东北一带，并在这两个地区停滞一段时间，于 8 月中旬又南撤。一般在 5 月和 6 月上旬，由于西太平洋副热带高压偏南，因此，季风达到华南和江南地区，这正是华南和南岭一带的前汛期；到了 6 月上中旬，西太平洋副热带高压的脊线北移到 20°N 附近，季风雨带北跳到长江、淮河流域，这时长江和淮河流域的梅雨、日本梅雨(Baiu)和韩国梅雨(Changma)就开始了；雨带在这些地区停滞一段以后，到了 7 月中旬，由于西太平洋副热带高压脊线又北移到 30°N，使得季风雨带又北跳到华北、东北一带，这时中国北方雨季开始，而江淮流域梅雨结束；到了 8 月中旬左右，此雨带又迅速南撤到长江流域(孙力 等,2003)。

3　东亚夏季副热带季风影响范围的边界研究

EASSM 的活动不仅受到西太平洋副热带高压和西风急流的作用，还受到高纬度北极涛动(Arctic Oscillation,AO)的作用(主要是冬春季 AO 的活动，对紧接着的夏季风的滞后作用)。龚道溢(2003)指出，春季北极涛动强时，随后的夏季急流位置通常偏北，雨带位置也北移，从而造成梅雨——Changma—Baiu带降水量减少，反之亦然；并进一步指出，较强的AO异常对降水的影响更明显，而较弱的AO与降水异常的对应关系并不显著。对于中高纬度地区来说，其北面受到以西风带系统为主导的影响（西风槽、大陆阻高和冷涡），南面受到北进的季风的影响，要判断该区域有多大的范围可以受到北上季风的影响，或者说季风的北边界在哪个纬度，到目前为止还没有一个共同认可的结论。最近，孙力等（2003）和廉毅等（2003）研究表明，季风不仅影响中国东北地区而且有的年份可以到达50°N附近。因为东北地区的降水主要集中在夏季（6—8月），夏季降水主要分为两个时段，入夏初期的冷涡降水（6月至7月上旬）和后期的季风降水（廉毅 等，2003）。另外，中国东北地区夏季降水量距平分布与中国夏季降水量距平自北向南的“－＋－＋”波列分布休戚相关（黄荣辉 等，2005）。所以，确定EASSM影响范围，不仅对中国东北地区的夏季降水量而且对全国范围的夏季降水量的多少都有举足轻重的作用。EASSM北边界对于研究中国东北地区夏季降水和北方荒漠化都有积极的意义，而它的西（北）边界一般认为是位于中国农牧交错带上（汤绪 等，2006），该边界的探讨对于研究东亚季风边缘后退所造成的该区域生态恶化、土地荒漠化以及气候干旱化方面均有着重要的作用。

EASSM系统是一个很复杂的体系，影响它活动的因素很多，其不仅受到印度季风、西太平洋副热带高压、中纬度的大气扰动影响，还受到西太平洋暖池、赤道太平洋区域的ENSO循环、青藏高原以及极冰等众多因素的影响。此外，夏季副热带对流层中高层的西风急流也对季风雨带的推进有重要影响。研究表明，夏季急流的位置偏南，则雨带偏南，长江流域降水偏多；急流的位置偏北，则雨带偏北，长江流域降水偏少，华北降水偏多（沈柏竹 等，2004）。还有，极涡的强弱都对季风北上产生影响。这些众多因子的作用再加上季风的年际波动，使得研究季风北上的问题变得比较困难。在有的年份季风对某些区域有影响，而在有的年份对该区域并没有影响。这些由于季风的年际波动而形成的具有很强的年际变率的区域，称为季风边缘活动区（沈柏竹 等，2005；竺可桢，1934）。中国地处东亚季风区，EASM带来的季风雨对中国东部大陆意义重大（Tao et al.，1987），中国北方很多地区属于干旱、半干旱区，位于EASM作用的北部边缘地带，受EASM影响相对较弱，而夏季风北边缘带的活动对该地区的降水、沙尘暴等重要天气气候现象的影响也是很重要的。

季风的边缘区是一个带状区域，这是一个多年气候平均的概念，其范围在年际上和年代际变化上都有很大的不确定性，该区域的确定与季风指数的定义密切相关（Ramage，1971）。如果指数定义的指标不同，由此表示的夏季风活动的北界也存在较大差异。降水量多寡是影响季风边缘区环境发生变化的主要因子，但是，由于年与年之间受夏季风的影响不尽相同，从而降水等气候指标变化就比较大。中国北方地区的降水量本来就少，其雨量变化对植被的影响就更强烈，既是半干旱带，也是气候和生态系统的过渡地带。已有的研究表明，很多的自然灾害，特别是旱涝的发生与这条季风北边缘带的位置变化有关（黄荣辉 等，2002）。

张兰生等（1993）将平均年降水量250～450 mm等值线之间的地区称为北方季风尾闾区，并将年降水量250 mm等值线的地区称为北方季风闾区的北界。史正涛（1996）将年降水量等值线200～450 mm（或500 mm）的区域称为季风边缘区的范围。富元海等（2007）在研究中国北方季风边缘区边界位置的年代际变化时，将夏季风边缘带分为东北、华北和西北东部3个区，其中，将东北区的范围取为黑龙江、吉林、辽宁和内蒙古的东北部，而西北东部则包括了甘肃、陕西和宁夏部分地区。

汤绪等（2007）将每年7 d滑动平均的5 mm日降水量等雨量线推进所能达到的位置定义为EASM的最北边缘线，并指出位于黄河上游到华北北部的最北降水线密集带为气候上的夏季风最北边缘线。胡豪然等（2007）综合降水、风场以及假相当位温场，定义了表征EASM北边缘的新标准，研究了110°—120°E这一区域的纬向平均夏季风最北边缘的位置，利用1961—2001年共41 a的资料，分析发现在这41 a中，夏季风在这一区域最北位置低于40°N的年份只有9 a且都集中在1979年以后。

WNPSM和ISM都是典型的热带季风，在WNPSM、EASM和ISM之间有一个较宽区域的缓冲带即中印半岛、云贵高原这一带。由于青藏高原和云贵高原的地形阻挡作用，在EASM指数的定义时，有很多学者就以100°—110°E的范围作为所定义区域的西边界，把青藏高原和云贵高原相连的这一带作为EASM的边界即西边界。以降水量作为指标来研究东亚季风进退的手段有一定的局限性，中国地形复杂，在研究季风降水的过程中，往往很难准确判断在季风影响期内，有多少降水量是由季风造成的。因此从大尺度着眼，以大气环流角度来定义季风边缘应该会有很好的结果。廉毅等(2003)定义的EASM在中国东北地区建立的标准中所给出的1980—2000年的建立、持续和撤退的时间年历表表明，EASM北界超过45°N的有9 a。Sun等(2007)以水汽输送和收支为主要条件来研究EASM对中国东北地区影响的时间和范围时指出，在110°—115°E范围内，东亚季风影响的时间为7月第1候至8月第3候(7月1日至8月15日)，影响的最北界为42°N；在115°—120°E范围内，东亚季风影响的时间为7月第3候至8月第4候(7月25—31日)，7月第6候(7月25—31日)南风水汽输送最强，影响的最北界为46°N。

廉毅等(2007)用850 hPa候平均图上假相当位温(θ_{se})达到336 K和南风≥4 m/s为阈值来描述EASSM的活动。以东亚—西北太平洋(EA-WNP)的EASSM建立的概率等值线的空间分布，设阈值确定EA-WNP的EASSM活动范围。定义概率＜35％等值线的地区为其影响的边缘地带，其西边界由黑龙江省北部48°N附近向西南方向沿中蒙边境止于青藏高原的东北边界；北边界始于黑龙江省北部48°N附近向东南经由黑龙江省南部和俄罗斯滨海省穿过日本海，止于日本北海道南部(42°N，145°E)，建立概率≥35％和≤65％地区为EA-WNP的EASSM的中度影响地区，或称拉锯式影响地区，主要位于东北地区中部；建立影响概率＞65％的地区为EA-WNP的EASSM的主要影响地区。

杨建平等(2005)研究得出中国半干旱区、农牧交错带的地理位置和空间分布，可以发现廉毅等(2007)划分的季风边缘带的范围基本上同杨建平等(2005)给出的中国的半干旱区、农牧交错带的位置一致。

汤绪等(2006)通过对多年平均降水量场、水汽场、风场、水汽输送场的研究发现，东亚季风边缘带的大致位置沿西南—东北走向，自西向东经过黄河上游—河套—华北北部—东北，这与生态系统中植被的分区也比较一致。有的学者将该边界进一步扩大，徐袁等(2003)的研究指出，东亚季风边缘带是指东亚季风年际波动在内陆一侧所影响的区域，大致与中国北方农牧交错带一致。王静爱等(1999)的研究指出，中国北方农牧交错带核心区域大致为内蒙古高原东南边缘和黄土高原北部的半干旱气候区域。中国北方农牧交错带与非洲的萨哈尔地带都处于季风边缘地带，这两个地区主要受季风强弱及其南北摆动的控制，除温度条件外，在千年尺度和世纪尺度的波动变化规律上基本一致，且都在波动中趋于干旱化(史培军 等，2002)。现在对中国北方干旱化问题已经有了不少的研究，它与东亚季风边缘的研究既有联系又有区别。可以说中国北方干旱化的发展是环境系统整体的异常变化，很可能是全球变化，特别是全球增暖，对区域生存环境影响的综合表现，其中包括温度升高、蒸发加大、降水量减少、河网干枯、湖泊萎缩、雪线抬升、土地荒漠化和生态系统退化等。因此，北方干旱化是一个典型的交叉科学问题(符淙斌 等，2002)。

东亚季风边缘研究着重于大尺度的气候研究，且其重点放在了季风强弱变化、影响范围的变化等方面，其研究结果可以作为北方干旱化问题的背景之一，是引起北方干旱化的原因之一。中国东北地区位于EASM的最北端，年降水量特别是夏季降水量的多少，在很大程度上决定于夏季风活动的强弱(Qian et al.，2002)。由于受到东亚季风年际变化大的影响，导致该地区夏季洪涝和干旱等气候灾害异常活跃，特别是近些年来旱涝发生的频率明显增加。这些频繁出现的旱涝灾害给当地的国民经济特别是农业生产造成了很大的损失。虽然人们从不同的角度曾对东北旱涝问题做过一些研究，但是对东亚季风与东北地区夏季降水量异常的关系还缺乏认识，这可能与东亚季风系统本身所具有的复杂性和东北地区所处的地理位置等因素有关。正如上面所说的那样，EASM可达的北界与季风的强弱有关，因此准确地来说，EASM北界是一个区域的范围。确定EASM的北边界对于进一步了解其对东北地区降水异常的影响以及该地区夏季旱涝的短期气候预测都具有一定的意义。

EASM在中国大陆的西边界，一方面决定大地形的高度——大兴安岭—太行山系的作用，这些山脉的走向大致为东北—西南走向，这些山脉的高度在1000～2000 m，除了气温随高度上升而下降的效应之外，从东南面来的气流在沿山脉爬升时由于降水造成的水汽损失对降水也有显著的影响(Ye et al.，1979)。另一方面是青藏高原的阻挡，青藏高原对季风环流的影响，既有热力作用又有地形动力作用。边界层受到地面的影响最大，对于其上空的大气来说，青藏高原在夏季是个热源，冬季是个冷源(Li et al.，1996)，与周围的大气有比较大的东西向和南北向的温度梯度，除了北美西部山地所造成的比较弱的类似效应之外，其他地方都没有该特点(Webster et al.，1998；卓嘎 等，2002)，因此，作为世界上最大最高的高原，青藏高原在季风的形成中起到了重要的作用，对于季风边缘地区来说就更是如此。深厚的高原边界特征将使河套地区与黄河流域的夏季云量及降水有所减少(朱乾根 等，1989)。青藏高原热力作用的变化与季风边缘的关系是值得研究的一个问题。还有一方面是取决于南海热带季风和越过菲律宾海共同向中国大陆西北地区扩张影响的强度，由于每年季风环流系统的强度变化不尽相同，其到达的边界位置有异，所以，EA-WNPM 的西边界是一个动态变化较大的半干旱区域。一般来说，季风定义决定着季风影响的大致区域，而有关季风边界的确定则是一件比较复杂且细微的研究工作。

目前对 EASSM 边界的确定大体可以分为 3 种：一种是动力边界，以风的变化为主；一种是热力边界，以气团的性质，如用 θ_{se} 的某一阈值作为气团属性外包线抵达的范围，来表明季风可达到的范围；一种大都以降水量作为指标对雨季的开始时间给予定义。

4 东亚夏季副热带季风指数的相关研究

季风指数是通过数学表达式定量确定季风强度的一种无量纲化的标量，并且对于研究季风的年际变化有很重要的意义，朱乾根等(1989)提出用等露点线来判断季风影响的观点，他们使用 850 hPa 的 14 ℃等露点线作为标准。季风是季节性海陆热力差异导致的大气环流现象，因此有些学者单从热力学方面来定义夏季风指数。郭其蕴(1983)认为，东亚地区纬向海陆热力差异是季风在该地区产生的主要原因，他利用 10°—50°N 各纬度上海平面气压资料，计算出 110°E 气压减去 160°E 气压的差(月平均)值，然后对差值 ≤−5 hPa 的各种纬度值累加起来取绝对值然后标准化，首次提出一种夏季风指数(SMI)。施能(1996)和施能等(1996)为了消除不同格点均方差不均匀带来的影响，对郭其蕴(1983)定义的指数做了部分改进，取 20°—50°N 范围内 5 个纬度的 110°、160°E 的标准化的海平面气压差之和的序列的标准化值作为东亚季风指数。陈隆勋等(1999)和孙秀荣等(2002)从季风形成的本质出发，定义了一个东亚海陆热力差指数。孙秀荣等(2001)用东亚季风区(27°—35°N，105°E 以东的大陆)范围内的地表温度(T_{EC})和副热带西太平洋(15°—30°N，120°—150°E)海表温度(SST_{STNWP})之差表示东西向海陆热力差异，用华南地区(27°N 以南、105°E 以东的大陆)的地表温度(T_{SC})和南海(5°—18°N，105°—120°E)海表温度(SST_{SCS})之差表示南北向海陆热力差，从而定义出一个东亚海陆热力差指数。

为了表征副热带夏季风进退活动，不少学者都着眼于寻找季风在副热带活动的指标，如汤明敏等(1983)使用 1000 hPa 上 344～348 K 等 θ_{se} 线，或 850 hPa 上 336～340 K 等 θ_{se} 线为副热带夏季风进退指标；施尚文等(1982)用 850 hPa 上 332 K 等 θ_{se} 线作为北方地区夏季风影响期的临界值等。也有学者利用高、低空风场变化为指标来定义季风指数。祝丛文等(2000)将东西向海平面气压差与低纬度高、低层纬向风切变相结合，定义了一个 EASM 指数，较好地反映了东亚冬、夏季风变化；张庆云等(2003)将东亚热带季风槽(10°—20°N，100°—150°E)与东亚副热带地区(25°—35°N，100°—150°E)6—8 月平均的 850 hPa 风场的纬向风距平差，定义为东亚季风指数。周兵等(2003)将(10°—25°N，100°—125°E)区域的平均经向风垂直切变 I_{es} 定义为东亚副热带季风指数。Lau 等(2000)利用不同地区的平均纬向风之差定义了一个东亚季风指数 RM2，即 200 hPa 副热带激流两侧平均纬向风切变。廉毅等(2003)依据 850 hPa 候平均南风风速和 θ_{se} 的阈值定义了 EA-WNP 的 EASSM 建立和撤退时间(候)的标准，计算了 102.5°—147.5°E 每隔 5 个经度，20°—50°N 每隔 5 个纬度的东亚和西北太平洋的 EASSM 建立、持续和撤退时间(候)，为 EA-WNP 的 EASSM 活动范围确定了观测计算的一个标准。

中纬度与中高纬度地区影响季风进退的主要系统不同，如果只用一个指数来描述东亚的季风进退难免有些困难，如何定量准确地描述季风的活动是今后要加强研究的一个课题。

5 结语

EASSM 活动范围和强度对中国大陆的中东部、朝鲜半岛、日本和西北太平洋地区的降水量有重要影响。因此，研究 EASSM 活动范围的年际和年代际变化，对于夏季风的进退、季风边缘带的气候变化、中国北方干旱化和生态环境的演变等科学问题的研究都有重要的借鉴作用。

EASSM 活动的范围是一个动态区域，因此，东亚夏季副热带季风活动的边界用概率来划分可能更切合实际一些，决定 EASSM 活动的范围和可能的边界，不仅与当年的热带季风（南海季风）向北爆发的强弱有关，还与极地和中高纬度环流系统有关。已有研究表明，AO/NAM、NAO、NPO 都与 EASSM 活动相关显著，同时，青藏高原对其的动力和热力作用也不可忽视，今后要加强这方面的研究工作。

进一步加强定量描述 EASSM 强度变化的研究，将有助于进一步科学地界定 EASSM 活动范围。只用一个季风指数很难描述 EASSM 在幅员辽阔东亚地区的进退，季风指数的确定仍是今后有关季风气候和动力学研究的前沿和热点课题。

参考文献

陈际龙，黄荣辉，2006. 亚洲季风各子系统气候学特征的异同研究 I. 夏季风流场结构[J]. 大气科学，30(6)：1091-1102.

陈隆勋，1984. 东亚季风系统的结构及其中期变化[J]. 海洋学报，1(6)：744-785.

陈隆勋，刘洪庆，王文，等，1999. 南海及其邻近地区夏季风爆发的特征及其机制的初步研究[J]. 气象学报，57(1)：16-29.

符淙斌，安芷生，2002. 我国北方干旱化研究——面向国家需求的全球变化科学问题[J]. 地学前缘，9(2)：271-275.

富元海，刘宜飞，2007. 中国北方季风边缘区边界位置的年代际变化[J]. 南京气象学报，30(1)：94-100.

龚道溢，2003. 北极涛动对东亚夏季降水的预测意义[J]. 气象，29(6)：3-6.

郭其蕴，1983. 东亚夏季风强度指数及其变化的分析[J]. 地理学报，38(3)：207-217.

胡豪然，钱维宏，2007. 东亚夏季风北边缘的确认[J]. 自然科学进展，17(1)：57-65.

黄荣辉，周连童，2002. 我国重大气候灾害特征、形成机理和预测研究[J]. 自然灾害学报，11(1)：1-9.

黄荣辉，顾雷，许予红，等，2005. 东亚夏季风爆发和北进的年际变化特征及其与热带西太平洋热状态的关系[J]. 大气科学，2(1)：20-36.

金祖辉，陈隆勋，1983. 夏季东亚季风环流系统的中期变化及印度季风系统的相互关系[C]//全国热带夏季风学术会议文集编写组. 全国热带夏季风学术会议文集. 昆明：云南人民出版社：204-215.

廉毅，沈柏竹，高枞亭，等，2003. 东亚夏季风在中国东北区建立的标准、日期及其主要特征分析[J]. 气象学报，61 (5)：548-558.

廉毅，沈柏竹，高枞亭，等，2007. 确定东亚—西北太平洋地区夏季副热带季风建立和活动范围的一种方法[J]. 气象学报，65 (4)：503-510.

沈柏竹，廉毅，2004. 青藏高原积雪与东北地区夏季冷涡雨季降水关系的初探[J]. 吉林大学学报（地球科学版），34 (7)：111-118.

沈柏竹，廉毅，2005. 东亚夏季风与同期及前期海陆热状况异常的联系[J]. 气象学报，63(Z1)：191-199.

施能，1996. 近 40 年东亚冬季风强度多时间尺度变化特征及其与气候的关系[J]. 应用气象学报，7(2)：175-182.

施能，朱乾根，官彬贵，1996. 近四十年东亚夏季风及中国夏季大尺度天气气候异常[J]. 大气科学，20(5)：573-583.

施尚文，巢俊民，1982. 中国北方地区的夏季风[C]//全国热带夏季风学术会议文集编写组. 全国热带夏季风学术会议文集. 昆明：云南人民出版社：72-85.

史培军，哈斯，2002. 中国北方农牧业交错带与非洲萨哈尔地带全新世环境变迁的比较研究[J]. 地学前缘，9(1)：121-128.

史正涛，1996. 中国季风边缘带自然灾害的区域特征[J]. 干旱区资源与环境，10(4)：1-7.

孙力，安刚，唐晓玲，2003. 东亚区域夏季 850 hPa 南风异常与东北旱涝的关系[J]. 大气科学，3(11)：425-434.

孙秀荣，陈隆勋，何金海，2001. 东亚海陆热力差指数与东亚夏季风强度关系讨论[C]//何金海，丁一汇，高辉，等. 南海夏季风建立日期的确定与季风指数. 北京：气象出版社：96-108.

孙秀荣，陈隆勋，何金海，2002. 东亚海陆热力差指数及其与环流和降水的年际变化关系[J]. 气象学报，60(2)：164-172.

汤明敏，黄士松，1983. 1979 年中国东部夏季风的进退[C]//全国热带夏季风学术会议文集编写组. 全国热带夏季风学术会议文集. 昆明:云南人民出版社:311-317.
汤绪,钱维宏,梁萍,2006. 东亚夏季风边缘带的气候特征[J]. 高原气象,25(3):375-381.
汤绪,孙国武,钱维宏,2007. 亚洲夏季风北边缘研究[M]. 北京:气象出版社:122.
王安宇,尤丽钰,1990. 对流层低层的季风[J]. 地理学报, 45(3):302-310.
王静爱,徐霞,刘培芳,1999. 中国北方农牧交错带土地利用与人口负荷研究[J]. 资源科学,21(5):19-25.
徐袁,钱维宏,2003. 东亚季风边缘活动带研究综述[J]. 地理学报,58(Z1):138-146.
杨建平,丁永建,陈仁升,等,2003. 近 50 年中国干湿气候界线波动及其成因初探[J]. 气象学报,61(3):364-373.
杨建平,丁永建,陈仁升,等,2005. 亚洲季风与中国干湿、农牧气候界线之关系[J]. 地理科学,25(4):441-447.
曾庆存,张邦林,1998. 大气环流的季节变化和季风[J]. 大气科学,22(6):805-813.
张家诚,1983. 大气环流的季节变化和季风的科学概念[C]//全国热带夏季风学术会议文集编写组. 全国热带夏季风学术会议文集. 昆明:云南人民出版社:1-9.
张兰生,史培军,侯乐峰,等,1993. 中国北方季风尾闾区全新世不同时期降水变化及其区域分异规律的研究[C]//张兰生. 中国生存环境历史演变规律研究. 北京:海洋出版社:147-154.
张庆云,陶诗言,陈烈庭,2003. 东亚夏季风指数的年际变化与东亚大气环流[J]. 气象学报,61(5):559-568.
赵哈林,赵学勇,张铜会,等,2002. 北方农牧交错带的地理界定及其生态问题[J]. 地球科学进展,17(5):739-747.
周兵,何金海,吴国雄,等,2003. 东亚副热带季风特征及其指数的建立[J]. 大气科学,27(1):123-135.
朱乾根,何金海,1985. 亚洲季风建立及其中期振荡的高空环流特征[J]. 热带气象,1(1):9-18.
朱乾根,杨松,1989. 东亚副热带季风的北进及其低频振荡[J]. 南京气象学院学报,1(3):249-257.
朱乾根,刘宣飞,2000. 大气环流的正压斜压流型特征与季风类型[J]. 气象学报,58(2):194-201.
竺可桢,1934. 东亚季风与中国之雨量[J]. 地理学报,1(1):1-27.
祝从文,何金海,吴国雄,2000. 东亚季风指数及其与大尺度热力环流年际变化[J]. 气象学报,58(4):391-402.
卓嘎,徐祥德,陈联寿,2002. 青藏高原边界层高度特征对大气环流动力学效应的数值试验[J]. 应用气象学报,13(2):163-169.
DING Y H,1992. Summer monsoon rainfalls in China[J]. J Meteor Soc Japan,70(1B):373-396.
DING Y H,WANG Q Y,YAN J Y,1996. Some aspects of climatology of the summer monsoon over the South China Sea [C]//IAP/ CAS. Atmospheric Circulation to Global Change. Beijing:China Meteorological Press:107-117.
KRISHNAMURTI T N,BHALME H N,1976. Oscillation of a monsoon system. Paurt Ⅰ:observational aspects[J]. J Atmos Sci,33(10): 1937-1954.
LAU K M,KIM K M,YANG S,2000. Dynamical and boundary forcing characteristions of regional components of the Asian summer monsoon[J]. J Clim,13(14):2461-2482.
LAU K M,YANG S,1997. Climatology and interannual variability of the Southeast Asian summer monsoon[J]. Adv Atmos Sci, 14(2):141-162.
LI C,YANAI M,1996. The onset and interannual variability of the Asian summer monsoon in relation to land-sea thermal contrast[J]. J Clim,9:358-375.
MURAKAMI M,MATSUMOTO J,1994. Summer monsoon over the Asian continent and Western North Pacific[J]. J Meteor Soc Japan,72(5):719-745.
QIAN W H,KANG H S,ICE D L,2002. Distribution of seasonal rainfall in the East Asian monsoon region[J]. Theor Appl Clim,73(3):151-168.
RAMAGE C S,1971. Monsoon Meteorology [M]. New York: Academic Press:4-6,296.
SUN E,SHEN B Z,GAO Z T,et al,2007. The Impacts of Moisture Transport of East Asian monsoon on Summer Precipitation in Northeast China[J]. Adv Atmos,24(4):606-618.
TAO S Y,CHEN L X,1987. A review of recent research on the east Asian summer monsoon in China[C]//Chang C P, KRISHNAMURTI T N. Monsoon Meteorology. London:Oxford University Press:60-92.
WEBSTER P J,MAGANA V O,PALMER T N,et al,1998. Monsoon:Processes,predictability and the prospects for prediction[J]. J Geophys Res,103(C7):14451-14510.
WEBSTER P J,1987. The Elementary Monsoon [M]//Fein J S,Stephens P L. Monsoons. New York:A Wiley－Interscience Publication:3-32.

YE D Z,CAO Y X,1979. Meteorology of Tibetan Plateau[M]. Beijing:Science Press.

ZHANG R H,SUM A,KIMOTO M,1996. Impact of El Niño on the East Asian Monsoon;A Diagnostic study of the 86/87 and 91/92 events[J]. J Meteor Soc Japan, 74(1):49-62.

ZHU Q C,HE J H,WANG P X,1986. A study of circulation differences between East Asian and Indian summer monsoon with their interaction [J]. Adv Atmos Sci,3(3):446-477.

The Impacts of Moisture Transport of East Asian Monsoon on Summer Precipitation in Northeast China*

SUN Li[1,2] SHEN Baizhu[1,2] GAO Zongting[1,2] SUI Bo[1] BAI Lesheng[3]
WANG ShengHung[2] AN Gang[1] LI Jian[1]

(1 Institute of Meteorological Sciences of Jilin Province, Changchun 130062; 2 Wuhan Institute of Heavy Rain, Wuhan 430074; 3 Polar Meteorology Group, Byrd Polar Research Center, The Ohio State University, Columbus, Ohio 43210)

Abstract: By using the ECMWF reanalysis daily data and daily precipitation data of 80 stations in Northeast China from 1961 to 2002, the impacts of moisture transport of East Asian summer monsoon on the summer precipitation anomaly in Northeast China, and the relationship between the variation of moisture budget and the establishment of East Asian summer monsoon in this region are studied. The results demonstrate that the moisture of summer precipitation in Northeast China mainly originates from subtropical, South China Sea, and South Asia monsoon areas. East China and its near coastal area are the convergent region of the monsoonal moisture currents and the transfer station for the currents continually moving northward. The monsoonal moisture transport, as an important link or bridge, connects the interaction between middle and low latitude systems. In summer half year, there is a moisture sink in Northeast China where the moisture influx is greater than outflux. The advance transport and accumulation of moisture are of special importance to pentad time scale summer precipitation. The onset, retreat, and intensity change of the monsoonal rainy season over Northeast China are mainly signified by the moisture input condition along the southern border of this area. The establishment of East Asian summer monsoon in this area ranges from about 10 July to 20 August and the onset in the west is earlier than that in the east. The latitude that the monsoon can reach is gradually northward from west to east, reaching 50°N within longitude 120°—135°E. In summer, the difference of air mass transport between summers with high and low rainfall mainly lies in whether more air masses originating from lower latitudes move northward through East China and its coastal areas, consequently transporting large amounts of hot and humid air into Northeast China.

Keywords: Northeast China; East Asian monsoon; moisture transport; source sink structure; air mass trajectory

1 Introduction

Northeast China is located in the northern end of the East Asian monsoon area, where annual

* 本文原载于 *Advances in Atmospheric Sciences*, 2007, (4): 606-618

precipitation is mainly determined by the intensity of summer monsoon activity. Influenced by the large annual variation of East Asian monsoon, meteorological and climatic disasters, such as torrential rain and floods, as well as durative drought, occur in this area frequently. The frequency of drought and flood events are evidently an increasing trend since the 1990s (Sun et al., 2002a). Although some studies of the summer precipitation anomaly in this area have been carried out in many ways, such as through climatic rules of precipitation (Sun et al., 2000), characteristics of atmospheric circulation anomalies (Sun et al., 2002b), and impacts of sea surface temperature (Sun et al., 2003a, 2004), the relationship between moisture transport of East Asian monsoon and the summer precipitation anomaly in this area remains undiscovered. In fact, from this point of view, the study is very important to Northeast China, not only because—compared with other areas of East China—the air mass in this area does not often meet with hot and humid conditions, but also because the strength or weakness of summer monsoon in this area often corresponds with special circulation patterns (Sun et al., 2003b). Thus, it is possible that characteristics of monsoonal moisture transport and budget in this area are significantly different from those in other areas of China.

Local precipitation anomalies mainly result from circulation and moisture change. Moisture transport comprehensively reflects the impacts of both parts. In the last few years, more and more attention has been paid by studies analyzing the impacts of East Asian monsoon on drought and flood in China from the viewpoint of moisture transport and its recycling. Huang et al. (1998) studied the characteristics of moisture transport in the East Asian monsoon area and its difference from that of the South Asian monsoon area. They believed that the moisture convergence in the East Asian monsoon area was mainly caused by advection processes, and that its longitudinal transport is very important, and thus it is very different from the Indian monsoon area. Through analyzing the characteristics of moisture transport and budget of two torrential rains in the Changjiang River and Changjiang-Huaihe River drainage area in 1998 and 1991, Ding et al. (2003) and Hu et al. (2003) revealed the relationship between these two torrential rains and large scale moisture transport, as well as East Asian monsoon systems. Xu et al. (2002, 2003a, 2003b, 2003c) also discussed the main characteristics of the areas influenced by monsoonal moisture transport and presented a teleconnection model and a physical image of moisture transport in the monsoonal Meiyu belt. All these studies suggest that it is an indispensable direction for analyzing precipitation anomalies to study moisture transport and budget, which is essential for further understanding precipitation mechanisms and forecasting drought and flood processes.

Since the 1980s, many significant advancements in studying the East Asian monsoon have been achieved [see Tao et al. (1988) and Chen et al. (1991) for summaries and generalizations of such studies before the 1990s]. In recent years, many considerable developments have been made in the aspects of variability of East Asian monsoon (Huang et al., 2003; Wu et al., 2000), the relationship between East Asian monsoon and the ENSO phenomenon (Wang et al., 2000, 2001; Wu et al., 2002; Huang et al., 2004), and the precipitation characteristics of East Asian monsoon (Wang et al., 2002; Yu et al., 2001). Ding et al. (2004a, 2004b) described the research results in this field for the past ten and more years in detail. On the whole, compared with other areas of China, relatively few studies have been carried out with regard to the determination of the activity region (borders that the monsoon can reach), the influence duration, and intensity of the East Asian monsoon to the north of Huaihe River. Many aspects should be investigated further.

In this study, by using the ECMWF (European Centre for Medium-range Weather Forecast) reanalysis daily data and daily precipitation data of 80 stations in Northeast China from 1961 to 2002, the impacts of moisture transport and its source-sink structural characteristics on the summer precipitation anomaly in Northeast China, as well as the relationship between the variation of moisture budget and the establishment of East Asian summer monsoon (including the onset, retreat, and the impact areas of the monsoonal rainy season) in this region are studied. It is helpful for discussing the mechanism of drought and flood formation in Northeast China, as well as for enhancing the understanding of the interaction between middle and low latitude systems and the impacts of East Asian monsoon on this region. The data and analysis method are described in section 2; the analysis results are presented in section 3; and section 4 summarizes the main conclusions.

2 Data and method

Moisture flux vector Q is defined as:

$$Q(\lambda,\varphi,t) = Q_u(\lambda,\varphi,t)i + Q_v(\lambda,\varphi,t)j, \tag{1}$$

where λ, φ and t are longitude, latitude and time, respectively; i and j are meridional and longitudinal unit vectors, respectively. $Q=(Q_u^2+Q_v^2)^{1/2}$ is the value of Q.

The two moisture flux components, Q_u and Q_v are defined as follows:

$$Q_u(\lambda,\varphi,t) = \frac{1}{g}\int_{p_s}^{p_t} q(\lambda,\varphi,p,t)\, u\,(\lambda,\varphi,p,t)\mathrm{d}p \tag{2}$$

$$Q_v(\lambda,\varphi,t) = \frac{1}{g}\int_{p_s}^{p_t} q(\lambda,\varphi,p,t)\, v\,(\lambda,\varphi,p,t)\mathrm{d}p \tag{3}$$

where Q_u and Q_v are the eastward and northward moisture fluxes, respectively; q, u and v are specific humidity and horizontal wind velocity components, respectively; g is the gravity constant; p is pressure; p_s is surface air pressure; and p_t is top air pressure.

The estimation of the trajectory of two-dimensional air parcel transport is based on the backward-trajectory model of the semi-Lagrangian integration scheme, which searches for the upstream point (Robert, 1982). In a given wind field, the upstream point of the air parcel that reaches the target point is searched for through an iteration procedure. The iteration formula is as follows:

$$\alpha_{ijk}^{n+1} = \Delta t(m^2U)(X_i - \alpha_{ijk}^n, Y_j - \beta_{ijk}^n, \sigma_k, t) \tag{4}$$

$$\alpha_{ijk}^{n+1} = \Delta t(m^2V)(X_i - \alpha_{ijk}^n, Y_j - \beta_{ijk}^n, \sigma_k, t) \tag{5}$$

where α_{ijk} and β_{ijk} are the position changes along the x and y directions in a map projection plane; U and V denote x and y components of wind velocity in a map projection plane; m denotes the map magnification coeffcient; polar stereographic projection of the Northern Hemisphere is adopted; i, j and k represent x, y and vertical direction; $\sigma=(p-p_t)/(p_s-p_t)$; n denotes the iteration number; and X and Y are the position of the air parcel along the x and y directions respectively.

Data for wind field evolution are ECMWF reanalysis data, which are available four times a day (0000, 0600, 1200, and 1800 UTC). Data for other times are attained through interpolation. In order to overcome topographical effects, the trajectory is modeled in a σ coordinate system. Considering the moisture and its transport are mainly concentrated in the lower layers, the trajectory is estimated in the two-dimensional wind field of the layer with $\sigma_k=0.9$, with a time step of one hour. When the position in a previous time point is determined, the iteration method can be applied at this point to continue to search for the upstream point backwardly. Thus, through repeating the procedure, the transport trajectory of the two-dimensional air parcel that reaches the target point at a certain time point will be attained.

In order to understand the climatic characteristics of moisture transport over Northeast China, the moisture fluxes of the four borders (east, south, west, and north) of the region (40.0°—52.5°N, 117.5°—135.0°E) are calculated.

Moisture transport through a certain border is defined as:

$$\frac{1}{ga}\int_{p_s}^{p_t}\int_L V_n q\,\mathrm{d}l\mathrm{d}p \tag{6}$$

where L is the horizontal border; l is the length of the horizontal border; V_n is normal velocity perpendicular to the border surface; and a is the region area. Positive and negative values represent outflux and influx, respectively. Moisture flux of each of the four borders is, respectively:

East border:

$$\frac{1}{ga}\int_{p_s}^{p_t}\int_L qu\,\mathrm{d}l\mathrm{d}p \tag{7}$$

south border:

$$-\frac{1}{ga}\int_{p_s}^{p_t}\int_L qv\,\mathrm{d}l\mathrm{d}p \tag{8}$$

west border:

$$-\frac{1}{ga}\int_{p_s}^{p_t}\int_L qu\,\mathrm{d}l\mathrm{d}p \tag{9}$$

north border:

$$\frac{1}{ga}\int_{p_s}^{p_t}\int_L qv\,\mathrm{d}l\mathrm{d}p \tag{10}$$

The data used in this paper mainly include daily precipitation of 80 stations uniformly distributed in Northeast China (Fig. 1) and the ECMWF reanalysis daily data (ERA-40). The dataset is a global atmospheric analysis of many conventional observations and satellite data streams for the period September 1957 to August 2002 (Uppala et al., 2005). The dataset used here contains surface and upper level variables on a global 2.5°×2.5° latitude-longitude grid. All variables are reported four times a day. The upper level variables, geopotential height, temperature, specific humidity, and latitudinal and longitudinal components of wind velocity (1000, 925, 850, 700, 500, 400, 300, 250, 300, 150, 100, 70, 50, 30, 20, 10 hPa) are used for this study. Specific humidity is the primary focus between 1000 hPa and 300 hPa.

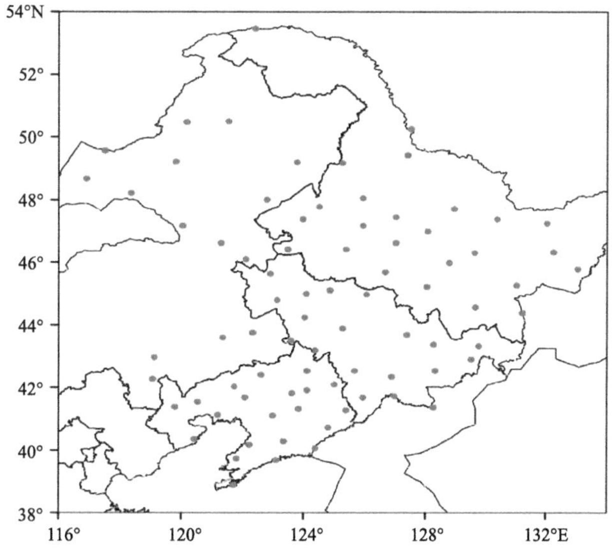

Fig. 1 Geographical distribution of 80 stations in Northeast China. The stations selected are labeled as maculae

3 Results

3.1 Source-sink distribution characteristics of water vapor flux for summers with high and low rainfall

Fig. 2 shows the time series of area average summer (July—August) precipitation anomaly percentage from 1961 to 2002 in Northeast China. Taking the summer (July—August) precipitation anomaly percentage as standard, the six highest (1962, 1964, 1966, 1985, 1994, and 1998) and six lowest (1968, 1976, 1980, 1989, 1992, and 1999) rainfall summers have been chosen as cases of high and low precipitation in this study. The average value of the precipitation anomaly percentage with high rainfall summers is 31.71%, and −24.11% with low rainfall summers. The departure composite maps of water vapor flux (Q) both for high and low rainfall summers over Northeast China from 1961 to 2002 and their difference map (water vapor flux anomaly for high rainfall summers minus water vapor flux anomaly for low rainfall summers) are shown in Fig. 3.

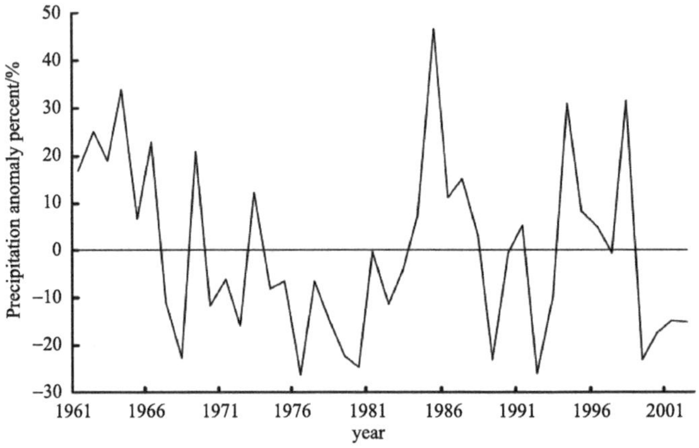

Fig. 2　Time series of area average summer (July—August) precipitation anomaly percentage in Northeast China from 1961 to 2002

In high rainfall summers, it can be seen from the map (Fig. 3a) that a large and notable positive value area covers from Northeast China to the Korean peninsula, where the water vapor flux is above normal, meaning this region is a water vapor transport "sink" (in this paper, the positive value area of water vapor flux anomalies is defined as a water vapor transport "sink" and, conversely, the negative value area is called a water vapor transport "source"). The Sea of Okhotsk, the western Pacific Ocean, the South China Sea, and the Indian Ocean to the south of the Bay of Bengal, are the notable areas of negative anomalies, and are therefore the "sources" of water vapor transport. In low rainfall summers, the features of water vapor flux anomaly distributions are the opposite to high rainfall summers. The region from Northeast China to the Korean peninsula is a notable negative departure distribution, and the Sea of Okhotsk, the western Pacific Ocean, the South China Sea, and the Indian Ocean to the south of the Bay of Bengal, are the notable positive departure distributions (Fig. 3b). The source-sink structural characteristics of water vapor transport in summers with high and low rainfall over Northeast China are even more clearly displayed in Fig. 3c. Both the positive value distribution from Northeast China to the Korean peninsula, and the negative value distributions over the western Pacific Ocean, the South China Sea, the Indian Ocean to the south of the Bay of Bengal, and Sea of Okhotsk, are all statistically significant at the 95%

confidence level (t-test). This indicates that the differences of water vapor transport between high and low rainfall summers are very remarkable over the above mentioned areas.

Fig. 4 displays the composite maps of the water vapor flux vector (Q) anomaly both for summers with high and low rainfall. It can be seen from the Fig. 4a that the anomalous moisture currents (water vapor flux vector anomaly) which influence the precipitation of Northeast China can be tracked southward up to the lower latitudes of the western Pacific Ocean and South China Sea, and northward up to the east of Lake Baikal in high rainfall summers. There are three different branches of anomalous moisture current (water vapor flux vector anomaly) which converge in Northeast China, forming a cyclonic moisture anomaly circulation system, leading to a great increase in convergence, and consequently influences precipitation in this area. The first current originates from the lower latitudes of the western Pacific Ocean, moves westward, and then turns right. The second current originates from the South China Sea and moves directly northward. The two currents assemble over East China and its coastal areas, where they are strengthened and then continue to extend northward. The third current originates from the Sea of Okhotsk, propagating westward up to the east of Lake Baikal, and then turning southward before entering into Northeast China. It can also be observed from Fig. 4a that the anomalous moisture currents originating from the East Asian monsoon region are much stronger than those coming from the middle and high latitudes. In low rainfall summers, the characteristics of the water vapor flux vector anomaly are the opposite to the above (Fig. 4b).

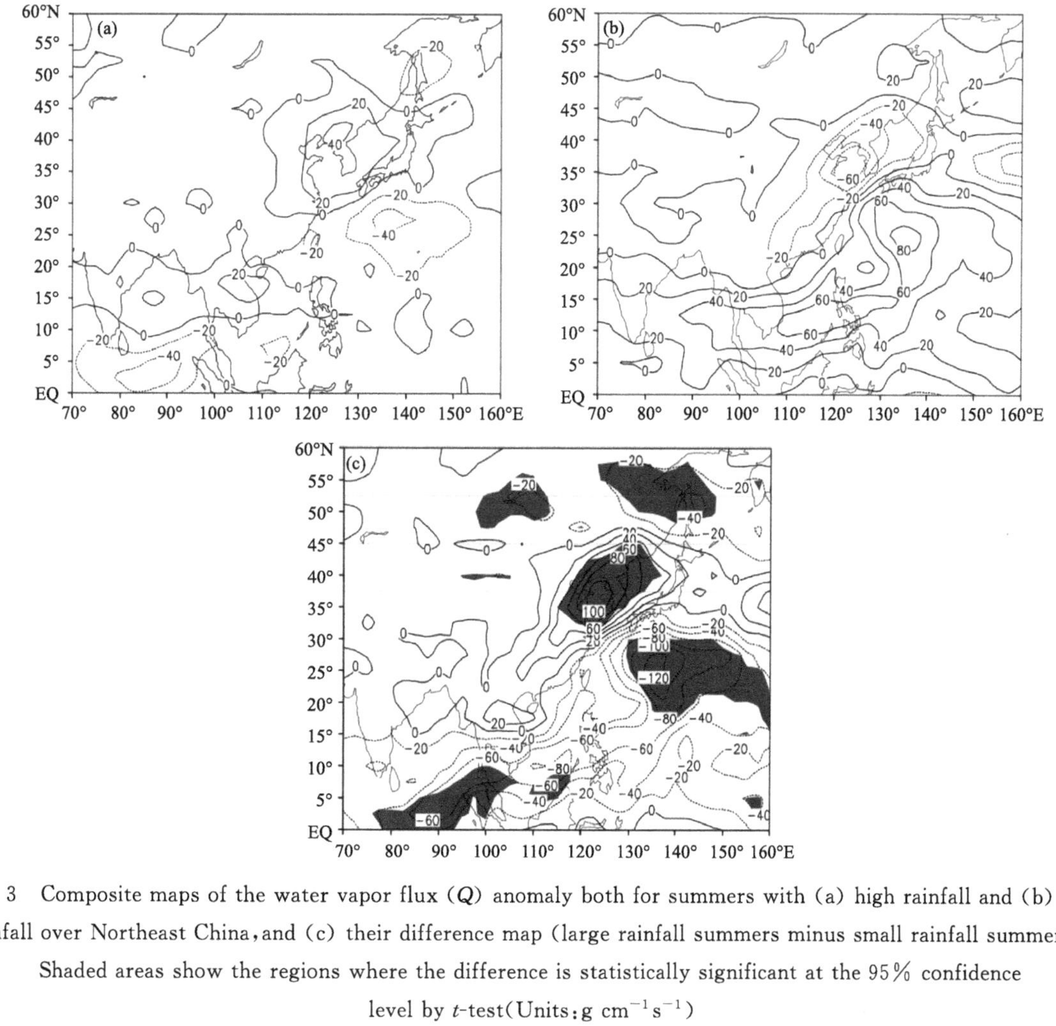

Fig. 3　Composite maps of the water vapor flux (Q) anomaly both for summers with (a) high rainfall and (b) low rainfall over Northeast China, and (c) their difference map (large rainfall summers minus small rainfall summers). Shaded areas show the regions where the difference is statistically significant at the 95% confidence level by t-test(Units: g $cm^{-1}s^{-1}$)

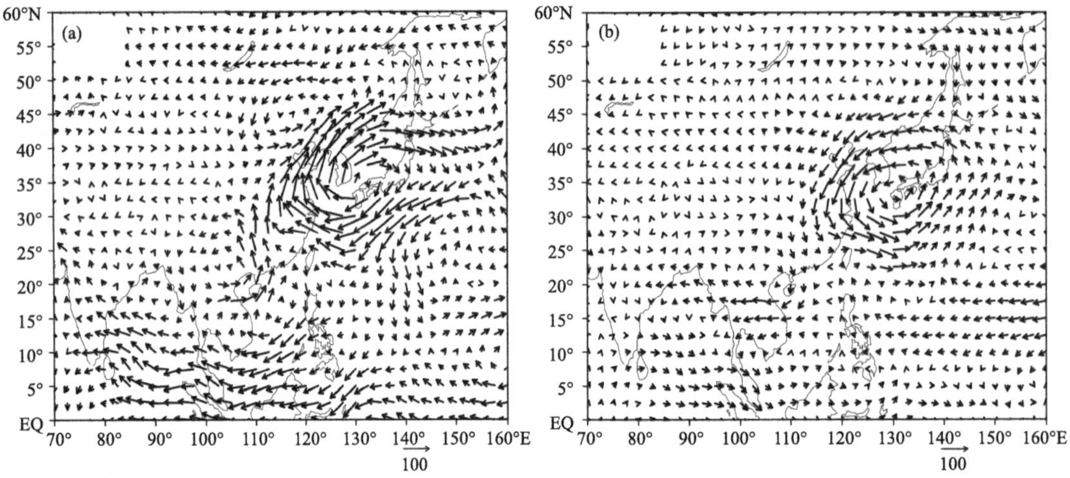

Fig. 4　Composite maps of the water vapor flux vector (Q) anomaly both for summers with (a) high rainfall and (b) low rainfall(Units: g $cm^{-1}s^{-1}$)

It is evident from the above analyses that the moisture of summer precipitation in Northeast China mainly originates from the East Asian monsoon areas, including subtropical monsoon areas and the South China Sea monsoon area. East China and its coastal areas are the convergent regions of the monsoonal anomalous moisture currents (water vapor flux vector anomaly), as well as the "transfer station" for the moisture continuing to transport northward. Although this precipitation anomaly of Northeast China has a complicated climatic background, and specific characteristics of anomalous atmospheric circulation, as well as an existence of evident multi-scale interactions, in fact, the monsoonal moisture transport, as a main link or bridge, connects the interactions between middle and lower latitude systems and, as an important form or media, realizes the impacts of East Asian monsoon on Northeast China.

3.2　Correlation analysis between the summer precipitation anomaly and moisture transport of East Asian summer monsoon

Fig. 5 demonstrates the correlation distribution between summer (July—August) precipitation and the two moisture flux components, Q_v (Fig. 5a) and Q_u (Fig. 5b). The results shown in Fig. 5a suggest that the significant positive correlation area is mainly located in East China and its coastal areas. This positive correlation area is belt-shaped in the southnorth direction and can extend to the north part of the South China Sea, and the sea to the east of Taiwan in the southerly direction. In particular, the northward moisture transport over the area between the middle and lower reaches of the Changjiang River and Huanghe River, and sea to the east of this area, are vital to the precipitation of Northeast China. This indicates that if the northward moisture transport coming from the East Asian monsoon area is stronger (or weaker) than normal, then it is favorable for the occurrence of more (or less) precipitation in Northeast China. It can be seen from the map (Fig. 5b) that Northeast China and the majority of northern China forms the notable positive distribution, and the region from the Sea of Okhotsk up to Lake Baikal, westward, forms the notable negative correlation area. This shows that the eastward moisture transport coming from Northwest China and North China, and the westward moisture transport in the region from the Sea of Okhotsk up to Lake Baikal are favorable to more precipitation in Northeast China.

Furthermore, in the pentad time scale, precipitation in Northeast China also tightly relates to the previous regional atmospheric precipitable water (water content), the amount of transported moisture, and divergence of moisture flux. For example, the precipitation of the sixth pentad of July (the average of 25—31 July) (Fig. 6) has a positive correlation with water content of the immediately previous pentad in Northeast China (the largest coefficient with the confidence criteria better than 0.01), and with longitudinal moisture transport in the lower latitudes of the western Pacific Ocean and East China, plus its coastal area up to Northeast China (the largest correlation coefficient with confidence criteria better than 0.05～0.01). On the contrary, the precipitation has a negative correlation with the divergence of moisture flux of the immediately previous pentad in Northeast China (the largest correlation coeffcient with the confidence criteria better than 0.01). The aforementioned results indicate that, although Northeast China is not short of atmospheric disturbance, because of relative high latitude and long distance from the ocean, the air mass in this area does not often meet with hot and humid conditions. Therefore, in summer, advanced moisture transport and accumulation is of special importance to pentad time scale precipitation.

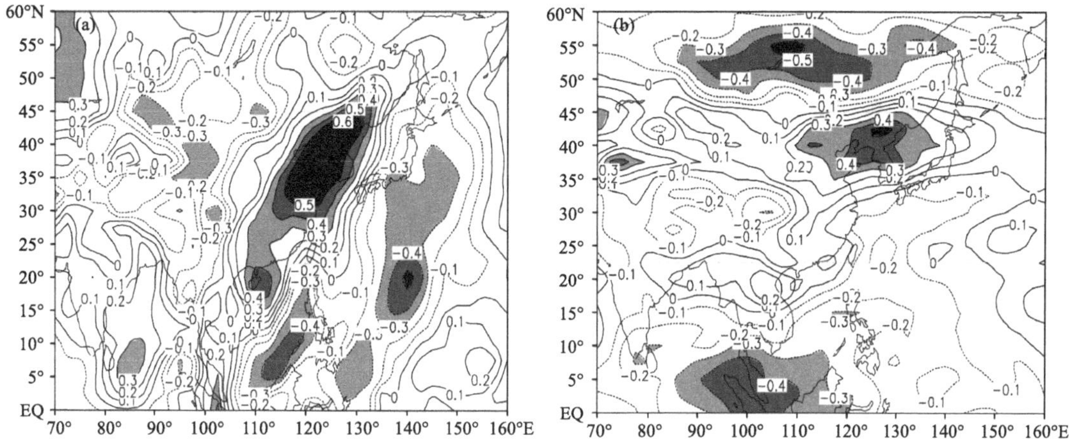

Fig. 5　Correlation distribution between summer (July—August) precipitation in Northeast China and the two components of moisture flux of the whole atmosphere layer, (a) Q_v and (b) Q_u. The shaded areas from light to heavy show the regions where the correlation is statistically significant at the 95%, 99% and 99.9% confidence levels

3.3 Climatic characteristics of the regional moisture budget and the impacts of East Asian

Fig. 7 shows the pentad-to-pentad moisture budget along every border of Northeast China (40.0°—52.5°N, 117.5°—135.0°E) from April to September. Positive and negative values represent outflux and influx, respectively. It can be seen from the map that moisture is always in a state of outflux along the east border from April to September. During this period, the outflux is relatively stable, without any fluctuation from April to June, and begins to increase rapidly from the second pentad of July, reaching a maximum in the first pentad of August, before beginning to slowly decrease soon after. Moisture is always in a state of influx along the west border from April to September, meaning that moisture influx here is always of special importance to summer precipitation in Northeast China. During this period, the influx from April to June is similar to, and therefore counteracts, the outflux along the east border, and then increases slightly by a low magnitude from July to September. Thus, the net moisture budget is a positive value from April to September in the westeast direction. This represents net outflux, especially significant in July and August. Moisture transport along the north border is basically a positive value from April to September. During this period, the absolute value is very low and fluctuates near to zero from April to May, before increasing from June to reach a maximum value in July, and then decreasing from

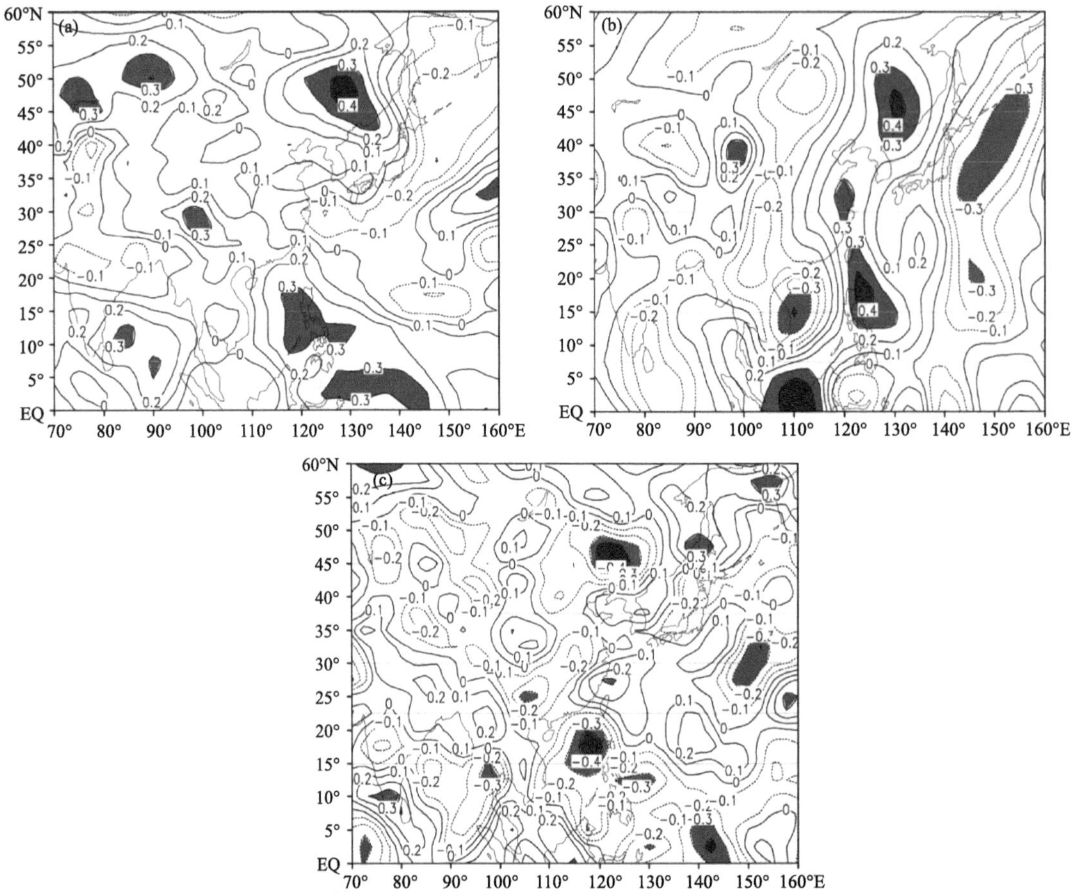

Fig. 6　Correlation distribution between the precipitation of the sixth Pentad of July in Northeast China and (a) the previous pentad's precipitable water, (b) the previous pentad's component of moisture flux Q, and (c) the previous pentad's component of moisture flux divergence. Shaded areas from light to heavy show the regions where the correlation is statistically significant at the 95% and 99% confidence levels

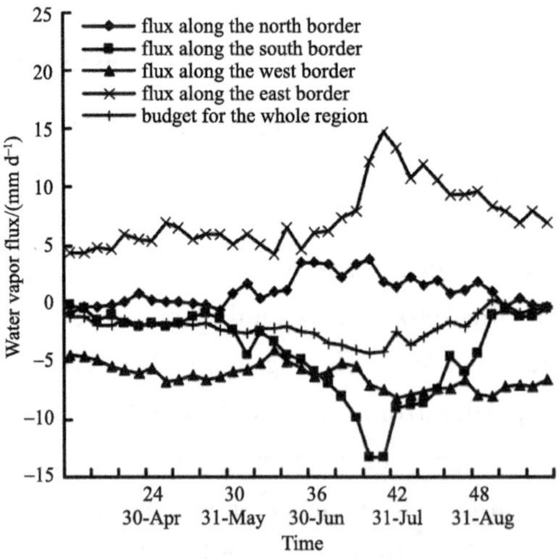

Fig. 7　Pentad-to-pentad moisture budget along every border in Northeast China(40.0°—52.5°N, 117.5°—135.0°E) (average from 1961 to 2002) from April to September, with positive and negative values representing outflux and influx, respectively. The x-axis represents the pentad number and its corresponding month and day (Unit: mm d^{-1})

August. On the whole, the magnitude of moisture outflux along the north border from April to September is not high, even in July and August. This indicates that the moisture transported by the south wind has a limited impact upon the area to the north of Northeast China. Compared with Fig. 8 (pentad-to-pentad rainfall distribution in Northeast China from April to September (average from 1961 to 2002)), moisture transport along the south border is analyzed in detail. It can be found that moisture transport along the south border is always a negative value from April to September; in other words, a moisture influx. During this period, the influx along the south border, with a small change from the first pentad of April to the second pentad of June, is evidently smaller than that along the west border, and the corresponding rainfall of each pentad is smaller than 2 mm d^{-1}. From the third pentad of June to the second pentad of July, moisture influx along the south border begins to increase, and its value is still smaller than, but similar to, that of the west border. During this phase, the rainfall of each pentad also increases up to around 4 mm d^{-1}, and the moisture influxes in Northeast China are influenced by the combined forces of moisture transport, both along the west and south borders. From the third pentad of July to the fourth pentad of August (10 July to 20 August), the moisture influx along the south border increases significantly and rapidly exceeds that along the west border (notably, the value is more than two times that along the west border from the fifth pentad of July to the first pentad of August). During this phase, the corresponding rainfall of each pentad reaches a peak for the whole year: around 6 mm d^{-1}. This means the establishment of the monsoonal rainy season (major flood period) in this area. After the fourth pentad of August, the moisture influx along the south border decreases rapidly, and in the third pentad of September, reverts to a level similar to that of April and May. From April to September, moisture is in a state of net influx from the viewpoint of the moisture transport budget in the north-south direction, which is most significant in July and August. Synthesizing the latitudinal and longitudinal moisture budget, it can be found that, in the summer half year, moisture influx is greater than outflux in Northeast China, with a significant net moisture increase from June to August, and thus this area can be classed as a "moisture sink" in summer.

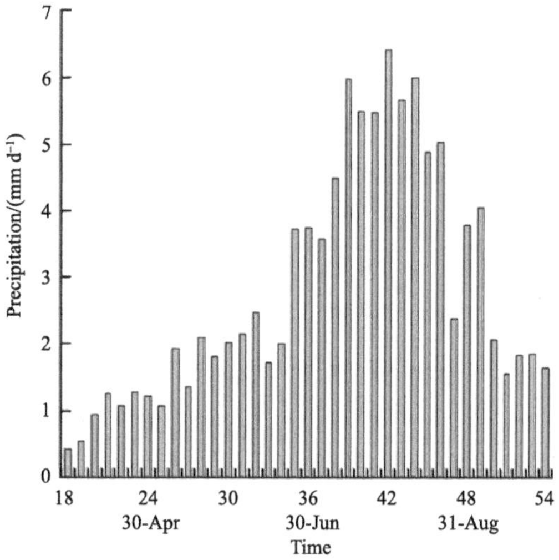

Fig. 8 Pentad-to-pentad rainfall distribution in Northeast China from April to September (average from 1961 to 2002). The x-axis represents the pentad number and its corresponding month and day(Units: mm d^{-1})

The abovementioned analysis indicates that in Northeast China the monsoonal rainy season phenomenon (establishment of East Asian summer monsoon in this area) corresponds to a significant reinforcement of moisture influx along the south border, eventually exceeding that along the west border, and developing into the primary moisture source for this region. This implies that the onset, retreat, and intensity change of the monsoonal rainy season in Northeast China are signified by moisture input conditions along the south border. Therefore, the impacts of East Asian summer monsoon on this area should be measured mainly by northward monsoonal moisture transport.

Fig. 9 shows the temporal-latitudinal profile of vertical integrated moisture transport by south

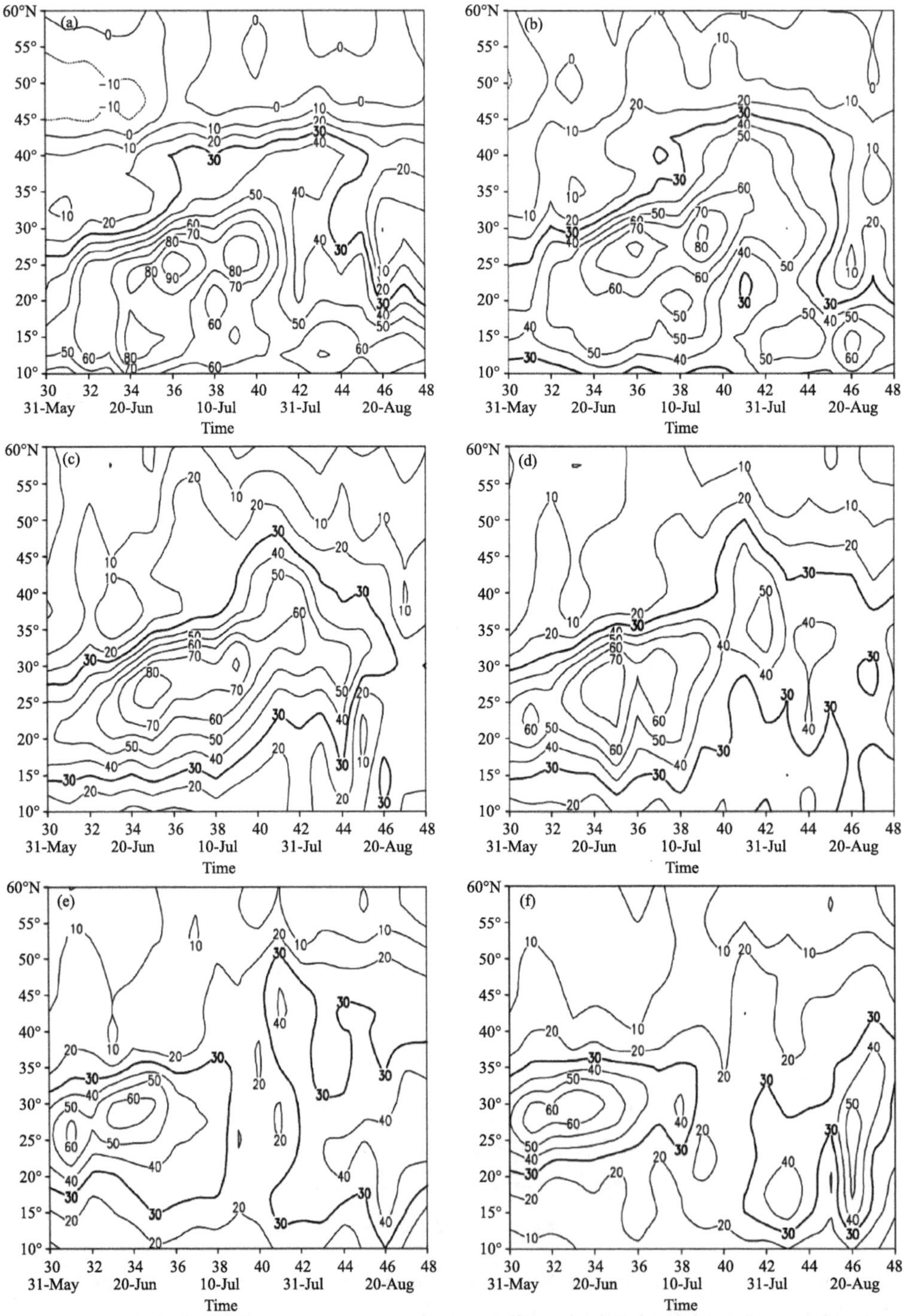

Fig. 9　Temporal-latitudinal profile of vertical integrated moisture transport by south winds averaged by every five longitudes along 110°—140°E (average from 1961 to 2002). The x-axis represents the pentad number and its corresponding month and day. Along (a) 110°—115°E, (b) 115°—120°E, (c) 120°—125°E, (d) 125°—130°E, (e) 130°—135°E, and (f) 135°—140°E (Units: mm d^{-1})

winds, averaged by every five longitudes along 110°—140°E (average from 1961 to 2002). It can be seen from this map that, after summer begins, the area with the greatest value of moisture transport by south winds gradually extends northward, but in different longitudes, and the latitudes which the area with high moisture transport values by south winds can reach are different, as are the times at which they reach them. If we define the time (pentad) that the isoline 30 mm d^{-1} in the map moves across latitude 40°N as the onset date (pentad) of East Asian summer monsoon in Northeast China, the durative (accumulated) numbers of pentad as the impact duration, the time that the isoline retreat to the south of latitude 40°N as the end date (pentad) of the impact of East Asian summer monsoon in this area, and the most northerly latitude that the isoline can reach as the north border of impact of East Asian summer monsoon, then the duration and extension of the impacts of East Asian summer monsoon in Northeast China can be determined. Based on the above definition, and combined with Fig. 9, the following facts can be found. Within longitudes 110°—115°E, the impacts of East Asian summer monsoon begins from the first pentad of July and finishes in the third pentad of August (1 July to 15 August), and the north border is 42°N. Within longitudes 115°—120°E, the impacts of East Asian summer monsoon begins from the third pentad of July and finishes in the fourth pentad of August (11 July to 20 August), with the strongest moisture transport by south winds in the sixth pentad of July (26 July to 31 July), and the north border is 46°N. In this longitude range, the duration of the impacts of East Asian summer monsoon is well matched to the conclusion drawn by Lian et al. (2004) that along 117.5°E the average onset and retreat date of East Asian summer monsoon at 40°N are, respectively, pentad 37.65 (around 10 July) and pentad 46.76 (around 20 August). Within longitudes 120°—125°E, the impacts of East Asian summer monsoon begins from the fourth pentad of July and finishes in the third pentad of August (16 July to 15 August), with the strongest moisture transport by south winds in the sixth pentad of July (26 July to 31 July), and the north border is 49°N. Within longitudes 125°—130°E, the impacts of East Asian summer monsoon begins from the fifth pentad of July and finishes in the fourth pentad of August (21 July to 20 August), with the strongest moisture transport by south winds in the sixth pentad of July (26 July to 31 July), and the north border is 51°N. Within longitudes 135°—140°E, the northward moisture transport of East Asian summer monsoon weakens sharply and the isoline 30 mm d^{-1} can not move across 40°N. On the basis of above discussion, we demonstrate the northernmost border that East Asian summer monsoon reaches in Northeast China. The impacts of East Asian summer monsoon are relatively weak in areas to the west, north, and east of the border.

Synthesizing the above analysis, the following statements can be reached. The active duration of East Asian summer monsoon in Northeast China is from around 10 July to 20 August. However, in the same latitude, the summer monsoon onset dates are different and, specifically, the onset in the west side of the area is earlier than that in the east side, with an interval of around one or two pentads every five longitudes. The retreat dates, however, are similar. East Asian summer monsoon has the strongest impact on Northeast China from 25 July to 5 August. The latitude that the monsoon can reach is gradually northward from west to east in this area and, notably, it is up to around 50°N within longitude 120°—135°E. This is also the area in Northeast China where the most significant impacts of summer monsoon occur.

3.4 Trajectory characteristics of air mass transport in summers with high and low rainfall

According to the departure percentage of monthly regional average precipitation, four typical rainy months and four typical rainless months are picked out from the months of July and August in the years since 1980. The years with a rainy July were 1985, 1986, 1991, and 1994, and the years with a rainless

July were 1982, 1997, 1999, and 2000. The average percentage value of the precipitation anomaly with a rainy July is 32.25%, and −33.5% for a rainless July. The years with a rainy August were 1985, 1987, 1997, and 1998, and the years with a rainless August were 1980, 1989, 1991, and 1992. The average percentage value of the precipitation anomaly with a rainy August is 54.5%, and 44.0% with a rainless August. In order to compare the trajectory characteristics of air mass transport in the summers with high and low rainfall, one backward trajectory model is used to calculate trajectories of the two-dimensional air mass that can reach Changchun (43°54′N, 125°13′E) for four rainy July and August months, as well as four rainless July and August months. The trajectory of air mass that can reach Changchun is calculated for each rainy (or rainless) July and August with an interval of six hours from 0000 UTC of the first day of every month to 1800 UTC of the last day of every month. As a result, 124 trajectories are attained for each rainy (or rainless) July and August. Thus, there are 496 trajectories for four rainy (or rainless) July and August months. The time span of a trajectory is set as 10 days and the point step in the same trajectory is set as one hour. In order to overcome topographical effects, trajectory is modeled in a sigma (σ) coordinate system, and considering the moisture and its transport are mainly concentrated in the lower layers, trajectory is estimated in the two-dimensional wind field of the layer with σ=0.9, around 900~850 hPa. For the sake of saving space, Fig. 10 only provides the frequency distribution of trajectory dots of air mass that reaches Changchun for four rainy (or rainless) July and August months. The results

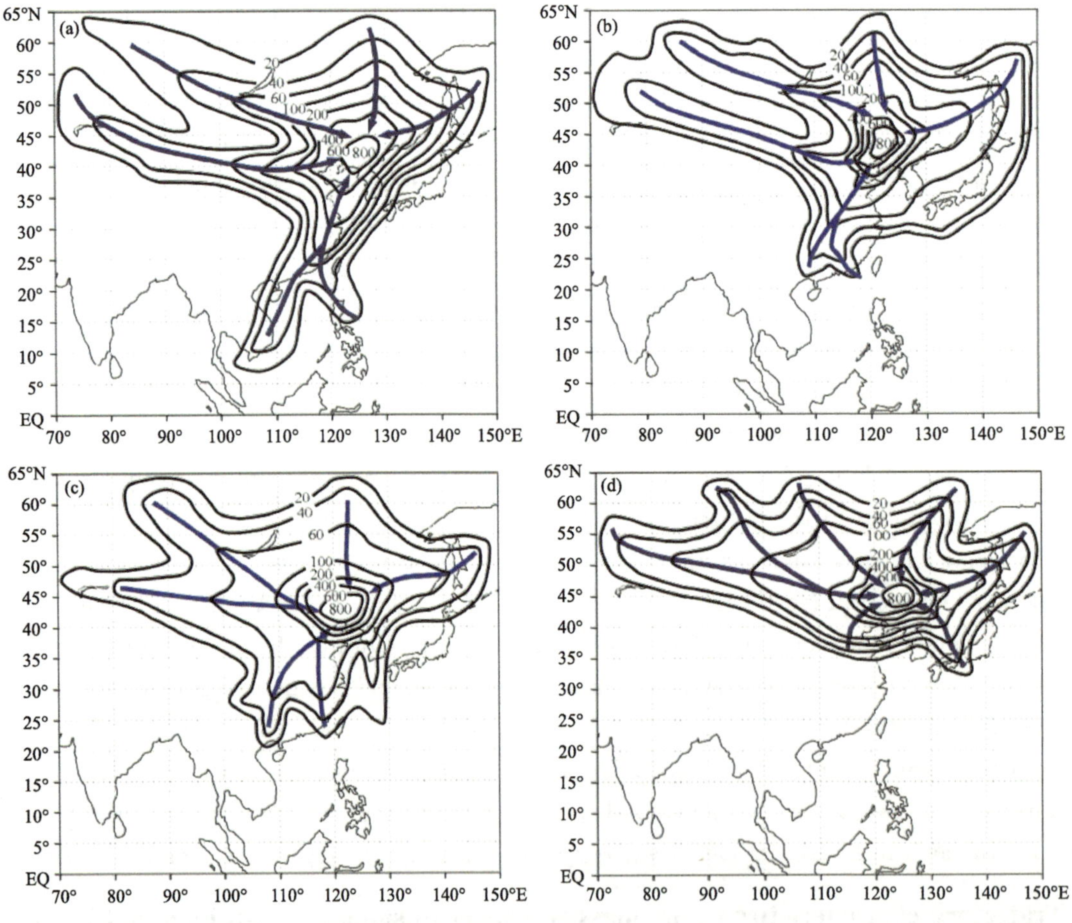

Fig. 10　Frequency distribution of trajectory dots of the air masses that can reach Changchun (43°54′N, 125°13′E) for the four rainy (or rainless) July and August months in Northeast China. (a) Four rainy July months (1985, 1986, 1991, and 1994); (b) four rainless July months (1982, 1997, 1999, and 2000); (c) four rainy August months (1985, 1987, 1997, and 1998); (d) four rainless August months (1980, 1989, 1991, and 1992)

shown in Fig. 10 suggest that the air masses which reach Changchun in the four rainy July months mainly come from near east, near north, northwest, near west, southwest, and southeast paths. The air masses with near south paths (southwest and southeast paths) can transport large amounts of warm and moist air into Northeast China and the air masses with near west and near north paths mainly provide the cold air for the local precipitation. The air masses that reach Changchun in the four rainless July months also mainly come from the near east, near north, northwest, near west, and near south paths, among which the first four paths are similar to those of the rainy July months. The main difference lies in the near south paths. Specifically, the trajectories of air masses in East China and its coastal area in rainy July months are evidently denser than those in rainless July months, and, in the meantime, the isolines with the same frequency can extend southward into much lower latitudes. For example, the isoline with frequency 20 in the southwest path can reach 10°N, the south end of Indochina, and in the southeast path it can reach 15°N, near the Philippines. In contrast, the isoline in the rainless July months is not able to reach 20°N. This indicates that the difference between rainy and rainless months mainly lies in whether more air masses originating from lower latitudes move northward along East China and its coastal area, consequently transporting large amounts of hot and humid air into Northeast China. The trajectory characteristics of air masses in rainy (or rainless) August months is similar to that in July months, only the air mass transport from the near south path in August months is weaker than that in July months.

4 Conclusions

In this paper, the monsoonal moisture transport and its source-sink structural characteristics of summer precipitation in Northeast China, as well as the variation of moisture budget and the impacts of East Asian summer monsoon on this area are analyzed. The results are summarized as follows:

(1) The moisture of summer precipitation in Northeast China mainly originates from the subtropical monsoon area, the South China Sea monsoon area, and even the South Asian monsoon area. The precipitation anomaly is mainly determined by the moisture change and its northward transport characteristics of these monsoon regions. East China, plus its coastal area (especially the area between the middle and lower reaches of the Changjiang River and Huanghe River, and the sea to the east of this area) is the convergent region of the monsoonal moisture currents and the "transfer station" for the moisture continuing to transport northward. In fact, the monsoonal moisture transport, as a main link or bridge, connects the interactions between middle and lower latitude systems and, as an important form or media, realizes the impacts of East Asian monsoon on Northeast China.

(2) In the pentad time scale, precipitation in Northeast China tightly relates to previous regional atmospheric water content, the amount of transported moisture, and the convergence of regional moisture flux. This means the advanced transport and accumulation of moisture are of particular importance to summer precipitation in the pentad time scale.

(3) In the summer half year (April to September), Northeast China, with the main moisture outflux occurring along the east and north borders, and in contrast, influx along the south and west borders, is a "moisture sink" (influx greater than outflux).

(4) The monsoonal rainy season phenomenon in Northeast China (establishment of East Asian summer monsoon in this area) corresponds to a significant reinforcement of moisture influx along the south border, and eventually exceeds that along the west border, developing into the primary moisture source of this area. This implies that the onset, retreat, and intensity change of the monsoonal rainy season in

Northeast China are signified by the moisture input conditions along the south border. Therefore, the impact of East Asian summer monsoon on this area should be measured mainly by northward monsoonal moisture transport.

(5) The active duration of East Asian summer monsoon in Northeast China ranges from around 10 July to 20 August. However, in the same latitude, summer monsoon onset dates are different and, specifically, onset in the west side of the area is earlier than that in the east side, with an interval of around one or two pentads every five longitudes. The retreat dates, however, are similar. East Asian summer monsoon has its strongest impact upon Northeast China from 25 July to 5 August. The latitude that the monsoon can reach is gradually northward from west to east in this area and, notably, it is up to around 50°N within the longitude 120°—135°E. This is also the area in Northeast China where the most significant impact of summer monsoon occurs.

(6) The transport trajectory characteristics of air masses in summers with high rainfall is similar to that in summers with low rainfall in the near east, near north, northwest, and near west paths. The main difference lies in the near south path, i. e., whether more air masses originating from lower latitudes move northward along East China and its coastal area, consequently transporting large amounts of hot and humid air into Northeast China.

Acknowledgements. This study was jointly supported by the National Natural Science Foundation of China under Grant Nos. 40633016 and 40575047, the Special Social Public Welfare Foundation of the Ministry of Science and Technology of China, the Open Foundation of Wuhan Institute of Heavy Rain under Grant No. IHR2006K05, and the Special Foundation of Climate Change of China Meteorological Administration under Grant CCSF2006-18.

References

CHEN L, ZHU Q, LUO H, 1991. The Study of East Asian Monsoon[M]. Beijing: China Meteorological Press: 362. (in Chinese)

DING Y H, HU G, 2003. A study on water vapor budget over China during the 1998 severe flood periods[J]. Acta Meteor Sinica, 61(2): 129-145. (in Chinese)

DING Y H, LI C Y, HE J H, et al, 2004a. South China Sea monsoon experiment (SCSMEX) and the East Asian monsoon [J]. Acta Meteor Sinica, 62(5): 561-586. (in Chinese)

DING Y H, LI C, LIU Y, 2004b. Overview of the South China Sea Monsoon experiment[J]. Adv Atmos Sci, 21(3): 343-360.

HU G, DING Y H, 2003. A study on the energy and water cycles over Changjiang River—Huaihe River basins during the 1991 heavy rain periods[J]. Acta Meteor Sinica, 61(2): 146-163. (in Chinese)

HUANG R, ZHOU L, CHEN W, 2003. The progresses of recent studies on the variabilities of the East Asian Monsoon and their causes[J]. Adv Atmos Sci, 20(1): 55-69.

HUANG R, ZHANG Z, HUANG G, et al, 1998. characteristics of the water vapor transport in East Asian Monsoon Region and its difference from that in South Asian Monsoon region in summer[J]. Sci Atmos Sinica, 22(4): 460-469. (in Chinese)

HUANG R, CHEN W, YAN B, et al, 2004. Recent advances in studies of the interaction between the East Asian winter and summer monsoon and ENSO cycle[J]. Adv Atmos Sci, 21(3): 407-424.

LIAN Y, SHEN B Z, GAO Z T, 2004. An exploration on the determination of East Asian Summer Monsoon index[J]. Acta Meteor Sinica, 62(6): 782-789. (in Chinese)

ROBERT A, 1982. A semi-Lagrangian and semi-implicit numerical integration scheme of the primitive meteorological equations[J]. J Meteor Soc Japan, 60: 319-325.

SUN L, AN G, 2003a. The effect of north Pacific Sea surface temperature anomaly on the summer precipitation in Northeast China[J]. Acta Meteor Sinica, 61(3): 346-353. (in Chinese)

SUN L, AN G, TANG X L, 2003b. Relationship between the Northeast Asian summer south wind anomaly and the precipitation in Northeast China[J]. Chinese J Atmos Sci,27(2):186-196.

SUN L,AN G,DING L,et al,2000. A climatic analysis of summer precipitation features and anomaly in Northeast China[J]. Acta Meteor Sinica,58(1):70-82. (in Chinese)

SUN L,AN G,DING L,2002a. The characteristics of summer drought and flood in Northeast area of China[J]. Sci Geographica Sinica,22(3):311-316. (in Chinese)

SUN L,AN G,LIAN Y,et al,2002b. A study of the characteristics of the general circulation anomaly of the flood and drought in Northeast China[J]. Clim Environ Res,7(1):102-113. (in Chinese)

SUN L,SHEN B Z,AN G,et al,2004. A study on the relationship between the summer precipitation in Northeast China and the global sea surface temperature anomaly (SST) in preceding seasons[J]. J Tro Meteor,10(1):43-52.

TAO S,CHEN L,1988. A Review of Recent Research on then East Asian Summer Monsoon in China[M]// Chang C P, Krishna-murti T N. Oxford:Oxford University Press:60-92.

UPPALA S M,KLLBERG P W,SIMMONS A J,et al,2005. The ERA-40 reanalysis[J]. Quart J Roy Meteor Soc,131:2961-3012.

WANG B,LIN H,2002. Rainy seasons of the Asian-Pacific monsoon[J]. J Clim,15:386-398.

WANG B,WU R,FU X,2000. Pacific East Asian teleconnection:How does ENSO affect Asian climate? [J]. J Clim,13:1517-1536.

WANG Y,WANG B,OH J H,2001. Impact of preceding El Niño on the East Asian summer atmospheric circulation[J]. J Meteor Soc Japan,79:575-588.

WU R, WANG B,2000. Interannual variability of summer monsoon onset over the Western North Pacific and the underlying processes[J]. J Clim,13:2483-2501.

WU R,WANG B,2002. A contrast of the East Asian Summer Monsoon and ENSO relationship between 1962—1977 and 1978—1993[J]. J Climate,15:3266-3279.

XU X,TAO S,WANG J,et al,2002. The relationship between water vapor transport features of Tibetan Plateau-monsoon "Large Triangle" affecting region and drought-flood abnormality of China[J]. Acta Meteor Sinica,60(3):257-266. (in Chinese)

XU X,CHEN L,WANG X,et al,2003a. The source-sink structure of water vapor transport of Meiyu belt in Changjiang River basin[J]. Chinese Sci Bull,48(21):2288-2294. (in Chinese)

XU X,MIAO Q,WANG J,et al,2003b. The water vapor transport model at the regional boundary during meiyu period[J]. Adv Atmos Sci,20(3):333-342.

XU X,ZHOU L,CHANG S,et al,2003c. characteristics of the correlation between regional water vapor transport along with the convective action and variation of the Pacific Subtropical High in 1998[J]. Adv Atmos Sci,20(2):269-283.

YU R,ZHANG M,YU Y,et al,2001. Summer monsoon rainfalls over mid-eastern China lagged correlated with global SSTs [J]. Adv Atmos Sci,18:179-196.

2011年春夏季长江中下游地区旱涝急转特征分析*

沈柏竹[1,2] 张世轩[1] 杨涵洧[1] 王阔[1] 封国林[3]

(1 兰州大学大气科学学院,兰州 730000;2 吉林省气象科学研究所,长春 130062;
3 国家气候中心中国气象局气候研究开放实验室,北京 100081)

摘 要:选用美国国家环境预报中心(NCEP)/国家大气研究中心(NCAR)提供的全球再分析资料,国家气候中心提供的中国758站降水资料及国家气象信息中心提供的中国区域2011年降水格点资料,对2011年长江中下游地区6月初旱涝急转现象及环流背景特征进行简要分析,结论如下:(1) 2011年1—5月长江中下游地区降水较气候态偏少且降水变率较小,6月长江中下游地区降水量急剧增多,发生旱涝急转现象,其转折点为第31候(6月第1候)。(2)旱涝急转前后,水汽通量及其经验正交函数分解的第1、第2模态的时间和空间演变均表现出弱水汽输送向强水汽输送转变的特征。(3)转折前后大气环流场差异显著。急转前,北半球冬季风偏强,南半球夏季风偏弱,在春季,东亚大槽偏强导致季风的转换延迟,不利于南方暖湿气流的北上,南方降水偏少,并发生大范围持续性的严重干旱。6月初(急转期)环流迅速调整,西太平洋副热带高压突然西伸北跳,东亚大槽异常偏西偏强、鄂霍次克海阻高减弱,使冷暖空气在长江中下游地区汇合,有利于该地区的降水发生并持续,是长江中下游地区由严重干旱向洪涝急转的主要原因。

关键词:旱涝急转;水汽输送;大气环流

1 引言

长江中下游地区地处东亚,受季风影响其降水时空变化很大,成为旱涝灾害最严重的区域之一,旱涝异常发生的规律、成因和预测问题也一直是短期气候研究的一个热点和中心课题(吴志伟 等,2006;Dai et al.,1998)。早在20世纪40年代,涂长望等(1994)从季风的角度研究了中国东部雨带进退与季风之间的联系。之后文献(陶诗言,1980;陶诗言 等,1997;林贤超 等,1989;Ding,1992;Shi et al.,1993;施能 等,1996;戴新刚 等,2003,2002)进一步深入分析了夏季风对中国地区夏季旱涝的影响。戴新刚等(2002)通过研究印度季风与东亚夏季环流的遥相关特征发现,当印度夏季风强盛时,江淮等地易旱;反之,易涝。文献(陈汉耀,1957;陶诗言 等,1962;张庆云 等,1998)对夏季江淮流域旱涝期间的大气环流特征进行分析,指出江淮流域旱涝的发生与环流异常变化有关。Nan等(2003)对夏季长江流域降水研究发现其与春季南半球环状模存在密切的关系。文献(黄荣辉,1990;吴国雄 等,1995;张琼 等,2003;李琰 等,2007;Shi et al.,2005;Shi et al.,2006)研究表明,中国东部夏季风降水及异常与海温异常以及海-陆-气相互作用相联系,并通过不同统计方法对其特征进行了分析。这些研究增进了对长江流域旱涝成因及机制的认识,对实际预测具有十分重要的意义,但大都着眼于夏季平均降水异常的空间分布特征及成因等分析。对长江中下游降水季节内尺度变化,尤其是季节内旱涝的时间非均匀分布,如季节内旱涝急转的研究尚未见报道(Wu et al.,2006)。

2011年6月长江中下游地区发生了一次快速、剧烈旱涝急转现象。6月之前,该地区发生了近60 a

* 本文原载于《物理学报》,2012,61(10):530-540

来最严重的冬春持续气象干旱，而进入 6 月，接连 4 轮强降水过程又使该区遭受严重暴雨洪涝灾害。因此，本文通过对此次长江中下游旱涝急转现象的特征进行分析，为进一步研究夏季旱涝急转的成因提供一定的参考。

2 资料及分析方法

本文研究选用的资料主要包括美国国家环境预报中心(NCEP)/国家大气研究中心(NCAR)提供的 1948—2011 年全球日平均、月平均再分析资料(Kalnay et al.,1996)，其中包括垂直方向 1000 hPa 到 10 hPa共 17 层等压面的风场 (u,v 分量)、温度场(t)、高度场(h)资料及垂直方向 1000～300 hPa 共 8 层等压面的比湿场(q)，其水平空间分辨率为 2.5°×2.5°的格点资料；国家气候中心提供的中国 758 站逐日降水量资料 (1951—2010 年) 以及国家气象信息中心提供的中国区域 2011 年逐日 08 时降水格点资料 (0.25°×0.25°)(http://cdc.cma.gov.cn/shishi/pre grid 0.25.jsp)。在分析过程中，对个别缺测值用前后两天的资料进行了线性插补，由于缺测值很少，通过比较表明对分析结果影响很小。

利用丁一汇(2005)水汽输送和收支计算方法，进行水汽特征分析，其中大气中水汽输送向量场 F 为

$$F(\lambda,\varphi,p,t) = \frac{1}{g}qV \tag{1}$$

对单位面积空气柱，总的水汽水平通量为：

$$Q(\lambda,\varphi,t) = \int_{p_t}^{p_s} F\mathrm{d}p = \int_{p_t}^{p_s} qV\frac{\mathrm{d}p}{g} \tag{2}$$

式中，q 为比湿；$V=(u,v)$；u,v 分别为纬向和经向风速；g 为重力加速度；p_s为表面气压；p_t为顶部气压，这里取 300 hPa。并利用经验正交函数分解(EOF)方法对其进行分解分析。

本文计算均取 1981—2010 年 30 a 平均为气候态。分析过程中对降水做日、候(1 年 72 候，每月 6 候)、月、季平均处理。

3 2011 年长江中下游地区降水异常特征分析

3.1 降水变率的空间分布特征

2011 年 1—6 月降水变率较大的区域分布于长江流域及其以南的华南地区，最大变率中心位于长江中下游地区，中心值可达 20 mm 以上。另一个范围稍小的中心位于华南地区，中心值在10 mm左右(华南南部)，表明长江中下游及华南地区是 2011 年 1—6 月中国降水变化较显著的两个地区。中国大部分地区 1—4 月的逐月降水变率较小，月内标准差的变化范围大都在 1～5 mm，而进入 5 月，长江以南的大部分地区降水变率突然增大，标准差变化范围为 10～15 mm，中心位于华南地区，最大值为 25 mm；至 6 月，华南一带降水减弱，降水变化剧烈的地区北移至长江中下游的湖南省、湖北省、安徽省、江西省、浙江省的大部分地区，其标准差达到 20 mm 以上，中心值高达 40 mm 以上，表明长江中下游地区 5—6 月降水异常发生了较大的转变。

3.2 降水异常的时间演变特征

为了更好地反映长江中下游地区的降水异常演变的特征，选取能够较好代表长江中下游地区降水变化特征的上海、南京、芜湖、九江、汉口 5 站平均候降水量的时间变化图(图 1)进行分析。

从图 1 可以看出，1—5 月长江中下游的候平均降水量偏少，最大为 4 mm 以上，其中第 18～28 候长江流域干旱持续并有所加重，直至第 29 候(5 月第 5 候)华南地区和长江中下游局部地区出现强降水过程，但旱情仍未缓解，到第 30 候(5 月第 6 候)降水又急剧转少，而从第 31 候 (6 月第 1 候) 开始降水形势再一

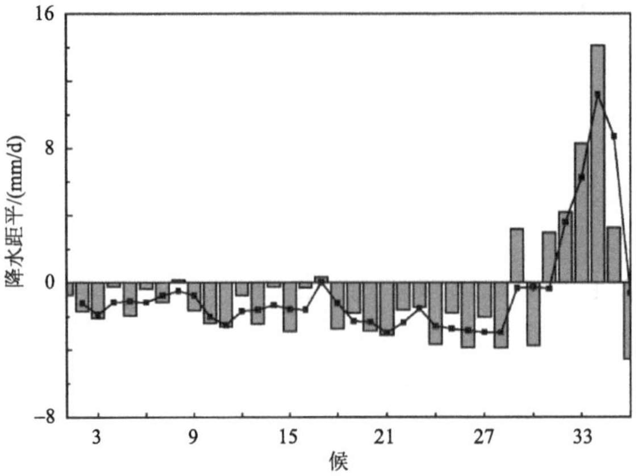

图 1　2011 年第 1～36 候长江中下游 5 站平均降水距平时间序列图
（柱状为候平均；曲线为 10 天滑动平均）

次发生转变，表现为由第 30 候（5 月第 6 候）的明显偏少急剧转为明显偏多，且一直持续到第 35 候（6 月第 5 候）。其中第 33 候（6 月第 3 候）和第 34 候（6 月第 4 候）的候平均降水量分别比气候态偏多 8.3 mm 和 14.1 mm，整个降水偏多形势持续时间为 5 候（25 d 左右）。这一持续性的强降水过程是引发长江中下游涝灾的主要原因。10 年滑动平均降水距平曲线变化形势与候降水距平异常相似，但略有区别。2011 年 1—5 月整个长江中下游地区降水均为负距平，第 29～30 候受华南、长江中下游地区的一次强降水过程影响，降水负距平减小至 0 mm 附近。从第 31 候开始，降水距平值迅速增加，并持续到第 35 候左右，其中第 34 候达到 11 mm 左右。

2011 年 1—5 月长江中下游和南方地区持续性干旱，降水量偏少且降水变率相对较小，5 月下旬华南前汛期降水开始，但相对较晚，6 月初长江中下游大部分地区强降水突然开始，并提前入梅。综合上述两方面的分析，可以确定长江中下游的旱涝急转发生于第 31 候（6 月第 1 候）左右，其原因一是降水有明显的负距平转为明显的正距平，二是转折后降水正距平持续较长时间（5 候），且降水变率相对较大。

4　水汽输送异常及其特征分析

4.1　水汽输送的 EOF 分布特征

由于水汽输送是产生降水的直接和必要条件，因此研究此次旱涝急转，有必要考察其水汽输送特征。取 5—6 月降水变率差异较大的（0°—60°N，70°—140°E）区域为长江中下游地区降水异常分析区，分别对该区 2011 年和 1981—2010 年多年平均的第 1～36 候（1—6 月）逐候水汽通量场进行 EOF 分解，着重分析长江中下游地区旱涝急转前后的水汽输送异常及转折特征。

表 1 为 EOF 分解前 10 个模态的方差贡献率，1981—2010 年气候态平均水汽通量 EOF 分解的前 10 个模态占总方差贡献率的 99.5%。其中第 1、第 2 模态分别占方差贡献率的 59.7%和 34.8%（累积方差贡献率达 94.5%），而 2011 年前 10 个模态占总方差贡献率的 90.8%，其中第 1、第 2 模态分别占方差贡献率的 39.2%的 30.1%（累积方差贡献率达 69.3%）左右。因此，EOF1 和 EOF2 是最主要的水汽输送型，能够表征水汽活动的整体特征。以下着重分析 EOF1 和 EOF2 模态的变化特征。

表 1　第 1～36 候水汽输送 EOF 分解分析累积方差贡献率　　单位：%

项目	EOF1	EOF2	EOF3	EOF4	EOF5	EOF6	EOF7	EOF8	EOF9	EOF10
气候态	59.7	94.5	97.3	98.3	98.8	99.0	99.0	99.3	99.4	99.5
2011 年	39.2	69.3	74.3	78.3	82.2	84.7	86.6	88.2	89.6	90.8

气候态平均与 2011 年第 1～36 候的水汽输送 EOF 分解空间分布十分相似，故以 2011 年 EOF 分解的空间分布图进行说明。EOF1 空间分布表现为，赤道西太平洋和印度洋分别有一支向西和向东传输的水汽输送带，在 100°E 中南半岛附近交汇后转向东北输送，为中国华南及长江中下游地区的降水提供水汽。与之对应的时间系数标准化值变化则显示第 1～36 候时间系数呈逐渐上升趋势，但波动变化较大，除第 2 候、第 17 候外第 1～24 候时间系数值均在气候态以下，尤其在冬、春季节转换的第 18～21 候，水汽输送较气候平均明显偏弱，第 25～27 候水汽输送较气候态偏强，与华南前汛期几次强降水过程相对应。第 28～30 候水汽输送由强转弱，由第 31 候开始又陡然上升，使得向北输送水汽急剧增加，造成 6 月初长江中下游的强降水过程。

水汽输送通量的 EOF2 空间分布与 EOF1 略有不同。低纬度印度洋和孟加拉湾地区存在一支向东的强水汽输送带，并在 100°E 以东地区转向，与(10°N，140°E)附近太平洋西部的一个反气旋性向西北的水汽输送通量相结合，形成 110°—140°E 范围内的向北水汽输送区，沿东北方向输送到 30°N 附近的长江中下游地区。其时间系数变化与 EOF1 类似，第 1～36 候呈上升趋势，但波动变化较为平缓，其中第 1～25 候均为负或弱的正时间系数，大多比气候态偏小。表明在此期间水汽输送较弱，不利于中国南方降水的发生，第 26～27 候和第 31～34 候有两次较大的变化，分别对应华南地区和长江中下游流域的强降水过程。

由以上分析表明，2011 年第 1～30 候，中国南方地区整体水汽输送强度较气候态偏弱，而 5 月之后，第 25 候和第 31 候附近水汽输送出现 2 次较大的位相变换，尤其是 31 候的水汽输送强度较气候态显著偏强。因此，EOF1、EOF2 时间和空间分布均能较好地表现此次降水异常的时空变化特征。

4.2　水汽输送异常特征

为进一步分析此次旱涝急转的降水异常变化特征，分别对 2011 年长江中下游地区发生旱涝急转的第 30～31 候的候平均水汽输送通量进行分析。

从 2011 年第 27～30 候水汽输送整体呈一种相似变化特征，以第 30 候为例，中国东部均为向北和向西北的水汽输送异常分布，华南与长江中下游地区为显著的负水汽通量异常，不利于降水发生，造成该区域的干旱少雨形势。而在第 31 候中国东南部沿 120°E 附近向北的水汽输送突然增强，水汽主要来源于孟加拉湾南部向东南方向及西太平洋向西的输送，其在 100°E 附近汇合后向北输送，在华南和长江中下游地区为水汽汇区，为强降水提供充沛的水汽条件。而第 32～33 候，水汽直接由孟加拉湾向东北方向输送，到中国大陆东部，在(20°N，120°E)左右与西北太平洋的西南暖湿气流汇合，在长江流域一带汇合。虽然水汽输送来源有所变化，但始终维持一种向长江中下游的强输送形势，使该地区维持强降水，从而导致由旱向涝的急剧转折。

综上所述，实际的水汽通量变化与 EOF 分解的主要模态的变化一致，2011 年第 1～30 候西太平洋和印度洋向北的水汽输送偏弱，不利于中国华南及长江流域降水的发生，导致中国南方地区发生持续干旱；而第 31 候之后，印度洋(孟加拉湾)、西北太平洋水汽汇合在 110°—140°E 附近，形成较强的向北水汽输送大值带，为长江中下游地区提供了有利的持续性水汽输送，造成强降水，并使该区发生旱涝急转。

4.3　前期及同期大尺度环流特征分析

4.3.1　前期干旱期环流背景分析

在 2011 年前冬 500 hPa 位势高度场(图 2a)的中低纬度地区均为负距平，西北太平洋副热带高压偏低，大西洋副热带高压偏弱。极涡异常偏弱，格陵兰高压、冰岛低压异常偏强，而阿留申低压偏弱，北半球中高纬度呈显著的北极涛动(AO)负位相分布；在中高纬度地区，表现出冬季典型的 3 个低压槽，分别为欧洲浅槽、东亚大槽以及北美大槽，且三者强度偏强。

2011 年春季(图 2b)与前冬相比在极区和中纬度变化较大，表现出自冬季环流型向夏季环流型转变的环流特征。由沿 30°E—极区—130°W—120°E 有一条越极地的负距平带，极涡异常加强，格陵兰高压、冰

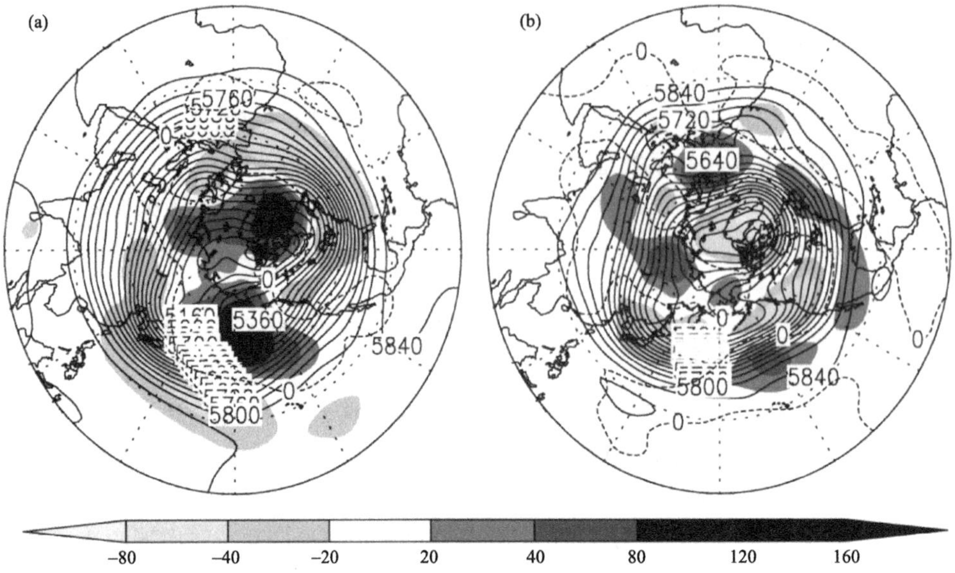

图 2　2011 年前冬(a)和春季(b) 500 hPa 位势高度场空间分布
(阴影为距平场;等值线为原场;粗虚线为位势高度距平 0 线,单位:gpm)

岛低压偏弱,而阿留申低压偏强。北半球呈 AO 正位相分布型。欧洲大陆北部正距平,欧洲浅槽偏弱且位置向东略为偏移。北美大槽在北美东、西海岸逐渐分离为两个低压槽,前者强度偏弱,而后者强度偏强。东亚大槽异常偏强,仍然维持在日本上空,并没有表现出东移的趋势。西北太平洋副热带高压位于正距平区,但强度稍弱,控制范围较小,西伸脊点偏东,位于西太平洋 150°E 上空。

2011 年前冬至 2011 年春季,500 hPa 环流场上表现出冬季风系统偏强的环流型。尤其在春季,东亚大槽维持在日本上空持续偏强;极涡由正距平转为负距平,强度偏强;在西伯利亚上空出现较强的高压脊。上述系统均有利于冷空气持续向南侵袭,阻碍冬季风向夏季风转换,这一点在 850 hPa 经向风(v 分量)上亦有很好的体现。

图 3a 和图 3b 分别为 2011 年前冬和春季 850 hPa 的经向风场。2011 年前冬 (图 3a),在 110°—150°E 范围内,除赤道部分地区外,30°S—60°N 地区几乎都为偏北风,且偏北风较强,尤其在中国南海一带存在正值中心,向西南延伸至赤道附近,在越赤道后向东南延伸,形成强烈的越赤道气流。在春季(图 3b),上

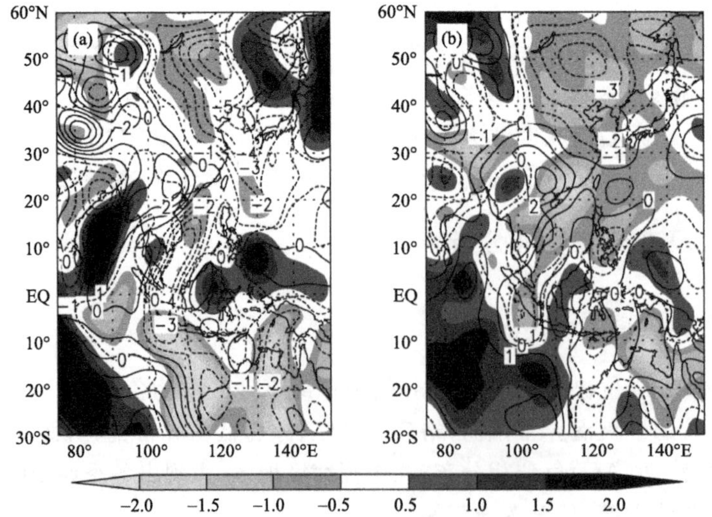

图 3　2011 年前冬(a)和春季(b) 850 hPa 经向风场(v 分量)空间分布
(阴影为距平场;等值线为原场;粗虚线为位势高度距平 0 线,单位:m/s)

述地区也同样多为 0.5 m/s 以上的负距平控制，110°E 附近由 5°S 至北半球中高纬度有一负值中心带，赤道附近地区，仍然存在偏北风的越赤道气流，表明春季南、北半球间的北风分量异常维持并加强，不利于南风向北的输送分量。因此，2011 年前冬至春季在中国东部，偏北风偏强、偏南风偏弱，与 2011 年前冬、春季东亚大槽异常偏南并持续相对应。此种北风偏强的分布型，也不利于南方暖湿空气的北上，2011 年前冬和春季中国南方地区均呈向西南的水汽输送异常，印度洋孟加拉湾及西北太平洋的水汽向西或向南输送。因此，经风向的异常亦为 2011 年前冬、春季中国南方地区水汽呈负距平分布的直接和主要原因。

4.3.2　旱涝急转期环流异常特征分析

图 4 为 2011 年长江中下游地区降水异常急转前后的 500 hPa 高度场和 850 hPa 风场候距平分布。由图 4a 可以看出，第 30 候，乌拉尔山、贝加尔湖阻高及鄂霍次克海阻高均偏强，而东亚大槽较弱，北方冷空气向南输送稍弱，西太平洋副热带高压虽然偏强，但其西伸脊点稳定在 130°E 附近。第 31 候(图 4b)，乌拉尔山阻高持续偏强，整个东亚地区和鄂霍次克海地区呈现与第 30 候相反的形势，东亚大槽由弱转强，鄂霍次克海阻高则由强变弱。西太平洋副热带高压区仍为正距平控制，但反气旋式的差值环流加强并向西移动，中心位于 100°—130°E，表明西太平洋副热带高压西伸。第 31 候西太平洋副热带高压西脊点(134°E)较气候态异常偏西，达 107°E 附近 (偏西 27 个经度)，较第 30 候突然西伸 20 个经度。32 候的环流形势图可以看到(图 4c)，东亚地区及鄂霍次克海地区的异常形势一直持续，高纬度地区的正距平显著区则持续向西减弱、收缩，至第 33 候(图 4d)乌拉尔山附近已由显著正距平转为负距平控制。从第 32～33 候，太平洋副热带高压虽然东撤 10 个经度左右，但始终维持在 115°—120°E，有利于副热带高压后部的西南暖湿气流向北输送，致使长江中下游地区持续性洪涝灾害发生。第 34 候副热带高压东撤，长江中下游地区降水减弱，整个强降水过程结束。

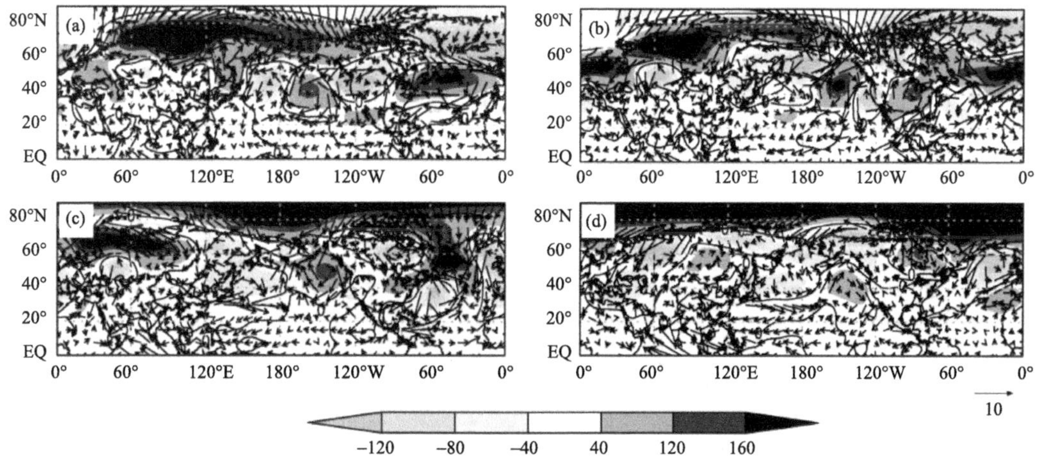

图 4　2011 年长江中下游地区 500 hPa 高度场和 850 hPa 风场候平均距平分布

a—d. 第 30～33 候(阴影：高度距平场，单位 gpm；矢量场：风场，单位 m/s)

与上述环流场异常相对应，第 30 候受中高纬阻塞高压和东亚大槽的影响，经向风场(图 5a)表现为除华北东北部、东北地区外，中国北方大部分地区都为负距平，尤其在长江中下游、华南等地受强偏北风影响，且西太平洋副热带高压并未西伸，其西侧较强的偏南风位于海洋上空。随着大气环流场的改变，第 31 候经向风场也随之调整(图 5b)。中高纬度正距平向西向北收缩及东亚大槽加强导致在 85°E 以西的中高纬度地区盛行偏南风，而 85°E 以东地区则受较强的偏北风控制，随着西太平洋副热带高压的西伸，受其影响，我国长江以南地区均转为受较强的偏南风控制。此种环流形势的配置，有利于北方冷空气南下，南方暖湿气流向北输送加强，冷暖空气在长江中下游一带交汇。从第 32～33 候经向风场(图 5c、5d)，也可以看到长江中下游及华南地区仍受持续较强的偏南风控制，华北、东北地区受偏北风控制。

因此，从环流形势的演变可以看出，造成长江中下游地区旱涝急转的一个主要因素是东亚大槽突然由弱转强，鄂霍次克海阻高突然由强变弱，加之西太平洋副热带高压的突然西伸，使得冷暖空气在长江中下游地区交汇。其中西太平洋副热带高压的这种突然西伸和长时期的维持，是导致长江中下游地区大量的降水，以及造成暴雨洪涝灾害的直接原因。

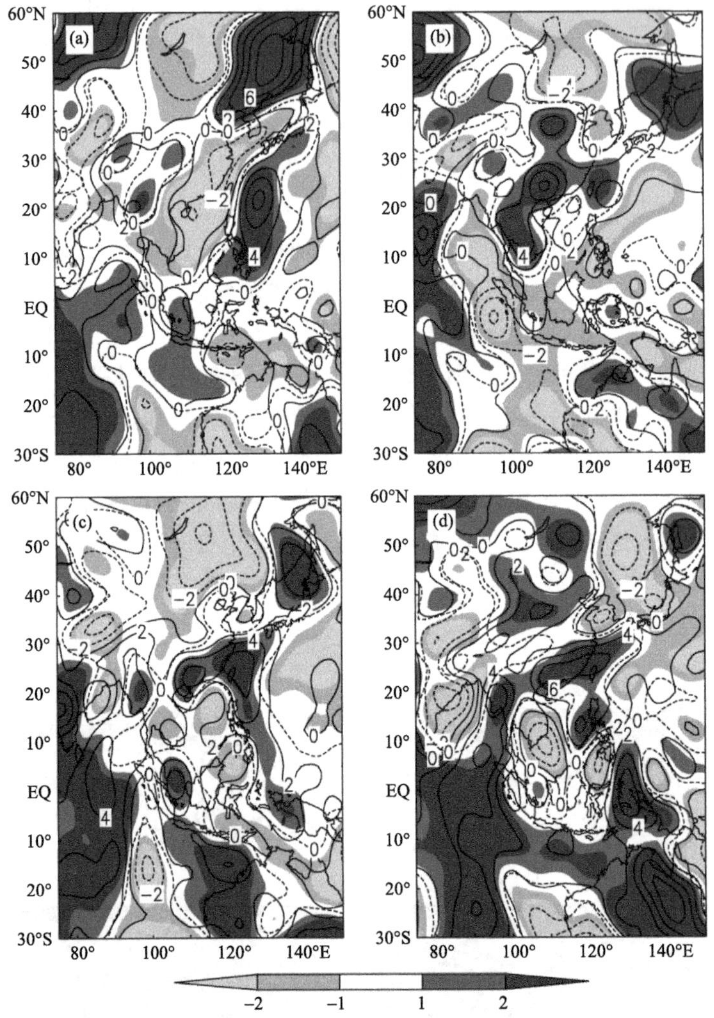

图 5　850 hPa 经向风场（v 分量）空间分布

a—d. 第 30～33 候（阴影为距平场；等值线为原场；粗虚线为位势高度距平 0 线，单位：m s^{-1}）

5　结论与讨论

选用 NCEP/NCAR 的 1948—2011 年全球日平均、月平均再分析资料，国家气候中心提供的中国 758 站逐日降水量资料以及国家气象信息中心提供的中国区域 2011 年逐日 08 时降水格点资料，对 2011 年春夏季长江中下游旱涝急转现象及急转发生前后环流场的变化特征等进行了简要分析，结论如下。

(1)前冬至春季长江中下游降水偏少，降水变化率较小，在第 31 候（6 月第 1 候）长江中下游地区发生旱涝急转现象，降水变化率急剧增长，整个降水过程持续 1 个月左右。

(2)旱涝急转前后，水汽输送通量差异较大。前期干旱发生时，印度洋、西太平洋向北的水汽输送偏弱，不利于中国华南及长江流域降水的发生，而第 31 候之后，印度洋（孟加拉湾）、西北太平洋水汽汇合在 110°—140°E 附近形成较强的向北水汽输送大值带，为长江中下游地区强降水提供了丰沛的水汽条件。

水汽通量的 EOF1、EOF2 模态可以较好地表现此种水汽输送的时间及空间异常特征，对降水异常的发生发展有较好的表征。

(3)旱涝急转前后大气环流场差异显著。前冬至春季，中低纬度西北太平洋和大西洋副热带高压均偏弱，中高纬度呈冬季典型的三槽分布且强度偏强，东亚大槽仍持续偏强，维持在日本上空。表现为北半球冬季风偏强，南半球夏季风偏弱，在春季导致季风的转换延迟。而 6 月初大气环流场迅速调整，东亚大槽加强，鄂霍次克海阻高偏弱，西太平洋副热带高压突然西伸并长时期维持，与之对应的长江中下游地区南风分量偏强。

(4)西北太平洋副热带高压的突然西伸和长时期的维持，夏季东亚大槽的异常偏西，使来自印度洋和西北太平洋的暖湿气流与北方冷空气在长江中下游地区汇合，是导致长江中下游地区旱涝急转的直接因素，而乌拉尔山、贝加尔湖阻高对此次降水异常的影响较弱。

综上所述，2011 年长江中下游地区由前冬至春季的异常干旱转变为 6 月的强降水异常，呈显著旱涝急转型，其成因不是单一的，而是多种因素综合互相配合、持续异常作用的结果。本文分析表明，西北太平洋副热带高压、东亚大槽的异常发展和鄂霍次克海阻塞高压等大气环流形势及孟加拉湾和西北太平洋(南海)暖湿空气的向北输送异常等均对其有重要影响。然而，气候系统是一个高度非线性、无标度、多层次、强迫耗散的复杂巨系统(封国林 等，2001，2006；龚志强 等，2006)，影响此次旱涝急转过程的因子绝不仅限于上述讨论的几类，例如大气驻波的传播与扰动(李建平 等，2003；Dai et al.，2005；Wang et al.，2004)。对此次旱涝急转的作用还有待于进一步研究，因此将在此基础上另文分析此次旱涝急转前期干旱的控制因素以及海-气环流的异常特征及其对长江中下游地区旱涝急转的影响机制。

参考文献

陈汉耀，1957. 1954 年长江淮河流域洪水时期的环流特征[J]. 气象学报，28(1)：1-12.

戴新刚，丑纪范，吴国雄，2002. 印度季风与东亚夏季环流的遥相关关系[J]. 气象学报，60(5)：544-552.

戴新刚，汪萍，丑纪范，2003. 华北汛期降水多尺度特征与夏季风年代际衰变[J]. 科学通报，48(23)：2483-2487.

丁一汇，2005. 高等天气学[M]. 北京：气象出版社：585.

封国林，侯威，董文杰，2006. 基于条件熵长江三角洲温度的非线性动力学特征分析[J]. 物理学报，55(2)：962-968

封国林，戴新刚，王爱慧，等，2001. 混沌系统中可预报性的研究[J]. 物理学报，2001，50(4)：606-611.

龚志强，封国林，万仕全，等，2006. 基于启发式分割算法检测华北和全球气候变化的特征[J]. 物理学报，55(1)：477-484.

黄荣辉，1990. ENSO 及热带海-气相互作用动力学研究新进展[J]. 大气科学，14(2)：234-241.

李建平，丑纪范，2003. 非线性大气动力学的进展[J]. 大气科学，2003，27(4)：653-673.

李琰，王亚非，魏东，2007. 前期热带太平洋、印度洋海温异常对长江流域及以南地区 6 月降水的影响[J]. 气象学报，65(3)：393-405.

林贤超，徐淑英，1989. 东亚季风强弱变化及其对初夏我国东部地区降水的影响[J]. 地理研究，8(2)：44-54.

施能，朱乾根，吴彬贵，1996. 近 40 年东亚夏季风及我国夏季大尺度天气气候异常[J]. 大气科学，20(5)：575-583.

陶诗言，徐淑英，1962. 夏季江淮流域持久性旱涝现象的环流特征[J]. 气象学报，32(1)：1-10.

陶诗言，李吉顺，王昂生，1997. 东亚季风与我国洪涝灾害[J]. 中国减灾，7(4)：17-24.

陶诗言，1980. 中国之暴雨[M]. 北京：科学出版社.

涂长望，黄士松，1944. 中国夏季风之进退[J]. 气象学报，18(1)：1-20.

吴国雄，刘还珠. 1995. 降水对热带海表温度异常的邻域响应 I. 数值模拟[J]. 大气科学，19(4)：422-434.

吴志伟，李建平，何金海，等，2006. 大尺度大气环流异常与长江中下游夏季长周期旱涝急转[J]. 科学通报，51(14)：1717-1724.

张庆云，陶诗言，1998. 亚洲中高纬度环流对东亚夏季降水的影响[J]. 气象学报，56(2)：199-211.

张琼，刘平，吴国雄，2003. 印度洋和南海海温与长江中下游旱涝[J]. 大气科学，27(6)：992-1006.

DAI A，TRENBERTH K E，KARL T R，1998. Global variations in droughts and wet spells：1900—1995 [J]. Geophys Res Lett，25(17)：3367-3370.

DAI X G，FU C B，WANG P，2005. Interdecadal change of atmospheric stationary waves and North China drought[J].

Chinese Physics,14(4): 850-858.

DING Y H,1992. Summer monsoon rainfalls in China[J]. J Meteor Soc Japan,70(1B):373-396.

KALNAY E,KANAMITSU M,KISTLER R,et al,1996. The NCEP/NCAR 40-year reanalysis project[J]. Bull Amer Meteor Soc, 77:437-471.

NAN S L,LI J P, 2003. The relationship between the summer precipitation in the Yangtze River valley and the boreal spring Southern Hemisphere annular mode [J]. Geophys Res Lett,30(24):2266.

SHI N,GU J Q,YI Y M,et al,2005. An improved south Asian summer monsoon index with Monte Carlo test[J]. Chinese Physics,14(4):844-849.

SHI N,YI Y M,GU J Q,et al,2006. On the correlation of nonlinear variables containing secular trend variations: numerical experiments [J]. Chinese Physics,15(9):2180-2184.

SHI N,ZHU Q G,1993. Studies on the northern early summer teleconnection patterns, their interannual variations and relation to drought / flood in China [J]. Adv Atmos Sci,10 (2): 155-168.

WANG P,DAI X G,2004. Numerical solution of Helmholtz equation of barotropic atmosphere using wavelets[J]. Chinese Physics,13(10):1770-1776.

WU Z W,LI J P,HE J H,et al. ,2006. Occurrence of droughts and floods during the normal summer monsoons in the mid- and lower reaches of the Yangtze River[J]. Geophys Res Lett,33(5):1-1. DOI:10. 1029/2005GL024487.

第三部分

东北生态环境与区域气候变化

Quantitative Assessment of Impacts of Regional Climate and Human Activities on Saline-Alkali Land Changes: A Case Study of Qian′an County, Jilin Province*

LIAN Yi[1,2] WANG Jie[1,2] TU Gang[1,2] REN Hongling[1,2] SHEN Baizhu[1,2]
ZHI Keguang[1,2] LI Shangfeng[1,2] GAO Zongting[1,2]

(Laboratory of Research for the Mid-High Latitude Circulation System and East Asian Monsoon, Institute of Meteorological Science of Jilin Province, Changchun 130062)

Abstract: Interdecadal and interannual variations of saline-alkali land area in Qian'an County, Jilin Province, China were comprehensively analyzed in this paper by means of satellite remote sensing interpretation, field flux observations and regional climate diagnosis. The results show that on the interannual scale, the impact of climate factors accounts for 71.6% of the total variation of the saline-alkali land area, and that of human activities accounts for 28.4%. Therefore the impact of climate factors is obviously greater than that of human activities. On the interdecadal scale, the impact of climate factors on the increase of the saline-alkali land area accounts for 43.2%, and that of human activities accounts for 56.8%. The impact of human activities on the variation of saline-alkali land area is very clear on the interdecadal scale, and the negative impact of human activities on the environment should not be negligible. Besides, changes in the area of heavy saline-alkali land have some indication for development of saline-alkali land in Qian'an County.

Keywords: saline-alkali land; regional climate change; human activities

1 Introduction

Along with the global change, global warming and the development of social economy, regional climate changes and human activities have greatly altered the natural environment, and correspondingly, land use and land cover have also changed. Webb (1931) considered that the land use was determined by local drought degree; afterwards, Kraenzel (1955) investigated the forest areas and agricultural areas in the South America, the Southeast Asia and Africa, and held that the land use was jointly affected by climatic, economic and social factors.

In the 1970s, the study on the global changes had become one of hot issues in the academic field, and as one of basic anthropogenic factors affecting the global changes, the land use/cover change (LUCC)

* 本文原载于 *Scientia Geographical Sinica*, 2010, 20(1): 91-97

was very actively studied. In 1995, the International Geosphere-Biosphere Programme (IGBP) and the International Human Dimensions Programme (HDP) on Global Environmental Change jointly presented the Land Use/Cover Change Programme, and the study on LUCC became a frontier and hot topic of global change research (Committee on Global Change, 1990; Turner et al., 1990; Meyer et al., 1994; Chen et al., 2001; Shi et al., 2002; Chen et al., 2003). In recent years, researchers in North America turned to grid cell automatic recognition models instead of concept models (Turner, 1997), Japanese researchers quantitatively investigated the regional LUCC using quantitative and economic models (Otsubo, 1999; Wu et al., 2007), and European researchers simulated the past and current LUCCs to reveal the social and economic driving force of the LUCC.

Studies on LUCC in China advanced rapidly, and some progresses have been achieved in the correlation among climate change, human activity and LUCC (Zhuang et al., 1997; Sun et al., 1998; Yu et al., 2002; Zheng et al., 2003). In the 1990s, the development of land desertification in the west of Northeast China drew much attention (Qiu, 1992; Huang et al., 1996; Lian et al., 1999, 2001; Lin et al., 1999; Sun et al., 2001a; Ren et al., 2002), and some studies qualitatively analyzed the driving forces of the LUCC (Huang et al., 2003; Liu et al., 2005, 2006; Zhang et al., 2006; Li et al., 2007). In recent years, assessments on the status of land desertification are very active due to the application of the remote sensing, geographic information system and global positioning system (3S) techniques, and the comprehensive assessment method has become the development direction of land desertification assessment, however, there are still some biases and unsolved scientific problems (Li et al., 2002; Yang et al., 2004; Chen et al., 2005). For instance, there are some difficulties in selecting satellite data. Most existing studies only selected two phase data in different years to reveal the development trend of land desertification, thus the impact of climate anomaly in the current years on the land desertification was often neglected. The randomness problem in selecting remote sensing data arose (Lian et al., 2007a, 2007b; Sun et al., 2001b). How to separate and quantify the impacts of climate change and human activity on land desertification is a hot spot in the research community (Wang, 2003). Recently, Gao et al. (2006) hold that in the last 20 years, the dual impacts by climate changes and economic development had led to a shift of Land Use Degree Weight Centre of China northeastward by 5 km; and with regard to Land Use Degree Excursion Intensity, in the eastwest direction, 81% is caused by climatic changes and 19% by anthropogenic activities.

Qian'an County in Jilin Province, which is located in the front edge of the soda saline-alkali land (SAL) in the west of Northeast China (Ren et al., 2002), was chosen as the study area in this paper. Using four phases data of Landsat TM, and adopting the "3S" System of Jilin Province Landsat Remote Sensing Information Processing and Application, the paper revealed the variation rates of SAL on the interdecadal and interannual scales in Qian'an and the conversion of major land cover types, and attempted to quantitatively assess the impacts of regional climate change and human excessive reclamation and overgrazing on SAL area variation by means of regional climate diagnosis, field flux observation and satellite remote sensing monitoring.

2 Study Area and Methodology

2.1 Study area

Qian'an County (44°38′—45°19′N and 123°25′—124°23′E) lies in a blind drainage area between the

Songhua River and the Nenjiang River in the west of Jilin Province, Northeast China, with an area of about 3616.48 km^2, and is in semiarid—semihumid continental monsoon climate region. The county, with less amount of seasonal surface runoff, is an unique county without rivers in Northeast China, and its waters consist of 74 saline-alkali ponds, wherein water is importable and also unsuitable for irrigation. Therefore, the variation of the water area is only related to the climatic factors, such as precipitation, temperature and evapotranspiration, which greatly simplifies the water budget estimation of Qian'an, as well as the quantitative assessment of impacts of regional climate changes and human activities on the changes of SAL area.

2.2 Data and method

The main data analyzed in this study were the Landsat TM data of the four phases on 2nd October 1988, 22nd September 1996, 25th September 2000, and 12th September 2001. And the monthly precipitation and average monthly temperature from January to December over Qian'an in 1961—2001 were also used in this paper, which were provided by Jilin Meteorological Bureau, China Meteorological Administration.

The selection of Landsat TM channels, the enhancement processing and the supervised classification of remote sensing images, the information extraction of SAL and other land types referred to Lian et al. (1999); the computation of annual and seasonal relative change rates of precipitation referred to Lian et al. (2001); and the time series of monthly precipitation and temperature from May to September were also standardized.

Evapotranspiration was estimated using the energy balance equation (Bowen, 1926; Tu et al., 2003):

$$R_n = H + LE + G + C \tag{1}$$

where R_n is the net radiation; H is the sensible heat flux; LE is the latent heat flux, L is the latent heat of vaporization of water (about 585 cal g^{-1}①), E is the evapotranspiration; G is the ground heat flux; and C is the outward advection loss. Variations of G and C over a long period, even several days or one day, can be neglected in comparison with the other terms. In such a case, Equation (1) can be reduced to

$$R_n = H + LE \tag{2}$$

R_n was measured by the TBB-1 net radiometer authenticated by China Meteorological Administration, and H was observed by a Large Aperture Scintillometer (LAS) of the Wageningen University. On the aspect of acquiring the large scale surface water and heat fluxes over the nonhomogeneous underlying surface, the LAS is to some extent able to solve the problem of the limited spatial representativeness of the Bowen ratio method and the eddy covariance measurement method (Hu et al., 2003; Lu et al., 2005). Details of Qian'an LAS observation refer to Zhi et al. (2002).

The regional water budget was estimated by using the soil moisture budget (Huang, 1963), which contains 13 parameters, such as precipitation, slope runoff, etc., some of them are difficult to be obtained. In the arid-semiarid area, some minor factors can be neglected, and furthermore, the runoff can also be neglected because Qian'an has no rivers, except some inner lakes, therefore, the soil moisture budget equation is reduced to the following form (Zhi et al., 2003):

$$W_2 - W_1 = R - E \tag{3}$$

where W_2 and W_1 are the total amounts of water at the end and beginning of a period, respectively; and R

① 1 cal=4.186 J。

is the precipitation in the period. That is to say, if precipitation is greater than evapotranspiration in a certain period, the water amount is surplus, otherwise it is deficit.

3 Results and Analyses

3.1 Variation in saline-alkali land area on interannual and interdecadal scales

Table 1 gives the variation of saline-alkali land (SAL) area in Qian'an from 2000 to 2001 and its causes. The SAL area in 2001 increased by 65.83 km^2 (8.5%) compared with that in 2000, in which 71.6% came from the drying of saline-alkali waters and wetlands, and 28.4% from the degradation of grasslands. The interannual variation of farmlands was ignored, because it was very small.

Table 1 Variations of SAL area in 2000—2001 and its causes in Qian'an

SAL	Area in 2000	Area in 2001	Variation in 2000—2001		Cause for area variation
	km^2	km^2	km^2	%	
Heavy	696.11	712.84	+16.73	+2.4	
Medium	76.54	125.65	+49.10	+64.1	
Total area	772.65	834.49	+65.83	+8.5	40.8% from reduction in saline-alkali waters 30.8% from reduction in wetland 28.4% from degradation of grassland

It can be seen from the comparison between the interpretation results of 1988 and 1996 (Lian et al., 1999; Ren et al., 2002) that during the 8 years (the period more than 8 years generally represents the interdecadal scale), land cover types with a larger area variation are orderly grassland (−63.1%), farmland (+26.2%), SAL (+51.5%), and saline-alkali waters (−44.5%). Table 2 shows the variation in SAL area in Qian'an from 1988 to 1996 and its causes. The reduction in waters and wetland contributed 43.2% to the total increment of SAL area, while the degradation of farmland and grassland accounted for the rest 56.8%.

It can be found from the comparison of the variations between medium SAL and heavy SAL in Table 1 and Table 2 that the interannual variation of medium SAL area (64.1%) was much more distinctive than that of heavy SAL area (2.4%); while on the interdecadal scale the relative variation in heavy SAL area reached 59.1%, far greater than that (19.2%) in medium SAL area. Apparently, whether on the interannual scale or the interdecadal scale, the variation rates of the heavy SAL area were about the same order of magnitude as that of the total SAL area. Therefore, monitoring variation of heavy SAL area has indication for the development of land salinization in Qian'an.

Table 2 Variations of SAL area in 1988—1996 and its causes in Qian'an

SAL	Area in 1988	Area in 1996	Area variation in 1988—1996		Cause for area variation
	km^2	km^2	km^2	%	
Heavy	423.76	674.28	+250.52	+59.1	
Medium	100.22	119.47	+19.25	+19.2	
Total area	523.98	793.75	+269.77	+51.5	37.4% from reduction in saline-alkali waters 5.8% from reduction in wetland 56.8% from degradation of grassland and expansion of farmland

3.2 Estimation of water budget in growing season

Winter precipitation in Qian'an is about 50 mm, accounting for about 10% of the annual precipitation, therefore the water budget only in the growing season (May—September) was considered. The water budget in 2000—2001 was estimated using the field observation data (net radiation, sensible heat flux, etc.) and Equations (1)—(3) (Table 3). The water budget in 1988—1996 was not estimated because Qian'an Large Aperture Scintillometer (LAS) station was established in 1999.

Table 3 Water budget in growing season in 2000—2001

	2000/mm	2001/mm	Variation in 2000—2001	
			mm	%
Precipitation	434.7	209.2	−225.2	−51.9
Evapotranspiration	411.3	373.4	−37.9	−9.2
Water budget	+23.4	−164.2		

It can be seen from Table 3 that on the one hand, the interannual variation in evapotranspiration was smaller than that in precipitation, the difference of precipitation between 2000 and 2001 was 225.5 mm, while that of evapotranspiration was only 37.9 mm; and on the other hand, the difference in the water budget between the two years was large, the budget in 2000 was 23.4 mm, while the budget in 2001 was 164.2 mm. Therefore, even if the precipitation in growing season decreased, the total evapotranspiration did not declined in proportion, so the water deficit could happen and the environment drought might be increasing. This is consistent with the 8.5% increment of the SAL area in 2001 (Table 1). Therefore, as far as interannual variation of SAL area is concerned, the increase/decrease in precipitation, as well as the surface evapotranspiration are both very important.

3.3 Impacts of regional climate and human activity on land salinization

The precipitation in growing season, SAL and waters areas in Qian'an in 1988, 1996, 2000 and 2001 were compared successively (Table 4). For the purpose of comparison convenience, the standardized time series of the total precipitation and average temperature in growing season from 1961 to 2002, and variations of the saline-alkali waters and SAL areas during the periods of 1988—1996, 1996—2000, and 2000—2001 were showed in Fig. 1 and Fig. 2 respectively.

It can be seen from Fig. 1 and Fig. 2 that the climate in Qian'an has become warm/dry since the mid-1990s, which accords with the expansion of SAL area. Variations in the saline-alkali waters area and the precipitation in growing season were in phase, but variations in SAL area and the average temperature in growing season were out of phase. The mean temperature in growing season had some impact on variations in saline-alkali waters area and SAL area, but its impact was much less than that of the precipitation in growing season.

The time series of precipitation in growing season in Fig. 1 also shows a typical wavy pattern. The years when the anomaly of the precipitation exceeded $\pm 1.2\sigma$ were defined as a positive/negative anomalous years respectively. There were 9 anomalous years in total, accounting for 21.4% of the 42 years (1961—2002), therefore, the positive and negative anomalous years of the precipitation were basically within 10% respectively. Therefore, the driving forces of extreme precipitation events for saline-alkali waters and SAL area variations were very robust.

Table 4 shows that the interdecadal and interannual variations of the saline-alkali waters area and total precipitation in growing season in Qian'an are both larger, but their trends are in a similar phase (Fig. 1). The area of saline-alkali waters decreased 44.5% from 1988 to 1996, but the SAL area increased 51.5%, which was consistent with the warm/dry climate trend of the west of Northeast China in the mid-1990s (Lian et al., 2001, 2005; An et al., 2002; Liu et al., 2003; Lian et al., 2007b). In 1998, an exceptionally serious flood happened in the Songnen Plain, directly resulting in the increase of 84.72 km^2 (+65.2%) in saline-alkali waters area in 2000 compared to that in 1996, and the waters area basically returned to that of 1988. However, due to the exceptional severe drought, the standardized growing season precipitation decreased 3.0σ (Fig. 2) (Table 3), and the deficit of water budget reached 164.2 mm (+44%), resulting in the reduction of 29.8% in waters area and the increase of 8.5% in SAL area in 2001 in comparison with those in 2000 (Table 4).

Table 4 Variations in precipitation, saline-alkali waters and saline-alkali land area in Qian'an

Year	Precipitation in growing season		Saline-alkali waters area		Total SAL area	
	mm	%	km^2	%	km^2	%
1988	466.70		226.92		523.98	
1996	242.70		125.95		793.75	
2000	434.70		210.67		772.65	
2001	209.20		147.87		839.49	
1988—1996	−224.00	−48.0	−100.97	−44.5		+51.5
1996—2000	+192.00	+79.1	+84.72	+65.2	−21.10	−2.7
2000—2001	−225.50	−51.9	−62.80	−29.8	+65.84	+8.5

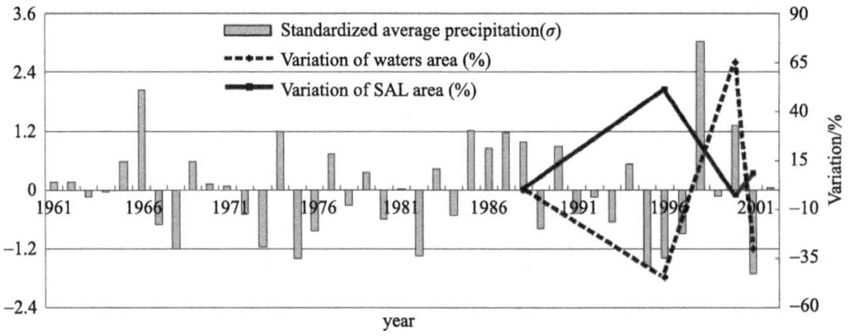

Fig. 1 Total precipitation in growing season and variations of waters and SAL areas after 1988 in Qian'an

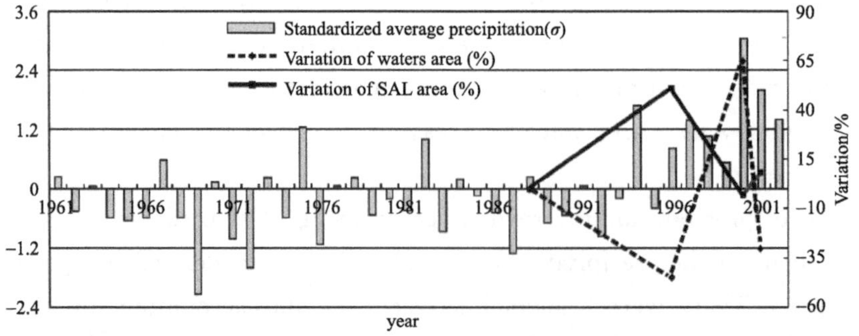

Fig. 2 Average temperature in growing season and variations of waters and SAL areas after 1988 in Qian'an

It can be held that the interdecadal and interannual variations of waters (saline-alkali waters) and wetland, major types of land cover in Qian'an, are mainly affected by climate. So the change in SAL area associated with the changes of saline-alkali waters and wetland should be attributed to climatic factors. The increase of 43.2% in the SAL area of Qian'an from 1988 to 1996 was transformed from waters and wetland (Table 2), and the increase of 71.6% in SAL area from 2000 to 2001 came from the reduction of waters and wetland (Table 1), indicating that the impact of the climatic factors on the interdecadal scale accounted for 43.2% of variation in SAL area, while the impact on the interannual scale accounted for 71.6%. Variations in farmland and grassland areas were mainly affected by human activities, and the related variation in SAL area was mainly caused by excessive reclamation and overgrazing. Indeed, the impact of the warm/dry trend of climate also accounted for the variation to some extent. It can be found from Table 1 and Table 2 that human activities accounted for 56.8% of the interdecadal variation, and 28.4% of the interannual variation in SAL area, respectively.

Therefore, for the variation in SAL area on the interdecadal scale, the effect of farmland expansion and grassland reduction was approximately equivalent to that of the reduction of waters and wetland due to the warm/dry trend of climate, while, on the interannual scale, the impact of the warm/dry trend of climate was obviously greater than that of human activities.

4 Conclusions

The specific geologic and geographic environment of Qian'an County, Jilin Province, provides the simplified condition for understanding the mechanism of land salinization. The interdecadal and interannual variations in SAL area have been analyzed comprehensively by the way of satellite remote sensing interpretation, field flux observation, and regional climate diagnosis. It can be concluded that for the interannual variation of land salinization, the impact of climate was the main factor (71.6%), and that of human activities was the second (28.4%), while for the interdecadal variation, the impact of human activities (56.8%) was greater than that of climate (43.2%). The negative impact of human activities on the environment is not negligible. Qian'an is one of areas where land salinization distinctively developed due to excessive reclamation and overgrazing in Jilin Province, and lies in the west of the Songliao Plain, a national main grain production region. Therefore, the comprehensive management and protection of ecological environment in Qian'an is important for national grain safety.

The relative variation of heavy SAL area on interannual and interdecadal scales were more close to the relative variation of total SAL area, therefore, the variation of the heavy SAL area has some indication for the development of land salinization in Qian'an.

This paper is only a preliminarily attempt to quantitatively assess the effect of regional climate and human activities on SAL area variation. The impact of climate is only partly separated from that of human activities, meanwhile the impact of excessive reclamation and overgrazing on SAL area variation still contains the effect of the warm/dry trend of climate. To solve this problem, numerical simulation should be adopted in future research work.

References

AN G, SUN L, LIAN Y, et al, 2002. The climatic changes analysis of Qian'an in the last 40 years[J]. Clim Environ Res, 7(3): 370-376. (in Chinese)

BOWEN I S, 1926. The ratio of heat losses by conduction and by evaporation from any water surface[J]. Phys Rev, 27(6): 779-787.

CHEN B M, LIU X W, YANG H, 2003. Review of most recent progresses of study on land use and land cover change[J].

Prog Geo,22(1):22-29. (in Chinese)

CHEN H L,XU X D,LIU Y J,2005. Review of researches on remote sensing monitoring and impact on environment of land use/cover change[J]. Meteor Sci Technol,33(4):289-294. (in Chinese)

CHEN Y Q,YANG P,2001. Recent progresses of international study on land use and land cover change (LUCC)[J]. Econ Geo,21(1):95-100. (in Chinese)

Committee On Global Change,1990. Research Strategies for the U. S. Global Change Research Program[C]. Washington, DC:National Academy Press:112,117-118.

GAO Z Q,LIU J Y,2006. The LUCC responses to climatic changes in China from 1980 to 2000[J]. Acta Geograghica Sinica, 61(8):865-872. (in Chinese)

HU L Q,WU R Z,FANG Z Y,2003. Large aperture scintillometer and its application to surface energy balance monitoring [J]. J Appl Meteor Sci,14(2):197-205. (in Chinese)

HUANG B W,1963. Physical geography and soil moisture balance,with special reference to four recent papers by J. C. Wilcox[J]. Acta Geograghica Sinica,29(4):331-352. (in Chinese)

HUANG F,LIU X N,WANG P,et al,2003. Land use/cover change and its driving forces of west of Songnen Plain[J]. J Soil Water Conser,17(6):14-17. (in Chinese)

HUANG X C,MENG X X,1996. Study on Ecological Environment Fragile Zone in the West of Northeast China[M]. Beijing:Science Press:1-20. (in Chinese)

KRAENZEL C F,1955. The Great Plain in Transition[M]. Norman:University of Oklahoma Press:428.

LI B L,ZHOU C H,2002. A study on monitoring sandy desertification in sandy land of west Northeast China Plain[J]. J Remote Sens,6(2):117-122. (in Chinese)

LI Y C,LIU C X,2007. Land use/cover dynamic changes in Northern China[J]. Scientia Geographica Sinica,27(1):45-52. (in Chinese)

LIAN Y,GAO Z T,REN H L,et al,1999. A monitoring analyse of desertification development in western Jilin of China by Landsat TM data[J]. Quaternary Sci,57(6):662-667. (in Chinese)

LIAN Y,GAO Z T,REN H L,et al,2001. Desertification development and regional climatic change in Northeast China in the 1990s[J]. Acta Meteor Sinica,59(6):730-736. (in Chinese)

LIAN Y,GAO Z T,SHEN B Z,2007a. Climate change and its impacts on grain production in Jilin Province[J]. Adv Clim Change Res,3(1):46-49. (in Chinese)

LIAN Y,SHEN B Z,GAO Z T,et al,2005. Correlation of drought development tendency on China's climate transition zone and activity of East Asia summer monsoon and polar vortex[J]. Acta Meteor Sinica,63(5):740-749. (in Chinese)

LIAN Y,WANG Y J,2007b. Research advance on land desertification and climate change of Northeast China[J]. Sci Geographica Sinica,27(suppl):97-102. (in Chinese)

LIN N F,YANG J,BIAN J M,et al,1999. The Quaternary environmental evolution and the problem of desertification in Northeast Plain[J]. Quaternary Sci,20(5):448-455. (in Chinese)

LIU D W,SONG K S,WANG D D,et al,2006. Dynamic change of land-use patterns in west part of Songnen Plain[J]. Sci Geographica Sinica,26 (3):277-283. (in Chinese)

LIU X H,WANG J F,LIU J Y,et al,2005. Quantitative analysis approaches to the driving forces of cultivated land changes on a national scale[J]. Transactions of the CSAE,21(4):56-60. (in Chinese)

LIU Z X,LIAN Y,SHEN B Z,et al,2003. Seasonal variation features of 500 hPa height in North Pacific Oscillation region and its effect on precipitation in Northeast China[J]. J Appl Meteor Sci,14(5):553-561. (in Chinese)

LU L,LIU S M,SUN M Z,et al,2005. Advances in the study of areal surface fluxes with large aperture scintillometer[J]. Adv Earth Sci,20(9):932-938. (in Chinese)

MEYER W B,TURNER B L Ⅱ,1994. Change in Land Use and Land Cover:A Global Perspective[M]. London:Cambridge University Press.

OTSUBO K,1999. Land Use for Global Environmental Conservation (LU/GEC):Final Report of the LU/GEC First Phase (1995—1997)[R]. Tokyo:National Institute for Environmental Studies, Environmental Agency of Japan.

QIU S W,1992. Serious hazards of sandy desertification,salinization and grassland degradation in the west of Northeast China, and their comprehensive treatment[M]//SHI Y F, HUANG D C, CHEN P Q. The Analysis of Disaster Situation and the Countermeasures to Reduce Natural Hazards in China. Wuhan: Hubei Science and Technological Press: 468. (in Chinese)

REN H L, LIAN Y, GAO Z T, et al, 2002. Using remote sensing technique to study the desertification in the west of Northeast China[J]. Quaternary Sci, 22(2): 136-140. (in Chinese)

SHI P J, SONG C Q, JING G F, 2002. Strengthening the study of land use/cover change and its impact on ecoenvironmental security—The trend of the study of the dynamics of human-nature system based on "Global Change Open Science Conference 2001" in Amsterdam, Netherlands[J]. Adv Earth Sci, 17(2): 161-168. (in Chinese)

SUN H Y, WANG C Y, NIU Z, et al, 1998. Analysis of the vegetation cover change and the relationship between NDVI and environmental factors by using NOAA time series data[J]. J Remote Sens, 2(3): 204-210. (in Chinese)

SUN J M, LIU T S, 2001a. Desertification in the Northeastern China[J]. Quaternary Sci, 22(1): 72-78. (in Chinese)

SUN L, AN G, 2001b. A diagnostic study of northeast cold vortex heavy rain over the Songhuajiang—Nenjiang River Basin in the summer of 1998[J]. Chinese J of Atmos Sci, 25(3): 342-354. (in Chinese)

TU G, ZHI K G, SUI C Y, 2003. Estimation of evapotranspiration of Qian'an area using energy balance method[M]//SUN L, LIAN Y, GAO Z T. Climate and Ecological Environment. Beijing: China Meteorological Press, 143-146. (in Chinese)

TURNER B L Ⅱ, KASPERSON R E, MEYER W B, et al, 1990. Two types of global environmental change: Definitional and spatial-scale issues in their human dimension[J]. Global Environ Change, 1(1): 14-22. DOI: 10. 1016/09-59-3780(90)90004-S.

TURNER B L Ⅱ, 1997. Socializing the pixel in LUCC[J]. LUCC Newsletter, 1: 10-11.

WANG T, 2003. Study on sandy desertification in China—2. Contents of desertification research[J]. J Desert Res, 23(5): 428-477. (in Chinese)

WEBB W P, 1931. The Great Plain[M]. Boston: Ginn and Company: 525.

WU W B, YANG P, SHIBASAKIP Y, et al, 2007. Agent-based model for land-use/cover change: A review[J]. Sci Geographica Sinica, 27(4): 573-578. (in Chinese)

YANG X H, ZHANG K B, CI L J, 2004. Desertification assessment in China: The state-of-art, problems and solution measures[J]. Sci of Soil and Water Conserv, 2(1): 22-28. (in Chinese)

YU X X, YANG G S, 2002. The advances and problems of land use and land cover change research in China[J]. Prog Geo, 21(1): 51-57. (in Chinese)

ZHANG Y H, LUO Y, LIU J Y, et al, 2006. Driving forces of land use changes in irrigated agriculture[J]. Resour Sci, 28(1): 81-86. (in Chinese)

ZHENG H J, HUA L, OU L Y, 2003. Summarization of the study of land use/land cover change in China[J]. Journal of Capital Normal University (Natural Science Edition), 4(3): 89-95. (in Chinese)

ZHI K G, TU G, LIAN Y, et al, 2002. Measuring sensible heat flux over saline-alkali land area in Qian'an[J]. Acta Meteor Sinica, 60(6): 780-785. (in Chinese)

ZHI K G, TU G, LIAN Y, 2003. Environmental drought in the west of Jilin Province[M]//SUN L, LIAN Y, GAO Z T. Climate and Ecological Environment. Beijing: China Meteorological Press: 147-150. (in Chinese)

ZHUANG D F, LIU J Y, 1997. Study on the model of regional differentiation of land use degree in China[J]. J Natural Resour, 12(2): 105-111. (in Chinese)

吉林省西部荒漠化发展的陆地卫星遥感监测分析*

廉毅　高枞亭　任红玲　唐晓玲　安刚　张文哲　孙力

（吉林省气象科学研究所，长春 130062）

摘　要：依据1996年8—10月和1988年8—10月两个时期的Landsat TM遥感信息静态反演，对两个时期TM资料采用图像增强处理和计算机监督分类进行荒漠化（侧重乾安县盐渍化）解译和动态量变分析，并与1976年和1981年的卫片进行对比分析，研究结果表明，吉林省西部的生态环境在恶化，主要表征为荒漠化发展并向东扩张，已波及吉林省中部松辽平原。大安县是吉林省盐渍化面积最多的县份，盐碱地已占全县总面积的50%以上；乾安县是盐渍化面积漫延较快的县份，盐渍化面积8 a增加了269.77 km^2，增加量占县域面积的7.87%。

关键词：吉林省西部；陆地卫星资料；生态环境；荒漠化发展

1　引言

吉林省西部位于(43°—47°N，122°—125°E)，处于中国湿润的东亚季风区和干旱内陆之间的气候过渡带（年降水量300～500 mm），是生态系统从半湿润森林草原向半干旱草原和沙漠之间的过渡带，也是中国历史上农牧交错带的位置（符淙斌 等，1995）。该区受气候的自然变化和人类活动共同作用，其环境变化比较敏感，是全球变化响应比较突出的区域，已引起国内学者的关注，对近代该区荒漠化还在向东发展，是生态脆弱带和气候脆弱区等特征已形成共识（《东北平原》第四组自然环境形成与演化基金课题组，1990；肖荣寰，1995；方修琦，1993；吴正，1987；董光荣 1994）。文中采用吉林省陆地资源卫星遥感信息处理与应用的3S系统，利用1996年8—10月和1988年8—10月两个时期的Landsat TM遥感信息静态反演，对两个时期TM资料采用图像增强处理和计算机监督分类进行荒漠化（侧重乾安县盐渍化）解译和动态量变分析，再与1976年和1981年不同时期的卫片相对照，揭示了吉林省西部的生态环境进一步恶化，其主要表征——荒漠化发展并向东扩张的量变证据。

2　TM资料荒漠化的反演与解译

2.1　吉林省陆地卫星影像地图的制作和解译

选用的陆地卫星资料为1996年8—10月间的Landsat 5 TM资料。吉林省及其邻域范围共有19帧资料。由于TM资料受天气等因素的影响，不可能选择同一天的资料，另考虑到东北地区地物以秋季对比较为明显，故所选时间大多数为8—10月，以增加可比性。

TM资料选取7、4、3这3个波段资料进行合成，形成假彩色图像。在有关软件平台的基础上，开发和研制了吉林省陆地资源卫星遥感信息处理与应用的3S系统，利用此系统通过各帧的拼接、各帧色调差异的调整、几何纠正、几何配准和增加各种注记等处理，最后制作出了吉林省陆地卫星影像地图（图略）。

* 本文原载于《气象学报》，1999，57(6)：662-667

吉林省中西部地区陆地卫星遥感影像图中的地面景观特征表明了西部地区荒漠化相当严重的态势(图略)。荒漠化所包含的盐渍化和沙漠化区分别在图中显示为以深蓝色的碱泡(水面)为核心的浅黄白色区和黄褐色的带状或成片状区,呈椭圆形的盐渍化区,区中由数十个大小碱泡构成,其外围浅黄白色区或连成条状区,或呈星罗棋布,为松嫩沙地、嫩江下游西侧的大片盐渍化区(以吉林省大安县为中心)。沿45°N 纬圈(124°—125°E)漫延了约 20 km,正开始侵蚀第二松花江下游著名的稻米产区——前郭灌区,盐渍化区显然已扩展连成带状区;43°—45°N 的与季风盛行风走势一致的西北—偏西—西南走向的沙漠带变宽,东辽河两岸的沙地从内蒙古东部和辽宁省北部境内,呈扫把状明显向东和东北方向迅速伸展,10 a来已东扩 20 km,直接威胁着吉林省的产粮大县梨树县,基本上已与北面的沙带相融为一区;位于松辽平原商品粮基地中部的中国产粮二连冠的农安县,荒漠化亦在发展,盐渍化区分别向东北、西北和南部 3 个方向延伸。吉林省西部地区荒漠化已对中国的松辽平原商品粮基地构成严重的威胁。

2.2　盐渍化区发展和向东扩展的动态量变分析

2.2.1　信息源

以吉林省乾安县盐渍化区动态分析为例,乾安县位于科尔沁草原东端,在沙垄间平地与波状平原中共有 72 个大小湖泡。国内以往在提取植被信息处理上,选择 5、4、3 通道,植被解译为红色(图略)(东北师范大学 等,1989),采用 1996 年 9 月 22 日和 1988 年 10 月 2 日两景秋季的 TM 遥感资料,选择其中的可见光的 3 波段(因为它受大气的阴霾影响最小)、近红外的 4 波段和中红外的 7 波段,将 3、4、7 波段组成的假彩色图像作为解译主要信息源,并分别选择同一景 TM 图像采用 K-T 变换后得到的土壤特征伪彩色图作为辅助信息源。

2.2.2　遥感图像的增强处理

图像增强处理的重点是在图像中突出所需的信息特征或模式,减弱和去掉不相干信息,使图像适合人们视觉系统的习惯和屏幕目视解译特点。主要采用了灰度变换(即对比度增强)和直方图均衡变换两种处理方法。

灰度变换是通过修正图像的灰度等阶实现图像对比度增强。现采用线性变换:

$$g(x,y)=\frac{f(x,y)}{a_2-a_1}(b_2-b_1)+b_1 \tag{1}$$

假定原图像灰度值变量动态范围为$[a_1,a_2]$,待扩展的范围为$[b_1,b_2]$,$b_1<a_1$,$a_2<b_2$,原图像元的灰度值为 $f(x,y)$,扩展后的灰度值为 $g(x,y)$,可采用多个分段线性函数分别对图像的暗区、中等亮度区或亮区进行线性变换,突出感兴趣的目标或灰度区间,尤其是盐碱地的遥感信息分布于图像的亮区,因此突出亮区而抑制暗区,处理后的图像增强了盐碱地信息质量。

对 7、4、3 波段图像分别进行一系列图像处理后赋予红、绿、蓝三基色合成一幅假彩色图像。此影像以绿色为主,河流沟渠的走向、道路以及城镇居民地分布都清晰可见,耕地呈绿色或黄绿色,水体呈蓝黑色,沼泽地为黑绿色。盐碱地土层薄或土壤侵蚀较严重,显示浅黄色或浅白色;重盐碱化地(碱巴拉)呈白色不规则形状分布在湖泡边缘及各种低洼地,有些盐碱地因返盐重,白色盐斑特征明显,沙地反射率也强,呈蓝白色,伴有浅蓝色条纹,显示其中有轮耕地,草地退化严重的撂荒地呈淡红色。

2.2.3　遥感图像计算机分类和盐碱地信息的提取

进行盐碱地信息的识别和提取与遥感图像的计算机分类密不可分。分类是利用遥感图像的光谱信息、空间信息以及多时期信息对目标进行识别并归类。我们采用最大似然分类进行计算机监督分类(林培,1995),就是用 Bayers(贝叶斯)定理求出每一类别的判别函数,为

$$G_i(x)=\frac{1}{2}(x-M_i)\sum^{-1}(x-M_i)-\ln P_i(W_i)+\frac{1}{2}\ln\left|\sum\nolimits_i\right| \tag{2}$$

式中，i 为类别序号；W_i 为类别标志；$P_i(W_i)$ 为第 i 类的先验概率；x 为所要判别的像元点的多光谱亮度值；M_i 为第 i 类 W_i 的均值向量；Σ_i 为协方差矩阵。当要判别任意像元属于哪一类时，可将 x 代入各类别的判别函数进行归类。然后对分类结果进行校验，以便决定是否进行分类训练区更正或重新分类。

上述的整个过程都是在地理信息系统 GIS 中完成的，对复杂的难以区分的自然景观，以计算机分类为基础结合同一景经 K-T 变换后得到的土壤特征图像，在图像处理软件支持下，采用人机交互方式对初始分类结果多次修正，确保分类结果数据的精度，比如，盐碱土表层容易结盐，地表光滑坚实而发白，因此可以根据影像白色色调的多少来区分出不同程度的盐碱化土壤。但是，在一些排水不通畅或土壤质地较黏重的地段，在图像中呈暗色；还有一些中度和轻度的盐碱土上覆盖着许多草类，不易区分。因此选用 K-T 变换后得到亮度、绿度和湿度这 3 个主分量，将其中亮度与湿度分量组合形成一幅土壤特征信息图像作为辅助判别信息，这幅图像突出了土壤盐碱化信息，便于区分各种等级的盐碱土。

2.2.4 大安县是吉林省盐渍化面积最多的县份

用以上方法解译出大安县 1996 年重中度盐碱地为 1915 km^2，占全县总面积的 39.25%；轻度盐碱地的面积为 391.7 km^2，占全县总面积的 8%；碱泡及部分水体为 181.5 km^2，总计盐碱地范围占全县总面积的 50%以上。

2.2.5 乾安县盐渍化发展并向东扩展

将 1988 年 10 月 2 日和 1996 年 9 月 22 日两景 TM 图像处理分类计算后得出：1988 年乾安县重盐碱地 423.76 km^2，占全县总面积的 12.35%；中度盐碱化面积 100.22 km^2，占全县总面积的 3%；碱泡及部分水体面积 205.63 km^2，占全县总面积的 6%，总计盐碱地的范围占全县总面积的 21.35%。1996 年乾安县重盐碱地 674.28 km^2，占全县总面积的 19.66%；中度盐碱化面积 119.47 km^2，占全县总面积的 3.5%；碱泡及部分水体面积 104.66 km^2，占全县总面积的 3.05%，总计盐碱地的范围占全县总面积的 26.21%。经过 8 a 的变化，重盐碱地增加了 250.52 km^2，中度盐碱地增加 19.25 km^2，两者合计，增加范围占全县总面积的 7.87%，其中因水面减少了 100.97 km^2，而使盐碱地增加范围占全县总面积的 2.9%，占盐渍化面积增加量的 37.4%；原有陆地盐碱地增加占全县总面积的 4.93%，占总增加量的 62.6%。

乾安县盐碱地多分布于碱泡(呈深蓝色的水面)边缘及漫岗间的碟形洼地和带状封闭洼地带(图略)。可明显看出，盐碱地的变化主要表现为碱泡面积缩小而边缘盐碱地明显扩大；乾安县东部边界中段的呈电视塔状的大片绿地，后已蜕变成浅白色外围和浅灰色的盐碱地，并明显地往西北方向延伸一直达县界，其东西跨度约为 8 km，只剩下少部分核心绿地。由此可认定，乾安县盐碱地向东扩展明显。乾安县盐碱地发展有两个直接原因，一是水面减少而使湖泡周围的盐碱地裸露扩大，二是东部草场退化使盐碱化面积扩大。

3 结论和讨论

(1)依据 1996 年 8—10 月的吉林省及邻域 Landsat TM 资料，经用吉林省陆地资源卫星遥感信息处理与应用的3S系统反演，从研制成的大型、彩色和高分辨率的吉林省陆地卫星遥感影像图的解译表明，吉林省西部的生态环境在恶化，主要表征为荒漠化(盐渍化和沙漠化)区域发展并向东扩张；松嫩沙地、嫩江下游西侧的大片盐渍化区(以吉林省大安县为中心，包括乾安县)，经与 1976 年和 1981 年的卫片相比照，20 a 来，沿 45°N 纬圈(124°—125°E)蔓延扩张 20 km，已开始侵蚀第二松花江下游著名的稻米产区——前郭灌区，大安县的盐渍化区明显发展连成带状区；43°—45°N 的与季风盛行风走势一致的西北—偏西—西南走向的沙漠带变宽，这是中国中高纬度沙地由于季风风成作用的明显痕迹，并与北面的盐渍化区连成一片，已逼近到长春市所属的吉林省中总商品粮基地之一的农安县边界；位于吉林省西南部的东辽河两岸的沙地，从内蒙古东部和辽宁省北部境内呈扫把状明显向东和东北方向迅速伸展，10 a 来已东扩20 km，直接威胁着吉林省的产粮大县梨树县，亦与北面的沙带相融为一区。荒漠化已从3个方向逼近，对中国的商品粮基地——吉林省中部松辽平原产粮区构成严重威胁。而商品粮基地中的农安县盐渍化亦在发展，生态环境不容乐观。所有这些都表明前郭、农安和梨树县是荒漠化发展前沿区。

(2)选用1996年9月22日和1988年10月2日两个秋季TM的7、4、3通道资料，经盐碱地信息的增强提取以及计算机监督分类，植被呈绿色、盐碱地呈浅黄白色，与国内以往的5、4、3通道处理方法相比较，既符合自然景观，也符合人们的视觉习惯。结果表明，大安县是吉林省盐渍化面积最多的县份，盐碱地范围占全县总面积的50%以上；经过8 a的变化，乾安县重盐碱地增加250.52 km^2，中度盐碱地增加19.25 km^2，二者合计增加量占全县总面积的7.87%，其中因碱泡水面减少而露出盐碱化土地达100.97 km^2，而使盐碱地增加占总面积的2.9%，其余是盐碱土地增加占全县总面积的4.93%；由乾安县东部边界大片绿地蜕化为以盐碱地为主，且东西跨度约为8 km。这表明大安县是吉林省西部荒漠化的严重区，乾安县东部是其盐渍化前沿发展区。

(3)以卫星遥感和计算机软硬件为前沿工程，积极应用20世纪90年代国际上异军突起的3S一体化高新技术，可以实现对生态环境研究不再因循由点到面的地学研究的传统演绎方法，而是从宏观到微观细部的拓展(陈述彭，1997)。吉林省大型、彩色和高分辨率的陆地卫星遥感影像可从宏观目视、计算机监督分类的宏观和微观相结合的分析，揭示东北西部，特别吉林省西部是欧亚大陆和中国土地沙漠化发展的东缘，是荒漠化向东发展的显著区，是中国重要的松辽平原商品粮生产基地的屏障地区，是农牧渔业发展潜力较大的区域，该区生态环境恶化情况比较严重，加强对该区荒漠化变化成因、监测和预测研究，不仅对大气科学和有关学科的发展和应用有着重大现实意义，也将对世界和中国生态环境整治、建设具有重大示范意义。

(4)本文所做的研究工作只是刚刚开始，还有许多问题，如生态环境正逆过程与气候变化的相关性、如何区分自然与人为对环境的作用等都有待于今后做坚持不懈和深入一步的研究。

参考文献

陈述彭，1997．遥感地学分析的时空维[J]．遥感学报，1(3)：161-171．

《东北平原》第四组自然环境形成与演化基金课题组，1990．中国东北平原第四组自然环境形成与演化[M]．哈尔滨：哈尔滨地图出版社：185-200．

东北师范大学，中国科学院长春地理研究所，吉林农业大学，1989．吉林省陆地卫星影像图集[M]，长春：东北师范大学出版社：12．

董光荣，1994．科尔沁沙地沙漠化的几个问题[J]．中国沙漠，14(1)：1-9．

方修琦，1993．中国北方农牧交错带环境问题分析[C]//中国环境科学学会．中国青年环境论坛首届学术年会走向21世纪——中国青年环境论坛首届学术年会论文集．北京：中国环境科学出版社：423-428．

符淙斌．叶笃正，1995．全球变化和我国未来的生存环境[J]．大气科学，19(1)：116-126．

林培，1995．农业遥感[M]．北京：北京农业大学出版社：119-121．

吴正，1987．风沙地貌学[M]．北京：科学出版社：316．

肖荣寰，1995．松嫩沙地的土地沙漠化研究[M]．长春：东北师范大学出版社：168-169．

中国东北西部地区荒漠化发展前沿区域的遥感研究*

任红玲[1] 廉毅[1] 高枞亭[1] 张文哲[1] 钟林生[2] 丁立[1]

(1 吉林省气象科学研究所,长春 130062;2 中国科学院沈阳应用生态研究所,沈阳 110000)

摘　要:依据 1996 年 7—9 月和 1988 年 8—10 月两个时期的 Landsat TM 遥感信息,分别对中国东北西部地区荒漠化发展前沿的典型区域吉林省乾安县的盐渍化、吉林省梨树县土地沙化变化和乌尔吉木伦河与新开河之间的土地利用状况,进行遥感解译和实地调查。分析研究结果表明,东北西部这三个区域的生态环境 1996 年比 1988 年有所恶化,主要表征为盐渍化和沙化土地向东有较强发展和蔓延,撂荒地显著扩大,与沙化土地面积量级接近。由此东北西部应该成为中国荒漠化和生态环境治理的重点区域。

关键词:东北西部;荒漠化发展;典型区域;生态环境

1　引言

中国东北西部地区系指辽宁、吉林、黑龙江三省的西部和内蒙古自治区东部的四盟区域,地理范围跨 42°—50°N,117°—125°E,总面积约为 3.2×10^5 km^2。该区气候基本上处于中国湿润的东部季风向内陆干旱、半干旱的过渡区域,自然植被上横跨森林草原或草甸草原、典型草原乃至荒漠化草原地带,生态环境较为敏感和脆弱。由于自然和人为因素两方面的原因,该区土地和草地盐渍化、沙化、贫瘠化发展较为严重,近年来并有向东发展的趋势,对东北地区中部我国重要的粮食生产基地黑土地已形成较大威胁(裘善文,1990;肖荣寰,1995;吴正,1987;董光荣,1994;廉毅 等,1999;孙继敏 等,2001)。廉毅等(1999)的研究强调该区生态环境恶化情况比较严重,应加强对该区荒漠化成因、监测和预测的研究,特别是孙继敏等(2001)的研究表明,倘若不及时采取措施的话,中国东北的黑土地区将变为类似于西北的荒漠区。

本文运用吉林省陆地卫星遥感信息与应用的“3S”系统,利用东北西部 1996 年的陆地卫星遥感影像图,并应用 GPS 实地验证对照,对处于该区盐渍化发展前沿的吉林省乾安县的盐渍化变化、处于沙漠化发展前沿的吉林省梨树县的土地沙化变化以及科尔沁沙地腹地的乌尔吉木伦河与新开河之间的土地利用变化分别进行了解译,从而获得 20 世纪 90 年代中后期东北西部地区生态环境不断恶化、荒漠化在加速的若干证据,为在该地区开展荒漠化治理工作提供直接依据。

2　资料和处理方法

采用 1996 年 7—9 月的 16 景 Landsat TM 数据资料。首先对各景资料选用 TM7、4、3 通道合成影像,其次对各影像进行几何纠正和图像配准,然后对影像进行了一系列图像处理,镶嵌合成了东北西部地区 1996 年夏季陆地卫星遥感影像图(图 1)。

选择 1988 年 10 月 2 日、1996 年 9 月 22 日两个时相的 TM 遥感信息,经过图像处理并配准制成(乾安县)1∶100000 的工作底图。根据该研究区域下垫面的特点和 TM 原始数据各波段的相关性,选用 K-T

---

* 本文原载于《第四纪研究》,2002,22(2):136-140

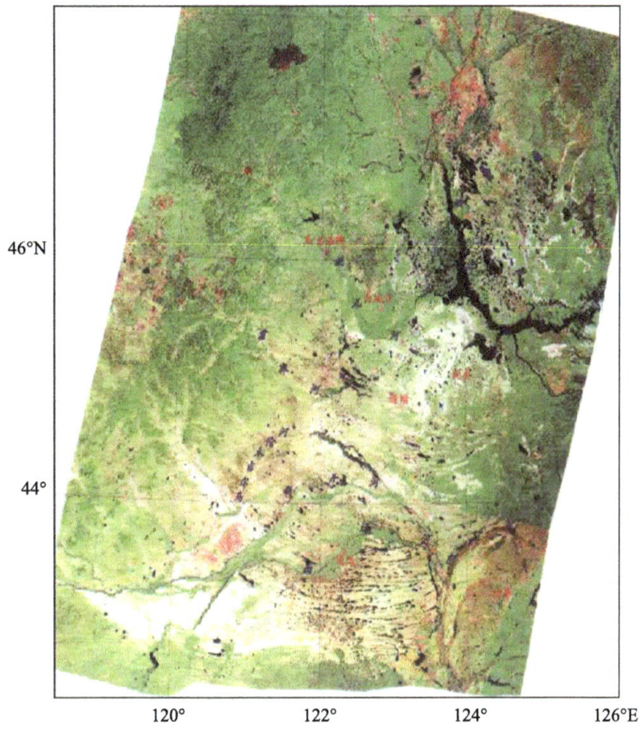

图 1　东北西部地区 1996 年夏季陆地卫星遥感影像图

变换后的 3 个主分量组成的图像，辅以 NDVI 植被指数数据进行最大似然法分类，以分类结果的伪彩色图为主，利用 GIS 中的地形图、有关土壤资料对分类结果进行修正，分别制作出吉林省乾安县两个时相的土地利用图(图 2)。将 1989 年乾安县土地利用图和野外调查数据作为真实数据，用基于误差矩阵的精度评价方法，对分类结果进行评价，得到 1988 年分类总精度 87.3%，Kappa 系数 0.86；1996 年分类总精度 90.7%，Kappa 系数 0.89。而后采用分类后比较法对这两个时相中盐渍化土地的变化信息进行专题提取。

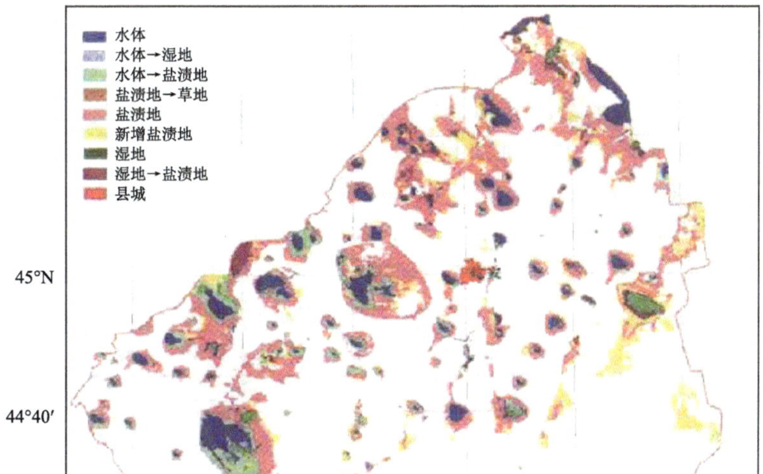

图 2　乾安县 1988 年与 1996 年盐渍化土地动态变化图

选择1988年11月3日、1996年5月15日的Landsat-5 TM遥感信息和2000年6月5日的Landsat-7 ETM遥感信息对吉林省梨树县沙化土地进行解译。由于梨树县的地理环境和物候特征以及春季和秋末地面植被覆盖较少，是提取沙化土地的最佳时间。分别对3个时相的TM数据进行相关分析，选取TM5、4、2三波段合成影像，以影像为主，将经K-T变换后得到的亮度与湿度分量组合形成的土壤特征信息作为辅助判别信息建立解译标志，分别进行沙化土地信息的提取。

选用1996年9月4日TM遥感信息对乌尔吉木伦河与新开河之间的撂荒地进行解译。选取TM7、4、3波段数据经图像处理并几何配准制作成1：100000的影像图，该图沙地呈白色或淡黄色，耕地呈浅绿色，撂荒地呈浅粉色。图像色彩丰富，反差大，各类别层次清晰，容易区分，因此对研究区域中的耕地、撂荒地和大片的沙地进行目视判读。

3 结果分析

3.1 东北西部地区陆地卫星遥感影像图

从东北西部地区1996年夏季陆地卫星遥感影像图可见(图1)，44°N以北，嫩江下游两岸的松嫩草地盐渍化土地景观十分显著，呈浅灰白色，部分还以蓝色的碱泡沼(即湖沼)为其核心。裘善文(1990)、李建东等(1997)的研究表明，松嫩苏打盐碱区面积大约为2.3×10^4 km^2，其中2/3分布在吉林省；44°N以南则是我国北方五大沙地之一——横跨内蒙古东部和东北西南部的科尔沁沙地，呈现出大面积的黄色和白色荒漠化景观，仅在河流狭窄河道两侧和大兴安岭南麓的山地地区，才显现相对较明显的绿色农田、草地与森林景观，这与我国西北地区的景观颇有相似之处。

3.2 东北西部地区荒漠化证据

3.2.1 吉林省乾安县盐渍化监测

从1988年10月2日与1996年9月22日两个时相陆地卫星遥感解译的乾安县土地利用现状对比(表1)可知：耕地、草地、水域和盐渍地8 a间变化较大；耕地和盐渍地增加，草地和水域减少。1988年盐渍化土地面积为523.98 km^2，其中重度盐渍地423.76 km^2，中度盐渍地100.22 km^2；1996年盐渍化土地已达793.75 km^2，其中重度盐渍地674.28 km^2，中度盐渍地119.47 km^2。1996年比1988年盐渍地增加了269.77 km^2，重度明显增加，许多地方已发展成寸草不生的荒漠化土地。在增加的面积中因碱泡沼面积(水域)减少而增加的盐渍化土地约占总增加面积的37.4%，因湿地(未利用地中的部分)面积减少而增加的盐渍化土地约为15.63 km^2，占总增加面积的5.79%，两项相加达43.19%，因耕地和草地退化而导致盐渍化土地增加占总增加面积的56.81%。仅仅8 a时间，乾安县中重度盐渍化土地面积增加已达该县总面积的7.49%，其速度不能不令人惊异。乾安县1988年与1996年盐渍化土地动态变化如图2所示。图中蓝色泡沼周围的浅绿色图斑就是因水面减少而增加的盐渍化土地，黄色的图斑为因耕地和草地退化增加的盐渍化土地。

表1 乾安县1988年与1996年土地利用面积动态监测表 单位：km^2

土地利用现状	耕地	林地	草地	园地	居民地	水域	交通用地	盐渍地	未利用地
1988年	1527.41	323.60	777.92	2.11	118.47	226.92	42.72	523.98	73.35
1996年	1928.25	270.04	286.75	6.31	123.85	125.95	42.89	793.75	38.68
变化	+400.84	−53.56	−491.17	+4.20	+5.38	−100.97	+0.17	+269.77	−34.67

3.2.2 吉林省梨树县东辽河流域沙化动态

图3为吉林省梨树县1988、1996、2000年3个时相的沙化土地遥感解译示意图。从图中可见，1996

年比 1988 年沙化土地迅速向东或东北方向扩张且面积增加，1996 年的最东端比 1988 年的最东端沿经向向东扩展了近 9 km，廉毅等(2001)揭示了 1996 年比 1988 年沙化土地向东发展的事实。2000 年沙化面积比 1996 年又有较快发展(图中绿色的图斑)，原有的 5 条近东北—西南走向的沙带变宽，沙化土地有连成一片的发展趋势，东辽河两岸的沙地明显增加。

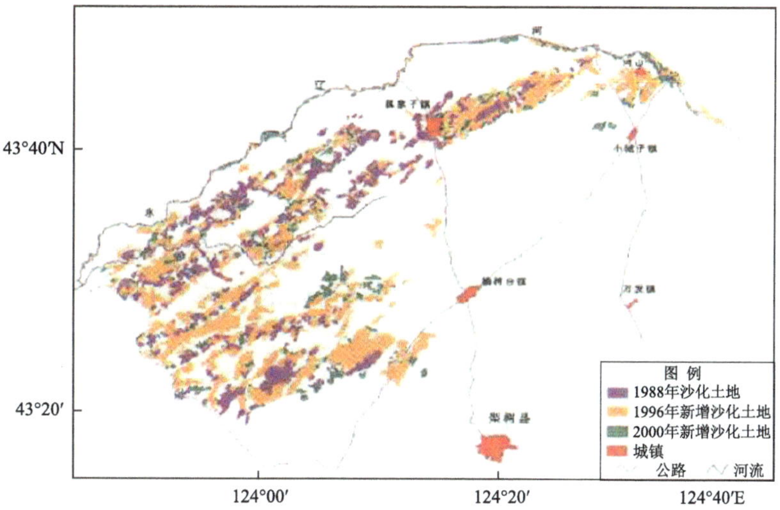

图 3　梨树县 1988—1996 与 2000 年沙化土地分布遥感解译示意图

3.2.3　科尔沁沙地腹地的乌尔吉木伦河与新开河间撂荒地现状

从图 1 可见，乌尔吉木伦河与新开河之间流域位于西辽河南侧的大范围沙地的北侧，跨度范围为(43°20′—44°35′N，119°55′—122°15′E)，是原哲里木盟新开河与西辽河之间农牧业生产区的重要西北屏障，面积约为 8320 km^2。经对该研究区域判读解译，1996 年耕地约为 1024.1 km^2，大片的沙地约为 793.72 km^2，撂荒地(有被人开垦过的痕迹，但当年未耕种，基本没有植被覆盖)约占该区总面积的 6%(图 1 影像中粉红色的景观)，达 498.64 km^2，几乎与该区沙地的面积处于同一量级。这种因为人类过垦和干旱化造成的土地利用荒芜现象，在 2000 年 6 月 2—7 日赴原哲里木盟和赤峰市实地考察时，得到证实。

4　结论与讨论

Nicholson 等(1988)曾指出，“自 20 世纪 70 年代荒漠化引起关注开始，争议就相当大。一个主要分歧是气候和人的因素在荒漠化变程中的贡献量”。本文亦认为关于生态环境荒漠化和干旱化成因的争议是当代科学研究的难点之一。本文就东北西部地区荒漠化发展的几个典型区域应用卫星遥感、社会和气候调查做一尝试性研究。

(1)乾安县属半干旱、半湿润的大陆季风气候，境内无江河，仅有少量的季节性地表径流，是东北西部(也是吉林省西部)唯一没有河流的县，只有 74 个盐碱性泡沼，泡沼中的水既不能人、畜饮用又不能用来灌溉，因而其水面面积仅与自然降水、气温和蒸发量的大小有关。经过 1988 年 10 月 2 日和 1996 年 9 月 22 日卫星遥感解译分析，1996 年比 1988 年盐渍化面积增加、盐碱性泡沼水面积减少而裸露出盐碱地和盐碱性沼泽(湿地)，并进一步演变成干性盐渍化土地的面积约占盐碱地总增加面积的 43.19%。由于两个时相遥感资料获得前的短期气象条件相似，而乾安县夏季气温 20 世纪 90 年代的均值比 80 年代的均值增加了 0.5 ℃，同期降水则减少了 77 mm，蒸发量增加了 55 mm。因此可以肯定地指出，这部分盐渍化面积增加是由于自然气候变化的暖干化倾向并和特定的地理环境背景有直接关系。而剩下的盐渍化增加部分占总面积的 56.81%，是由耕地和草地退化造成的，应该说主要因为人类过垦、过牧所致。因此可以推断乾安县 1996 年比 1988 年盐渍化面积的增加，是干旱化和荒漠化发展的重要标志之一，其中气候变化和地理环境与人类活动的消极因素影响近于等量齐观。所以在今后的生态环境整治过程中，除了必须对人为活动造成的生态环境恶化加以治理，也许还需对自然因素引致的环境恶化加以治理和恢复。

(2)东北西部地区沙化的前沿区域发展强烈，以及科尔沁沙地覆盖区撂荒地的大量增加，区域内河流及其流域的干旱化是其重要的背景。如从图 1 的 46°N 以南地区可以明显看出，东北西部所有的河流及流域不是伴随着沙地、土地沙化就是其下游的盐渍化(44°—46°N)。经实地调查，1986 年东辽河主河道宽 80 m，现在已超过 200 m。在现场看见主河道的水流宽 10 m，离之不远的水稻灌渠却宽 30 m，而且纵横交错。由于主河道水流量的大量减少和频繁改道，丢弃了大量沙滩、决口扇、天然堤和阶地陡坎，从而在风成作用下形成移动沙滩，可见人类对水资源的不合理利用是造成主河道干旱化的重要原因。尽管 44°N 以南科尔沁沙地的气候 1996 年比 1988 年变得相对暖湿，1989—1996 年比 1981—1988 年的年平均降水量增加了 44 mm，但撂荒地仍在大量增加，并与沙地融为一片，这些都说明人类活动对土地沙化的影响是相当严重的。上述分析说明，在大规模治理生态环境、强调退耕还林(草)的同时，不应忽视对河流及流域干旱化的系统整治。

(3)东北西部地区是气候和生态的双重过渡带(符淙斌 等，1995)，对气候异常变化和人类活动作用比较敏感，它应该成为我国荒漠化和生态环境治理的重点区域。自然界本身引起的生态环境变化在一定程度上带有循环性，亦即正逆过程交替，而人类活动的无序和消极影响一旦大于积极影响作用于环境，将加速向恶化单一方面发展。因此只有当气候变化存在着趋利避害的可行性条件时，再通过人类活动的积极参与，人类制定的荒漠化治理和生态环境建设规划才具有真正科学性，才能逐步达到预期目的。东北西部地区属半干旱地区，年平均降水量约为 350～450 mm，土地的荒漠化与西北固定戈壁沙漠景观相比，具有一定的可变性。从这个意义上说，为保护我国仅有的黑土地、保障我国东北粮食生产基地的安全，东北西部地区应该成为我国荒漠化和生态环境治理的重点区域，与大西北的开发和生态环境治理同时进行并形成掎角之势。

参考文献

董光荣，1994. 科尔沁沙地沙漠化的几个问题[J]. 中国沙漠，14(1)：1-9.

符淙斌，叶笃正，1995. 全球变化和我国未来的生存环境[J]. 大气科学，19(1)：116-126.

李建东，郑慧莹，1997. 松嫩平原盐碱化草地治理及其生物生态机理[M]. 北京：科学出版社：271.

廉毅，高枞亭，任红玲，等，1999. 吉林省西部荒漠化发展的陆地卫星遥感监测分析[J]. 气象学报，57(6)：662-667.

廉毅，高枞亭，任红玲，等，2001. 中国东北区 20 世纪 90 年代荒漠化的发展与区域气候变化[J]. 气象学报，59(6)：720-726.

裘善文，1990. 科尔沁沙地形成与演变研究[M]//裘善文. 中国东北平原第四纪自然环境形成与演化. 哈尔滨：哈尔滨地图出版社：185-200.

孙继敏，刘东生，2001. 中国东北黑土地的荒漠化危机[J]. 第四纪研究，21(1)：72-78.

吴正，1987. 风沙地貌学[M]. 北京：科学出版社：316.

肖荣寰，1995. 松嫩沙地的土地沙漠化研究[M]. 长春：东北师范大学出版社：168-169.

NICHOLSON S E，TUCKER C J，BA M B，1998. Desertification，drought and surface vegetation：An example from the West African Sahel[J]. Bull Amer Meteor Soc，79(5)：815-830.

吉林省中西部地区生态环境整治的紧迫性与经济开发的协调性分析*

廉毅 齐艳霞 唐晓玲

（吉林省气象科学研究所，长春 130062）

摘　要：依据陆地卫星遥感信息，研究了吉林省西部荒漠化在发展的反演和解译结果，强调对吉林省中西部地区生态环境整治的紧迫性；通过分析其整治和经济开发的协调性问题，初步提出了系统整治的策略、原则和对策。

关键词：吉林省；中西部地区；生态环境整治；对策；经济开发

1 引言

松辽平原大部分在吉林省境内，位于 42°—46°N，121°—131°E，是驰名中外的玉米带，是中国的重要商品粮基地。随着人类社会的发展和科学技术的进步，人类对自然界及其规律性的认识有所加深，然而，人们对自然的认识总是远远落后于对自然的破坏。廉毅等(1999)指出，依据两个时期 Landsat TM 资料，采用图像增强处理和计算机监督分类进行荒漠化解译和动态量变分析，再与 1976 年和 1981 年不同时期的卫片相对照，揭示了吉林省西部的生态环境进一步恶化，其主要特征——荒漠化继续发展并向东扩张的量变证据。该区生态脆弱，以及受气候和人类活动共同作用环境变化是比较敏感的，已引起国内学者的关注。对近代该区荒漠化继续发展已形成共识(东北平原第四纪自然环境形成与演化基金课题组，1990；肖荣寰，1995；方修琦，1993；吴正，1994；董光荣，1994)，并已引起吉林省政府的高度重视。

面向 21 世纪，人类与环境要协调发展的科学论断和社会发展热点，进一步增强了吉林省中西部地区生态环境整治的紧迫性。分析其整治和经济开发的协调性问题，初步提出了系统整治的策略、原则和某些对策。

2 吉林省中西部地区生态环境整治的紧迫性分析

生态环境是当代全球变化研究的核心和热点，它可横跨数省区，亦超越了国界，从整体上来了解人类赖以生存的环境变化规律和原因，并监测和预测其发展趋势，是世界当代科学和可持续发展的一个前沿领域。显然，只用局部利益的观点去掳夺生态环境，往往会适得其反。比如，咸海地区的乌兹别克斯坦、土库曼斯坦和哈萨克斯坦三国，1960—1994 年，由于大量从阿姆河和锡尔河两大内陆河引水扩大灌溉，浪费水源及单一农作物(如棉花)种植，使这两条河流入咸海水量剧减，使世界第四大湖的咸海蓄水量损失了约 3/4(杨小平，1998)。在干枯了的湖底和周围耕地上荒漠化迅速发展，1992 年引起国际和三国的重视，将咸海地区定为生态环境灾害区，制定了可行的治理方案，计划到 2005 年将上述两河入咸海湖总量达到阻止咸海水面下降的水平。中国东北区与三国的咸海地区面积大体相当，可以说也面临着这样尖锐的同一性质问题。

---

* 本文原载于《吉林农业大学学报》，2000，(Z1)：17-21

从文献(廉毅 等,1999)可见,20 世纪 90 年代与 80 年代相比较,吉林省西部盐渍化沿 45°N 已向东蔓延了 8 km,平均每年 1 km;而位于吉林省西南部四平地区的产粮大县梨树县,沿 43°N 有 2 条沙带,8 a 来向东北、东方向分别移动了 14.3 km 和 8.9 km;吉林省西部有关县份,已相继发生荒漠化,造成移民现象;曾获全国产粮两连冠的农安县,盐渍化亦在发展。如果未来 100 a 内气候振动和 20 世纪相比幅度并不很大,尽管期间会有正逆过程发生,但如果不抓紧进行系统整治,东北区中部的黑土带将被侵蚀很大,国家的重要商品粮基地很有可能出现拦腰截断的后果。由此可见,东北西部,特别是吉林省西部是中国荒漠化发展的显示区,是世界和中国生态环境整治、建设有示范意义的典型区,也是中国重要的松辽商品粮生产基地的屏障地区。为此建议应把东北西部,特别是吉林省中西部生态环境的监测、预测和治理作为中国北方中高纬度地区的重中之重。

3　整治的策略、原则和对策分析

3.1　吉林省中西部生态环境治理的战略要点

(1)制定一个比较详细周密的系统工程规划。其要点是否可以考虑为“遥感监测为先、战略分割为主、先易后难切块、短期由中向西、长期从源抓起、水资源重新分配、国家和省级立项并重”的治理策略。其内涵是:以建立气候变化对吉林省西部荒漠化影响的遥感立体监测系统为前沿工程,加强对其宏观监测;以其反演的荒漠化状况为背景与吉林省以往的科研成果相结合,把西部划分为 6 个需要治理的区域——“梨、怀、双”沙化区,乾安—前郭盐渍化和土地退化区,农安—扶余部分荒漠化区,镇赉湿地,大安为核心的重盐碱和沙化区,向海—乌兰图嘎弧形隆起大沙带。本着先易后难切块,短时期(5 a 内显效的)由中部向西部推进治理,长期难以治理的从影响源地着眼规划;嫩江、洮儿河、霍林河、辽河水重新分配,使西部水分收支趋于良性循环;长期的大工程采取由国家立项为主、省级为辅的治理策略。

(2)选择 2 个气候过渡和生态环境恶化特征不同的前沿地带,作为建设的试验和示范区,力争在 5 a 内有明显的突破和好转,以此带动全盘,充分体现战略分割、先易后难、由中部向西推进、遏止荒漠化向东发展的治理策略。

(3)实行三个并重的方针。监测与成因研究、综合治理和科学评估并重;整顿治理、改善保护和经济开发并重;纵向、横向和交叉并重。因为人类社会的机构、体制、规范都是按纵向设计的,而人类面临的包括重大生态环境问题在内的种种棘手问题都是横向蔓延的。

(4)坚持“四边”技术路线。坚持“边监测分析、边研究创新、边规划治理、边评估反馈”的弹动思维技术路线,以利于人类认识更符合生态变化的规律,达到人类与生存环境和谐共存的目的。

(5)贯彻“五个相结合”的原则。长、中、短期目标相结合,以长期为背景,以中短期治理为主攻方向;跨省、跨区、跨县治理相结合,尤其要重视跨省治理的立项;国家、省级和地方投入相结合,长期的、跨省的和省级大项目以争取国家投资为主;政府行为、专家建议和增强当地干部群众生态意识相结合;扶贫开发、科学治理和生态保护有机结合,是生态环境建设系统工程能否持久的重要保证。

(6)如何区分气候自然变化、地质环境和人类活动对荒漠化发展正逆过程的不同影响度,是科学制定治理荒漠化计划的关键所在。如果在人类活动的积极影响参与下,而气候带变化又存在有趋利避害的可行条件,这时人们制定治理荒漠化计划才具有科学性、可行性,并能逐步达到预期的目的。另外,吉林省西部沙漠化和盐渍化发展的机制既有相同之处,也存在较大差异,如何区分两者的差异,是制定行之有效的治理方案的关键之一。比如,植树造林和改变植被是抗御沙漠化强有力的对策,而对盐碱地的治理则不然,尚须有其他对策才能奏效。

(7)只有坚持生态环境研究和治理的全球和全局性观点,才能使之达到利大于弊的效果。

3.2 两个试验示范生态区治理构想

3.2.1 乾安—前郭盐渍化、草原和耕地退化治理要点

该区为南北走向的长方形，东西向为 50 km，南北向为 65 km，总面积为 3250 km^2。西侧以乾安县城为界，向南至查干花镇以北，向北延至乾安县界止，东侧以前郭的长龙乡为界。

该区大部为波状冲积平原区，湖泡较多，呈花斑影像，大部以草甸草原为主，地带性土壤为淡黑钙土，广大甸地为盐渍化草甸土，局部地区为苏打盐土，呈白色的斑块。

根据 1988 年和 1996 年美国 Landsat TM 资料的解译，盐渍化、草原和耕地退化面积见表 1。

表 1 乾安—前郭盐渍化、草原和耕地退化面积

年份	盐渍化面积/km^2			变化/%	草原和耕地退化面积/km^2
	严重	中度	合计		
1988	168.1	104.4	272.5		
1996	253.4	175.6	429.0	+57.4	180

除去区内的水面外，需要治理的面积占陆地面积的近 1/3，其中严重区占 1/6。有关研究表明，该区的盐渍化的景观是属不稳定的(即可变性较大)，而相邻的大安县的盐渍化景观是稳定型(即多年形成的固定景观)(刘惠清 等，1997)，与陆地卫星的动态监测分析结果一致。正因为是近期变化，增加了短期治理的可行性和易变性。另外，气象卫星监测结果表明，1998 年 5 月底比 1996 年 6 月初草场面积减少 8%。

首先要进行该区盐渍化、草原和耕地退化的监测分析和对策研究的详细评估，进一步搞清现状，揭示其成因；其次有针对性和分门别类提出综合治理的对策；最后对对策可能产生的利弊进行科学评价，做经费预算和可行性的定量分析，所需时间 1 a；2000 年开始实施，大约 2005 年完成。可以从改善植被、排碱、水利工程和化学应对等诸方面综合治理盐碱(裘善文，1996)，草原和耕地退化区可以大规模种草；充分利用该区约 200 km^2 的泡沼大兴养鱼，发展渔业、加工业。

3.2.2 "梨、怀、双"东辽河区域防治沙化东北移治理要点

该区近似为东北—西南走向的长方形，东侧以四平—范家屯的铁路为界，西界从吉林省的东辽河南起点至公主岭市的双城堡乡，面积近 5000 km^2。由于东辽河和招苏太河的下切，地面破碎，该区南部形成淡黄色起伏台地。

根据 1988 年和 1996 年美国 Landsat TM 资料解译，该区以沙化迅速向东和东北部推进为荒漠化发展的主要特征。

首先进行该区沙化迅速向东和东北发展的年际间和年代间量化监测和成因分析。据初步判断和 1999 年 6 月 4 日的实地调查，该区的沙化迅速往东北方向发展，不仅是因为地形和季风盛行风成风蚀作用，还因为东辽河的流量减少和频繁改道，丢弃大量沙滩、决口扇、天然堤和阶地陡坎，在风成作用下形成移动沙丘。所以，不仅要植树造林、改良土壤，还需对东辽河以系统整治，防治水土流失和河沙移动。另外防风林带在哪儿造，究竟造多长，要造几道，坐落在什么方向，都需对该区的风成规律和河流河滩的演变特征加紧研究，这有利于完善生态林工程的实施方案。为此前期研究仍需约 1 a 的时间。2005 年前示范初具规模并争取整治东辽河的国家立项。表 2 为"梨、怀、双"区域沙化面积表。

表 2 "梨、怀、双"区域沙化面积

年份	沙化面积/km^2	变化/%	向东推进/km	向东北推进/km
1988	70	—	—	—
1996	198	+182.0	8.9	14.3

3.3 关于湿地的保护问题的新建议

吉林省领导在整治西部生态环境的思路中紧紧抓住湿地保护问题，切中要害。目前，过去传统的防洪技术许多发展中国家已经放弃了，美国又重新运用。1993 年美国密西西比河泛滥时，沿岸的村镇便迁移到山上，把低地留为湿地，放弃了以往筑堤建坝的现代防洪办法。发大洪水在某种意义上说，是气候的自然变化，是自然界生态平衡规律性的突出表现。比如，吉林省白城地区嫩江和洮儿河流域发大洪水，自 1949 年以来基本呈现约 12 a 为 1 个周期。如 1957 年、1969 年发大洪水，20 世纪 70 年代有所中断，到 1986 年、1987 年恢复，1998 年又引发了百年不遇的特大洪水，这对湿地的保护和半干旱地区水面的恢复有着人类所不可替代的作用。为此，我们建议，在洪水不对人口稠密的大中城镇构成严重威胁时，应该在嫩江下游和洮儿河、霍林河设置必要的洪泛区，这是人对气候自然变化趋利避害的一种巧力。在干旱之年，引水浇灌湿地、泡沼及尽可能增加洮儿河、霍林河和东辽河上游流量亦是十分重要的举措。

另外，对向海自然保护区的治理，是否首先要考虑恢复霍林河，减少霍林河上游水库的蓄水量，其次考虑“引嫩济霍工程”的国家立项问题。

4 关于生态环境整治和经济开发协调性的思考

是否可以将生态环境的整治和经济开发视为一个现代社会发展的有机整体和系统，如果是，可遵循下列 10 项基本原则协调中西部地区两个方面工作：整体性、联合性、预测性、开放性、层次性、可靠性、定量化、综合性、协调性和最优性。为此，前述已提到中西部地区分 6 个战略分割区域进行生态环境治理，现在加上一些与经济开发(特别是产业结构调整)协调性的思考，借以抛砖引玉，为各级政府领导提供某些参考意见。

4.1 联合治理乾安—前郭盐渍化和土地退化区工程

前郭比乾安地理自然环境优越，其突出表现在拥有得天独厚的查干泡和松花江流经县境，查干泡面积约为 260 km^2，是吉林省最大的淡水湖，与嫩江水似连非连，而乾安县无江无河，只有 70 余个泡沼。前郭应该为乾安县提供淡水资源。一方面扶植该县的牧草业和农业，另一方面引灌泡沼，增加其淡水含量，养鱼和种植芦苇。两县都应大兴水利工程，在中重度盐渍化区挖深沟引水排碱，可在退化耕地和草地(在两县交界处及其南面)引进羊草良种，重点发展牧草出口业，组建和壮大饲草集团公司，迅速增加牧业的比例，力争在 5 a 内与种植业并驾齐驱。

如果乾安县的盐渍化得不到治理，会更加迅速向前郭蔓延，所以两县联手，对荒漠化进行重点整治，一荣共荣的局面会逐渐显现的，经济开发前景亦是可观的。建议该区可作为吉林省生态环境治理的一期和重点工程。

4.2 “梨、怀、双”沙带区的东辽河治理和植树造林工程

前述已提到与植树造林工程并列考虑的应是东辽河治理工程。首先是梨树和公主岭市是否共同考虑调整原有的灌区用水量，采取滴灌等节水措施，提高水的利用率(应考虑沙土壤的漏水率较大)；应因地制宜，调整农业水旱田的面积(减少水稻面积)，以达到恢复东辽河主河道的流水量，从根本上遏止和减少沙源，再与植树造林工程相匹配，防止沙化向东和向东北方向发展。

梨树的建筑材料业已有基础，是否考虑提高科技含量，使之工业升级换代。三县亦要逐渐增加牧业的比例，尤其是原双辽县畜牧业要超过种植业。

乾安—前郭盐渍化和土地退化区与“梨、怀、双”沙带区两个区的治理，是能否把西部荒漠化向东发展势头首先遏制住的关键所在，可列为吉林省中西部地区生态环境整治的首选区域。

4.3　大安为核心的重盐碱和沙化区治理工程

该区包括大安县、洮南县和通榆县的东北部分，是吉林省荒漠化的严重区，其根治要花费大力气，要高投入，要打持久战，还要搞好跨省协调（主要与邻省协调水资源的重新分配），应以国家立项为主，以兴修大型水利工程为重点，也是吉林省西部生态环境治理的重中之重。

工程的核心是：从月亮泡引嫩江水，形成人工渠纵横网，畅通洮儿河，恢复大安东部和南部的泡沼湿地；适当减少察尔森水库的库容量（暂时可考虑缩小 50％的存量保持约 5×10^{8} m^{3}），增加春季洮儿河的流量；设置 10 a 一遇的洪泛区，拦贮洪水和丰水年自然降水；修筑排碱深沟，引水补充碱泡沼的淡水。实施大规模的排碱、治碱和发展渔业和芦苇业，建议大安和洮安调整大田耕种面积，大兴种草，对原有草原定期实行草库伦保护措施。同样应加强饲草集团公司的建设，形成各县有力的牧草买方市场，扩大出口份额，以调动广大农民种草的积极性。

4.4　农安—扶余部分荒漠化区的治理要点

该区是吉林省中部商品粮基地重要组成部分，农安县有少部分盐渍化区分布在泡沼的干涸边缘区，可采取植树造林和种耐碱的良草，发展林牧业，经过数年可遏止其发展；远期尚须考虑引水源灌入波罗泡、前巴吉垒泡、敖宝图泡和元宝泡，制止泡沼水位下降，乃是防治农安县中心地带荒漠化发展的根本工程。

原扶余县虽是产粮大县，但沙化面积带有全境性，其中北部盐碱化的发展亦是显而易见。其中南部以种植红皮花生闻名国内外。土壤的改良将是该县治理荒漠化的长期和核心措施；松花江环绕本县，对生态环境治理利大于弊，但亦仍须防治因频繁改道的二松所丢弃的沙滩、陡坎和不断拓宽的河道，以及在风成和风蚀作用下，由盛行的西南风对全县境内的沙化侵蚀。虽然该区荒漠化没有西部市县那么重，但千万不可掉以轻心，该区的生态环境治理是吉林省西部荒漠化向东发展而进行先易后难遏止治理战略成败与否的关键所在。

4.5　保护镇赉湿地工程要点

镇赉县是吉林省西部水域面积最多的县份，如果连同沼泽大约已达该县面积的 1/3，是吉林省，也是东北区西部重要的湿地集中区，主要分布在镇赉的东部靠近嫩江西岸区，从 1998 年大洪水的淹没面积为全省最多的县份得到证明。

为了吉林省西部生态环境治理的大局，必须恢复其自然湿地的本来面目，以成为对东北地区西部水分平衡的重要因子。镇赉是否可以考虑由省级立项，实施保护湿地工程。将该区发展为吉林省渔业、水产品养殖和芦苇业为主的生产加工基地，以此改变吉林省中部大中城镇主要采购辽宁水产品的局面，增加吉林省长白铁路线的物资流量。

4.6　向海—乌兰图嘎弧形隆起大沙带水利和造林种草工程

该区以沙化为主要特征的荒漠化是地质、地理环境和气候过渡带共同促成的景观。据地质专家介绍，每年仍以毫米的高度在缓慢隆起，这是不以人的意志为转移的。

我们利用卫星监测到该地植树造林是卓有成效的，比较成功地制止了大部分移动沙丘。要仍以原有的治理方法，继续扩大植树造林和扩大种草的面积，发展林牧业。

该区是吉林省免受内蒙古沙漠侵袭的重要屏障区，是非常重要的；向海是世界级的自然保护区，目前尚存约 70 km^{2} 的水域面积，20 世纪 70 年代原有的水渠已消失了。由于地处半干旱地区和干旱区的过渡边界上，年降水量不足以弥补向海水位下降，必须引邻域的水给予必要的补充。霍林河已日益缩小，在吉林省西部几乎绝迹，如果能上项目使之恢复则是很好的措施，是否可行，还须做科学研究和论证。

通榆县可迅速重振吉林省牧业县的雄风，通榆的红牛是省内外闻名的肉食牛，应当加强肉类加工工业的建设。

长岭县近几年盐碱化面积在部分地区蔓延得比较快，如县城南和新安镇以北的东西长近 30 km，南北宽近 20 km，约 600 km^2 地带大片盐碱地在卫星影像上清晰可见。该县重点要放在遏止草地和耕地的退化治理上，同样应调整作物结构，发展油料作物和畜牧业，迅速增加其比重。

参考文献

东北平原第四纪自然环境形成与演化基金课题组，1990.中国东北平原第四纪自然环境形成与演化[M].哈尔滨：哈尔滨地图出版社：185-200.

董光荣，1994.科尔沁沙地沙漠化的几个问题[J].中国沙漠，14(1)：1-9.

方修琦，1993.中国北方农牧交错带环境问题分析[C]//中国环境科学学会.中国青年环境论坛首届学术年会走向21世纪：中国青年环境论坛首届学术年会论文集.北京：中国环境科学出版社：423-428.

廉毅，高枞亭，任红玲，等，1999.吉林省西部荒漠化发展的陆地卫星遥感监测分析[J].气象学报，57(6)：662-667.

刘惠清，龙花楼，1997.为生态建设服务的吉林省西部景观类型研究[J].地理研究，17(4)：389-397.

裘善文，1996.松辽平原低洼易涝盐碱地综合治理与农业发展[M].北京：科学出版社：189.

吴正，1987.风沙地貌学[M].北京：科学出版社：310.

肖荣寰，1995.松嫩沙地的土地沙漠化研究[M].长春：东北师范大学出版社：168-169.

杨小平，1998.中亚干旱的荒漠化与土地利用[J].第四纪研究，18(2)：119-127.

Climate Change and Its Impacts on Grain Production in Jilin Province*

Lian Yi[1,2] Gao Zongting[1,2] Shen Baizhu[1,2] Ren Hongling[1,2]
Tang Xiaoling[1] Liu Yuying[3]

(1 Institute of Meteorological Science of Jilin Province, Changchun 130062; 2 Laboratory of Research for Middle-High Latitude Circulation System and East Asian Monsoon, Changchun 130062; 3 Climate Center of Jilin Province, Changchun 130062, China)

Abstract: The climate observation data, reanalysis data, and grain/soybean yields per unit area were used to analyze and interpret the impact of climate change on grain production. The results show that Jilin Province was located in a remarkable increase area of temperature during the growing season (May—September) from 1948 in the middle latitudes of the Northern Hemisphere. The midwest and south of Jilin Province and Liaoning Province were located in a clear linear decrease tendency area of annual precipitation, wherein a warm/dry tendency of climate change dominated, while the east of Jilin Province lay in a clear linear decrease tendency area of annual precipitation. The climate warming played an important role in continuous increase in the grain yield per unit area since the 1980s in the main grain production areas of Jilin Province, however, from the end of the 20th century to the beginning of the 21st century, the beneficial effect seemed to be not obvious any longer, the grain yield per unit area fluctuated with annual precipitation.

Keywords: Jilin Province; climate change; grain production

1 Introduction

Jilin Province is located in the middle of the Northeast Plain, i. e. one of the three biggest black soil zones of the world, covering an area of 1.874×10^5 km^2, which accounts for 1.95% of the whole country's territory. The arable land area is about 3.955 million hm^2, and especially suitable for the cultivation of temperate zone grains such as corn, soybean, rice, and some other economic crops. The mid-west region of Jilin Province is one of the major bread baskets, and the rates of grain output, grain export and grain commodity all always headed the list in the past 10 years.

Jilin Province is very important for both the climate and the ecological environment of China. The dense natural forests, rich black soils and available climate resources make it quickly develop into one of the provinces that play a very important strategic role. Many researchers have revealed regional response characteristics of obvious winter warming in Northeast China (Qiu, 1992; Huang et al., 1996; Lian et al., 1999, 2001; Sun et al., 2001; Ren et al., 2002; Cramer et al., 1999; Jiang et al., 1999; Chen et al., 1998).

* 本文原载于 *Advances in Climate Change Researon*, 2008, 4(Suppl.): 56-59

In order to take scientific and feasible measure against the climate warming, we should know how to quantitatively assess the regional response. It is very important not only for the economy revitalization of Northeast China but also for the ecological environment protection of Northeast Asia. And it is also meaningful for the economy construction and sustainable development.

In this paper, we just give a brief analysis result about the regional climate distribution characteristics of Jilin Province, and reveal the fact of warming during the growing season and its effect on the grain production.

2 Data and methods

In this study, the global surface temperature data from 1948 to 2005, the 80 stations' monthly precipitation data in Northeast China from 1961 to 2000, the 41 stations' monthly precipitation and average temperature data of Jilin Province from 1961 to 2004, and some kinds of the grain yield data of Jilin Province from 1949 to 2004, were used.

The method of linear trend analysis (Lian et al. ,2005) for precipitation analyses, the method in the reference (Wang, 1995) for separating the grain/soybean yield fluctuations, and the time series analysis for analyzing other data, were employed.

3 Characteristics of climatic change in Jilin Province

3.1 Obviously increasing temperature regions during crop growing season in the Northern Hemisphere

Fig. 1 shows the time series of average temperatures during the growing season (May—September). It shows that there was an obvious linear rising trend in Jinlin Province, the trend line has already reached the area above the average temperature in the early 1980s, and especially after the middle 1990s, when persistent higher temperature occurred. The linear rising trend of the average temperature of American corn belt (40°—43°N, 100°—85°W) was relatively more gentle. However, the average temperature over the 45°N latitude band showed a generally reducing linear trend. There was an obvious temperature increasing after the middle 1990s, whereas, in fact, it was still lower than that in the early 1950s.

3.2 Linear trends of annual precipitation in Jilin Province

Fig. 2 shows the distribution of linear trends of annual precipitation in Northeast China. There were clear reduction regions of annual precipitation in the west (43°—46°N, 122°—124°E) of Jilin Province, the main crop production area in Songliao Plain in the middle (43°—46°N, 124°—126°E) of Jilin Province, and the Changbai Mountain (west of 128°E except for Yanbian Prefecture) in the mid-south of Jilin Province. All these regions and Liaoning Province formed the major area wherein precipitation was linearly reduced in the mid-south of Northeast China.

In addition, Yanbian Prefecture in the east of Jilin Province, and Mudanjiang and Jiamusi prefectures in the east of Heilongjiang Province were all obviously positive trend areas of precipitation. The probability of the annual total precipitation exceeding 600 mm in Yanbian is 97%. This brings not only rich climate resource for medicine industry and forestry, but also plentiful surface runoff for many big rivers in Northeast China. Due to this kind of climate, this regional climate's warm/moist trend is self-evidence.

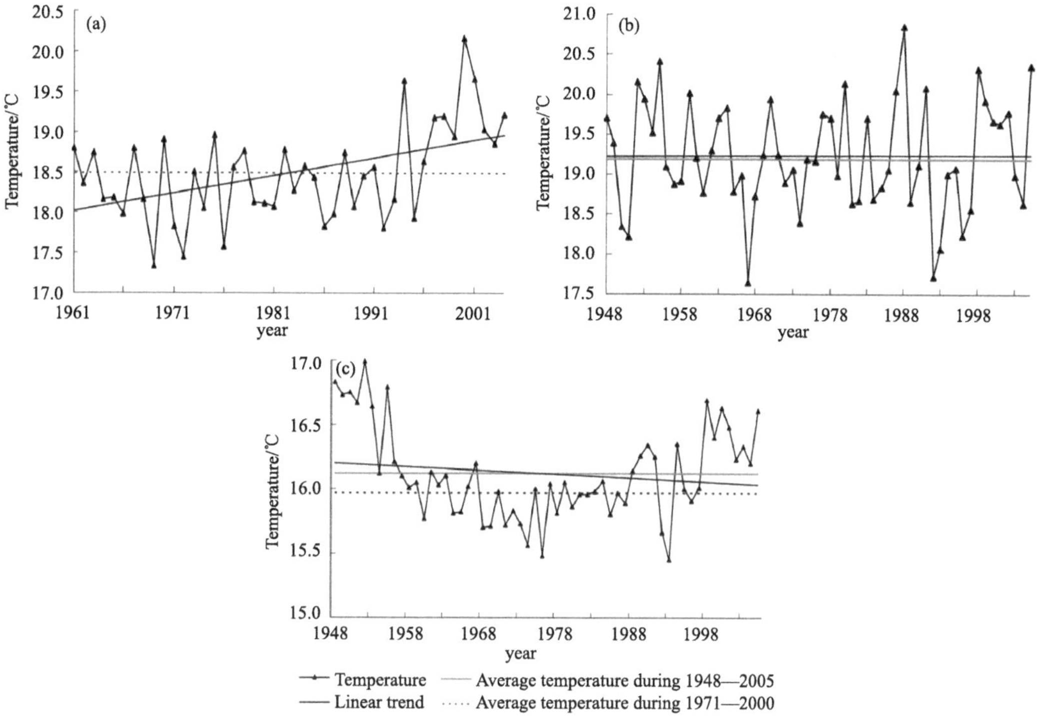

Fig. 1　Time series of average temperatures over May—September (a) Jilin Province, (b) the corn-growing belt of USA, (c) the Northern Hemisphere latitude band along 45°N

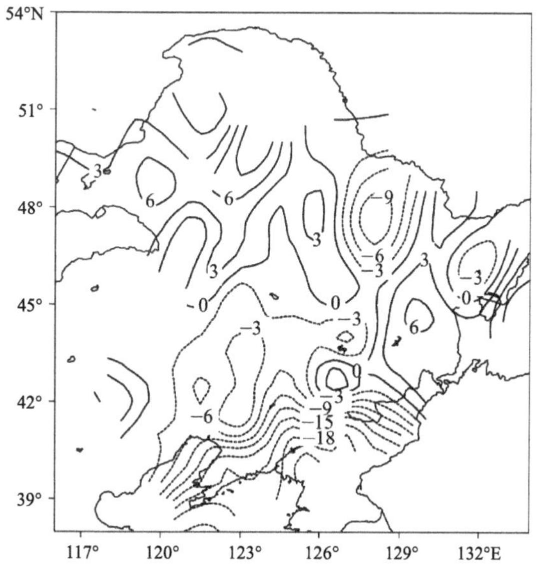

Fig. 2　Linear trends of annual precipitation in Northeast China during 1961—2000 (solid: positive; dashed: negative; unit: mm a^{-1})

4　Effects of warm/dry climate trend on grain production

Fig. 3 shows time series of precipitation and average temperature in crop growing season (May—September), and the time series of the grain/soybean yield per unit area in Jilin Province. The grain/soybean yield per unit area in Jilin Province in 1949 was 1000 kg/hm^2, and the yield still hovered 1000—2000 kg/hm^2 until the 1970s, but since the 1980s the yield reached 4000—5000 kg/hm^2. There may be two reasons: one is the adjustment of agriculture policy, and the other is that after the 1980s, Jilin Province agriculture got rid of the damages caused by low temperature like those in the 1960s and 1970s, and the average temperature of Songliao Plain in crop growing season came into a relatively warm stage. After the 1990s, the yield reached 5000—6000 kg/hm^2, but its fluctuation amplitude also increased. The total precipitation from May to September had a downtrend after the year 1998 when it reached the second peak, this downtrend continued up to the year 2004. There was a similar phenomenon in the crop production after the per unit area grain/soybean yield reached the second peak in the 1990s, showing a declining trend and being below 6000 kg/hm^2 for several years.

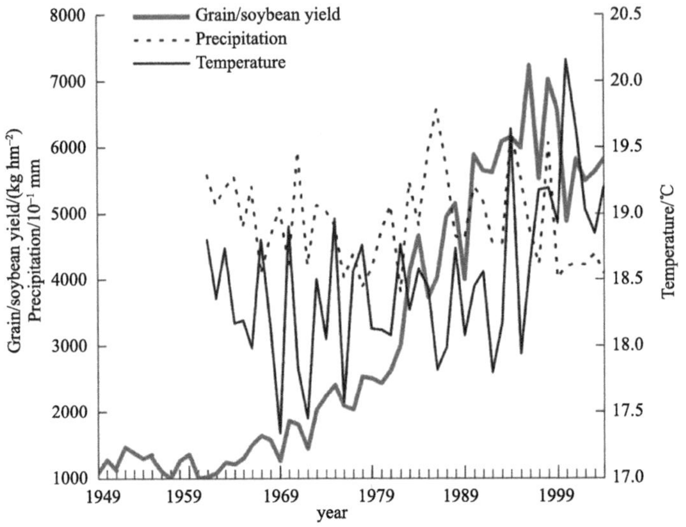

Fig. 3　Time series of grain/soybean yield per unit area, precipitation and average temperature in the growing season from May to September in Jilin Province during 1949—2004

Fig. 4 shows the time series of the per unit area grain/soybean yield fluctuation (meteorological yield) separated from trend yield (3-year running mean), and the temperature and precipitation fluctuations in crop growing period (May—September). It is clear seen from Fig. 4 that the fluctuation of meteorological grain/soybean yield is similar to that of the temperature from May to September during 1986—1996, and is similar to that of precipitation during 1997—2003. Analyzing the fluctuation ampli-

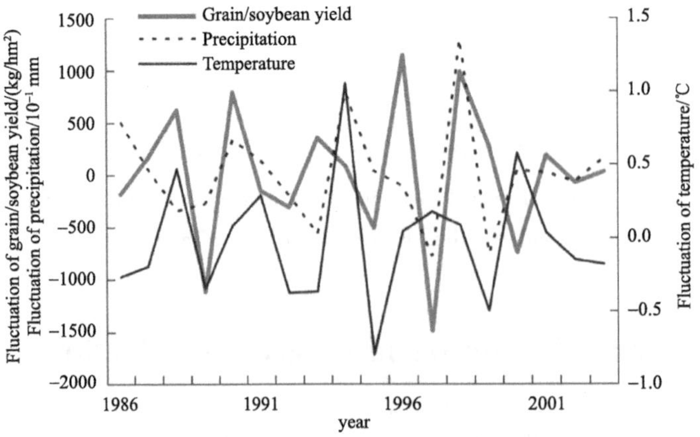

Fig. 4　Time series of fluctuations of the grain/soybean yield per unit area separated from trend yield and the temperature and precipitation in the growing season in Jilin Province during 1986—2003

tude of grain/soybean yield, we found that the Jilin grain/soybean yield fluctuation was related to the drying trend of climate, showing that the grain production of Jilin Province was closely related to the climate condition of current year.

5 Conclusions and discussion

(1) It is well known that temperature is increasing in the middle-high latitude regions in winter. The temperature in Jilin Province is increasing not only in winter but also in crop growing season from May to September. After the 1980s, the average temperature in the growing season was continuously increasing, and after the 1990s, the rising trend was especially significant. Under the background of global warming, regional precipitation response characteristics were not the same in many regions of Northeast China during the past 40 years. For example, Songliao Plain, the south central of Jilin Province and Liaoning Province all showed an significant reducing trend in the annual precipitation, but the north central and the east of Northeast China displayed a significant increasing trend.

(2) In the main grain production regions of Jilin Province, climate warming played a very important role in pushing a sustained increase in the grain yield since the 1980s. Unfortunately, when the time came to the late 20th century, the benefit no longer existed. The trend of annual grain production was mainly coincident with that of the annual precipitation. By using the wavelet analysis of the average temperature during crop growing season, we found that the temperature in Changchun may change from a warm period to a comparatively cold period in the year 2010. However, under the background of global warming, some questions are worthy of our attention and further investigation, for example, whether the regional temperature change period can switch or not, how much the amplitude will be, and whether the phenomenon that the low-temperature breaks out in many regions at the same time will appear again or not. It is a significant subject closely related to the grain production change of Northeast China and the grain safety of our country.

References

CHEN L X, ZHU W Q, WANG W, et al, 1998. Studies on climate change in China in recent 45 years[J]. Acta Meteorologica Sinica, 56 (3): 257-271. (in Chinese)

CRAMER W, KICKLIGHTER D W, BONDEAU A, et al, 1999. Comparing global models of terrestrial net primary productivity (NPP): overview and key results[J]. Global Change Biology, 5(S1): 1-15.

HUANG X C, MENG X X, 1996. Research on Ecological Environment Fragile Zone in the West of Northeast China[M]. Beijing: Science Press: 1-20. (in Chinese)

JIANG H, APPS M J, ZHANG Y L, et al, 1999. Modeling the spatial pattern of net primary productivity in Chinese forests [J]. Ecological Modeling, 122(3): 275-288.

LIAN Y, GAO Z T, REN H L, et al, 1999. A monitoring analysis of desertification development in western Jilin of China by landsat TM data[J]. Acta Meteorologica Sinica, 57 (6): 662-667. (in Chinese)

LIAN Y, GAO Z T, REN H L, et al, 2001. Desertification development and regional climatic change in Northeast China in the 1990s[J]. Acta Meteorologica Sinica, 59 (6): 730-736 . (in Chinese)

LIAN Y, SHEN B Z, GAO Z T, et al, 2005. Correlation of drought development tendency on China's climate transition zone and activity of East Asian summer monsoon and polar vortex[J]. Acta Meteorologica Sinica, 63 (5): 740-749. (in Chinese)

QIU S W, 1992. Three Kinds of Serious Damages of Soils in the West of Northeast China, Comprehensive Treatment is Imperative under the Situation[M]// SHI Y F, HUANG D C, CHEN P Q. Analysis of the Nature Disasters in China and Their Countermeasures. Wuhan: Hubei Technology and Science Publishing House: 387-391. (in Chinese)

REN H L, LIAN Y, GAO Z T, et al, 2002. Using remote sensing technique to study the desertification in the west of Northeast China[J]. Quaternary Sciences, 22 (2): 136-140. (in Chinese)

SUN J M, LIU T S, 2001. Desertification of black soil land in Northeast China[J]. Quaternary Sciences, 21 (1): 72-78. (in Chinese)

WANG S Y, 1995. Study on Crop Chilling Damages[M]. Beijing: China Meteorological Press: 1-246. (in Chinese)

中国东北地区土地荒漠化与气候变化研究进展*

廉毅[1,2]　王雅君[1,2]

(1 中高纬度环流系统与东亚季风研究开放实验室，长春 130062；2 吉林省气象科学研究所，长春 130062)

摘　要：综述了近十几年中国东北地区西部土地荒漠化这一领域研究进展，侧重介绍了土地荒漠化与区域气候变化的研究成果，并对将来研究中应思考的问题提出某些建议。指出：在全球变暖的背景下，对东北地区西部生态环境的恶化趋势，如何采取系统和大规模的工程整治和保护措施，不仅对保障国家重要的东北地区中部产粮区的安全生产有重要战略意义，而且对人类适应、减缓和应对气候变暖具有典型的示范作用；进一步鼓励和支持开展东北地区土地荒漠化与气候变化的研究是十分必要的，建议加强有关气候变化、环境干旱化、土地覆盖变化的固定和野外观测系统建设，加强多要素的气候系统观测工作，加强多学科交叉研究立项工作，不仅要掌握卫星遥感信息，更要掌握器测的地表有关要素的信息，从中准确得出该区土地覆盖变化的频繁区和敏感区的分布，为研究土地覆盖动力学和制定对策奠定科学基础。

关键词：东北地区；荒漠化；区域气候变化；研究进展

1　引言

中国东北区地处东北亚的核心区，中国东北平原在全新世气候时为草原或森林草原环境，并广泛发育了富含有机质和养分的黑土层，有的地域在黑土之下还有厚沙层，从而造成该区具有潜在的沙漠荒漠化风险。中国东北平原西部沙地位于欧亚大陆中纬度巨型沙带的东部边缘，为中国沙质荒漠化土地的东缘，该区基本上处于中国湿润的东部季风和内陆干旱、半干旱气候的过渡区域，是生态系统从半湿润森林-草原向半干旱草原和沙漠之间的过渡带，也是中国的农牧交错带。在全球变暖的背景下，区域气候变化的主要响应为暖干倾向，生态环境亦在恶化，主要表征为荒漠化（盐渍化和沙漠化）区域的发展，已引起国内学者广泛关注（孙力 等，2000；杨建平 等，2003；廉毅 等，2005；孙继敏 等，2001；廉毅 等，2004；裘善文，1992；黄锡畴 等，1996；廉毅 等，2001；任红玲 等，2002；周成虎 等，2002；包慧娟 等，2003；杨晓军 等，2004；慈龙骏 等，1997；王涛，2004；王涛 等，1998）。分析气候变化对自然生态系统的可能影响，正确把握气候变化对中国各方面影响的深度和广度，提出相应减缓对策，是中国在可持续发展道路中面临的和必须要认真对待并加以解决的重大课题之一。本文侧重综述和评价近十几年土地荒漠化与区域气候变化研究方面的成果，提出未来在东北地区土地荒漠化研究中应思考和注意的问题。

2　中国东北地区西部土地荒漠化发展趋势

在近代历史时期，特别是清朝光绪年间以及民国时期，中国东北地区平原西部的黑土地被大量放垦，草场退化和土地沙化加剧，长期的垦荒已经使东北的黑土地在风力的吹扬下，被剥蚀殆尽，其下的厚层第四纪沙源活化，从而使东北黑土地成为中国目前荒漠化危害最为严重的地区（孙继敏 等，2001）。有研究

* 本文原载于《地理科学》，2007，27(S1)：97-102

表明，吉林省西部耕地1996年为12472 km^2，比1930年增加近3倍，1996年的林地和草地比1930年减少了近30%（廉毅 等，2004）。

裘善文（1992）利用20世纪80年代陆地卫星照片数据和土壤情况调查资料，得到的东北西部土地沙漠化面积约为10.25万 km^2，占东北西部土地面积的24.35%，其中科尔沁沙地沙漠化面积为53283.44 km^2，松嫩沙地沙漠化面积为18716.84 km^2，呼伦贝尔沙地沙漠化面积为30493.74 km^2，并继续向东侵入蔓延，正在形成从南到北全线压向东北中部商品粮基地的险恶形势；东北西部碱化土地面积为37160.33 km^2，主要分布在松嫩平原，碱化是草原的主要灾害，草原退化是土地沙化、碱化恶性循环的必然结果（黄锡畴 等，1996）。黄锡畴等（1996）研究表明，东北地区西部为生态环境脆弱带。

东北平原西部沙地位于欧亚大陆中纬度巨型沙带的东部边缘，为中国沙质荒漠化土地的东缘，沙质荒漠化景观日益明显，生态环境在恶化，原哲里木盟西部开河与乌尔吉禾伦河间的撂荒地明显增加，荒漠化从3个方面逼近中国的商品粮基地——吉林省中部松辽平原产粮区，其中吉林省大安县是盐渍化严重区，乾安县东部是其盐渍化发展的前沿地区（廉毅 等，2001；任红玲 等，2002）。20世纪90年代东北平原西部沙地的沙质荒漠化发展速度很快，但区域差异很大（周成虎 等，2002）。科尔沁沙地植被盖度小于30%的较严重沙漠化土地从1995年至2000年以61万 hm^2/a 的速度扩展，景观破碎度变大（包慧娟 等，2003）。

3　关于土地荒漠化评价的现状、问题和发展趋势

中国荒漠化评价现状存在的主要问题是评价基准的不确定性（杨晓军 等，2004）。比如，目前对中国的荒漠化土地面积大体存在两种估计，一种根据大规模的普查和调研工作，按1994年6月通过的《联合国防治荒漠化公约》规定范围（UNCCD，1994），初步得出中国的荒漠化面积为262万 km^2，占中国陆地国土总面积的27.2%（慈龙骏 等，1997；CCICCD，1996；慈龙骏，1998）。还有一种根据若干典型地区不同时期航片解译以及地面调查资料，得到中国的荒漠化土地面积图，统计出中国的荒漠化土地面积为86.16万 km^2，占中国陆地总面积的8.97%（朱震达，1998）。目前，对于荒漠化评价过程中量化指标综合方法，多数学者都采用根据经验给定每一等级中的指标阈值的方法，确定评价单元的荒漠化等级。有的通过给定风蚀地表形态占地面积（%）、植被盖度和土地生物生产量下降数（%）3个指标各等级下的阈值，再结合地表景观综合特征来制定风蚀作用下土地荒漠化的程度（高志海 等，1998）。罗札诺夫等（1983）通过给定不同自然带植被退化中植物生物量贮藏、生物产量和植被盖度3个指标的变化阈值来确定评价单元的荒漠化程度等级。联合国粮食及农业组织（FAO）和联合国环境规划署（UNEP）发表的《荒漠化评价与制图方案》，虽然各评价方法所选的指标不同，但评价的方法的原理大同小异（王涛，2004）。上述的一系列方法的可能缺陷是，对特定的评价单元，各指标的变化可能不一致，轻重度不一样，不好综合，而且多数指标体系没有考虑尺度问题（王涛 等，1998），大部分定量指标看起来更适合传统的小尺度（通过人的眼睛）地面调查，而不适合在尺度较大的荒漠化遥感评价中应用。

近10 a来，在荒漠化的现状和危险性评价方面，基于“3S”（地理信息系统，全球定位系统，遥感）技术的荒漠化综合评价方法研究相当活跃，似乎已成为荒漠化评价技术研究的发展方向（廉毅 等，2001；任红玲 等，2002；周成虎 等，2002；王涛，2004；王涛 等，1998；张玉贵 等，1998；廉毅 等，1999）。但是，其中也存在一些偏差和尚未解决的科学问题。例如，因为陆地卫星遥感资料较为昂贵，已发表的论文当中一般只用不同年代的2个时相来表明荒漠化的变化，而往往忽视了气候异常变化对土地荒漠化当年变化的影响，试想一下，用一个降水十分充沛的气候条件下的土地覆盖类型的遥感信息解译为背景，然后再用一个降水异常偏少，十分干旱的气候变化条件下的土地覆盖类型遥感信息解译为动态变化样本，这一前一后两个时期的比较，肯定会得出荒漠化发展比较明显的结论，因此，产生了遥感数据选用的随机性，因为，两个时期的土地覆盖的变化不仅有人类活动的影响，还包括气候变化的影响，如何区别两者对土地荒漠化的影响，二者在土地荒漠化中的贡献及其耦合效应是国际该领域悬而未决的研究问题。王涛等（2004）提出如何

复原特征时期沙漠与沙漠化土地空间格局的问题，是必须解决的关键问题，这个问题不是短时间可以解决的，要花费大量资金、人力和物力。

4　区域气候变化与土地荒漠化的可能驱动力

中国东北地区西部位于生态区和气候区的过渡带上，受气候变异与人类活动共同作用，其环境变化是比较敏感的，亦是全球变暖响应比较突出的区域，该区现代荒漠化发展较显著，是生态脆弱区，但又具有一定的可恢复性等特征(孙继敏 等，2001；廉毅 等，1999，2001，2004；裘善文，1989，1992，2003；黄锡畴 等，1996；任红玲 等，2002；周成虎 等，2002；包慧娟 等，2003；王涛 等，2004；李取生 等，1998)。东北地区西部土地荒漠化的变化过程，与中国西北地区有类似的方面，如荒漠化的正、逆过程往往与气候变化的干旱化紧密相关，亦与人类社会的消极活动(过垦、过牧、过度开采地下水等)休戚相关。

关于该区对全球变暖的气候变化响应特征有些学者进行了研究。关于气温变化的响应特征大都达成共识，即中国东北地区同北半球中高纬度地区一样，冬季显著变暖(丁一汇 等，2006；陈隆勋 等，2004；廉毅 等，1997)，不仅冬季变暖明显，农业生长季(5—9 月)的平均温度自 20 世纪 80 年代以来进入一个相对暖的阶段(廉毅 等，1997，2007)，20 世纪 90 年代中期以后出现持续高温。与中国东北地区同一纬度位于西半球的美国玉米带(80°—100°W，40°—43°N)农业生长季平均温度趋势变化比较平稳，而其 45°N 的平均温度则呈线性下降趋势，这表明吉林省农业生长季平均温度是北半球中纬度升温显著的地区(廉毅 等，2007)。还有研究表明，东北地区春季气温日距平值基本上是由高纬度向低纬度、由东向西增加的，1957—2000 年气温日距平强度呈正趋势增长(王冀 等，2007)。关于东北地区对全球变暖的区域降水响应特征的分析，由于采用方法不同，研究结果有较大差异。孙力等(2000)用服从泊松(Poisson)Ⅱ型分布将不满足正态分布的夏季降水量换算为降水 Z 指数，以此来评估夏季旱涝等级，结果表明，20 世纪 80 年代初至 80 年代后期，雨涝较为集中，90 年代则进入一个旱涝交替变化的阶段。杨建平等(2003)采用降水量观测数据，利用彭曼公式计算了潜在蒸发和干燥度指数，分析了中国东北地区近 50 年干湿气候界线波动，得出 20 世纪 50—90 年代东北地区由湿—干—湿的变化规律。廉毅等(2005，2007)采用线性趋势分析方法，揭示了 1961—2000 年东北地区年降水量变化区域的不均一特征，吉林省中西部、南部和辽宁省为东北地区年降水量线性减少较显著的区域，气候变化以暖干倾向为主，而黑龙江的中北部和东部的牡丹江与佳木斯市、吉林省东部、原内蒙古的呼伦贝尔市为年降水量明显增加的区域。由 88 位中国科学家编制的《气候变化国家评估报告》中指出，“中国东北北部和内蒙古大部地区 1956—2000 年的年降水量有一定程度的增加。但是，华北、西北东部、东北南部等地区年降水量出现下降趋势；中国华北和东北地区干旱趋势严重”(丁一汇 等，2006)。由于东北地区西部每年仅有 300 mm 左右的降水量，而且降水量的年际变率很大(沈柏竹 等，2000)，必然会对土地荒漠化和土地覆盖变化有较大影响。所以，首先要准确地评估东北地区西部降水对全球变暖的区域响应特征，如果单纯地只用一个 10 年的降水量与另一个 10 年的降水量简单的数量上的加减，而忽视其趋势和年际变率的分析，往往会掩盖干旱发展的趋势。以吉林省西部为例，1995—2004 年的 10 年中，6 年的降水量仅为 200 mm，2 年为 300 mm 左右，只有 2 年大于 400 mm。在过去的 50 年里，沙漠化土地蔓延的主要部分是反映在人类活动和脆弱生态环境相互作用的土地退化区(王涛，2004)，东北地区西部与中国西北地区荒漠化变化明显有别，该区土地荒漠化特别是盐渍化面积增长较快(廉毅 等，1999)，而且分布比较复杂，比较突出的一点是，相当数量的荒漠化土地与农田和草地交织在一起，不如土地覆盖类型界面比较清楚的区域，增加了对荒漠化监测和发展趋势判断的困难，所以，不能仅凭某一局域荒漠化发展而判断整个或较大区域荒漠化亦发展(肖荣寰，1995)。

慈龙骏等(2002)在研究未来气候变化对中国荒漠化的潜在影响时指出：“目前国内有关气候变化的研究多以气候变化与农业生产和陆地生态系统间相互作用为主，而有关气候变化与荒漠化间反馈关系的研究也多集中在历史上的气候变化对荒漠化的影响上。”王涛(2003)指出：“荒漠化过程的起因主要包括气候变化和人类活动两个方面，二者在土地荒漠化中的贡献及其耦合效应是国际科学界长期悬而未决的问题，是必须解决的关键问题。”

土地荒漠化在很多区域被公认是全球变化研究中的一个关键因素，当前，国际上又提出一个抓重点敏感区的研究方法，很值得借鉴。由于目前人们缺乏对土地覆盖变化动力学的认识，尚未确定土地变化的频繁和敏感地区，这些地区被称之为临界地带或土地覆盖变化的热点区。利用遥感数据和由其他手段获得的指标的测定相结合，寻找土地覆盖变化与影响因素的相关性，当这些直接的诱发过程突出表现出来的时候，它们的结果就会被用来识别高频变化区今后变化的基础(孙成权 等，2003)。

因此，如何准确地评估区域气候变化对全球变暖的响应特征，如何进行有效的监测和准确判断荒漠化变化的主要趋势，准确获取土地变化的频繁和敏感地区的分布，是如何评估气候变化和人类活动对荒漠化影响的核心问题，也是为进一步研究土地覆盖变化的动力学，以及如何制定切实可行整治和恢复生态环境恶化地区的措施奠定科学基础。

5　中国东北地区西部气候变化与土地荒漠化发展趋势研究的讨论

根据国内外的发展趋势，在今后的研究中，必须加强固定的和野外的观测系统建设，在东北地区西部加强从空间向地面的空基遥测和以地面现代化监测为主的地基观测系统建设。目前，东北地区西部有关气候变化的干旱化和土地荒漠化监测面临的困境为：①至今还未建立该区协调一致的现代化监测体系，只有 EOS 和 MODES 是不够的，还必须具有时次较密和足够的陆地卫星遥感信息数据库，至少有 20 a 的春、秋两个时期的时相资料；仅基于空间遥感监测干旱化和荒漠化的动态发展还是很不够的，一方面空基遥感信息还必须经过地面的校正，以便提高监测的精确度，另一方面空间遥感还存在一定的局限性，对地表水资源的变化目前只能获取河流的长与宽和湖泊面积，还不能监测到水资源的总量变化，更重要的是对干旱化的监测仅限于地表植被指数(NDVI)和土地覆盖度，而对环境的蒸散量、水分盈亏、地表和大气低层的水分和能量交换是无法取得精确的估算。另外，仅基于空间遥感和监测会忽视地面野外观测站建设(孙成权 等，2003；叶笃正 等，2003)，忽视土地利用、土地覆盖度变化分析的诸多信息，特别是人口统计、收入、生产方式(包括农事活动)、经营政策、工程措施的建设等方面的信息，也就是忽视人类活动对干旱化和荒漠化影响的监测。②现代化监测系统未建立亦制约着荒漠化预警系统的建立，直接关系到该区的经济、社会环境和谐与可持续发展，亦直接影响到对生态环境恶化程度的认知和采取必要工程措施的投资。③东北地区西部对气候变异和人类活动作用都比较敏感，尚未确定东北地区西部土地覆盖变化的频繁区和敏感地区的分布(只获得部分地域)，这些地区被称之为临界地带或土地荒漠化变化的热点地区，自然界本身引起生态环境变化在一定程度上带有循环性质，亦即变坏之后，还会变好。而人类活动的消极影响一旦加进来，影响往往是单一的，不是循环的，随着人类社会的发展和科学技术的进步，人类对自然界及其规律性的认识有所加深，然而到目前为止，大量事实表明人们对自然的认识总远远落后于对自然的破坏(叶笃正 等，2003；陈敏豪，1995)。④如何准确地评估和判断生态环境、土地覆盖受气候变化和人类活动消极影响到不可恢复而发生强烈发展的阈值，正是干旱化和荒漠化预警系统的任务，只有通过该系统较长时间的边检测、边调查、边试验、边研究的技术路线才能逐渐逼近并获得这个阈值，这个阈值应该是在多学科的交叉研究过程中获得，而且还要不断采用现代化的器测手段，比如，最近由支克广等研制的《土壤荒漠化观测》的国家标准，已经国内著名专家审定并由国家有关部门机构正式颁布(中华人民共和国国家质量监督检验检疫总局 等，2006)。该阈值的获得关系到该区人类社会的生态安全、经济安全的可持续发展的核心问题。

目前，国内外有关气候变化与土地覆盖变化的研究，另一个值得注意的动向是，已从简单地分析气候主要要素气温和降水量的变化对土地覆盖影响的阶段，进入到以诸多气候要素的监测和各种变异分析与突出陆面过程的气候数值模拟相结合的方法，研究气候变化与土地覆盖相互作用的科学问题。符淙斌等(2002)通过卫星遥感植被和地表覆盖信息与气候要素关系的诊断分析，以及生态和气候模式的数值模拟，揭示了季风气候变化(季节、年际和长期)对生态系统的影响及大范围植被覆盖状况变化对季风区域气候的作用。研究表明，受季风气候季节跳跃特征的影响，植被的生长状况也呈现突变症状；在年际和更长时间的尺度，季风气候大的变率也强烈地影响着生态系统的变化。在此基础上提出了季风生态系统的概念(Monsoon-driven ecosystems)(符淙斌 等，2002)。Ji(1995)将陆表物理过程和植被生物学过程相结合，使大气对植被的影响和植被对大气的作用处于一个动态的相互作用过程中(AVIM)，通过气候模式的结合不仅能较好地模拟东亚季风的气候特征，而且可以模拟植被与大气之间的碳交换，对植物初级生产力的模拟有良好的结果。丁一汇等(2005)利用高分辨率区域气候模式(REGCM-NCC)模拟了中国区域植被发生改变后引起的局域或区域气候变化，结果表明，大范围区域植被变化对区域降水、温度的影响非常显著，内蒙古地区土地荒漠化可导致中国北方大部分地区降水减少，尤其加剧了华北、西北地区的干旱。还有学者用该模式模拟了中国近代历史时期土地利用/覆盖变化对中国区域气候的影响，认为由于植被变化改变了地面温度，使海-陆温差进一步增大，从而使区域大气环流流场发生变化(李巧萍 等，2006)。中国有些学者近几年来，不仅对干旱区，还注重对东北地区及半干旱地区的区域气候变化和环境演变，特别是荒漠化的成因进行研究，并且取得进展(陈杰 等，2000；林年丰 等，2001；朱金花 等，2001；闫敏华 等，2003；张耀存 等，2005)。

6　结论

在全球变暖的大背景下，已有的研究表明，地处中高纬度的东北地区区域气候变化响应特征为：东北地区是中国显著变暖的地区，不仅表现在冬季变暖，农业生长季(5—9月)变暖也比较明显；东北地区降水量变化趋势的空间分布并不一致，东北地区北部近50 a来是增加的趋势，而中西部和南部都是呈下降趋势；西部45°N以南地区气候变化主要特征为暖干倾向。

中国东北平原西部沙地位于欧亚大陆中纬度巨型沙带的东部边缘，为中国沙质荒漠化土地的东缘，在全球变暖的背景下，生态环境亦在恶化，主要表征为荒漠化(盐渍化和沙漠化)区域的不断扩展。如果不注意采取系统和大规模工程整治和保护措施，继续任荒漠化发展下去，不仅对国家重要的产粮区——东北中部构成严重威胁，还有可能使东北平原西部成为第二个西北。

进一步开展中国东北地区气候变化与土地荒漠化的研究是十分必要的。东北地区中部的粮食主产区和东部长白山森林区是防止欧亚大陆中纬度巨型沙带的沙质荒漠化向东扩展的最后一道绿色屏障，因此，该区必须采取科学的、系统的战略性整治措施，建立人类适应、减缓和应对气候变暖和荒漠化的典型示范区。

应加强该区有关气候变化、干旱化、土地覆盖变化的固定和野外观测系统建设，要加强多要素的气候系统观测工作，要加强多学科的交叉研究。因此，在各级政府的支持下，各学科积极联合争取到国际或国内的有关科研项目，不仅要掌握时次较密的卫星遥感信息，更要掌握用现代化仪器测得的地表有关要素的信息，特别要不断掌握东北地区西部土地覆盖变化的频繁区和敏感区分布的准确信息，与获取的多样本的要素变异建立关系，为研制东北区域的植被-土地覆盖-大气相互作用的数值模拟系统奠定坚实基础。

参考文献

包慧娟，姚云峰，张学林，等，2003. 科尔沁沙地景观格局变化的研究[J]. 干旱区资源与环境，17(2)：83-88.
陈杰，龚子同，高尚玉，2000. 干旱地区草场荒漠化及其评价[J]. 地理科学，20(2)：176-181.
陈隆勋，周秀骥，李维亮，2004. 中国近80年来气候变化特征及其形成机制[J]. 气象学报，62(5)：634-646.
陈敏豪，1995. 生态文化与文明前景[M]. 武汉：武汉出版社：365.
慈龙骏，1998. 我国荒漠化发生机理与防治对策[J]. 第四纪研究(2)：97-107.
慈龙骏，吴波，1997. 中国荒漠化气候类型划分潜在发生范围的确定[J]. 中国沙漠，17(2)：107-111.
慈龙骏，杨晓晖，陈仲新，2002. 未来气候变化对中国荒漠化的潜在影响[J]. 地学前沿，9(2)：287-294.
丁一汇，李巧萍，董文杰，2005. 植被变化对中国区域气候影响的数值模拟研究[J]. 气象学报，63(5)：612-613.
丁一汇，任国玉，石广玉，等，2006. 气候变化国家评估报告(1)：中国气候变化的历史和未来趋势[J]. 气候变化研究进展，2

(1):3-8.
符淙斌,温刚,谢力,等,2002.东亚季风区气候和生态系统相互作用的诊断和模拟研究[J].南京大学学报,38(5):281-294.
高志海,魏怀东,丁峰,1998.TM影像荒漠化土地特征信息的提取[J].中国沙漠,18(Z1):42-45.
黄锡畴,孟宪玺,1996.东北地区西部生态环境脆弱带研究[M].北京:科学出版社:1-20.
李巧萍,丁一汇,董文杰,2006.中国近代土地利用变化对区域气候影响的数值模拟[J].气象学报,64(3):257-270.
李取生,裘善文,邓伟,1998.松嫩平原土地次生盐碱化研究[J].地理科学,18(3):268-272.
廉毅,安刚,倪超玉,等,1997.吉林省四十年气温和降水的变化[J].应用气象学报,8(2):197-204.
廉毅,高枞亭,任红玲,等,1999.吉林省西部荒漠化发展的陆地卫星遥感监测分析[J].气象学报,57(6):662-667.
廉毅,高枞亭,任红玲,等,2001.20世纪90年代中国东北地区荒漠化的发展与区域气候变化[J].气象学报,59(6):730-736.
廉毅,任红玲,高枞亭,等,2004.中国东北西部气候变化与荒漠化发展及某些对策建议[C]//全国政协人口资源环境委员会,中国气象局.气候变化与生态环境研讨会论文集.北京:气象出版社:387-393.
廉毅,沈柏竹,高枞亭,等,2005.中国气候过渡带干旱化发展趋势与东亚夏季风、极涡活动相关研究[J].气象学报,63(5):740-749.
廉毅,高枞亭,沈柏竹,等,2007.吉林省气候变化及其对粮食生产的影响[J].气候变化研究进展,3(1):46-49
林年丰,汤洁,2001.中国干旱半干旱地区的环境演变与荒漠化的成因[J].地理科学,21(1):24-29.
罗札诺夫,1983.根据苏联的经验评沙漠化的定义、判断及评价[J].世界沙漠研究(4):21-25.
裘善文,1989.浅论科尔沁沙地的形成与演变[J].地理科学,9(4):317-328.
裘善文,1992.东北西部土地"三化"灾害严重,综合治理势在必行[C]//施雅风,黄鼎成,陈泮勤.中国自然灾害灾情分析与减灾对策.武汉:湖北科学技术出版社:468.
裘善文,张柏,王志春,2003.吉林省西部陆地荒漠化现状、特征与治理途径[J].地理科学,23(2):188-192.
任红玲,廉毅,高枞亭,等,2002.中国东北西部地区荒漠化发展前沿区域的遥感研究[J].第四纪研究,22(2):136-140.
沈柏竹,廉毅,高枞亭,2000.中国东北地区西部降水变化敏感带及干旱趋势的某些特征分析[C]//廉毅,孙力,王书裕.气候与生态环境.北京:气象出版社:60-65.
孙成权,林海,曲建生,2003.全球变化与人文社会科学问题[M].北京:气象出版社:198.
孙继敏,刘东生,2001.中国东北黑土地的荒漠化危机[J].第四纪研究,21(1):72-78.
孙力,安刚,丁立,2000.中国东北地区夏季降水异常的气候分析[J].气象学报,58(1):70-82.
王冀,江志红,张海东,2007.1957—2000年东北地区春季极端气温变化及其与北极涛动的关系[J].气候变化研究进展,3(1):41-45.
王涛,2003.我国沙漠化研究的若干问题:2.沙漠化的研究内容[J].中国沙漠,23(5):477-482.
王涛,2004.我国沙漠化研究的若干问题:3.沙漠化研究和防治的重点区域[J].中国沙漠,24(1):1-9.
王涛,吴薇,王熙章,1998.沙质荒漠化的遥感监测与评估:以中国北方沙质荒漠化区内的实践为例[J].第四纪研究,18(2):108-118.
王涛,朱震达,赵哈林,2004.我国沙漠化研究的若干问题:4.沙漠化的防治战略与途径[J].中国沙漠,24(2):115-123.
闫敏华,邓伟,陈泮勤,2003.三江平原气候突变分析[J].地理科学,23(6);661-667.
肖荣寰,1995.松嫩沙地的土地沙漠化研究[M].长春:东北师范大学出版社:1-20.
杨建平,丁永建,陈仁升,等,2003.近50年中国干湿气候界线波动及其成因被初探[J].气象学报,61(3):364-373.
杨晓军,张克斌,慈龙骏,2004.中国荒漠化评价的现状、问题及其解决途径[J].中国水土保持科学(1):22-28.
叶笃正,符淙斌,董文杰,等,2003.全球变化科学领域的若干研究进展[J].大气科学,27(4):435-450.
张耀存,张录军,2005.东北气候和生态过渡区50年来降水和温度概率分布特征变化[J].地理科学,22(5):561-566.
张玉贵,BEEMAERT F R,刘华,1998.TM影像的计算机屏幕解译和荒淇化监测[J].林业科学研究,11(6):599-606.
中华人民共和国国家质量监督检验检疫总局,中国国家标准化管理委员会,2006.土地荒漠化监测方法:GB/T 20483—2006[S].北京:中国标准出版社.
周成虎,李宝林,2002.东北平原西部沙地沙质荒漠化的遥感监测研究[J].遥感学报,6(2):117-122.
朱金花,黎劲松,2001.基于遥感与GIS技术的吉林省西部土地沙漠化现状及发展趋势研究[J].地理科学,21(2).140-144.
朱震达,1998.关于中国土地荒漠化的概念、成因和防治[J].第四纪研究,18(2):145-155.
CCICCD,1996.China Country Paper to Combat Desertification[R].Beijing:China Forestry Publish House:18-31.

JI J J,1995. A climate vegetation interaction model:Simulating physical and biological processes at the surface[J]. Joumal of Biogeography,22:445-451.

UNCCD, 1994. United Nations Convention to Combat Desertification in Those Countries Experiencing Serious Drought and/or Desertification, Particulary in Africa[R]. https://catalogue. unccd. int/936_UNCCD_Convention_ENG. pdf.

附录 A　廉毅先生及团队主要论文年表

1981

[1] 廉毅.热成风适应原理在大雪短期分析预报应用中的某些尝试[J].气象学报,1981,39(2):182-191.

1983

[2] 廉毅,胡建勋,娄安良.吉林省白城地区一次强飑线“下曳风暴”初步分析//强对流天气文集编辑组,强对流天气文集[G].北京:气象出版社,1983:67-73.

1986

[3] 廉毅.我国大雪预报研究概况[J].气象科技,1986(3):35-37.

1992

[4] 孙力,廉毅,李东平.东北低压爆发性发展过程的诊断分析[J].气象学报,1992,50(3):301-309.

1996

[5] 高枞亭,廉毅,王洪奇.1995 年 7 月下旬东北地区暴雨的数值分析[J].吉林气象,1996(2):42-46.

[6] 廉毅,常宗治.吉林省圈中主要灾害——气象灾害链简介及对策建议//吉林地球物理学会,吉林地球物理[M].长春:吉林科学技术出版社,1996:15-19.

1997

[7] 薛东有,廉毅.吉林省气象灾害链简介及对策建议[J].吉林气象,1997(4):2-3,45.

[8] 廉毅,安刚,王琪,等.吉林省 40 年来气温和降水的变化[J].应用气象学报,1997,8(2):70-77.

1998

[9] 廉毅,安刚.东亚季风 El Niño 与中国松辽平原夏季低温关系初探[J].气象学报,1998,56(6):724-735.

[10] 安刚,廉毅,王琪.近九十年吉林省松辽平原作物生长季气温变化的小波分析[J].气象学报,1998,56(4):458-466.

1999

[11] 廉毅,高枞亭,任红玲,等.吉林省西部荒漠化发展的陆地卫星遥感监测分析[J].气象学报,1999,57(6):662-667.

[12] 支克广,涂钢,廉毅,等.使用 LAS 系统测量显热通量[J].吉林气象,1999(4):6-7.

2000

[13] 廉毅,孙力,王书裕.气候与生态环境——吉林省气象科学研究所成立四十周年文集《科研报告》[G],北京:气象出版社,2000:180.

[14] 孙力,安刚,廉毅,等.夏季东北冷涡持续性活动及其大气环流异常特征的分析[J].气象学报,2000,

58(6):704-714.
[15] 廉毅,齐艳霞,唐晓玲.吉林省中西部地区生态环境整治的紧迫性与经济开发的协调性分析[J].吉林农业大学学报,2000(S1):17-21.

2001

[16] 廉毅,高枞亭,任红玲,等.20 世纪 90 年代中国东北地区荒漠化的发展与区域气候变化[J].气象学报,2001,59(6):730-736.

2002

[17] 孙力,安刚,廉毅,等.中国东北地区夏季旱涝的大气环流异常特征[J].气候与环境研究,2002,7(1):102-113.
[18] 任红玲,廉毅,高枞亭,等.中国东北西部地区荒漠化发展前沿区域的遥感研究[J].第四纪研究,2002,22(2):136-140.
[19] 安刚,孙力,廉毅,等.乾安县 40 年气候变化的分析[J].气候与环境研究,2002,7(3):370-376.
[20] 刘宗秀,廉毅,高枞亭,等.东北冷涡持续活动时期的北半球 500 hPa 环流特征分析[J].大气科学,2002,26(3):361-372.

2003

[21] 刘宗秀,廉毅,沈柏竹,等.北太平洋涛动区 500 hPa 高度场季节变化特征及其对中国东北区降水的影响[J].应用气象学报,2003,14(5):553-561.
[22] 孙力,廉毅,高枞亭.气候与生态环境科研报告[G].北京:气象出版社,2003:184.
[23] 廉毅,沈柏竹,高枞亭,等.东亚夏季风在中国东北区建立的标准、日期及其主要特征分析[J].气象学报,2003,61(5):548-558.

2004

[24] 廉毅,沈柏竹,高枞亭.关于确定东亚夏季风强度指数的探讨[J].气象学报,2004,62(6):782-789.
[25] 廉毅,任红玲,高枞亭,等.中国东北西部气候变化与荒漠化发展及某些对策建议//董文杰,江静,等,人类北方干旱化的有序适应——观测、虚拟试验和实验研究[M].北京:气象出版社,2004:79-84.

2005

[26] 安刚,孙力,廉毅.东北地区可利用降水资源的初步分析[J].气候与环境研究,2005,10(1):132-139.
[27] 廉毅,沈柏竹,高枞亭,等.中国气候过渡带干旱化发展趋势与东亚夏季风、极涡活动相关研究[J].气象学报,2005,63(5):740-749.
[28] 杨青,廉毅,何金海.利用奇异值分解方法预测东北地区夏季气温[J].气象,2005,31(3):31-35.

2007

[29] 廉毅,高枞亭,沈柏竹,等.吉林省气候变化及其对粮食生产的影响[J].气候变化研究进展,2007,3(1):46-49.
[30] 廉毅.北太平洋涛动与东亚夏季风的相关研究[J].地理科学,2007,27(Z1):19-27.
[31] 廉毅,王雅君.中国东北地区土地荒漠化与气候变化研究进展[J].地理科学,2007,27(Z1):97-102.
[32] 杨青,廉毅,卢秉红.极涡面积指数与北太平洋海温季节变异相关分析[J].地理科学,2007,27(Z1):78-83.
[33] 支蓉,廉毅,封国林.基于幂律尾指数研究不同尺度系统对降水的影响[J].物理学报,2007,56(3):

1837-1842.

[34] 廉毅,沈柏竹,高枞亭,等.确定东亚—西北太平洋地区夏季副热带季风建立和活动范围的一种方法[J].气象学报,2007,65(4):503-510.

[35] 侯威,廉毅,封国林.基于搜索平均法的气象观测数据的非线性去噪[J].物理学报,2007(1):589-596.

[36] 李尚锋,姚志平.东亚夏季副热带季风活动范围研究进展[J].地理科学,2007,27(S1):111-117.

[37] 沈柏竹.春季海陆热状况异常与东亚夏季副热带季风活动的相关分析[J].地理科学,2007,27(S1):61-68.

[38] SUN L,SHEN B Z,GAO Z T,et al. The impacts of moisture transport of East Asian monsoon on summer precipitation in Northeast China[J]. Advances in Atmospheric Sciences,2007,24(4):606-618.

2008

[39] 沈柏竹,李尚锋,廉毅.初夏东北冷涡活动的谐波分析[J].吉林大学学报(地球科学版),2008,38(S1):204-208.

[40] 刘平,廉毅,曹玲,等.全球变暖东北区域响应对吉林省旅游业的影响[J].吉林大学学报(地球科学版),2008,38(S1):191-192,195.

[41] LIAN Y,GAO Z T,SHEN B Z,et al. Climate Change and Its Impacts on Grain Production in Jilin Province[J]. Advances in Climate Change Researon,2008,4(Suppl):56-59.

[42] 刘刚,沈柏竹,廉毅,等.东亚夏季副热带季风主要活动路径的特征分析[J].吉林大学学报(地球科学版),2008,38(S1):196-199.

[43] 廉毅.东北西部沙尘暴的形成和发展//裘善文.中国东北西部沙地与沙漠化[M].北京:科学出版社,2008:197-209.

[44] 廉毅.东北西部气候对沙漠化形成的影响//裘善文.中国东北西部沙地与沙漠化[M].北京:科学出版社,2008:225-238.

2010

[45] 李尚锋,沈柏竹,廉毅,等.东北典型冷、暖夏年的北太平洋地区海-气系统季节演变合成分析[J].吉林大学学报(地球科学版),2010,40(S1):127-132.

[46] 沈柏竹,廉毅,李尚锋,等.北半球对流中、上层及平流层极涡特征初步分析[J].吉林大学学报(地球科学版),2010,40(S1):140-145.

[47] 刘刚,沈柏竹,廉毅,等.北极涛动与北半球海表温度联系的某些特征[J].吉林大学学报(地球科学版),2010,40(S1):122-126.

[48] LIU S,YANG S,LIAN Y,et al. Time-frequency characteristics of regional climate over Northeast China and their relationships with atmospheric circulation patterns[J]. Journal of Climate,2010,23(18):4965-4972.

[49] 廉毅,布和朝鲁,谢作威,等.初夏东北冷涡活动异常与北半球环流低频变化[J].大气科学,2010,34(2):429-439.

[50] LIAN Y,WANG J,TU G,et al. Quantitative assessment of impacts of regional climate and human activities on saline-alkali land changes—A case study of Qian'an County,Jilin Province[J]. Scientia Geographical sinica,2010,20(1):91-97.

[51] SUN L,SHEN B Z,SUI B. A study on water vapor transport and budget of heavy rain in Northeast China[J]. Advances Atmospheric Sciences,2010,27(6):1399-1414.

2011

[52] 沈柏竹，林中达，陆日宇，等. 影响东北初夏和盛夏降水年际变化的环流特征分析[J]. 中国科学：地球科学，2011，41(3)：402-412.

[53] 沈柏竹，刘实，廉毅，等. 2009 年中国东北夏季低温及其与前期海气系统变化的联系[J]. 气象学报，2011，69(2)：320-333.

[54] SHEN B Z，LIN Z D，LU R Y，et al. Circulation anomalies associated with interannual variation of early-and late-summer precipitation in Northeast China[J]. Science China(Earth Sciences)，2011，54(7)：1095-1104.

[55] WU R G，YANG S，LIU S，et al. Northeast China summer temperature and North Atlantic SST [J]. Journal of Geophysical Research：Atmospheres，2011，116(D16)：doi：10.1029/2011JD015779.

2012

[56] 沈柏竹，廉毅，杨涵洧，等. 中国东北夏季温度年代际变化特征[J]. 地理科学，2012，32(6)：739-745.

[57] 沈柏竹，廉毅，张世轩，等. 北极涛动、极涡活动异常对北半球欧亚大陆冬季气温的影响[J]. 气候变化研究进展，2012，8(6)：434-439.

[58] 裘善文，王锡魁，张淑芹，等. 松辽平原古大湖演变及其平原的形成[J]. 第四纪研究，2012，32(5)：1011-1021.

[59] 李尚锋，廉毅，陈圣波，等. 东北初夏极端低温事件的空间分布特征及其成因机理分析[J]. 地理科学，2012，32(6)：752-758.

[60] 刘刚，沈柏竹，廉毅，等. 亚洲阻塞高压分类及其与东北冷涡活动和东北夏季低温的联系[J]. 地理科学，2012，32(10)：1269-1274.

[61] 沈柏竹，张世轩，杨涵洧，等. 2011 年春夏季长江中下游地区旱涝急转特征分析[J]. 物理学报，2012，61(10)：530-540.

[62] SHEN B Z，LIU S，LIAN Y，et al. The 2009 Summer low temperature in Northeast China and its association with prophase changes of the Air-Sea System[J]. Acta Meteorologica Sinica，2012，26(4)：438-453.

2013

[63] 马梁臣，刘海峰，王宁，等. 2011 年长春市一次持续浮尘天气成因分析[J]. 气象与环境学报，2013，29(6)：24-30.

[64] LIAN Y，SHEN B Z，LI S F，et al. Impacts of Polar Vortex，NPO，and SST Configurations on unusually cool summers in Northeast China. Part Ⅰ：Analysis and diagnosis[J]. Advances in Atmospheric Sciences，2013，30(1)：193-209.

[65] JIANG X W，YANG S，LI J P，et al. Variability of the Indian Ocean SST and its possible impact [J]. Climate Dynamics，2013，41(7-8)：2199-2212.

2014

[66] 裘善文，王锡魁，MAKHINOV A N，等. 中国东北平原及毗邻地区古水文网变迁研究综述[J]. 地理学报，2014，69(11)：1604-1614.

[67] 李尚锋，应爽，姚耀显，等. 东北夏季月低温事件的定义及大气环流年代际特征分析[J]. 气象与环境学报，2014，30(3)：38-45.

[68] 李尚锋，孙钦宏，姚耀显，等. 东北夏季极端低温天气事件的定义及其冷空气路径分析[J]. 地理科学，

2014,34(2):249-256.
[69] LIAN Y,ZHAO B,SHEN B Z,et al. Numerical experiments on the impact of Spring North Pacific SSTA on NPO and unusually cool summers in Northeast China[J]. Advances in Atmospheric Sciences,2014,31(6):1305-1315.
[70] 乔少博,沈柏竹,王晓娟,等. 欧亚北部2004年以来频繁冷冬的特征分析及机理初探[J]. 气象学报,2014,72(6):1143-1154.
[71] 王宁,马梁臣,姚瑶. 2012年东北地区玉米黏虫爆发的气象成因分析[J]. 气象灾害防御,2014,21(4):33-36,48.
[72] 谢静芳,董伟,王宁,等. 吉林省冬季燃煤民居室内CO污染监测分析[J]. 气象与环境学报,2014,30(1):75-79.
[73] 杨旭,陈长胜,苏丽欣. 台风梅花在吉林省降水异常偏少的物理因子分析[J],吉林气象,2014(2):2-4,44.
[74] 高峰,王宝书,王宁,等. 东北地区自然采暖期气温及采暖指标的时空变化特征分析[J]. 气象灾害防御,2014,21(3):1-5,29.

2015

[75] 金顺梅,杨旭,马吉伟. 长白山景区6—10月降雨量对游客流量的影响特征分析[J],气象灾害防御,2015,22(1):9-13.
[76] 杨旭,刘刚,廉毅,等. 夏季鄂海阻高活动天数的年际变化及其前期征兆分析[J]. 地理科学,2015,35(2):224-230.
[77] 沈柏竹,封国林,廉毅. 1月北极涛动异常程度特征及其对北半球同期温度的影响[J]. 地理科学,2015,35(10):1299-1305.
[78] 苏丽欣,廉毅,李尚锋,等. 东北区夏季低温事件概率空间分布与亚洲阻塞流型域及其冷空气活动路径[J]. 地理科学,2015,35(2):251-258.
[79] 刘刚,廉毅,颜鹏程,等. 5—8月东北冷涡客观识别分类及北半球大气环流特征分析[J]. 地理科学,2015,35(8):1042-1050.
[80] ZHAO J H,YANG L,HOU W,et al. Extra seasonal prediction of summer 500 hPa height field in the area of cold vortices over East Asia with a dynamical-statistical method[J]. Chinese Physics B,2015,24(5):664-670.
[81] LIU G,FENG G L,et al. Activity of cold vortex in northeastern China and its connection with the characteristics of precipitation and circulation during 1960—2012[J]. Journal of Geographical Sciences,2015,25(12):1423-1438.

2016

[82] LIAN Y,SHEN B Z,LI S F,et al. Mechanisms for the formation of Northeast China cold vortex and its activities and impacts an overview[J]. Acta Meteorologica Sinica,2016,30(6):881-896.
[83] 苏丽欣,杨旭,李尚锋,等. 2014年东亚夏季副热带季风的季内活动特征分析[J]. 气象灾害防御,2016,23(3):6-8,48.
[84] LI S F,JIANG D B,LIAN Y,et al. Interdecadal variations of cold air activities in Northeast China during springtime[J]. Journal of Meteorological Research,2016,30(5):645-661,doi:10. 1007/s13351-016-5912-6.
[85] 王宁. 长春市城市热岛效应特征分析[J]. 农业与技术,2016,36(15):116-117.
[86] 孙力,马梁臣,沈柏竹,等. 2010年7—8月东北地区暴雨过程的水汽输送特征分析[J]. 大气科学,

2016,40(3):630-646.

[87] 刘刚,王宁,秦玉琳,等. 5—6 月东北冷涡持续活动及强度特征[J]. 应用气象学报,2016,27(1):47-55.

[88] 应爽,谢静芳,刘海峰,等. 基于 WRF 模式产品的长春市紫外线等级逐时预报模型的建立[J]. 气象灾害防御,2016,23(3):20-22.

2017

[89] 廉毅,沈柏竹,刘刚,等. 东北冷涡活动及其短期气候预测研究进展[J]. 气象科技进展,2017,7(1):33-37.

[90] LI S F,JIANG D B,LIAN Y,et al. Trends in day-to-day variability of surface air temperature in China during 1961—2012[J]. Atmospheric and Oceanic Science Letters,2017,10(2):122-129.

[91] SUN L,SHEN B Z,SUI B,et al. The influences of East Asian Monsoon on summer precipitation in Northeast China[J]. Climate Dynamics,2017,48:1647-1659.

[92] 刘刚,封国林,秦玉琳,等. 初夏东北地区冷涡降水"累积效应"[J]. 大气科学,2017,41(1):202-212.

[93] LIU GANG,WANG T,YANG X Y,et al. Climate characteristics of abnormal double-blocking activities over the Ural Mountains and Sea of Okhotsk[J]. Journal of Meteorological Research,2017,31(4):694-707.

[94] 马梁臣,孙力,王宁. 东北地区典型暴雨个例的水汽输送特征分析[J]. 高原气象,2017,36(4):960-970.

[95] 王耸,高西宁,肖瑶,等. 1961—2010 年中国降水量变化分区及其区域特征[J]. 干旱地区农业研究,2017,35(6):284-293.

2018

[96] 李尚锋,姜大膀,廉毅,等. 冬季中国东北极端低温事件环流背景特征分析[J]. 大气科学,2018,42(5):963-976.

[97] ZHENG X H,LIAN Y,WANG Q G. The long-range correlation and evolution law of centennial-scale temperatures in Northeast China[J]. Plos one,2018,13(6):e0198238.

[98] 马梁臣,王宁,于月明,等. 长春市两次严重 PM_{10} 污染天气对比分析[J]. 气象灾害防御,2018,25(2):9-13,27.

2019

[99] 刘刚,徐士琦,廉毅. 夏季亚洲阻塞高压识别及其对中国东北气候异常的可能影响:不同再分析资料对比[J]. 气象学报,2019,77(2):303-314.

[100] LIU G,QU M H,FENG G L,et al. Application study of monthly precipitation forecast in Northeast China based on the cold vortex persistence activity index[J]. Theoretical and Applied Climatology,2019(135):1079-1090.

[101] 王宁. 吉林省旅游气候舒适度分析[J]. 吉林农业,2019(20):104.

[102] 王宁,马梁臣. 长春市现代设施农业直通式气象服务思考[J]. 农业与技术,2019,39(19):127-128.

2020

[103] 杨旭,钟中,廉毅,等. 两种微物理方案对夏季东北亚阻塞过程影响的数值模拟结果分析[J]. 气象学报,2020,78(4):593-607.

[104] 石晨,廉毅,杨旭,等. 东北亚和北半球冬季高空切断冷涡与中国极端低温事件的联系[J]. 气象学报,

2020,78(5):778-795.

[105]徐士琦,刘刚,廉毅,等.近40年5—6月东北亚和北半球其他区域阻塞高压活动特征及其联系[J].气象学报,2020,78(2):221-236.

[106]王宁,马梁臣,刘琳.吉林省旅游气候资源区划[J].气象灾害防御,2020,27(2):34-39.

[107]王冬妮,唐晓玲,廉毅,等.气候增暖对3种典型落叶乔木物候的影响——以长白山区为例[J].东北林业大学学报,2020,48(9):51-55.

2021

[108]苏丽欣,廉毅,高枞亭,等.东亚—西北太平洋地区初夏副热带季风季内纬向变化的主要特征分析[J].气象灾害防御,2021,28(4):1-4.

[109]李建平,马云飞,马艳敏,等.基于自动土壤水分观测数据的吉林省西部干旱预报方法研究[J].气象灾害防御,2021,28(3):24-28.

2022

[110]徐士琦,刘刚,杨雪艳,等.中国东北地区春季透雨早晚与2—3月热带印度洋海温异常的联系[J].大气科学,2022,46(4):873-885.

[111]李尚锋,任航,高枞亭,等.1981—2019年吉林省暖季冷涡降水时空变化特征[J].大气科学,2022,46(1):1-14.

[112]王宁,马梁臣,霍也.一次爆发性气旋引发东北地区暴雨成因分析[J].气象与环境学报,2022,38(3):19-28.

[113]马梁臣,王宁,朱刚,等.张家口赛区冬奥会和冬残奥会同期湿度变化特征[J].气象灾害防御,2022,29(4):1-5.

2023

[114]SHI C,LIN Z D,LIAN Y,et al. Summer Onset in Northern East Asia:Feature,Mechanism and Variability[J]. Atmosphere,2023(14):187.

[115]李尚锋,邵祺多,尹路婷,等.东北地区入春日期变化特征及前期信号场[J].大气科学,2023,48(4):1-13. doi:10.3878/j.issn.1006-9895.2212.22139

[116]杨旭,涂钢,李尚锋,等.基于人工识别的1961—2020年东北冷涡活动特征分析[J],气象灾害防御,2023,30(1):25-30.

附录B 廉毅先生主要活动年表

1965年7月，从北京气象专科学校大专三年预报班毕业，9月分配至吉林省气象局工作。

1965年10月至1966年12月，参加吉林省农村社会主义路线教育工作队，赴怀德和伊通县工作。

1969年12月至1972年5月，走“五七”道路，到吉林省延边朝鲜族自治州珲春县马滴达公社三道沟大队，当“五七”战士，曾任大队党支部副书记、书记。

1972年6月，任吉林省气象台预报员。

1977—1978年，于北京大学地球物理系暴雨进修班进修。

1978年，任吉林省气象台预报科副科长。

1979年，为保证我国洲际弹道导弹装备部队试验，现场做出较准确的7天预报，受到试验基地领导的通信表扬。

1980年10月—1983年9月，任吉林省白城地区气象局副局长(主持工作)。

1981年，受聘工程师。在北京大学陈秋士教授指导下，在《气象学报》发表了有关东北10月气旋大风雪数值试验结果的论文。

1983年4月28—30日，准确预报出白城地区暴风雪天气，受到白城行署的通信表扬。

1983年10月—1991年5月，任吉林省气象科学研究所副所长。

1989年10月，受聘副研究员。

1991—1993年，任吉林省气象局科技服务处副处长(主持工作)、处长。

1992年，获国家教育委员会科技进步奖二等奖(完成人排名第三名)。

1993年3月，任吉林省气象科学研究所所长、党支部书记。

1997年2月，经丁一汇教授推荐参加在印尼巴厘岛举办的WMO第一届国际季风会议。

1997年10月，受聘正高级工程师。

1997年，在《科研报告》第二期发表《关于面向“九五”和21世纪开展大气科学应用研究的几点战略思考》一文，提出“必须在立项等级和涉及的领域上有新的突破”，即只有争取进入国家课题层面，省级研究才能继续发展。

1997年10月至1998年2月，向北京著名教授求教如何选题，1998年9月，吉林省气象科学研究所首次获得国家自然科学基金面上项目的资助。

1998年，科技部授予吉林省气象科学研究所为“98科技界抗洪救灾先进集体”。

1998年10月，经丁一汇教授推荐，作为中国气象代表团成员，赴加拿大埃德蒙顿参加第三次中加双边气候与气候变化对农业影响的研讨会，作了《中国东北气候及异常变化成因和对生态、农业影响分析》报告。

1999年1月，经吉林省委省政府批准为吉林省第五批有突出贡献的中青年专业技术人才。

1999年6月22日，应吉林省省长指示和批示，在吉林省省长常务办公会议上，作了题为《关于〈吉林省资源卫星遥感影像地图〉反演东北地区西部生态环境恶化的情况和建议》的报告。主要负责制作遥感图的高枞亭、负责后勤的朱长生列席了会议。

1999年6月24日，吉林省省长到吉林省气象科学研究所视察，廉毅做汇报。

1999年10月，获吉林省劳动模范称号。

1999年，获吉林省科学技术进步奖二等奖(完成人排名第一名)。

1999年，应林学钰院士邀请，兼任长春科技大学博士生导师。

1999 年，享受国务院政府特殊津贴。

1999 年，参加中国科学院大气物理研究所主持的“973”项目。

1999 年，与国家卫星气象中心合作，建立了我国第一个大孔径闪烁仪(LAS)乾安观测站。

2000 年 5 月 1 日，参加中共吉林省委书记主持的劳模座谈会，提供遥感卫星和气候分析报告图，获时任省委书记首肯，该报告图成为省委省政府被国务院办公厅批准为生态试点省建设的主要依据之一。

2000 年 11 月，在中俄双边气候研讨会上，作了题为“20 世纪 80—90 年代全球增暖欧亚大陆显著区域特征分析”报告。

2001 年，获全国“五一”劳动奖章。

2002 年 10 月，建议并协助中国科学院大气物理研究所与吉林省气象局，共建吉林通榆国际野外生态试验站。

2001—2003 年，作为中国气象局总体规划研究设计室和中国气象局政策法规司咨询顾问，撰写了气象部门生态与环境研究总体设计和建议报告。

2001 年 12 月，作为吉林省代表团成员，赴韩国首都参加了韩中荒漠化防治国际学术研讨会，作了特邀报告。

2002 年 2 月，经丁一汇教授邀请，赴澳门参加 APN、ENSO 和西太平洋暖池监测与预测项目会议，作了大会报告，并且报告全文被收录为会议成果之一。

2002—2009 年，受聘吉林省委省政府经济社会发展与环境咨询委员会委员，完成了 20 余份决策咨询报告，获省委书记和省长批示 10 余份。获省长基金近 10 项，达 500 多万元。其中提出造福本省人民的建议对策多项。

2003 年 9 月，经教育部函审备案，吉林省气象科学研究所与吉林大学地球探测科学与技术学院合办硕士学位授予权的培养点，为第二研究方向领衔教授，培养了沈柏竹、刘刚和李尚锋等吉林省气象局 3 名硕士和外单位 5 名硕士；培养点后升格为博士点。

2003—2004 年，吉林省气象科学研究所连续两次在中国气象局科技与气候变化司组织的专家评估中，获全国省所第一名。

2003、2006、2012 年，分别获批国家自然科学基金面上项目各 1 项，廉毅为主持人。

2004 年 11 月 2—6 日，在中国杭州参加第三届国际季风会议。

2005 年 11 月，经中国气象局人事司批准，缓退一年。因承担国家自然科学基金面上项目 2 项。

2006 年，被吉林省政府表彰为优秀咨询专家。

2006 年，《土地荒漠化检测方法》(GB/T 20483—2006)项目，获国家质量监督检验检疫总局、国家标准化管理委员会批准发布。

2006 年 2 月，在北京通过以丁一汇院士为首的十几位专家论证，成立了中高纬度环流系统与东亚季风研究开放实验室，任室主任至今。

2008 年 5 月，应邀在东亚副热带季风研讨会上，作了题为《东亚夏季副热带季风 5—7 月主要活动路径初探》报告。

2008 年 9 月，经中国气象局国际合作司批准，在长春市举办了“东亚－北太平洋－北美气候与气候变化国际学术研讨会”。

2004、2008 年分别参加 WMO 在中国杭州和北京举办的第 3、第 4 届国际季风会议。

2008 年 7 月 4 日，参加吉林省省长组织召开的决策咨询专家座谈会，作了题为“吉林省气候变化对粮食生产的影响”发言，省长在会议总结时指示：请廉毅研究员将吉林省是否会发生低温的资料送给我。

2009 年 6 月 11 日，将《关于 2009 年 6 月以及夏季出现低温机率增大的可能性分析和对策建议》咨询报告呈送吉林省省长，立即得到省长批示。国家气候中心还派专家到吉林省调研，之后形成调研报告呈送国务院。

2009 年，参加了由科技部、中国气象局和中国科学院牵头组织的第二次《气候变化国家评估报告》东北区域的撰稿。

2009 年，参加了科技部“21 世纪工程有关气候变化应对战略”的部分研究工作。

2009 年，受聘二级正研（高级工程师）。

2010 年 12 月，协办了中国科学院大气物理研究所王会军所长在长春召开的“973”项目“全球变暖背景下东亚能量和水分循环及其对我国极端气候影响”的学术研讨会。

2010—2011 年，在封国林教授的指导和帮助下，起草了公益性行业（气象）科研专项指南。吉林省气象科学研究所获 3 项“公益性行业（气象）科研专项”（主持）。

2013 年 1 月，在长春召开了首届亚洲阻塞流型与极涡活动及其天气气候效应论坛，论坛由中高纬度环流系统与东亚季风研究开放实验室、中国科学院大气物理研究所大气科学和流体力学数值模拟国家重点开放实验室和国际气候与环境科学中心联合举办。

2013 年 12 月 21 日，在北京召开了中高纬度环流系统与东亚季风研究开放实验室近 8 年的学术和工作总结会议。

2013、2014 年，在 ADV ANCES IN ATMOSPHERIC SCIENCES 发表 2 篇文章，有关北太平洋海-气耦合系统的 SVD 模态特征分析；经数值模拟试验，揭示春季赤道中太平洋 El Nino\La Nina 位相衰减，中高纬度环流系统响应的非对称性。

2013 年 6 月，沈柏竹获兰州大学博士学位，导师为封国林、廉毅。

2014 年 4 月 15 日，在扬州参加封国林教授团队学术会议。

2014 年 5 月，应杨崧教授邀请，在广州中山大学作了《关于热带、副热带与极地、中高纬度海-气系统基本模态间联系研究简况》的学术报告。

2014 年 5 月，参加在北京香山召开了由大气科学和地球流体力学数值模拟国家重点开放实验室陆日宇主任承办的论坛。丁一汇院士、何金海教授等到会作了学术报告。

2014 年 7 月，在北京参加了国家重点基础研究发展计划（“973”计划）“气候变暖背景下我国南方旱涝灾害的变化规律和机理及其影响与对策”项目中期评估会，为项目咨询专家。

2014 年 8 月，在吉林省陪同王会军院士野外考察气候与中国科学院生态观测站。

2015 年 9 月 5 日，在长春召开了中高纬度阻塞系统与东北天气气候异常学术研讨会，邀请黄思训、何金海和封国林教授作学术报告。

2016 年 2 月，主持的国家自然科学基金面上项目“中国东北低温前期征兆与北太平洋海-气耦合过程变化的联系”结题。

2016 年，在 Journal of Meteorological Research，发表了题为《东北冷涡的形成和活动异常机理研究进展概述》论文。

2016 年 6 月，刘刚获兰州大学博士学位，导师为封国林、廉毅。

2016 年 9 月 25 日，经丁一汇院士提议，在南京召开弘扬陶诗言学术思想学术活动筹备组第二次会议，何金海、琚建华、廉毅、黄思训与会。

2016 年 9 月，中国科学院大气物理研究所布和朝鲁研究员主持的国家自然科学基金重点项目获得批准，吉林省气象科学研究所为协作单位。

2017 年 1 月，在北京参加了布和朝鲁教授主持的亚洲阻塞流型与极涡活动及其天气气候效应论坛第四次会议。

2017 年 6 月 1 日上午，协助吉林省气象局、吉林省气象科学研究所，在长春承办了弘扬陶诗言学术精神研讨会，作为来宾代表发言。下午召开了中高纬度环流系统与东亚季风研究开放实验室 2016 年学术和工作总结会议。

2017 年 8 月，在长春先后召开了东北亚环流系统低频变化特征及其短期气候预测学术研讨会和东北亚气候变异的非线性特征学术研讨会，邀请黄思训教授、梁湘三教授、冯爱霞与达朝究等专家，作了非线性专题报告。

2005—2017 年，一直参加徐祥德院士主持的中国现场统计研究会气象、地质统计委员会学术会议，被聘为副主任委员。

2013—2017 年，参加了中国科学院大气物理研究所的硕士、博士论文通信和会议评审。

2017 年，参加了中国科学院大气物理研究所李尚锋博士论文答辩会议。

2018 年 4 月，在上海参加了张人禾院士主持的国家自然科学基金重大项目——北极海冰-气系统对冬季欧亚大陆极端天气、气候事件影响及机理启动会；在南京参加丁一汇院士负责的第一课题启动会。

2018 年 7 月，邀请复旦大学武炳义教授来长春作学术报告，这是武炳义教授第二次来长春讲学。

2018 年 8 月，在长春召开亚洲阻塞流型与极涡活动及其天气气候效应论坛第五次会议。

2019 年 5 月 25 日，应邀参加了北京大学大气科学学科 90 华诞庆典。

2020 年 10 月，在南京应邀参加了东亚季风与 ENSO 国际研讨会——庆祝何金海教授从事大气科学研究 60 周年。

2021 年 6 月 15 日，在吉林省气象部门青年职工的线上会议上作了题为《迎接党的二十大，继往开来，做有志向的气象科技服务人》报告。

2021 年 6 月 25 日，参加吉林省气象局召开的庆祝建党 100 周年纪念大会，接受光荣在党 50 周年纪念章。

2022 年 12 月，与石晨、林中达等共同在 *Atmosphere* 发表了题为《东亚北部夏季开始的特征、机制和变率》的论文。

2023 年 1—5 月，统稿《东亚中高纬度环流与季风及其区域气候变异论文选》，约 62 万字，作者：廉毅、沈柏竹、刘刚、李尚锋、杨旭、王宁。

2009—2023 年，每年参加封国林教授团队的学术会议，还参加了兰州大学和扬州大学硕士生和博士生毕业答辩。

附录C　廉毅先生荣誉称号和活动剪影

聘书

兹聘请 廉 毅 为吉林省人民政府经济社会发展与环境咨询委员会咨询委员，聘期三年。

特发此证。

二[illegible]四年十二月

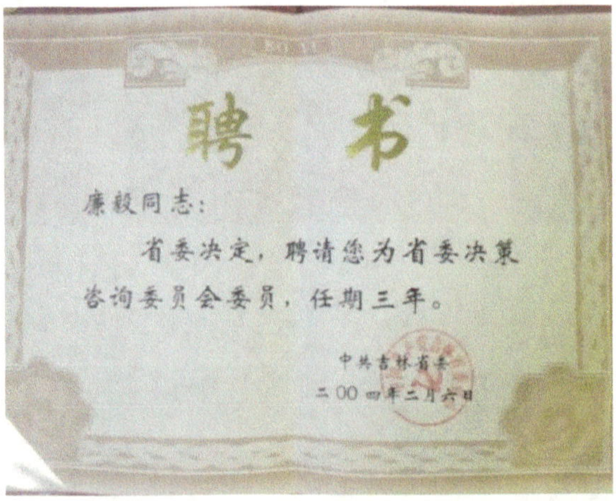

聘书

廉毅同志：

省委决定，聘请您为省委决策咨询委员会委员，任期三年。

中共吉林省委

二〇〇四年二月六日

1965 年 7 月毕业于北京气象专科学校大专班(最后排右六)

1978 年 6 月结业于北京大学地球物理系暴雨进修班(第四排右四)

1983—1991 年担任吉林省气象科学研究所副所长(右)

2008 年 9 月在长春市举办的国际会议专家合影

2014 年 5 月中国科学院大气物理研究所陆日宇教授组织的学术会议专家合影

1999 年 6 月 24 日吉林省省长到吉林省气象科学研究所视察，廉毅（前排右一）作汇报

2004 年 11 月 2—6 日在杭州参加第三届国际季风会议的全体代表合影

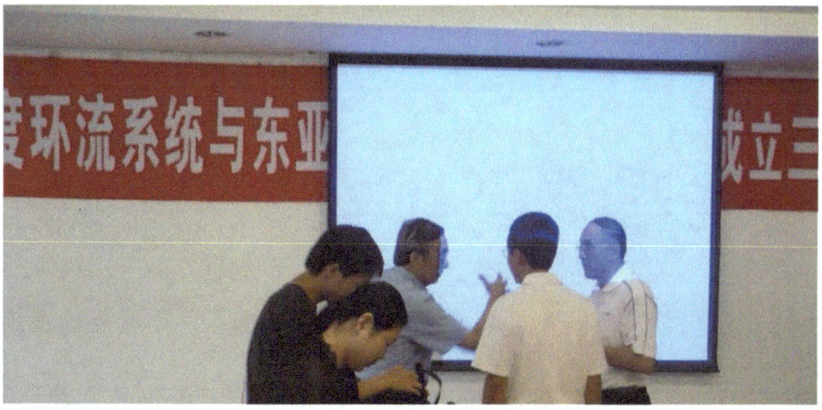

2009年在北京召开的中高纬度环流系统与东亚季风研究开放实验室成立三周年年会上，廉毅（右一）向丁一汇院士（右三）请教学术问题

2009年8月在西宁同徐祥德院士、丁一汇院士、张锦合影（左一为廉毅）

2007年2月，召开开放实验室第一次成立会议

2008 年 5 月丁一汇院士(左一)和王会军院士(左二)参加东亚副热带季风会议

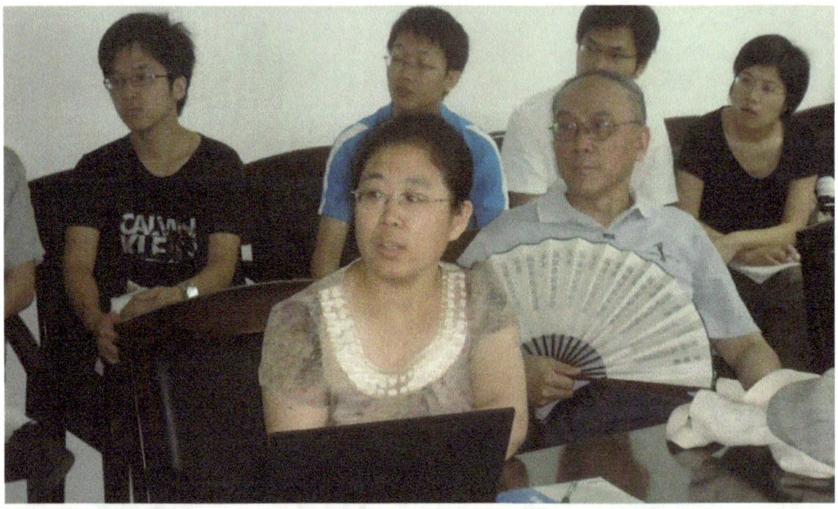

2010 年 7 月 8 日在兰州参加封国林团队课题会议(前排右一)

2008 年 5 月在北京参加东亚副热带季风研讨会

2014 年 6 月同王会军院士(右)在北京合影

2014 年 5 月在北京参加中国科学院大气物理研究所会议团队合影(左起刘刚、李尚锋、廉毅、杨旭和沈柏竹)

2010 年 7 月 10 日在兰州同封国林教授(右)、龚志强博士(左)合影

2014 年 5 月在北京香山参加会议与陆日宇教授(左一)、张朝林处长(左二)、布和朝鲁教授(右一)合影

2021年在南京参加梁湘三教授举办的“因果分析暨多尺度动力学暑期讲习班”全体专家合影
(从左到右:易斌教授、亓琳副教授、矫玉田教授、廉毅教授、黄思训教授、梁湘三教授、韩桂军教授、李威教授、吴拥政副教授、陶丽教授、王曰朋教授)